The Molecular Biology of Animal Viruses

Advisory Board

The Molecular Biology of Animal Viruses

VOLUME 2

edited by DEBI PROSAD NAYAK

Department of Microbiology and Immunology
School of Medicine
University of California
Los Angeles, California

MARCEL DEKKER, INC. New York and Basel

To my parents: Sarat Chandra and Durga Nayak

Library of Congress Cataloging in Publication Data
Main entry under title:

The Molecular biology of animal viruses.

Vol. 1 has title: RNA viruses.
Includes indexes.
1. Viruses. 2. Molecular biology. I. Nayak, Debi Prosad.
QR360.M55 576'.6484 76-29295
ISBN 0-8247-6534-6

MARCEL DEKKER, INC.
270 Madison Avenue, New York, New York 10016

Current printing (last digit):
10 9 8 7 6 5 4 3 2 1

PRINTED IN THE UNITED STATES OF AMERICA

Preface

In the last decade animal virology has become a major force in modern biology. During this period the major emphasis in animal virus research has been toward understanding the nature of animal viruses and the virus-host interaction at the molecular level. Furthermore, animal viruses have been used as a major tool in elucidating the basic regulatory mechanism involved in translation, transcription, and replication of infinitely more complex eukaryotic cells. It is not surprising to find that these basic regulatory processes of eukaryotic cells are essentially the same as those involved in the biology of animal viruses. After all, viruses must use the same host machinery, probably more efficiently, for their own survival.

Thus, the major assaults on both these fronts, i.e., viruses as pathogens and as probes in elucidating their host's function, have generated massive amounts of literature. Consequently, an ever increasing number of journals, specialized reviews, and monographs, as well as comprehensive multivolume treatises on virology are now being published.

This trend in modern biology and health care has created an increasing need for qualified virologists and has resulted in greater emphasis in graduate and postgraduate training in animal virology. Major universities now offer regular courses in the biology and chemistry of animal viruses. During the eight years I have been teaching courses in animal virology, I have felt the need of a suitable book covering the basic molecular biology of the major groups of animal viruses. The tremendous expansion of published literature in virology has made the need for such a book critical. The present treatise is an attempt to fulfill this need. This book (in two volumes) is therefore designed to serve as a text or a reference in advanced graduate courses in animal virology, with major emphasis on molecular biology.

Obviously, my experience in teaching animal virology has been a major influence in the organization of this book. I am fully aware that the design of each course and its emphasis will vary with individual instructors. This book is not an attempt to serve as a model for a course in animal virology, nor will it meet the specific needs of every instructor. Rather, the basic premise here is to present major groups of animal viruses. Each chapter is written by teacher(s), researcher(s) who are actively involved in their field of speciality. It is hoped that the basic information on the molecular biology of major groups of animal viruses presented here will be available and can be used by the students and the instructors regardless of their individual ways of synthesizing and formulating a course.

A few words are needed to explain the organization of the book's contents. As indicated above, the major emphasis is on elucidation of the nature of virus and host-virus interaction at the molecular level. Each chapter is not merely a chronicle of facts but represents an attempt to follow the logical progression of the life cycle of a major group of viruses. It includes the generality and uniqueness of their own group, as well as their relationships with other viruses outside the group. Chapters often present a hypothesis, raise critical questions about unsolved problems, and bring them to the attention of the readers, the future virologists. Although there is no attempt to explain viral pathogenesis in molecular terms, the relevance of each group of viruses to major animal and human diseases is indicated.

Some of the chapters are likely to become somewhat outdated soon. This is true for any fast-moving discipline, and even more so for animal virology. Indeed, this is a tribute to the dynamic nature of this field and to the unquenching curiosity of virologists. In spite of this inherently unsolvable problem, I hope this book will be of value in providing a source for basic information on the molecular biology of animal viruses to researcher and student alike. Newer information can always be found in recent journals to supplement the text as must be done in any advanced course in modern biology.

In addition, this book will serve as a reference for courses in biochemistry, molecular biology, genetics, or pathogenesis, where a basic understanding of the host-virus interaction at the molecular level is required. I hope it will be useful to present and future virologists as well as to other biologists.

I would like to thank the members of the advisory board for their help in reviewing one or more chapters, and to the contributors, who have done a remarkable job. Finally, I would like to thank my wife, Abantika, for her patience and understanding, and for help throughout this work.

Debi Prosad Nayak

Contents

Preface		*iii*
Contributors to Volume 2		*vi*
Contents of Volume 1		*vii*
Chapter 10	The Parvoviruses Lois Ann Salzman	539
Chapter 11	The Molecular Biology of the Papovaviruses Joseph Sambrook	589
Chapter 12	Adenoviruses William S. M. Wold, Maurice Green, and Werner Büttner	673
Chapter 13	The Herpesviruses Bernard Roizman	769
Chapter 14	Poxviruses Bernard Moss	849
Recent Developments		891
Author Index (Volumes 1 and 2)		901
Subject Index (Volumes 1 and 2)		995

Contributors to Volume 2

Werner Büttner, Institute for Molecular Virology, St. Louis University Medical School, St. Louis, Missouri

Maurice Green, Institute for Molecular Virology, St. Louis University Medical School, St. Louis, Missouri

Bernard Moss, Laboratory of Biology of Viruses, National Institute of Allergy and Infectious Diseases, National Institutes of Health, Bethesda, Maryland

Bernard Roizman, Departments of Microbiology and Biophysics/Theoretical Biology, University of Chicago, Chicago, Illinois

Lois Ann Salzman, Laboratory of Biology of Viruses, National Institute of Allergy and Infectious Diseases, National Institutes of Health, Bethesda, Maryland

Joseph Sambrook, Cold Spring Harbor Laboratory, Cold Spring Harbor, New York

William S. M. Wold, Institute for Molecular Virology, St. Louis University Medical School, St. Louis, Missouri

Contents of Volume 1

Chapter 1 Symmetry in Virus Architecture
Carl F. T. Mattern, Laboratory of Viral Diseases, National Institute of Allergy and Infectious Diseases, National Institutes of Health, Bethesda, Maryland

Chapter 2 Interferon
Hilton B. Levy, Freddie L. Riley, and Charles E. Buckler, Laboratory of Viral Diseases, National Institute of Allergy and Infectious Diseases, National Institutes of Health, Bethesda, Maryland

Chapter 3 The Molecular Biology of Picornaviruses
David M. K. Rekosh,* Department of Molecular Virology, Imperial Cancer Research Fund, London, England

Chapter 4 Togaviruses
James H. Strauss and Ellen G. Strauss, Division of Biology, California Institute of Technology, Pasadena, California

Chapter 5 Rhabdoviruses
David H. L. Bishop and M. S. Smith, Department of Microbiology, The Medical Center, University of Alabama in Birmingham, Alabama

Chapter 6 The Biology of Myxoviruses
Debi Prosad Nayak, Department of Microbiology and Immunology, School of Medicine, University of California at Los Angeles, California

Chapter 7 Paramyxoviruses
David W. Kingsbury, Laboratories of Virology, St. Jude Children's Research Hospital, Memphis, Tennessee

Chapter 8 Reoviruses
Robert F. Ramig and Bernard N. Fields, Department of Microbiology and Molecular Genetics, Harvard Medical School, Boston, Massachusetts

Chapter 9 Biology of RNA Tumor Viruses
Raymond V. Gilden, NCI Viral Oncology Program, Frederick Cancer Research Center, Frederick, Maryland

*Present affiliation: National Institute for Medical Research, London, England.

The Molecular Biology of Animal Viruses

Chapter 10

The Parvoviruses

Lois Ann Salzman

Laboratory of Biology of Viruses
National Institute of Allergy and Infectious Diseases
National Institutes of Health
Bethesda, Maryland

I. Introduction 540

II. Isolation of Parvoviruses 540

III. Characterization of the Virions 546

A. Particle Size 546
B. Capsid Configuration 547
C. Buoyant Density 547
D. Virion Sedimentation Coefficient and Molecular Weight 549
E. Hemagglutination 549
F. Antigenic Cross-Reactions of Parvoviruses 550
G. Host Range and Pathogenicity 551
H. Latency 553

IV. Structural Components of Parvoviruses 555

A. Structural Proteins 555
B. Nucleic Acid 560
C. Nucleic Acid Homology 565
D. Enzymatic Cleavage of Parvovirus DNA 565

V. Stimulation of Parvovirus Replication by Infected Cell Extracts or Helper Virus 566

VI. Parvovirus Replication 567

A. Cells Permissive for Viral Infection 568

B. Ultrastructural Studies of Infected Cells 569
C. Infectious Cycle 570
D. Viral Protein Synthesis 572
E. Viral DNA Synthesis 574
F. Viral RNA Synthesis 576
G. Interference 577

VII. Conclusions 579

References 579

I. Introduction

The family Parvoviridae contains small (15-24 nm) icosahedral particles composed of 1-3 proteins and a molecule of single-stranded DNA [1]. The virus particles, which have a relatively high specific density in cesium chloride (1.38-1.46 g/cc), are resistant to inactivity by heat (56-80°C for 30 min) and a variety of chemicals. The family is divided into 2 genera based on the ability of the virions to replicate autonomously in cell culture. Genus A, the larger group, consists of those autonomous or nondefective parvoviruses capable of replicating independently in cells, although replication requires dividing cells. Genus B contains the defective or satellite parvoviruses. These depend for complete or partial replication on coinfection of the host cell with an additional "helper virus" such as adenovirus or herpesvirus. Members of genus B are for this reason referred to as the adenosatellite or adenossociated viruses (AAV). The ICNV has recently proposed a third parvovirus genus, C or densovirus [2]. This group would contain the autonomous densonucleosis viruses of arthropods which may replicate their DNA in a manner analogous to the adenoassociated virions.

Parvoviruses are ubiquitous, having been isolated in a wide variety of hosts, including humans, animals, insects, and possibly bacteria. Other reviews of the parvovirus group are referenced at the end of the chapter [3-8]. This review will concentrate on the parvoviruses isolated from vertebrates. It will touch on those closely related members of the genus isolated from nonvertebrate hosts only when pertinent. Included in this review is information currently available on the virion's characteristics, pathology, and replication, and their suspected role in tumor interference and latent infections.

II. Isolation of Parvoviruses

In 1949, Schofield [9] isolated a filterable virus from the intestines of minks suffering from enteritis. This mink enteritis virus (MEV) filtrate was able to transmit the disease to other healthy minks. Wills [10] reported 3 years later that formalinized emulsion of infected tissues of minks suffering from mink enteritis could be used as an experimentally effective vaccine to protect the animals from contracting the disease. He also found an antigenetic relationship between MEV and feline panleucopenia virus of cats (FPV). Both MEV and FPV were later shown to be members of the parvovirus group [11].

In 1959, Kilham and Oliver [12] reported the first characterization of a parvovirus from rats (RV or KRV). Searching for an oncogenic papovavirus, they isolated a virus from several strains of metastasizing rat liver sarcomas and from the tissues of a rat bearing a transplantable leukemia [12]. The virus caused a cytopathogenic effect (CPE) in rat embryo tissue cultures and agglutinated guinea pig red blood cells (HA), and was heat stable and ether resistant. It did not produce tumors, however, and later it was shown to be smaller in size than members of the papovavirus group [13]. The following year, Toolan [14, 15] isolated a virus (H-1) from a human transplatable tumor (Hep-1). The H-1 virus, like KRV [16], may be pathogenic for hamsters. Depending on the age of the inoculated host and virus concentration, RV and H-1 produce antibodies, specific deformations, or death in hamsters. The viruses are, however, serologically distinct [17].

A steadily growing number of viruses have been identified as belonging to the parvovirus group. Isolated from a variety of hosts, they include (in order of isolation) the bovine hemadsorbing enteric virus (Haden) [18]; H-3 osteolytic virus from Hep-3 cells after blind passage in hamsters [3]; rat virus (L-2) [19]; rat virus X14 [20]; H-B and H-T virus from human placenta after blind passage in hamsters [21]; feline panleucopenia virus (FPV) [22, 23]; adenoassociated virus-1 (AAV-1) [24]; minute virus of mice (MVM) [25]; human adenoassociated viruses AAV-2 and AAV-3 [26]; porcine parvovirus (PPV) [27]; adenoassociated virus-4 (AAV-4) [28, 29]; avian adenoassociated virus (AAAV) or quail bronchitis virus [30]; hemorrhagic encephalopathy of rats (HER) [31]; minute virus of canines (MVC) [32]; the canine adenoassociated virus (CAAV) [33]; Kirk virus from human serum after passage through a Detroit 6 cell line [34]; bovine adenoassociated virus ($AAVX_7$) [35]; and KBSH, TVX, Lu III and RT, all from established continuous cell lines [36]. The viruses and the sources from which they were isolated are given in Table 10.1 (subgenus A, autonomous parvoviruses) and Table 10.2 (subgenus B, defective parvoviruses). In Table 10.1 the viruses from the same or closely related sources are grouped together.

The nonvertebrate members of the parvovirus group include the bacteriophage ϕX174 [37] and the densonucleosis virus (DNV) of the insect *Galleria mellonella* [38, 39].

There are several viruses which may possibly be members of the parvovirus group but which have not yet been sufficiently characterized to be

Table 10.1 Properties of Autonomous Parvoviruses

Virus	Source	Size (nm)	Density CsCl (g/cc)	Sedimentation coefficient ($S_{20,w}$)
KRV (RV)	Rat sarcoma, leukemia	15-25	1.38-1.41	110-122
X14	Mammary rat tumors	18-24	1.40	-
L-S	Choroleukemic rat	20-30		
HER	Rat central nervous system	20	1.37-1.39	
RT	Rat cell line			
MVM	Mouse adenovirus stock	19-28	1.41-1.43	110 ± 2
FPV	Cats leopard spleen	21-24	1.41	
MEV	Mink intestine, liver, spleen	21-24	1.41	
PPV	Hog cholera virus stock	20-28	1.31-1.39	108 ± 24
Haden (BPV)	Calf feces, bovine gastro-intestinal tract	22-28	1.38-1.425	
MVC	Dog feces	18-22	1.40	
H-1	Human tumor (Hep-1) and embryo	20-30	1.39-1.42	110 ± 2
H-3	Human tumor (Hep-3)	19-22		
H-T	Human placenta, and embryo	19-24		
H-B	Human placenta, and tumor	18-23		
Kirk	Human serum	18-20		
KBSH	Human cell line-KB	19-21	1.395	105 ± 10
TVX	Human tumor & cell line	19-21	1.395	
Lu III	Human cell line (Lu 106)	19-21	1.415	110

[a]Antiserum was prepared against H-1, H-3, KRV, X14, MVM, PPV, KBSH, TVX, Lu III, and RT and tested for hemagglutination inhibition (HI) for all indicated parvoviruses.

[b]Antiserum prepared against H-1, KRV, FPV, and PPV and tested for hemagglutination inhibition with Haden virus.

[c]Results of hemagglutination tests with red blood cells (RBC) from guinea pig (GP) human type O (HO) or rat (R); + = positive, 0 = no agglutination, +/0 = conflicting reports, – = not tested.

Serological cross-reaction (HI)	MW of Particle	RBC Agglutination[3]			References
		GP	H(O)	R	
KRV,[a] H_3,[a] X14,[a] HER	$6.6 \times 10^{+6}$	+	+/0	+	36,12,5,75,67,76,71
X14,[a] KRV,[a] H_3[a]		+	+/0	+	20,60,77,36,71
LVS, RV[c]		+	0	+	19,84
HER, KRV		+	–	–	75,31
RT[a]					36
MVM[a]		+	+/0	+/0	25,88,199,36,5,76,71
FPV	$5.9 \times 10^{+6}$				22,23,72,78,11,203,99
	$5.9 \times 10^{+6}$				10,72,203
PPV,[a] KBSH[a]		+	+	+/0	36,95,96,27,125,238,71
Haden[b]		+	+	0	18,5,79
MVC		0	0	0	32,80
H_1-1,[a] HT	$6.8 \times 10^{+6}$	+	+	+/0	3,6,21,76,36,103,128,71
H_3,[a] KRV,[a] X14[a]		+	+/0	+	77,36,18,71
H-1					21
					21
H_3		+	+	–	34,77
KBSH,[a] PPV[a]		+	+/0	+	36,123,77,71
TVX[a]		+	+	+	36,71
Lu III	5.7×10^6	+	+	+	36,186,190,171

Table 10.2 Properties of Defective Parvoviruses

Virus	Source	Size (nm)	Density CsCl (g/cc)	Sedimentation Coefficient ($S_{20,w}$)
CAAV	Infectious canine hepatitis stock	20-25		
AAAV	Quail bronchitis virus stock	18-20	1.42	
$AAVX_7$	Bovine adenovirus type I stock	22	1.35-1.38	135-147
AAV-1	Simian adenovirus virus 15 stock	21.8 ± 1.3	1.395	104,125
AAV-2	Human adenovirus 12 stock	23.8 ± 2.7	1.388	125,120
AAV-3	Human adenovirus 7 stock	21.4 ± 2.4	1.394	
AAV-4	Simian adenovirus (SV15) stock	22.0 ± 1.8	1.445	137

[a]Antiserum prepared against AAV-1, AAV-2, and AAV-3, incubated with various AAV types, and used to infect cells and helper virus. After cell lysis supernatant fluid was tested for presence or absence of AAV complement fixation antigen [64].

[b]Antiserum prepared against AAAV, AAV, AAV-2, AAV-3, and AAV-4. Aggregation of each virus was tested with homologous and heterologous antiserum [213].

Serological cross-reactivity	MW ($\times 10^{-6}$)	Virion DNA content (%)	RBC Agglutination[c] GP	Hu(O)	R	References
						33,81
AAAV[b]			0	0	0	73,30
$AAVX_7$			+	+	+	35,239,94
AAV-1[a]	5.4	18.9	0	0	0	5,179,8
AAV-2, AAV-3[a]	5.4		0	0	0	5,26,8
AAV-3,[a] AAV-2[a]			0	0	0	5,26
AAV-4	5.4	26.5	+	+	+/0	5,28,29,53

included definitely at this time. For example, the viruses isolated from human adult cases of acute infectious gastroenteritis [40] and hepatitis A [41] have several properties in common with the parvoviruses, including size (25-50 nm), heat and ether stability, and buoyant density in cesium chloride (1.38-1.41 g/cc). Other parvovirus-like particles with uncertain pathogenicity have been isolated from human sera [42, 43] and feces [44]. The particles are identical in size, morphology, and buoyant density to known parvoviruses. Antibodies to these viruses were found in 30-40% of the human population tested [42, 44]. The virus MVL 51, which infects mycoplasma cells, contains a single-stranded DNA molecule but the capsid does not appear to have icosahedral symmetry [45, 46]. The size, morphology, and density in cesium chloride of the virus associated with Aleutian mink disease suggest that it, too, may belong to the parvovirus group [47, 48].

III. Characterization of the Virions

A. Particle Size

Several techniques have been used to measure the size of virus particles. The most common are passage through filters of known pore size [49], ultrafiltration [50], and electron microscopy measurements [51]. Each of these techniques unfortunately has physical or biological factors which affect the measurements and values obtained. As a consequence, measurements can vary with the technique employed and even with the investigator using the technique. Difficulties encountered in the measurement of parvovirus size and structure are discussed in detail by Hoggan [5].

In 1961 Chandra and Toolan [52] found virus particles in the cytoplasm of Kupffer cells and in the interstitial cells of hamster kidneys infected with H-1 virus. The particles, visualized in thin sections in the electron microscope, consisted of a dense circular core approximately 15 nm in diameter. The core appeared to be encircled by a coat not made visible by their technique, but judged to be about 15 nm wide, thus giving the virus particle a diameter between 15 and 30 nm. Virions with diameters of 30 nm or less have the morphological subunits of their coat so close together that negative stain cannot penetrate between them satisfactorily [53]. Using negative staining of H-1 virus partially purified by absorption to guinea pig red blood cells, Toolan et al. [54] found virus particles with a mean diameter of 24.5 nm. Dalton et al. [241] found many intranuclear particles approximately 13 nm in diameter in kidney stromal cells infected with KRV. One difficulty with measurement of these small particles in cells is the problem of differentiating them from subcellular components such as ribosomes or cross-sectioned fibrils. Smith et al. [55] made 2 series of electron microscope measurements of purified adenoassociated virions (AAV), and found the mean major and minor dimensions to be 18 ± 1.1 and 21 ± 1.8 nm, respectively. Thus, the orientation of the particle in addition to the technique, method of measurement, stains, and instrument calibration can affect the size measurements [55, 5]. In Tables 10.1 and 10.2 the size variation reported for the parvoviruses is given. It is apparent from these tables that the reported virion diameters overlap; a mean value of 21 ± 3 nm would include most of the reported values.

B. Capsid Configuration

The principal technique used to determine the fine structure of the parvoviruses has been electron microscopy of the virions negatively stained with phosphotungstate. All of the measurement difficulties mentioned above also hamper the resolution of virion substructure. Perhaps for this reason the arrangement and number of the capsomeres in the protein coat of the virus has not been finally resolved. We know, however, from their resistance to lipid solvents (such as ether and chloroform), that parvoviruses do not possess an essential additional outer-lipid covering or envelope; thus they are considered "naked" particles [56,57,55].

Vasquez and Brailovsky [58] in 1965 observed the hexagonal outline of the KRV virion in the electron microscope. This led them to propose that the capsomere subunits of the virus protein capsid were arranged in icosahedral symmetry. The structural symmetry of the virus with a 20 nm diameter was represented by a pentagonal dodecahedron (12 sides) with pentagonal pyramids arranged on each face, giving a total number of 32 capsomers (Fig. 10.1). Karasaki [57] agreed that the KRV and also the H viruses contained 32 capsomers arranged in a 3-fold and 5-fold symmetry.

The fine structure of the capsid may vary with different members of the parvovirus group. Mayor et al. [53] found that selected AAV-4 virions had 6 capsomers along the edge with a central group of 3 capsomers in the form of a triangle. The concluded that the capsid was composed of 12 subunits in the form of an icosahedron as proposed for the bacteriophage parvovirus ϕX174 [59]. Smith et al. [55] found a resemblance between AAV and reovirus in that both appear to have capsids consisting of a net-like reticulum. Final resolution of the capsid configurations, however, as well as whether differences in configuration exist among the members of the parvovirus group must await future studies.

C. Buoyant Density

The parvoviruses have a relatively high buoyant density in cesium chloride, a factor attributable to their high DNA/protein ratio. This high specific

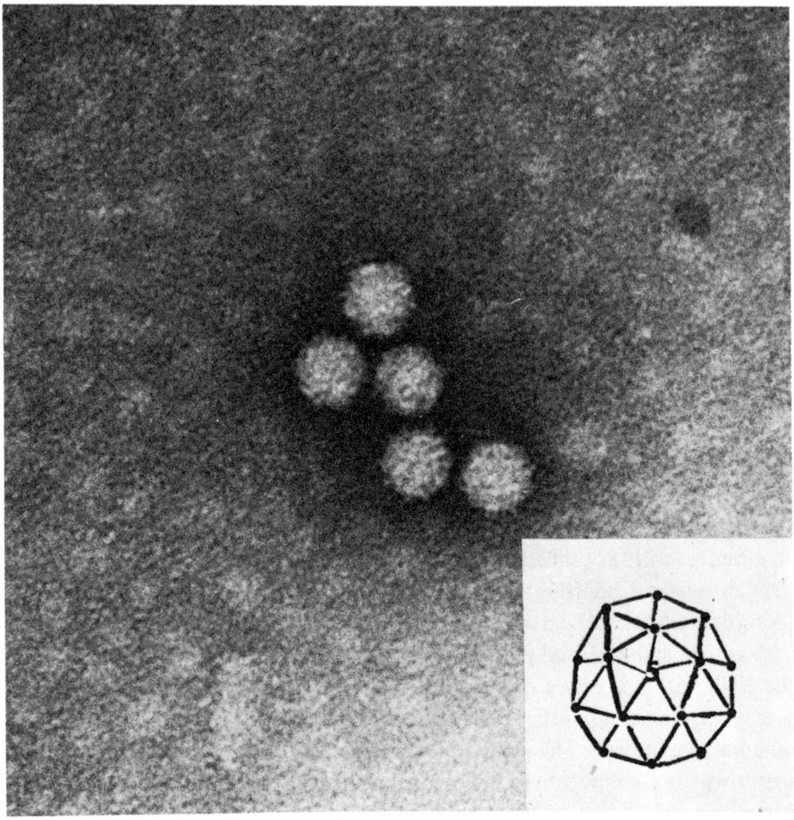

Fig. 10.1 Group of KRV particles negatively stained with phosphotungstic acid (X 300,000). The inset is a model of a pentagonal dodecahedron viewed along a 5-fold axis of symmetry.

density has been invaluable in grouping the viruses and in purifying them. Tables 10.1 and 10.2 give the buoyant density in cesium chloride for many parvoviruses. A majority of the parvoviruses have a buoyant density in cesium chloride of approximately 1.40 ± 0.02 g/cc. A range in values for the same virus in different laboratories has been reported. This may be attributable to differences in technique, variations in centrifugation times and speeds, and even differences in tube sizes. A given virus may also show a slightly different density when produced in different cells [5].

Isopycnic centrifugation of representative parvoviruses in cesium chloride has resulted in the detection of 1 or more minor bands in addition to the major infectious band discussed above. A virus band with very little if any infectivity can be detected at a lower cesium chloride density (1.31-1.36 g/cc) than the major band [60, 25, 61-67]. This band contains "empty particles" or virus capsids containing little or no DNA. An H-1, MVM, and AAV band with a cesium chloride density higher (1.44-1.47 g/cc) than the major virus band has also been described [5, 68]. These infectious particles, smaller in size than those in the primary band, may represent virions whose capsids are incomplete or altered [8, 69, 5]. Virions from this band, however, have also been reported as antigenically indistinguishable from the major infectious band [5].

D. Virion Sedimentation Coefficient and Molecular Weight

The sedimentation coefficient of a virus particle depends on the effective mass of virion in solution divided by its frictional constant [70]. The coefficient depends on the shape and size of the virion. It can be used to calculate the approximate molecular weight of a molecule and to compare virions within a group or between groups. The sedimentation coefficients of most parvoviruses were calculated by cosedimentation in a sucrose gradient with a virus of known sedimentation coefficient. Under well defined conditions, the distance migrated in the gradients is proportional to the relative sedimentation coefficients. The sedimentation coefficients can also be measured by sedimentation in an analytical ultracentrifuge. The known sedimentation coefficients and molecular weights of the virions are given in Tables 10.1 and 10.2. The molecular weights of some virions were calculated from the determined weight of their DNA and the percentage of the DNA in the intact virion. Again, reported differences may reflect differences in experimental conditions rather than significant differences in values.

E. Hemagglutination

Kilham and Oliver reported in 1959 [12] that KRV agglutinated guinea pig and rat erythrocytes but failed to react with human or chicken red blood cells. Since then, because of the ease and simplicity of the method, hemagglutination patterns with erythrocytes from a wide range of animal species have been used to characterize parvovirus group members and strains [3]. It is often difficult, however, to compare data from different laboratories. For this reason, where possible, the hemagglutination patterns from the same laboratory under identical conditions are used. Different results reported in another laboratory, or a borderline reaction, are indicated by the use of the symbol +/0 (Tables 10.1 and 10.2).

Hemagglutination by parvoviruses is prevented if the erythrocytes are treated with receptor destroying enzyme (RDE) which is a neuraminidase enzyme. Hemagglutination usually occurs equally well at 4° and 37°C except with MEV and FPV, and at variations in pH from 6.6 to 8.4. Virus can be eluted from the red blood cells at a pH of 9.0 without destruction of the erythrocyte receptor sites [7].

As seen in Table 10.1, the parvoviruses from rodents KRV, X14, L-S, HER, RT, and MVM all agglutinated rat and guinea pig red blood cells but not all agglutinate human type O cells. Hallauer et al. report that the KRV, MVM, and X14 will agglutinate human type O cells [71]. MVM can be distinguished from KRV and H-1 by MVM's ability to agglutinate mouse red blood cells [25]. FPV and MEV hemagglutinate pig erythrocytes at 4° but not at 37°C [72]. MVC does not agglutinate any of these 3 red blood cell types but will agglutinate pig erythrocytes [5]. In the cases tested, the parvoviruses from human sources, H-3 excepted, will agglutinate guinea pig and human type O erythrocytes but have either minimal or no reaction with rat red blood cells. H-3 is often considered a rodent virus derived from contamination of the Hep-3 cell line or from the isolation procedure.

The hemagglutination pattern of the defective parvoviruses is shown in Table 10.2. With the exception of AAV-4 and $AAVX_7$, they do not appear to agglutinate the erythrocytes tested [5, 35, 73]. AAV-4 does agglutinate human guinea pig and sheep erythrocytes at 4° but not at 37°C [74]. $AAVX_7$, the bovine parvovirus, can agglutinate the erythrocytes of the 3 sources tested.

F. Antigenic Cross-Reactions of Parvoviruses

The relationship of the parvoviruses to each other has been studied by several serological procedures. Included in Table 10.1 are the antigenic relationships of those autonomous parvoviruses which demonstrate hemagglutination. Antiserum was prepared against each virus. The ability of these specific viral antisera to inhibit hemagglutination of another parvovirus is an indication of a similarity between their hemagglutinating antigens. KRV demonstrates hemagglutination inhibition (HI) with antiserum prepared against X14, HER, L-S, Kirk, and H_3, all probably closely related rodent viruses [31, 75]. There was no hemagglutination inhibition of KRV with antiserum prepared against H-1, MVM, PPV, KBSH, TVX, Lu III or RT, according to tests conducted by Hallauer et al. [36]. Antiserum against MVM and RT did not prevent agglutination with any of the 10 viruses tested in Table 10.1, group 1. Some slight antigenic relationships among MVM, RV, and H-1 has been demonstrated by fluorescent antibody staining [5, 76], but these viruses are probably antigenically distinct. Antiserum prepared against Kirk and H-3 virus cross-reacted to give hemagglutination inhibition [77]. Hemagglutination of FPV was not inhibited when tested against antiserum prepared against KRV and

H-1 virus [78]. PPV and KBSH may be identical; they appear to have a close antigenic relationship as shown by hemagglutination inhibition [36]. The PPV could have been introduced as a contaminant of the KB cell line from which KBSH was isolated if the cells were dispersed with hog trypsin containing PPV. Haden virus does not appear to share hemagglutiantion antigens with H-1, KRV, FPV, or PPV [79]. MVC, when tested with antiserum against H-1, KRV, or MVM did not demonstrate hemagglutination inhibition [32, 80]. H-1 and H-T share common hemagglutination antigens. TVX and Lu III, isolated by Hallauer from continuous human cell lines [36], appear to be antigenically distinct from the rest of the viruses tested in Table 10.1, group 1. They may be cell line contaminants or distinct parvoviruses.

As most defective parvoviruses do not demonstrate hemagglutination, serological relationships have been demonstrated mainly by complement fixation antigen [1] or microagglutination as seen in the electron microscope [3]. Serum neutralization, fluorescent antibody, and immunodiffusion assays have also been used [5]. Only AAV-2 and AAV-3 have been shown to be serologically related by complement fixation [5]. AAAV [8] and $AAVX_7$ [35] do not appear to be related serologically to the simian and human AAVs. Onuma [81] reported that CAAV reacts antigenically with AAV-3. Other researchers, however, have not been able to detect this serological cross-reaction [8]. Hoggan [5] found no antigenic relationships among the 4 human and simian AAVs and the parvoviruses RV, H-1, MVM, and Haden. He also identified 3 strains of AAV-3 definable by serum neutralization and nucleic acid hybridization tests [5, 8]. The 2 strains of AAV-2 (H and M) have not been distinguished by serological tests [5].

G. Host Range and Pathogenicity

Detection of serum antibodies to a parvovirus indicates that the animals has had prior exposure either naturally or in the laboratory to the viral proteins. The presence of serum antibodies can reveal the range of hosts which are exposed to and infected by the virus. Specific antibodies to a virus can be detected by many techniques. Combination of antibody with the virus can be quantitated from the inhibition of viral hemagglutination, complement fixation, the failure of the virus to produce CPE or the reduction of plaques in tissue culture. When virus is injected into animals, specific antibodies, if present, may neutralize the virus and prevent the viral pathology if the virus in question produces disease. Antibody-virus interaction can also be seen in the electron microscope by immune election microscopy [40]. Kilham rat virus (KRV) was first isolated in rat embryo tissue cultures with material from tumor-bearing rats [12]. Antibodies to this isolated virus were found in a number of healthy laboratory and wild rats as well as in a germ-free rat.

These results suggested a wide dissemination of the virus and possible vertical transmission [12]. Naturally occurring antibodies to KRV, H-1, L-S, and MVM, indicating exposure to specific viral proteins, have been found in the rat [82-84]. Antibodies to PPV have been found in healthy pigs [85, 86], to MVM in normal mice [87, 88], and to Haden (bovine parvovirus) in a high proportion of cattle [17, 79]. Neutralizing antibody to MVC has been detected in a large proportion of healthy German shepherds and beagles [80].

Antibodies to AAV-1, -2, -3, and -4 have been found in both man and monkeys [89]. AAV-2 and -3 have been isolated from man [90, 91]. Blacklow et al. [90, 91] found that growth of the AAV-2 and -3 isolates in vivo requires the presence of a helper adenovirus. AAV-1 is believed to be of rhesus origin and AAV-4 is an African green monkey virus that occasionally infects man [92, 93]. Antibody to $AAVX_7$ has been reported both in human and cattle sera [5, 35, 94].

Although antibodies to parvoviruses can be detected in a wide variety of animals and humans, only a few of them, like FPV and MEV, have been shown to cause naturally occurring disease. Since the virions have a proclivity for the dividing cell, it is difficult to tell if the virus caused or influenced a disease, or if it is an adventitious association. Cartwright et al. [95, 96] have suggested that PPV infection might be related to porcine infertility and abortion. PPV has also been implicated in stillbirths [86, 97]. Haden virus may cause infertility and stillbirth in cattle [17, 79]. In kittens, FPV is believed to cause both cerebellar hypoplasia and the panleucopenia syndrome, the latter being a naturally occurring disease. Injection of the virus into cats can produce similar symptoms [98]. FPV infection in cats, mink ferrets, coatimundi, and raccoons may also result in anorexia, diarrhea, and even death [99, 22, 100, 101].

Toolan et al. [102, 103] noticed that newborn hamsters developed a peculiar deformity if injected at birth with fractions of Hep-1 cells containing H-1 virus. This abnormality was characterized by a flattened foreface with protruding tongue, absent or abnormal teeth, bone fragility, and small size. KRV was later shown to induce the same abnormal development in hamsters [15, 104]. The similarities between that parvovirus-induced, mongoloid-like deformity and Down's syndrome in man led Galton and Kilham [105] to look for chromosome abnormalities in deformed animals. The karyotypes of such animals are normal, the hamsters are fertile, and their offspring normal [3]. The deformities appear, therefore, to be due to a selective arrest of cells or specific cell functions in postnatal development. Another strain of the Kilham rat virus, designated L-S, and 2 strains originally associated with Maloney virus were not demonstrated to produce this osteolytic abnormality [106]. Attempts to demonstrate such activity with various AAVs have also been unsuccessful [5]. The ability of the KRV and H-1 to produce bone abnormality in hamsters is thus not a general property of the parvovirus group. Rat virus strains which had no osteolytic effect in newborn hamsters [106-108], as well as a virus of feline origin (FPV) [98], can produce a cerebellar hypoplasia and ataxia after intracerebral injection of the virus into hamsters, kittens, or ferrets. Histopathological studies of infected brains showed that the virus had a proclivity for the outer germinal layer of the cerebellum, a region active and developing in the neonatal period. Many nuclear inclusions were found in this area, which subsequently became hypoplastic. Kilham and Margolis [107, 108] concluded that KRV had an affinity for cells in active mitosis. Consequently, the parvoviruses would be found associated mainly in actively proliferating or regenerating cells; probably they require the milieu of such cells for their own replication. This is generally true (see Tables 10.1 and 10.2). Parvoviruses are isolated from infected and mitotically active cells. H-1 for example, is isolated from tumors, embryos, and placentas [3]. Growth of the virus in tissue culture also occurs best in cells from a malignant origin that are not contact-inhibited and continually divide.

Kilham and Margolis [109] have determined that KRV can be transmitted to rat sucklings through the milk of virus-inoculated mothers. The virus was believed to proliferate in the mammary tissue [109]. KRV thus might serve as a model for the study of transmission of virus in milk. Both H-1 and KRV [6, 110] can also cross the hamster placenta. When injected subcutaneously into pregnant hamsters, the offspring are either dead or deformed. Newborn hamsters injected intraperitoneally with KRV or MVM [87] can also develop hemorrhagic enteritis. The lethal effect of this enteritis decreases dramatically with an increase in the period between birth and inoculation. PPV can pass the placenta of pigs [97]. This results in stillbirth [86] with histopathology in the cerebral white and gray and leptominges [85]. H-3, a virus closely related to KRV, when injected subcutaneously following removal of a molar tooth or fracture, was found to localize wherever repair tissue proliferated [111].

Thus the paroviruses KRV, MVM, FPV, PPV, and the H viruses, all seem to share an affinity for the dividing cell. Diseases associated with the viral infection seem to be limited to tissues which at the time of infection are undergoing changes which involve cell proliferation [110]. Detailed information describing the pathology of parvovirus infections is available in other reviews and articles [3, 6, 112, 87].

H. Latency

Many studies have suggested the possibility of asymptomatic latent infections of animals and cells with several of the parvoviruses [113, 12, 114, 83, 115, 14].

Whether this latency is associated with incorporation of viral components into cell structures or the maintenance of a chronic cell infection at a level too low to detect is not certain.

A number of conditions have been reported to activate or induce the parvoviruses from a latent or chronic infection state. These conditions include manipulation of cells during cell passage; freezing and thawing cells [36]; changing cell conditions so that the cells begin to proliferate rapidly; x-irradiation of animals [20, 60]; and the presence of chemicals such as cyclophosphamide [75]. A chronic virus-cell relationship may be inferred from the observation that H-1 virus can cross the placenta from mother to embryo. It can remain latent in both for life. Presence of viral antigen is detectable only by serum antibody titers [3, 6]. Hamsters infected with H-1 only once at birth carry high antibody titers sometimes as much as 3 years later [6]. Accidental infection of man with H-1 can result in a change from a prior negative serum antibody titer to an elevated titer. This elevated antibody titer has been observed to remain constant for over 10 years, indicating the continued presence of the virus in some form. Melnick et al. [116] have reported 2 parvoviruses (HS-3) similar to both H-3 and KRV, and Kirk virus. Both were isolated from acute phase plasma of patients who had volunteered for studies of infectious hepatitis A. The authors believed that these viruses may have been latent agents activated by disturbance of the liver during hepatitic infection leading to reparative cell division. It is also possible that inoculation of the material from hepatitic patients leads to an activation of latent Kirk and HS-3 in the tissue cultures used to test for viral agents.

The defective parvovirus AAV has been reported to exist in a latent provirus state [117, 4]. Primary human embryonic kidney cells and African green monkey kidney cells have been found to be immunologically negative and not able to produce detectable infectious AAV-2 particles. However, when exposed to purified adenovirus, 1-2% of the human cells and up to 20% of the monkey cells released infectious AAV-2 or AAV-4 and became immunologically positive [4]. Since AAV in kidney cells in the absence of helper virus results in a rapid loss of ability to recover the virus [118, 119], it is possible that the virion under these conditions exists in a provirus state. Hoggan [113] has reported establishing AAV carrier lines of the 3 human AAV serotypes in Detroit 6 cells, in the absence of helper virus. For over 100 passages these abortively infected cells have been able to produce infectious AAV when exposed to helper virus. The cells may contain part or all of the AAV genome or there may be an undetected helper virus in the cell line.

A word of caution must be expressed in the interpretation of the reports on parvovirus latency. As previously discussed, the parvoviruses are stable, heat and chemical-resistant, and ubiquitous in the environment.

They could contaminate animals and cell cultures when their growth conditions are optimal. It is thus difficult to be certain of the origin of parvovirus isolates from tissues where recovery has been made.

IV. Structural Components of Parvoviruses

The parvoviruses are considered naked virus particles composed only of protein and DNA. They are resistant to lipid solvents, such as ether or chloroform, and are considered to contain no essential lipids [17, 24, 32, 120, 12]. They have been shown to contain DNA by isolation of DNA from purified virions [121, 122, 63, 123] and by prevention of virus replication with DNA inhibitors [20, 124-126, 80]. Viral proteins have also been isolated from the virus particles and studied further by gel electrophoresis. In order to examine the composition of parvoviruses, they must be extensively purified of any contaminating components of the cells, medium, or disrupted virions. Because many parvoviruses adhere readily to cells, membranes, and subcellular components, it is necessary to treat infected cells after lysis with receptor destroying enzyme, proteolytic and nucleolytic enzymes, or detergent [58, 127, 61, 128, 67]. The treated virus suspension is further purified by banding in cesium chloride [67, 25, 29] or potassium tartrate [13, 58] and by centrifugation in sucrose in a zonal rotor [94] by passage through millipore filters of known pore size [24, 26], or by attachment and elution from erythrocytes [54]. AAV particles are about 0.05 g/cc more dense in cesium chloride than its helper adenovirus and the 2 can be separated by this technique [8].

A. Structural Proteins

The amino acid composition of the total proteins of AAV-2 [129], H-3, KRV, and H-1 [130] shows that they are rich in acidic amino acids and have an overall negative charge. The mean ratios of acidic to basic amino acids for AAV, H-1, and KRV is 2.8, 1.7, 1.27, and 1.95, respectively. The knowledge of the polypeptide composition of the parvoviruses, however, is limited to but a few examples (Table 10.3). The viruses have generally been isolated, purified, and disrupted, and the polypeptide bands resolved by electrophoresis in a 5-7.5% polyacrylamide sodium dodecylsulfate (SDS) gel [131, 69, 130, 11, 129, 66]. The polypeptide bands can be made visible by staining the gels with coomassie blue. Virion proteins can be made radioactive by incorporating radioactive amino acids added during infection. The radioactive proteins can be added to polyacrylamide-SDS gels, electrophoresed, and the position of the protein in the gel determined. This is done by slicing the gels and assaying the

Table 10.3 Polypeptide Composition of Parvoviruses

		Polypeptides (MW X 10^{-3})[a] ± 10%			
Virus	(% protein)	A VP1	B VP2 [VP2′]	C VP3	Reference
KRV	74.4	76 (9)	*62 (60)	55 (10)	131
MVM		92	*69 [*72]	-	137
		82	*61.5 [*64]	-	138
FPL	71.5	73.1	*60.3	39.6	11
MEV	71.5	73.1	*60.3	39.6	11
Haden		85.5	76.8	*66.8	69
H-1		92 (8)	*72 (52)	56 (9)	130
		92 (12)	*69 [*72] (60)	-	135
Lu 3	72	75	*62	-	240
AAV-1	81.8	87 (.5)	73 (2.7)	*62 (52)	129
AAV-2		87 (3.5)	73 (2.7)	*62 (52)	129
AAV-3		87 (3.5)	73 (2.7)	*62 (52)	129
		91.6 (7)	79.3 (8)	*65.9 (72)	66

[a] An asterisk indicates the major polypeptide component of the virions. The Numbers in parentheses indicate the calculated vlaues for the number of polypeptide molecules per virion.

slices for radioisotope content [131]. As seen in Table 10.3, the parvoviruses contain 2 or 3 polypeptides. The major capsid protein (*) in each parvovirus has a molecular weight of 60,000-72,000. Whether the differences are actually significant or merely attributable to work done in different laboratories with slightly different techniques is not presently known since the molecular weight determination of the same protein can vary ± 10% on the basis of several gel determinations [131]. The major protein component of KRV (62,000 MW) has been reported by Salzman and White [131] as the only viral polypeptide to hemagglutinate guinea pig red blood cells and to demonstrate hemagglutination inhibition with antiserum prepared against KRV. It is, therefore, believed to be the capsid protein. The other 2 proteins in KRV, VP1 and VP3, can either be nonhemagglutinating or not present in sufficiently large amounts to react in the assay. An enzyme with DNA polymerase activity and a molecular weight close to that of VP1 has been found associated with KRV [132-134]. In KRV, VP1 is 13.1%, VP2, 75.5%, and VP3, 11.4% of the total radioactivity present. From the approximate amount of each protein

component and its molecular weight and the total molecular weight of the proteins in a parvovirus virion, one can calculate the number of units of each polypeptide component per virion. These values, as known, are given in parentheses in Table 10.3. If KRV contains 32 capsomeres [58] and approximately 60 units of VP2 per virion, there would be about 2 units of VP2 polypeptide per capsomere. The capsomeres may vary in composition; some might contain components VP1 or VP3, or these 2 components may also be internal. It is possible as well that the minor components could be cellular proteins, that the 3 proteins originated from some larger precursor molecule, or that they are precursor or degradation products of each other. The different molecular weight values reported by several investigators for MVM and AAV-3 (Table 10.3) may also reflect different growth conditions for the virions. Kongsvik et al. [130] found that H-1 virus grown in nonsynchronous hamster embryo cells contained 3 proteins of molecular weights 92,000, 72,000, and 56,000. When this virus was grown in synchronous hamster embryo cells, or synchronized SV40-transformed newborn human or embryo kidney, the 2 virion proteins found were 92,000 and 72,000 molecular weight [135]. The two invariant virion proteins of molecular weights 92,000 and 72,000 were designated VP1 or A and [VP2′] or B, respectively. The [VP2′] protein can be converted by proteolytic cleavage to a smaller VP2 protein of molecular weight 69,000 [135, 136]. The origin and significance of the protein with a molecular weight of 56,000 is not known. It is not believed to be coded for by the viral genome [136]. The investigators suggest that the capsid composition of H-1 would be 60 VP2 and 12 VP1 polypeptides. These data would fit a model for a parvovirus constructed of 12 capsomeres as suggested by Mayor et al. [53] better than one of 32 capsomeres [58].

One problem with the isolated virion proteins is that the combined molecular weights of the proteins are close to or greater than the coding capacity of the viral DNA [66, 131, 130, 129]. This fact suggests that the smaller viral proteins are subunits of the larger protein or one or more proteins may be partly or completely cellular in origin. As discussed in the preceding section on cesium chloride buoyant density, the parvoviruses can often be resolved by isopycnic centrifugation in cesium chloride into 3 groups: heavy particles (1.44-1.47 g/cc), light particles consisting of the major viral band (1.38-1.42 g/cc), and empty particles (1.31-1.36 g/cc). Clinton and Hayashi [137] found that mouse A-9 cells infected with MVM after 1 round of infection produced only heavy and empty particles which contained as the major protein [VP2′] (MW = 72,000). The heavy particle could be converted to a light particle when left in contact with the infected cell culture. The major capsid protein in the light particle was VP2 (MW = 69,000). [VP2′] appears to contain VP2 plus another, removable segment of protein. Tattersall et al. [138], also working with MVM, have found that the light particle (CsCl density 1.38-1.42 g/cc) contains 3 proteins but the amounts of VP2 and [VP2′] vary

Table 10.4 Properties of the Single-Stranded DNA from Parvoviruses

Source of DNA virus	MW[a] (× 10^{-6})	Length[b] (μm)	Sedimentation coefficient (S_{20w})
KRV	1.6	1.5 ± 0.2	27
MVM	1.5	1.2	–
FPV	1.7	–	23
MEV	1.7	–	24
H-1	1.7	–	27.8
Lu 3	1.59	1.61 ± 0.25	24 (14)
KBSH	1.4	–	24
AAV-1	1.35	1.38 ± 0.12	(15.5)
AAV-2[d]	1.35	1.38 ± 0.06	24 (15)
AAV-3	1.35	1.39 ± 0.06	(15)
AAV-4	1.5	1.5 ± 0.2	(15.7)

[a]Molecular weight of the DNA from a single virus particle.
[b]The length of AAV DNA molecules is given for double stranded DNA molecules.
[c]Sedimentation coefficients (S_{20w}) and density in cesum chloride (pCsCl) values are given for the defective AAV virions for both single-stranded and double-stranded (in parenthesis) DNA molecules.
[d]AAV-2 values for nucleotide composition given for (–)minus or heavy strand and (+)plus or light strand.

pCsCl[c] (g/cc)	Nucleotides (moles/100 moles)				References
	Adenine	Thymine	Guanosine	Cytosine	
1.726, 1.715	26.8, 26.7	29.6, 30.8	20.6, 20.0	22.9, 22.5	67,122,128 61,165
1.722	26.5	32.7	19.5	21.4	128,25,121
1.722					11
1.722					11
1.720	25.2, 25.5	33.1, 29.3	14.3, 22.6	27.4, 22.6	63,128
1.7254					155
1.724					64
1.729 (1.717)					121,8,156
1.726 (1.714)	(–)20.5 (+) 25.2	26.5 21.7	26.7 26.6	26.3 26.5	127,8
1.727 (1.715)					150,8
1.728 (1.720)					29,8

inversely from preparation to preparation. The amount of VP1 is constant in heavy, light, and empty particles. On the basis of peptide mapping, Tattersall et al. [138] concluded that VP2 is a cleavage product of [VP2′] which accumulates late in infection. VP2 may be a maturation protein which may be required for the binding of the virus to cell receptor sites. Antiserum prepared against SDS-dissociated VP2 [B] polypeptide of AAV [139] shows cross-reaction with VP3 [C] polypeptide, suggesting, if they are pure, a close relationship of the 2 proteins [139].

Further work needs to be done with the viral proteins. Determination of the polypeptides from many parvoviruses in the same laboratory at the same time would aid in comparison of the proteins. Use of radioisotopes to prelabel cellular proteins before infection could help determine the role (if any) the cell plays in viral protein replication. Further studies are also needed of the relationship of virion proteins to each other, the requirement of a specific protein for attachment to cell receptors, and possible enzymatic activity of the proteins.

B. Nucleic Acid

The characterization of the nucleic acid from parvoviruses presents technical problems and has been studied in relatively few cases. It is difficult to obtain large quantitites of the virions for study. In addition, the compact structure of the nucleocapsid makes it very resistant to treatments to extract DNA. DNA can be extracted with proteolytic enzymes or hot detergent [25, 140, 61]. Alkaline extraction of DNA and separation of the virion DNA and protein in cesium chloride solutions [128] or in sucrose gradients [140] have also been carried out. Recovery of virion DNA with enzymes and detergent can run as high as 60-80% [141]. Alkaline extraction followed by isopycnic or velocity sedimentation generally results in less broken and more homogeneous DNA molecules than the enzyme extraction procedure. Some of the properties of extracted parvovirus DNA are given in Table 10.4; additional information may be obtained from reviews by Rose [8], Hoggan [5], and Bèrns [142].

The early indications that KRV and H-1 contained DNA came from studies of infected cells. Those studies showed that intranuclear inclusions made up of virions could be stained with acridine orange and were Feulgen positive [143, 144]. Cheong et al. [145] found additional support for the DNA content of KRV in their discovery that [^{3}H]thymidine and not [^{3}H]uridine was incorporated into the virions. Final proof of the DNA content of a few parvoviruses came from studies isolating the DNA and identifying the separated nucleotides as shown in Table 10.4. It should be noted that the DNA of KRV, MVM, and H-1 contains a high percentage of thymidine (29-33%). This is a property shared with the bacteriophage ϕX174, which contains 32.8% thymidine [146].

The single-stranded nature of the virion DNA in the intact virus particles was demonstrated by several methods: staining techniques, nucleotide base ratios, enzyme treatments, and formaldehyde reaction. Formaldehyde can react with the free amino groups found only in single-stranded DNA and cause an increase in absorption at 260 nm. Because of hydrogen bonding between base pairs, double-stranded DNA polymers have no free amino groups to react with formaldehyde. H-1 [63], KRV [67], and MVM [121] were shown to contain single-stranded DNA using this procedure. A single-stranded structure for the DNAs of MVM and H-1 has also been confirmed by nearest neighbor analyses [122, 128], and for X14 by acridine orange staining [56]. Finally, the nucleotide analysis of several parvovirus virion DNAs (Table 10.4) demonstrated that the moles of adenine do not equal thymine and the moles of guanine do not equal cytosine, which would be the case in double-stranded DNA.

A controversy existed for several years over the DNA structure in the defective AAV parvoviruses. Although originally thought to be double-stranded in the virion [147], AAV DNA was subsequently shown by acridine orange staining to be single-stranded both in infected cells [148] and inside virus particles [56, 149]. However, the structural analysis of extracted AAV DNA was shown by Rose et al. [141] and Parks et al. [29] to be double-stranded with a molecular weight between 3 and 3.6 $\times$ 10^6. The average molecular weight estimate for KRV DNA is 1.6 $\times$ 10^6 [61, 67, 122], for H-1 DNA, 1.7 $\times$ 10^6 [63, 128], and for MVM DNA, 1.5 $\times$ 10^6 [121, 128]. The similarity in the sizes of AAV particles and the other parvoviruses, taken with the evidence that the DNA inside the AAV virions appeared to be single-stranded, prompted Crawford et al. [121] to suggest that AAV preparations contained 2 types of particles, one containing the (+) strand and the other the (–) strand. On extraction, these annealed to give a duplex of twice the molecular weight of the single-stranded DNA. This explanation has been confirmed. In a revealing experiment, Rose et al. grew virus in the presence of bromodeoxyuridine (BUdR) [150]. BUdR substitutes for thymidine in DNA. The BUdR containing DNA becomes denser in cesium chloride gradients and separable from unsubstituted light DNA. The extraction of DNA from a mixture of DNA labeled with BUdR and unlabeled AAV-3 virus resulted in formation of hybrid density molecules, suggesting that annealing occurred after extraction. Mayor et al. [149, 151], using AAV-4, and Berns and Rose [152], using AAV-2, showed that under conditions which did not denature the duplex DNA but minimized the association of complementary strands, the AAV virions contained single-stranded DNA. When BUdR was substituted for thymidine in duplex AAV-2 DNA, the 2 strands could be separated by equilibrium density centrifugation (heavy BUdR strand = 1.830 g/cc; light BUdR strand = 1.79 g/cc) [152]. The implication of this strand separation is that 1 of the viral strands had a high thymine content similar to that found with other parvoviruses. As shown in Table 10.4, when the base composition of AAV-2 was analyzed, the minus (–) strand, which banded at the heavy position in cesium chloride, contained 26.5% thymine and the (+) strand 21.7% thymine [127]. Berns and Adler [153] separated the 2 types of intact AAV virions by labeling the DNA with BUdR, banding the virions in CsCl, and extracting the DNA. Particles which banded at a heavier than average density in CsCl contained only a single heavy (–) strand of DNA. The interesting question of whether DNA from the AAV particles was double- or single-stranded was thus resolved. The production of virions containing either a single-stranded DNA (+) or (–) molecule may also occur with the defective parvovirus CAAV, as suggested with acridine orange staining [33] and with the densonucleosis parvovirus (DNV) from the insect *G.* mellonella [154, 38]. This property is not characteristic of any autonomous parvovirus.

The configuration of the parvovirus DNA was then studied to determine if it was linear or circular. Salzman et al. [122] exposed extracted KRV DNA to exonuclease I, an enzyme which specifically hydrolyzes linear single-stranded DNA. Within 4 hr of incubation the KRV DNA was 70-80% hydrolyzed. Further evidence of linearity was provided by electron microscopic examination of the DNA. Only linear KRV DNA strands were observed. This contrasted with the circular DNA configuration of the bacteriophage ΦX174. MVM [121] and Lu III [155] are also believed to have a linear structure after examination under the electron microscope. It remains possible, however, that the parvovirus DNA is a circular molecule linked by protein, hydrogen bands, or covalent bands which are very efficiently broken during extraction procedures.

The length of the linear DNA from the autonomous parvoviruses varies from approximately 1.2-1.61 as shown in Table 10.4. The length distribution of single-stranded DNA is generally reported to be more heterogeneous than that found with double-stranded molecules. Minor variations in conditions of ionic strength, cytochrome C, and surface forces during spreading and drying influence the length measurements, particularly of a single-stranded DNA. The nucleotide bases are not rigidly held and can stack, stretching and contracting in a nonuniform manner along the DNA molecule. Thus, the possibility exists that the apparent differences in measured length of DNA molecules from several parvoviruses may only be a matter of technique. A picture of the linear single-stranded DNA extracted from KRV is shown in Fig. 10.2 (A).

The configuration of the DNA from the defective AAV parvoviruses presented another dilemma. Vernon et al. [156] found that after osmotic shock of AAV-1 particles a very few circular duplex DNA molecules were seen. As discussed above, extraction of AAV plus and minus single strands

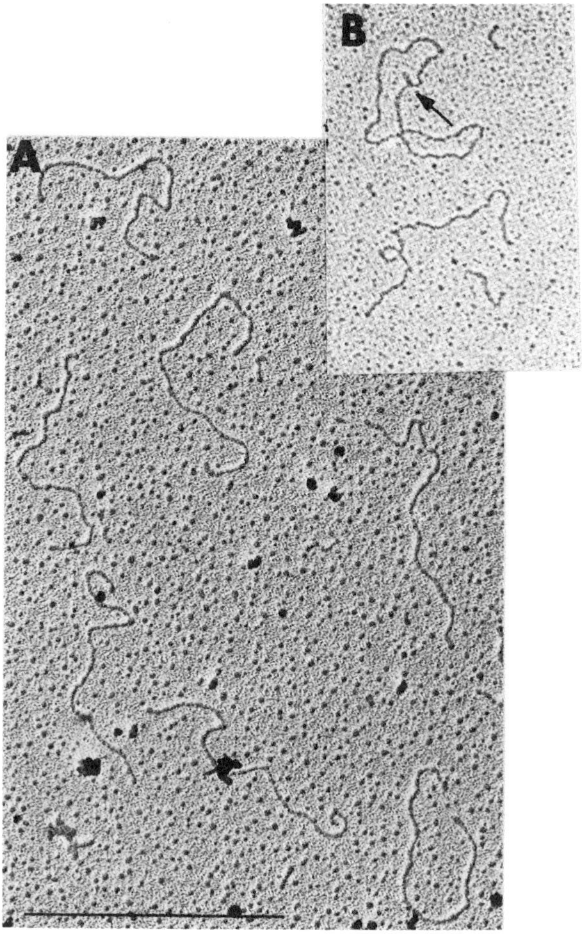
A
B

Fig. 10.2 Parvovirus single-stranded DNA (A) KRV-DNA: the line superimposed on the electron micrograph is equivalent to 1.0 μm. (B) AAV hydrogen-bonded circles: the arrow points to a pan handle type projection. (Picture courtesy of K. I. Berns and T. S. Kelly, Jr.)

usually results in the formation of a double-stranded DNA molecule. Gerry et al. [157], after neutral sucrose centrifugation of AAV DNA, isolated linear duplex circular molecules of unit length and linear duplex dimers. All duplex species are formed by linear single polynucleotide chains of unit length. Thus, duplex circles and linear dimers would be held together by short overlapping hydrogen-bonded regions [157]. Limited digestion of circular duplex monomers with 3′ exonuclease III disrupted one-half the circles and with 5′ T5 exonuclease the other half of the circles were opened. These experiments demonstrated that the overlap region may have 3′ or 5′ termini and that the overlap region is less than 6% of the length of the genome. Duplex linear monomers which do not form circles under annealing conditions will form circles or oligomers after a 1% degradation of the DNA with exonuclease III. Gerry proposed a model in which virion single-stranded, linear DNA consists of 2 or more permutations existing within a small segment, less than 6% of the genome length. The single DNA strands contain a terminal repetition (noninverted) representing 1% of the genome, similar to that found in linear bacterial DNAs [142, 157].

Koczot et al. [140] found that purified plus and minus strands of AAV-2 contain self-complementary sequences (inverted terminal repetitions). These terminal sequences comprise less than 10% of the strand length and can anneal to form single-stranded circular molecules which would be closed by realtively short hydrogen-bonded duplex segments or panhandles in an antiparallel configuration (Fig. 10.2, Part B). The hydrogen-bonded closure was demonstrated by its sensitivity to exonuclease III, its thermal stability comparable to double-stranded DNA, and its ability to regenerate circles after alkali denaturation [140]. The only other viral DNA known to contain this terminal sequence complementarity are the single strands of adenovirus DNA [158, 159]. As previously stated, adenovirus can be the obligatory helper virus of the defective AAV particles and this terminal sequence complementarity may be important in adenovirus-AAV interaction. If the base-paired regions holding the single-strand circles of AAV DNA together were sufficiently long, they could be observed under the electron microscope. Koczot et al. [140] found possible projections from single-stranded hydrogen-bonded circles in a panhandle-type formation. Berns and Kelly [160] found these projections at specific sites along the single-stranded circular molecules produced from linear plus AAV DNA chains. The projections corresponded to about 1.5% of the length of the DNA and may be the length of the region of the inverted terminal nucleotide sequence repetition. The panhandle in the circular AAV DNA is indicated by the arrow in Fig. 10.2 (B). The possible models to account for the data indicating the existence of both inverted and natural terminal nucleotide sequence repetition in a population of purified AAV DNA are discussed by several authors [142, 61, 140, 157]. It is not known if all the observed types of terminal sequences in AAV DNA are present in all the virion DNA strands. It is interesting, however, to observe that they all involve less than 10% of the genome length at either strand terminus. The single-stranded DNA from the human and simian AAV serotypes is believed to be a linear molecule in the virion and would be similar to the other parvoviruses in this respect. Circular DNA molecules have not been reported with the KRV, MVM, and Lu III autonomous parvoviruses [121, 122, 155].

In comparison, the DNA from the bacteriophage ΦX174 is a covalently closed single-stranded circular molecule [146, 161, 162].

C. Nucleic Acid Homology

The degree of homology between DNAs extracted from different parvoviruses has been studied only with a few of the defective parvoviruses. Rose [8] reports that in DNA-DNA competitive hybridization studies between AAV-1, -2, and -3 DNAs, 60-70% of the nucleotide sequences were found to be homologous. Studies with a technique employing synthetic RNA showed interserotypic homologies among the 4 human and simian AAV serotypes of 27-37%. It is possible that the latter estimates of homology are low due to incomplete copying of DNA template molecules [163].

Rose [8] reported the formation of heteroduplexes with AAV-1, -2, and -3 DNA. Under denaturing conditions, duplex DNA, with each strand of the duplex from a different AAV strain, shows a separation of the strands ("a melting out") in those regions where the base-pairs from the different DNA strands are imperfectly matched and hydrogen-bonded to each other (heterology region). Regions containing heterology melt out progressively as denaturing conditions are increased, suggesting that partial homology exists within the regions. Interserotypic homologies were found to map at similar locations in all 3 serotypes, suggesting that they all have conserved the same sequences. No melting was observed under similar conditions in duplex molecules with the plus and minus strands from the same strain. Nucleic acid hybridization experiments have not shown any DNA homology between AAV DNA and DNA from its adenovirus or herpesvirus helpers [8].

D. Enzymatic Cleavage of Parvovirus DNA

The enzymatic cleavage of AAV-2 DNA by the restriction endonuclease R·Eco R_1 has been reported by Carter and Khoury [164]. They found that linear monomeric AAV-2 DNA duplexes (MW 2.8×10^6) were cleaved at 2 specific sites yielding 3 fragments: A (57.2%), B (38.2%), and C (4.6%). Fragment C may consist of 2 similar fragments. The cleavage patterns of oligomeric forms of AAV were also determined, and the order of the fragments of the minus strand in the AAV genome was suggested to be 3′ OH *B C A* p5′ [166]. Separation of the complementary duplex strand of fragments A and B showed that A contained the 5′ terminus of the minus strand, and the 3′ terminus of the plus strand and vice-versa for fragment B [166]. The asymmetry in distribution of thymidine between the plus and minus strands is preferentially located in fragment A. Berns et al. [167] have also constructed a physical map of the AAV-2 genome on the basis of 5 fragments produced by the restriction endonucleases Hind II and III. They have ordered and oriented these fragments to those produced by the R·Eco R_1 enzyme. They also presented evidence that AAV DNA contains 2 nucleotide permutations, the starting points of which may be separated by 1% of the genome.

V. Stimulation of Parvovirus Replication by Infected Cell Extracts or Helper Virus

Chany and Brailovsky [168] found that coinfection of rat cells with adenovirus 12 and KRV led to considerable stimulation of KRV synthesis. Extracts of adenovirus 12-infected human cells could also elicit this stimulation. This stimulatory activity, named "stimulon," was thought to be an interferon antagonist. With other viruses, however, stimulon does not appear to reduce the protection conferred by interferon in infection. Coinfection of cells with infectious adenovirus can increase the parvovirus yield obtained with X14, H-1, and Kirk virus [77, 169, 115]. Mirkovic et al. [77] obtained a 4-fold increase in her yield of X14 when rat embryo cells were coinfected with adenovirus 7. H-1 will not grow autonomously in human embryonic lung cells (HEL). H-1 will, however, replicate in the HEL cells if they are coinfected with human adenovirus 12. Extracts of adenovirus-infected cells or ultraviolet (UV) inactivated virus did not substitute for the infectious adenovirus and did not permit the growth of H-1 [170, 169, 115]. A 100-fold increase in yield of Kirk virus is also reported where the Detroit 6 cells are coinfected with adenovirus 7 [77]. The autonomous growth of H-1 in NB cells transformed by simian virus 40 (SV40) may also depend on SV40 acting in some unknown way as a helper virus. The possible stimulation of the autonomous parvovirus growth by coinfection with another helper virus has not been studied either systematically or in detail. It could be important in understanding virus replication and control mechanisms. Because adenoviruses may act as helpers in some systems for the autonomous independent parvoviruses, Toolan and Ledinko [171] postulated that all parvoviruses are defective in some way. The adenoassociated viruses are defective and entirely dependent on coinfection with a helper virus, and other parvoviruses, such as KRV and H-1, are autonomous in some systems and defective and dependent on a helper adenovirus infection in others.

The members of subgroup B (Table 10.2) are the nonconditional defective viruses [172], including the adenoassociated virus. It is possible that these defective parvoviruses may yet be autonomous in some undiscovered cell line, as is the case with the members of subgroup A. The adenoviruses appear to be the only RNA or DNA viruses tested that support the production of complete infections AAV particles [24, 26, 173]. The production of AAV-1 immunofluorescent antigen, AAV DNA, RNA, and possibly empty capsids, but not infectious AAV, can occur after coinfection of cells with AAV and herpes simplex, or any of the following herpes viruses: infectious bovine rhinotracheitis virus, Epstein-Barr (EB) virus, cytomegalovirus, varicella-zoster virus or herpesvirus saimiri [173-178]. AAV replication may require the expression of a late event in the adenovirus replication cycle as adenovirus-transformed cells expressing only an early event in the adenovirus cycle (T-antigen synthesis) will not serve as host for AAV replication [26, 8]. However, AAV-1 has been reported to grow in cells where helper viruses replicate inefficiently, if at all [179, 117, 180], and AAV replication has been reported at nonpermissive temperatures in cells coinfected with DNA-minus temperature-sensitive adenovirus mutants [181, 179, 182]. Which step or steps in AAV replication are defective and must be supplied by helper virus is another unresolved question which promises a productive and informative area for future investigation.

VI. Parvovirus Replication

Studies of the intermediate steps in parvovirus replication have generally relied on growth of the virus in cell culture. The parvoviruses are generally fastidious in their requirements for a host cell. The host cells often must be derived from a specific species of animal and, sometimes, even specific tissue from the animal. Generally, the cell must be rapidly dividing (optimally in the S phase of the cell cycle) for virus infection and replication to occur [183-187]. In the S phase or DNA synthesis phase, the cellular enzymes and subcellular components are probably primed for the DNA-containing parvovirus to replicate. The effect of the parvovirus on the cell depends both on the virus and on the cell used. The virus infection can result in cell lysis or no demonstrable effect. Parvoviruses can cause temporary or permanent inhibition of cell mitosis [188] and CPE [100, 187-190]. The virus-cell interaction can be observed and monitored in the microscope either by immunofluorescence [191, 192, 139] production of inclusion bodies (FPV, MEV) or by the incorporation of specific radioisotopes into viral and subcellular components [193, 183, 177, 194, 195]. Detection of mature virions from infected cells has been carried out utilizing blind passages in vivo and in vitro [12, 14, 21, 106], by extraction of cell sheets with a glycine buffer [36], and from disrupted cell lysates [193, 183, 195, 177]. The amount of virus present has been quantified by inoculation into animals to determine lethal (LD_{50}) or pathologic dosage, tissue culture infective dose ($TCID_{50}$), hemagglutination (HA) and hemagglutination inhibition (HI), complement fixation (CF), fluorescent antibody titration (FA), and by plaque forming units (PFU) (Table 10.1).

Experiments concerned with precursors of virus components and the effect of the virus on cellular macromolecular synthesis have utilized radioisotopically-labeled thymidine for DNA studies, uridine for RNA studies, and amino acid hydrolysates for protein studies [193, 183, 177, 195, 129]. The advantage of the radioisotopic method is the small amount of radioactive material required for detection and study. DNA and RNA is determined as viral specific, generally by hybridization with viral DNA. Protein may be detected as viral specific or related by reaction with antibodies to intact or dissociated virion capsid proteins. Loss of a normal cellular component or function or the detection of a new cellular property may also occur in the infected cell and indicate viral regulation.

Most of the defective parvoviruses cannot be assayed by HA or HI (Table 10.2). The defective parvoviruses may produce a CPE. This is difficult to determine, however, because the required helper virus may also be the cause of the CPE. The defective parvoviruses are most often assayed by complement fixation antigen titration and fluorescent antibody titers.

A. Cells Permissive for Viral Infection

Although the parvoviruses in subgroup A (Table 10.1) are able to replicate autonomously in specific cell lines, they are dependent on the cell itself to supply most of the enzymes and subcellular structures required for virion replication. Some of the parvoviruses may be conditionally defective, growing autonomously only in certain cells under specific conditions. The cells used most often for parvovirus replication include actively dividing cells from nonconfluent monolayer cultures, cells derived from tumors that are not contact-inhibited, cells transformed by another virus (such as NB cells), and cells synchronized by either serum pulse [183, 185, 196] or DNA inhibitors [187, 188].

Rat embryo and hamster embryo cells in tissue culture will support the autonomous growth of many parvoviruses, including KRV, X14, L-S, HER, RT, KBSH, MVM, and H-1 [3, 82, 71, 25, 197, 198, 187, 196, 183, 192]. In addition, KRV has been grown in a rat nephroma cell line (RN) [68, 163]. MVM is the only known parvovirus that grows in mouse embryo (ME) and mouse L cells (MLC) [199, 88, 185]. FPV has been grown in feline kidney cells [22, 23, 100], PPV in pig kidney cells and continuous lines of human cells [125, 96, 71], Haden in fetal bovine cells [120], and MVC in dog epithelial cells as well as other primary and continuous cell cultures [80]. RT has been grown only in the continuous rat fibroblast cells from which they were isolated [71]. Lu III can replicate in HeLa cells and like TVX in other continuous human cell lines [71]. The parvovirus KBSH has been grown in KB cells [64]. H-1 virus has been grown in a newborn human kidney cell line transformed with the virus SV40 (NB). Incomplete replication of H-1 occurs in Salk monkey heart (SMH), a diploid human cell line (WI-26), and HEL [115]. The defective parvoviruses AAV-1, -2, and -3 appear to replicate in almost any cell line such as KB and HEK which can support complete or sometimes partial adenovirus replication [26, 118, 90, 200, 201]. $AAVX_7$ has been grown in bovine kidney cells with bovine adenovirus type 1 as a helper virus [35] and avian AAAV can be helped by avian adenovirus when grown in chicken fibroblasts [202].

B. Ultrastructural Studies of Infected Cells

Infecting cells with parvovirus may have profound effects not only on cell integrity but on the cell cycle as well. With the aid of the electron microscope, observers have noted morphological changes in the cell after infection. Late in infection mature virions are found in the nucleus, often in crystalline array. Virus is assembled in the nucleus and remains cell-associated long after infection [203, 125, 197, 190].

Al-Lami [197] found that the earliest effect of H-1 infection was localized in the nucleolus of H-1-infected nonsynchronized SMH and NB cells. Singer et al. [204, 205], studying synchronized cells, found that 12 hr after infection—immediately after the initiation of viral protein synthesis—chromatin condensation and vacuolation took place within the nucleolus. The negatively charged H-1 proteins showed a great affinity both for chromatin material [206] and the orthophosphate-binding nucleolar basic proteins. This suggests that the proteins could play a regulatory as well as a structural role in the kinetics of virus replication. By 24 and 36 hr after infection the nucleoli disintegrated. This was accompanied by extranucleolar chromatin condensation and accumulation of extensive amounts of interchromatin granules in the nucleus and cytoplasm. Current immunoelectron microscopic observations indicate that the specific binding of unassembled H-1 proteins to nucleolar-associated chromatin and nuclear heterochromatin 10 hr after infection are associated with their condensation, probably initiating nucleolar destruction and subsequent nuclear damage [204].

The early involvement of the nucleolus in parvovirus infection has also been reported during AAV infection in HEL cells in a herpes helper system [207]. The precise role that the nucleolus plays in virus synthesis and assembly, however, awaits further elucidation. Large aggregates of AAV capsids have been observed in close association with fragmented nucleoli [178]. Intranuclear densely stained aggregates attributed to AAV particles and empty particles appearing as honeycombed inclusions were reported in the nucleus of KB cells infected with AAV and adenovirus. Fibrillar inclusions attributed to AAV were also reported in the KB cell nuclei [208]. Thin sections of cells infected with AAV and adenovirus show that the nucleus contains predominately either AAV or adenovirus particles [209, 119, 208].

Difficulties in obtaining morphological data concerning parvovirus development in the infected cell are caused by its small size and densely staining core of the parvoviruses. This makes them difficult to distinguish from ribosomes with comparable morphology and stainability. In the case of the defective parvoviruses, electron microscope observations are further complicated by the presence of a helper virus and its effect on the cell.

C. Infectious Cycle

A single MVM virion is reported as capable of initiating a productive infection of a susceptible cell [185]. This finding may be true generally for autonomous parvoviruses. Defective parvoviruses, such as AAV, have 2 types of virions each containing either a plus or minus strand of DNA. It is not known whether 1 or both types of AAV virions are required for a productive infection in the presence of helper. Titrations of AAV carried out in the presence of excess adenovirus, however, suggest that a single AAV particle and a single adenovirus particle are required for infection [210, 175]. Infecting cells with physically separated plus and minus particles could answer the question. Infection multiplicities used have varied from 2-200 PFU or $TCID_{50}$ per cell. Approximately 90% of a 5 PFU/cell inoculum of KRV was taken up in 1 hr after infection [193]. Seventy-five percent of an 8 PFU/cell of Haden virus was adsorbed in 2 hr after infection [120]. Sixty to seventy percent of H-1 virus was adsorbed to human lung cells with or without the adenovirus 12 helper [169, 170]. Latent periods, during which very little infectious virus can be recovered, have been reported between 10 and 16 hr for Haden, KRV, FPV, H-1, HER, and Lu III [22, 197, 145, 193, 190, 119, 192, 184]. The production of new infectious virions is exponential for the autonomous virus beginning at 8-12 hr after infection and giving maximal yields by 22-30 hr after infection [100, 184, 189, 193, 196]. Parvoviruses are assembled in the nucleus [211, 197, 27, 64]. Final virus yields are 1-4 logs greater than that found in the latent period [193, 197, 184, 120, 196].

Little scientific attention has been accorded to the enzymatic changes that occur in the cell infected with a parvovirus. An enzyme associated with purified KRV has been shown to have DNA polymerase activity [132]. Highly purified preparations of the enzyme have been compared with the DNA polymerases of the cell in which the virus was grown [133, 134]. The viral-associated enzyme differs from the cellular enzymes in chromatographic properties, ion requirements, and substrate specificity. We do not yet know if the viral-associated enzyme is coded for by the virus, is a cellular enzyme which may have been modified, or is simply an adventitious association with the virus. Its role, if any, in viral DNA synthesis remains to be elucidated.

Cocuzza et al. [212] has described a decrease in thymidine kinase after infection of RE cells with H-1. This finding has been confirmed but with some reservations [213, 6]. Infection of SV40-transformed newborn human kidney resulted in a drop in radioactive thymidine incorporated into DNA, but this occurred late in infection. This incorporation, a measure of thymidine kinase activity, could be due simply to a decrease in the infected cell's capacity to synthesize DNA. The meaning of the decrease and its relationship to viral DNA synthesis is unknown.

The infectious cycle of the autonomous parvoviruses is dependent on the state of the cell at the time of infection. Irradiation of rat embryo cells with x-ray or UV light interferes with their capacity to synthesize KRV. This indicates a dependence of the virus on 1 or more cell functions [184, 185]. Treatment of the cells prior to infection with 5-fluorouracil, an inhibitor of DNA synthesis, also inhibited KRV replication. On the basis of these results, Tennant et al. [184, 185] concluded that KRV probably requires some cellular functions involving DNA synthesis in order to replicate. Use of synchronized cells have shown that the requirement or requirements for replication of KRV, MVM, H-1, and Lu III involve cellular functions during the S or G2 periods of cell growth [186, 188, 184, 187, 196]. Maximum yields and minimum latent periods of H-1 and Lu III are achieved if the cells are infected during early S phase [187, 187, 196]. Rhode reported that a DNA synthetic event (HA-DNA synthesis), which occurs in late S phase of the infected cell, is required for production of the viral mRNA for capsid protein (the viral hemagglutinin, HA) to occur. Viral HA synthesis follows this event by 1-2 hr [196, 68]. Virus production is, therefore, regulated by the cell synthesis of the component or components required for virus replication at a specific time in the cell cycle.

The adsorption, penetration, and uncoating of the defective parvoviruses occur to the same extent whether helper virus is absent or present. One to two hr after infection, 16-30% of AAV DNA radioisotopically labeled with tritiated thymidine was adsorbed to the cell [195, 142]. Within 2 hr after infection, radioactive DNA was found in the nuclear fraction of the cells. Using sensitivity to nuclease as a measure, 85% of the AAV DNA was uncoated by 16 hr after infection [214]. The remainder of the AAV infectious cycle is dependent on the time of helper virus infecton and the helper virus used. Simultaneous infection of KB cells with AAV-2 and adenovirus-2 [195] or AGMK cells with AAV-4 and helper SV15 [28] resulted in a AAV latent period of about 15-18 hr and an exponential synthesis of virus for an additional 12-15 hr. If the cells are infected with adenovirus 10-12 hr before AAV infection, the AAV latent period can be shortened to 4-6 hr [28, 215, 210]. With herpes simplex virus (HS-1) as a helper virus, the AAV latent period (no production of infectious virions) is about 6 hr [175]. In contrast to the observation with adenovirus helper, preinfection of cells with HS-1 before AAV infection delays the appearance of AAV fluorescent antigen [175]. The yields of AAV from infected cells are similar to those found with the H-1. Yields are much lower with the other autonomous parvoviruses. Each AGMK cell may yield more than 2.5×10^5 AAV particles [28]. Utilizing the average amount of AAV DNA synthesized per infected KB cell, Rose and Koczot [195] calculated that $2\text{-}4 \times 10^5$ AAV-2 genomes are synthesized per cell. As with autonomous parvoviruses, not all cells are competent for AAV infection and the particles synthesized per cell are probably even greater than reported [173, 215, 210].

D. Viral Protein Synthesis

The rate of protein synthesis is not significantly different between the infected and uninfected cell [188, 193]. Detection of viral specific protein has been accomplished most successfully by use of fluorescein-labeled antibody. The antibody is prepared against intact or dissociated viral capsid proteins labeled with fluorescein isothiocyanate and then added to infected cells. The antibody combines with the viral antigen in the cells. Cole and Nathanson [191], using $TCID_{50}$ per cell, studied the replication of the HER strain of KRV in rat embryo (RE), BHK-21, and L cells. They found specific fluorescence in the cytoplasm of infected RE cells as early as 5 hr after infection. The number of cells demonstrating cytoplasmic fluorescence increased to a maximum of 60% by 8 hr after infection, declining thereafter. Nuclear fluorescence was first observed 8 hr after infection; by 14 hr after infection it was present in 50% of the cells. About 20 hr after infection the cells were destroyed and extracellular virus hemagglutinin appeared. The virus cycle was approximately 20 hr in length in the RE cells. Cytoplasmic antigen was observed in BHK-21 and L cells but cell destruction occurred before nuclear involvement and no release of extracellular hemagglutinin or infectious virus was observed. Fields and Nicholson [192] confirmed the sequence of events in the growth cycle of KRV in RE, HE, and L-939 cells. They found that growth in HE cells lagged behind the growth cycle in RE cells and gave a decreased yield. The observation of cytoplasmic fluorescence followed by nuclear fluorescence has been also reported after infection of cells with KBSH and Lu III [64, 186]. The observation may indicate the migration of viral capsid proteins from the cytoplasm to the nucleus for incorporation into mature virions. The appearance of nuclear antigen correlates with the production of infectious virus. In KRV infection of L cells, it is possible that the viral capsid precursors are synthesized in the cytoplasm of the cells but they do not migrate to the nucleus and mature virions are not synthesized.

The order of appearance of cytoplasmic and nuclear antigen may vary with the parvovirus and the cell. Parker et al. [199] reported that MVM cytoplasmic antigen did not appear until 2 hr after the nuclear antigen. Kongsvik et al. [135] found that the 2 proteins of H-1 were synthesized concomitantly in the nucleus in about the same proportions as in the virion. As expected with viruses so dependent on the phase of the cell for virus replication, blocking cellular DNA synthesis with inhibitors also blocked nuclear antigen synthesis of MVM, Lu III and H-1 [199, 186, 135, 196]. In Lu III-infected cells, however, the cytoplasmic viral antigen appearance was blocked neither by DNA inhibitors nor by reduced mitotic activity of the cells [186]. Cytoplasmic antigen in this case may be independent of the cell physiological state; it may represent localized ingested viral antigen or the virus itself may be transcribed or act as messenger for this antigen. The cytoplasmic antigen also could be a modified cellular protein. More study is needed to unravel this mystery.

Further clarification of the steps in protein synthesis may develop from the use of temperature-sensitive (ts) mutants. Rhode [216] has isolated 2 ts mutants of H-1. Under restrictive conditions, both mutants synthesize a capsid protein which is defective in hemagglutination. The mutants are also distinguishable by the thermolability of the hemagglutinin.

The synthesis of fluorescent staining antigens of the defective parvoviruses (AAVs) produced with herpesvirus as a helper is morphologically similar to that produced with adenovirus as a helper [173, 210]. Herpesvirus helpers do not result in production of infectious AAV virions [173, 195, 210]. Johnson et al. [139] followed the kinetics of AAV-protein synthesis using labeled antiserum prepared against the 3 SDS-treated structural polypeptides VP1, VP2, and VP3. AAV proteins was detected 2-4 hr after the onset of AAV-specific DNA and RNA synthesis. Kinetic studies of antigen formation showed VP1 (main capsid protein) first stained the cytoplasm by 14 hr after infection. By 18 hr after infection, both cytoplasm and nucleus were stained. VP2 antiserum stained only discrete intranuclear areas and VP3 antiserum stained the entire nucleus. All 3 antigens appeared at about 14 hr postinfection, about 2 hr prior to the appearance of whole virion antigen. Antiserum against intact virion antigen stained only the nucleus [210]. The results suggested that at least 1 AAV polypeptide component, the major capsid component, (VP1) is synthesized in the cell cytoplasm. The minor capsid proteins may be present in the cytoplasm in quantities too low to detect [139].

Mayor et al. [217] investigated the synthesis of AAV proteins and DNA in the presence of herpesvirus and adenovirus helper. In the presence of a herpesvirus helper, AAV antigens and DNA are synthesized, but no complement-fixing antigens or infectious virus particles. The AAV DNA synthesized with a herpesvirus helper may occur in the form of a deoxynucleotide protein complex, which is a possible intermediate in DNA and protein synthesis. Preliminary studies with adenovirus as a helper suggest a very minor fourth capsid protein of AAV. This minor component may be a structural protein common to adenovirus and satellite virus, perhaps a common maturation protein.

E. Viral DNA Synthesis

In the KRV-infected RN cell, 6-7 hr after infection DNA synthesis decreased sharply in comparison to the uninfected cell [193]. Viral DNA synthesis started 7-8 hr after infection and the rate of DNA synthesis in the infected RN cell remained at a higher level than in the control cell for the rest of the 20-23 hr virus replication cycles. Differences in the effect of the virus infection on total cellular DNA synthesis may vary with the virus and the particular type of cell. In RE and rat kidney cells, total DNA synthesis decreased between 10 and 24 hr following KRV infection [188]. MVM infection of ME cells did not alter the rate of DNA synthesis [185]. KRV DNA synthesis began about 4 hr before the production of infectious virions and progeny production appeared to require concomitant viral DNA synthesis until RN cell lysis [193]. Synthesis of MVM [188], Lu III [190], and H-1 DNA [196, 68] are also reported to occur 8-17 hr postinfection. Lu III DNA synthesis precedes mature virion production by 2-4 hr and DNA synthesis ends with cell lysis [190].

Salzman and White [131] followed the fate of radioactive viral DNA after infection. They observed that within 60 min after infection 28-42% of the parental single-stranded linear DNA is converted to a new form. On the basis of its sedimentation in sucrose gradients, buoyant density, elution from hydroxyapatite, and appearance in the electron microscope, they judged the new form to be double-stranded and linear. It may be analogous to the double-stranded circular replicative form of the bacterial parvovirus ΦX174 [218]. A double-stranded form of viral DNA has also been reported after MVM [219, 220, 185, 190] and H-1 [68, 221, 216] infection of cells. Some of the linear double-stranded MVM viral DNA (20-40%) renatures intramolecularly, suggesting interstrand cross-linking or a hairpin structure. These self-complementary sequences have not been detected or looked for in other autonomous parvovirus DNA, but might play a part in parvovirus DNA replication. Rhode [68, 221, 216] has suggested that H-1 parental double-stranded form (RF) or another early DNA synthetic event (HA DNA) is synthesized shortly after infection [221]. Either this RF formation or HA DNA occurs near the end of S phase of the cell and is required for the synthesis of viral VP1 and VP2. Inhibition studies suggest that the synthesis of progeny RF DNA requires prior VP1 or VP2 synthesis. One would then hypothesize that progeny viral strand (V) DNA is copied from the synthesized complementary strand (C) DNA in the RF molecule [216]. Further confirmation of these steps in DNA synthesis may come from results with other parvoviruses and from the use of temperature-sensitive mutants [216]. Many steps in the use of RF as a template for single-stranded viral DNA synthesis, however, remain undiscovered.

The production of the defective parvovirus, DNA, RNA, and structural proteins are dependent on coinfection of a cell with an adenovirus or herpesvirus helper [173, 174, 195]. The kinetics of AAV nucleic acid synthesis depend on whether the cells are infected with helper virus simultaneously with or prior to infection. After simultaneous infection of KB cells with adenovirus, AAV DNA synthesis starts 6.5-7.5 hr and AAV RNA 10 hr after infection. If KB cells are infected with adenovirus 10 hr prior to AAV infection, AAV DNA and RNA are detected at about the same time, 3-4 hr after infection, and proceed together [222, 195]. Since the lag between AAV DNA and RNA synthesis is abolished by helper preinfection, AAV DNA and RNA synthesis may require separate helper functions appearing sequentially after adenovirus infection [222].

An attempt to pinpoint the effect of adenovirus DNA synthesis on AAV DNA synthesis has been made. Temperature-sensitive adenovirus mutants, unable to synthesize adenovirus DNA at the restrictive temperature, have been used as helper virus. The syntheses of AAV DNA, RNA, and infectious AAV are comparable to that when a non(ts)mutant was used. Thus, adenovirus DNA synthesis per se does not appear to be obligatory for satellite virus replication [181, 223, 224]. The ts mutants have also been shown to produce only early adenovirus RNA [223]. This suggests that the adenovirus factor or factors required for AAV multiplication are produced early in the adenovirus infectious cycle before adenovirus DNA synthesis. If AAV multiplication is linked to adenovirus DNA replication, AAV replication does not require all the factors needed for adenovirus DNA replication [223]. It is also possible that AAV replication does not depend on any function directly involved in adenovirus DNA synthesis.

There is little definitive information on the cycle of AAV DNA synthesis and the intermediates involved. Inverted terminal repetition and a concatameric AAV DNA form amy play a role in DNA replication. As discussed in the section on DNA, the AAV molecules can contain inverted terminal repetitions and can form panhandle-type molecules (Fig. 10.2). It is not surprising that they have been reported to form concatamers, 1 or more virus unit lengths joined together either by hydrogen bonding or covalent linkage. AAV DNA in the infected cell can be present in concatameric forms up to 4 times the AAV unit length [224]. The radioactive concatameric AAV DNA can be chased into single-stranded AAV DNA molecules, indicating that it may represent a replicative intermediate (RI). Some of the RI molecules contain covalently linked plus and minus strands. This suggests that AAV DNA replication may proceed from a terminal, self-priming sequence [224]. Because of the importance of these concatamers, the method of DNA synthesis required to form concatamers should continue to be a worthwhile area of study.

When Detroit 6 cells latently infected with purified AAV 2 are exposed to adenovirus, they are able to synthesize AAV proteins and infectious virus. The increased rate of annealing of [^{32}P] AAV DNA in 2 latently infected clones when compared to uninfected clones suggests the presence of 3-5 AAV DNA genome equivalents per diploid amount of cell DNA. This ability of AAV to infect cells latently in the absence of a helper function implies that the requirements for the establishment of latency are provided by the cell alone [225].

F. Viral RNA Synthesis

Little information is available about specific RNA synthesized after infection of the autonomous parvoviruses. Both Tenant and Salzman et al. have reported that in RE, rat kidney, and RN cells, the total RNA synthesized after KRV infection was not significantly changed [188, 193]. After H-1 infection of NP cells, there was a decrease in the number of cells synthesizing RNA. The 28S ribosomal RNA was selectively inhibited while the low molecular weight 4S RNA was selectively stimulated [226]. Salzman and Redler [227] studied the viral-specific RNA synthesized after infection of RN with KRV. The RNA was demonstrated to be viral-specific by analysis of its nucleotide base ratios and by hybridization with viral and cellular DNA. Viral RNA synthesis starts 2 hr after infection, preceding progeny DNA synthesis by 5-6 hr. The KRV RNA had a sedimentation coefficient of 18S in dimethylsulfoxide-sucrose gradients and a calculated molecular weight of 6.5-7.5 $\times$ 10^5. It would, if homogeneous, account for 40-50% of the viral genome. A small amount of 26S viral RNA with a molecular weight of 1.6-1.7 $\times$ 10^6 was also detected at 5-6 hr postinfection. It was not characterized but could represent a transcription of the entire KRV genome.

The synthesis of RNA by the defective parvovirus AAV-2 has been studied in detail by Carter et al. [177, 194, 222, 228, 166]. The AAV-2 RNA appears to be a single stable species of 20S with a molecular weight of 0.9-1.0 $\times$ 10^6. The RNA is synthesized in the nucleus and is not cleaved further during transport to the polysomes [194, 228]. A second heterogeneous population, ranging in size from 4-8S, is present in the nucleus and nonpolysomal regions of the cytoplasm. The heterogeneous AAV RNA may arise from incomplete transcription or degradation of the 20S RNA. The 20S RNA is present mainly in the cytoplasm in the polysome region and may be the only functional AAV messenger RNA. It may be monocystronic; if so, AAV DNA would contain 1 gene [228]. The stable 20S AAV-specific RNA is equivalent in size to approximately 70% of the DNA strand length [177, 228]. The terminally repeated nucleotide sequences, comprising 1.5-5% of the DNA strand length at each terminus, are not copied in the 20S AAV RNA [194, 140, 166].

As discussed in the section on viral DNA, AAV plus and minus DNA strands may be separated when BUdR is substituted for thymidine [151]. AAV-specific RNA was able to hybridize only to the heavy BUdR-substituted DNA strand containing more thymidine than its complement. The viral-specific RNA had an average base composition which was identical to that of the light strand [127]. Rose and Koczot concluded that AAV RNA is made only from the AAV DNA strand with the higher thymidine content, the minus strand [127].

Carter and Khoury [164] have cleaved the AAV-2 DNA with the restriction enzyme R·Eco R_1 and found that it yielded 3 fragments: B, C, and A. B is the 3′ terminus and A the 5′ terminus of the minus DNA strand. They found that transcription of AAV mRNA from the minus strand begins in fragment B and terminates in fragment A [166].

G. Interference

As discussed in preceding sections, parvoviruses have been shown to have antimitotic activity and to have profound effects on cellular macromolecular synthesis. They have also been demonstrated to interfere with the replication of coinfecting viruses, including obligatory helper viruses, and to cause oncolysis and suppression of tumor formation and leukemias.

Early studies of AAV replication showed that while AAV required co-infection with adenovirus, it also inhibited the replication of the helper virus and caused a reduction in adenovirus yield [26]. One AAV type could also interfere with other AAV types. Antiserum against the AAV prevented the interference [26]. Observing a reduction in infectious centers produced by adenoviruses and SV40 after AAV coinfection, Casto et al. [118, 229] felt that the coinfected cells did not yield any detectable adenovirus or SV40. In contrast to these results, Parks et al. [230] found that in coinfected single cells adenovirus was produced, but in reduced amounts. The AAV interference required infectious AAV in large amounts (> 40 AAV per cell). The interference was specific for helper viruses. It did not affect vaccinia virus, vasicular stomatitis, or poliovirus coninfection, and did not appear to involve interferon [230].

If AAV infection occurs after adenovirus DNA synthesis has started (10-12 hr after infection), the yield of adenovirus is not decreased [210]. It is possible that in the doubly infected cell nucleus there is a competition for a common function required for DNA synthesis [209, 16]. H-1, which requires adenovirus 12 to replicate in human embryonic kidney cells (HEK), inhibits the production of infectious adenovirus in the coinfected cells [170]. The H-1 and AAV interference are probably similar phenomena.

Viruses serologically similar to KRV have been isolated from rat leukemias of different origins [231, 232, 106, 19]. Berg [231, 232] reported that the KRV can act as an inhibitor of leukemogenic viruses. He found that rats inoculated intraperitoneally at birth with a mixture of the RNA mouse leukemia virus (MLV) and KRV developed fewer leukemias than rats inoculated with MLV alone. The HI antibody titer to KRV was higher in the nonleukemic than in the leukemic animals. This evidence suggested reciprocal interference between MLV and KRV in rats [231]. The parvovirus' role in suppression of the coinfecting virus is unknown.

H-1 virus and AAV have also been reported to inhibit oncogenesis. Hamsters injected with H-1 virus at birth were found, when aged, to have a low incidence of tumors. If H-1 and the oncogenic virus adenovirus 12 were injected together into newborn hamsters there was no apparent effect of adenovirus 12 on H-1 infection-induced deformities. H-1, however, reduced the number of visible tumors produced by adenovirus 12. The incidence of tumors was higher in females than in males. AAV has also been reported to inhibit adenovirus oncogenicity [233-237]. AAV reduced the frequency of tumors due to adenovirus 12 and delayed the appearance of those tumors which did occur when both viruses were injected into neonatal hamsters [233]. The AAV used had to be infectious to have this effect [234, 235]. When inoculated into hamsters, herpes simplex virus type 2-transformed tumor cells cause tumors and animal death. When these transformed cells are infected with AAV-1 prior to inoculation into hamsters there is a specific delay in the appearance of palpable tumors and an increased survival time of the animals [237].

VII. Conclusions

The parvoviruses present many unanswered questions: for example, they may have a unique method for single-stranded DNA replication and, possibly, may produce only a single messenger RNA species. The limited amount of information these viruses impart to an infected cell, however, may cause cell lysis and disease. The antimitotic action of autonomous parvoviruses is potentially of medical as well as biological significance. Of particular interest is their effect on the inhibition of coinfecting viruses and their influence on cancer expression. This interest is compounded by the fact that parvoviruses appear ubiquitous and extremely hardy, and may exist in a latent state. Thus it is entirely possible that the parvoviruses could influence the outcome of other viral diseases and impede rapid cellular growth, even in apparently healthy hosts. They certainly merit continued study.

References

1. C. Andrewes, Virology, 40, 1070 (1970).
2. P. A. Bachmann, M. D. Hoggan, J. L. Melnick, H. G. Pereira, and C. Vago, Intervirology, 5, 83 (1975).
3. H. W. Toolan, in International Review of Experimental Pathology, Vol. VI, Academic, New York, 1968, pp. 135-180.
4. M. D. Hoggan, in Progress in Medical Virology, Vol. 12 (J. L. Melnick, ed.), Karger, Basel, 1970, pp. 211-239.
5. M. D. Hoggan, in Comparative Virology (K. Maramorosch and E. Kurstak, eds.), Academic, New York, 1971, pp. 43-74.
6. H. W. Toolan, in Progress in Experimental Tumor Research, Vol. 16 (F. Homburger, ed.), Karger, Basel, 1972, pp. 410-425.
7. T. W. Tinsley and J. F. Longworth, J. Gen. Virol., 20, 7 (1973).
8. J. A. Rose, in Comprehensive Virology, Vol. 3 (H. Fraenkel-Conrat and R. Wagner, eds.), Plenum, New York, 1974, pp. 1-51.
9. F. W. Schofield, N. Am. Vet., 30, 651 (1949).
10. C. G. Wills, Can. J. Comp. Med., 16, 419 (1952).
11. R. H. Johnson, G. Siegl, and M. Gautschi, Arch. Ges. Virusforsch., 46, 315 (1974).
12. L. Kilham and L. J. Oliver, Virology, 7, 428 (1959).
13. S. S. Breese, Jr., A. F. Howatson, and C. Chaney, Virology, 24, 598 (1964).
14. H. W. Toolan, G. Dalldorf, M. Barclay, S. Chandra, and A. E. Moore, Proc. Natl. Acad. Sci. USA, 46, 1256 (1960).
15. H. W. Toolan, Science, 131, 1446 (1960).
16. L. Kilham, Proc. Soc. Exp. Biol. Med., 106, 825 (1961).
17. A. E. Moore, Virology, 18, 182 (1962).

18. F. R. Abinanti and M. S. Warfield, Virology, 14, 288 (1961).
19. G. S. Lum and A. W. Schriner, Cancer Res., 23, 1742 (1963).
20. F. E. Payne, C. J. Shellabarger, and R. W. Schmidt, Proc. Am. Assoc. Cancer Res., 4, 51 (1963).
21. H. W. Toolan, Proc. Am. Assoc. Cancer Res., 5, 64 (1964).
22. R. H. Johnson, Res. Vet. Sci., 6, 466 (1965).
23. R. H. Johnson, Res. Vet. Sci., 6, 472 (1965).
24. R. W. Atchison, B. C. Casto, and W. McD. Hammon, Science, 149, 754 (1965).
25. L. V. Crawford, Virology, 29, 605 (1966).
26. M. D. Hoggan, N. R. Blacklow, and W. P. Rowe, Proc. Natl. Acad. Sci. USA, 55, 1467 (1966).
27. A. Mayr and H. Mahnel, Zbl. Bakt. I. Abt. Orig., 199, 399 (1966).
28. W. P. Parks, J. L. Melnick, R. Rongey, and H. D. Mayor, J. Virol., 1, 171 (1967).
29. W. P. Parks, M. Green, M. Piña, and J. L. Melnick, J. Virol., 1, 980 (1967).
30. S. K. Dutta and B. S. Pomeroy, Am. J. Vet. Res., 28, 296 (1967).
31. A. H. El Dadah, K. O. Smith, R. A. Squire, G. W. Santos, and E. C. Melby, Science, 156, 392 (1967).
32. L. N. Binn, E. C. Lazar, G. A. Eddy, and M. Kajima, Abstract 68th Annual Meeting of the American Society of Microbiology, 1968, p. 161.
33. K. Domoto and R. Yanagawa, Jap. J. Vet. Res., 17, 32 (1969).
34. J. D. Boggs, J. L. Melnick, M. E. Conrad, and B. F. Felsher, JAMA, 214, 1041 (1970).
35. E. Luchsinger, R. Stobbe, G. Wellemans, D. Dekegel, and S. Sprecher-Goldberger, Arch. Ges. Virusforsch., 31, 390 (1970).
36. C. Hallauer, G. Kronauer, and G. Siegl, Arch. Ges. Virusforsch., 35, 80 (1971).
37. R. L. Sinsheimer, J. Mol. Biol., 1, 37 (1959).
38. E. Kurstak, in Comparative Virology (K. Maramorosch and E. Kurstak, eds.), Academic, New York. 1971, pp. 207-241.
39. G. Meynadier, C. Vago, G. Planteoin, and P. Atger, Rev. Zool. Agr. Appl., 63, 207 (1964).
40. A. Z. Kapikian, J. L. Gerin, R. G. Wyatt, T. S. Thornhill, and R. M. Chanock, Proc. Soc. Exp. Biol. Med., 142, 874 (1973).
41. P. J. Provost, O. L. Ittensohn, V. M. Villarejos, J. A. Arquedas, and M. R. Hilleman, Proc. Soc. Exp. Biol. Med., 142, 1257 (1973).
42. Y. E. Cossart, A. M. Field, B. Cant, and D. Widdows, Lancet, i, 72 (1975).
43. W. K. Paver, E. O. Caul, and S. K. Clarke, Lancet, i(7900), 232 (1975).
44. W. K. Paver, E. O. Caul, and S. K. Clarke, Lancet, i, 691 (1975).
45. A. Liss and J. Maniloff, Science, 173, 725 (1971).
46. A. Liss and J. Maniloff, J. Virol., 113, 769 (1974).
47. B. Chesebro, M. Bloom, W. Hadlow, and R. Race, Nature, 254, 456 (1975).

48. H. J. Cho and D. G. Ingram, J. Immunol. Meth., 4(2), 217 (1974).
49. F. L. Black, Virology, 5, 391 (1958).
50. M. K. Brakke, Adv. Virus Res., 7, 193 (1960).
51. R. C. Valentine, Adv. Virus Res., 8, 287 (1961).
52. S. Chandra and H. W. Toolan, J. Natl. Cancer Inst., 27, 1405 (1961).
53. H. D. Mayor, R. M. Jamison, L. E. Jordan, and J. L. Melnick, J. Bacteriol., 90, 235 (1965).
54. H. W. Toolan, E. L. Saunders, E. L. Greene, D. P. A. Fabrizio, Virology, 22, 286 (1964).
55. K. O. Smith, W. D. Gehle, and J. T. Thiel, J. Immunol., 97, 754 (1966).
56. H. D. Mayor and J. L. Melnick, Nature, 210, 331 (1966).
57. S. Karasaki, J. Ultrastruct. Res., 16, 109 (1966).
58. C. Vasquez and C. Brailovsky, Exp. Mol. Pathol., 4, 130 (1965).
59. W. J. Tromans and R. W. Horne, Virology, 15, 1 (1961).
60. F. E. Payne, T. F. Beals, and R. E. Preston, Virology, 23, 109 (1964).
61. D. M. Robinson and F. M. Hetrick, J. Gen. Virol., 4, 269 (1969).
62. K. Torikai, M. Ito, L. E. Jordan, and H. D. Mayor, J. Virol., 6, 363 (1970).
63. M. Usategui-Gomez, H. W. Toolan, N. Ledinko, F. Al-Lami, and M. S. Hopkins, Virology, 39, 617 (1969).
64. G. Siegl, Arch. Ges. Virusforsch., 37, 267 (1972).
65. E. L. Greene and S. Karasaki, Proc. Soc. Exp. Biol. Med., 119, 918 (1965).
66. F. B. Johnson, H. L. Ozer, and M. D. Hoggan, J. Virol., 8, 860 (1971).
67. L. A. Salzman and L. A. Jori, J. Virol., 5, 114 (1970).
68. S. L. Rhode, III, J. Virol., 13, 400 (1974).
69. F. B. Johnson and M. D. Hoggan, Virology, 51, 129 (1973).
70. C. A. Thomas, Jr., and L. A. MacHattie, in Annual Review of Biochemistry, Vol. 36, Part II (P. D. Boyer, ed.), Annual Reviews, Palo Alto, 1967, pp. 485-518.
71. C. Hallauer, G. Siegl, and G. Kronauer, Arch. Ges. Virusforsch., 38, 366 (1972).
72. R. H. Johnson and J. G. Cruickshank, Nature, 212, 622 (1966).
73. V. J. Yates, A. M. El Mishad, K. J. McCormick, and J. J. Trentin, Infect. Immun., 7, 973 (1973).
74. M. Ito and H. D. Mayor, J. Immunol., 100, 61 (1968).
75. N. Nathanson, G. A. Cole, G. W. Santos, R. A. Squire, and K. O. Smith, Am. J. Epidemiol., 91, 328 (1970).
76. S. S. Cross and J. C. Parker, Proc. Soc. Exp. Biol. Med., 139, 105 (1972).
77. R. R. Mirkovic, V. Adamova, D. W. Boucher, and J. L. Melnick, Proc. Soc. Exp. Biol. Med., 138, 626 (1971).
78. R. H. Johnson, G. Margolis, and L. Kilham, Nature, 214, 175 (1967).
79. J. Storz, R. C. Bates, G. S. Warren, and H. Howard, Am. J. Vet. Res., 33, 269 (1972).
80. L. N. Binn, E. C. Lazar, G. A. Eddy, and M. Kajima, Infect. Immun., 1, 503 (1970).

81. M. Onuma, Jap. J. Vet. Res., 19, 40 (1971).
82. L. Kilham, Natl. Cancer. Inst. Monogr. 20, 117 (1966).
83. R. E. Robey, D. R. Woodman, and F. M. Hetrick, Am. J. Epidemiol., 88, 139 (1968).
84. G. S. Lum, Oncology, 24, 335 (1970).
85. M. Narita, S. Inui, Y. Kawakami, K. Kitamura, and A. Maeda, Natl. Inst. Anim. Health Q. (Tokyo), 15(1), 24 (1975).
86. D. R. Redman, E. H. Bohl, and L. C. Ferguson, Infect. Immun., 10(4), 718 (1974).
87. L. Kilham and G. Margolis, Proc. Soc. Exp. Biol. Med., 133, 1447 (1970).
88. J. C. Parker, M. J. Collins, Jr., S. S. Cross, and W. P. Rowe, J. Natl. Cancer Inst., 45, 305 (1970).
89. N. R. Blacklow, M. D. Hoggan, and W. P. Rowe, J. Natl. Cancer Inst., 40, 319 (1968).
90. N. R. Blacklow, M. D. Hoggan, A. Z. Kapikian, J. B. Austin, and W. P. Rowe, Am. J. Epidemiol., 88, 368 (1968).
91. N. R. Blacklow, M. D. Hoggan, M. S. Sereno, C. D. Brandt, H. W. Kim, R. H. Parrott, and R. M. Chanock, Am. J. Epidemiol., 94, 359 (1971).
92. N. P. Rapoza and R. W. Atchison, Nature, 215, 1186 (1967).
93. W. P. Parks, D. W. Boucher, J. L. Melnick, L. H. Taber, and M. D. Yow, Infect. Immun., 2, 716 (1970).
94. E. Luchsinger, R. Stobbe, D. Dekegel, and G. Wellemans, Arch. Ges. Virusforsch., 33, 251 (1971).
95. S. F. Cartwright and R. A. Huck, Vet. Rec., 81, 196 (1967).
96. S. F. Cartwright, M. Lucas, and R. A. Huck, J. Comp. Pathol., 79, 371 (1969).
97. M. H. Lucas, S. F. Cartwright, and A. E. Wrathall, J. Comp. Pathol., 84, 347 (1974).
98. L. Kilham, G. Margolis, and E. Colby, Lab. Invest., 17, 465 (1967).
99. M. J. Studdert and J. E. Peterson, Arch. Ges. Virusforsch., 42, 346 (1973).
100. R. H. Johnson, Res. Vet. Sci., 8, 256 (1967).
101. R. H. Johnson and R. E. W. Halliwell, Vet. Rec., 82, 582 (1968).
102. H. W. Toolan, Science, 131, 1446 (1960).
103. H. W. Toolan, G. Dalldorf, M. Barclay, S. Chandra, and A. E. Moore, Proc. Natl. Acad. Sci. USA, 46, 1256 (1960).
104. L. Kilham, Virology, 13, 141 (1961).
105. M. Galton and L. Kilham, Proc. Soc. Exp. Biol. Med., 122, 18 (1966).
106. L. Kilham and J. B. Maloney, J. Natl. Cancer Inst., 32, 523 (1964).
107. L. Kilham and G. Margolis, Science, 143, 1047 (1964).
108. L. Kilham and G. Margolis, Science, 148, 224 (1965).
109. L. Kilham and G. Margolis, J. Infect. Dis., 129(6), 737 (1974).
110. L. Kilham and V. H. Fern, Proc. Soc. Exp. Biol. Med., 117, 874 (1964).
111. W. O. Engler, P. N. Baer, and L. Kilhom, Arch. Pathol., 82, 93 (1966).
112. G. L. Margolis, L. Kilham, and R. H. Johnson, in Progress in Neuropathology, Vol. 1 (H. M. Zimmerman, ed.), Grune and Stratton, New York, 1971, pp. 163-201.

113. M. D. Hoggan, in Proceedings of the Fourth Lepetit Colloquium Cocylic Mexico (L. G. Sylvestry, ed.), North Holland, Amsterdam, 1973, pp. 243-249.
114. Y. Matsuo and J. H. Spencer, Proc. Soc. Exp. Biol. Med., 130, 294 (1969).
115. H. W. Toolan and N. Ledinko, Nature, 208, 812 (1965).
116. J. L. Melnick, J. Infect. Dis., 124, 76 (1971).
117. K. O. Smith, W. D. Gehle, and H. Montes de Oca, Bacteriol. Proc., 52, 153 (1968).
118. B. C. Casto, R. W. Atchison, and W. McD. Hammon, Virology, 32, 52 (1967).
119. H. D. Mayor, M. Ito, L. E. Jordan, and J. L. Melnick, J. Natl. Cancer Inst., 38, 805 (1967).
120. R. C. Bates and J. Storz, Infect. Immun., 7, 398 (1973).
121. L. V. Crawford, E. A. C. Follett, M. G. Burdon, and D. J. McGeoch, J. Gen. Virol., 4, 37 (1969).
122. L. A. Salzman, W. L. White, and T. Kakefuda, J. Virol., 7, 830 (1971).
123. G. Siegl, C. Hallauer, and A. Novak, Arch. Ges. Virusforsch., 36, 351 (1972).
124. N. Ledinko, Nature, 214, 1346 (1967).
125. A. Mayr, P. A. Bachmann, G. Siegl, H. Mahnel, and B. E. Sheffy, Arch. Ges. Virusforsch., 25, 38 (1968).
126. J. Storz and G. S. Warren, Arch. Ges. Virusforsch., 30, 271 (1970).
127. J. A. Rose and F. Koczot, J. Virol., 8, 771 (1971).
128. D. J. McGeoch, L. V. Crawford, and E. A. C. Follett, J. Gen. Virol., 6, 33 (1970).
129. J. A. Rose, J. V. Maizel, J. K. Inman, and A. J. Shatkin, J. Virol., 8, 766 (1971).
130. J. R. Kongsvik and H. W. Toolan, Proc. Soc. Exp. Biol. Med., 139, 1202 (1972).
131. L. A. Salzman and W. L. White, Biochem. Biophys. Res. Commun., 41, 1551 (1970).
132. L. A. Salzman, Nature New Biol., 231, 174 (1971).
133. L. A. Salzman and L. McKerlie, J. Biol. Chem., 250(14), 5583 (1975).
134. L. A. Salzman and L. McKerlie, J. Biol. Chem., 250(14), 5589 (1975).
135. J. R. Kongsvik, J. F. Gierthy, and S. L. Rhode, III, J. Virol., 14, 1600 (1974).
136. J. R. Kongsvik and K. O. Ellem, unpublished work, 1976.
137. G. M. Clinton and M. Hayashi, Virology, 65, 261 (1975).
138. P. J. Tattersall, D. C. Ward, and A. J. Shatkin, in Abstract of the 3rd International Congress on Biochemistry, Madrid, Spain, September, 1975.
139. F. B. Johnson, N. R. Blacklow, and M. D. Hoggan, J. Virol., 9, 1017 (1972).
140. F. J. Koczot, B. J. Carter, C. F. Garon, and J. A. Rose, Proc. Natl. Acad. Sci. USA, 70, 215 (1973).
141. J. A. Rose, M. D. Hoggan, and A. J. Shatkin, Proc. Natl. Acad. Sci. USA, 56, 86 (1966).
142. K. I. Berns, Current Topics in Microbiology and Immunology, Vol. 65, Springer-Verlag, New York, 1974, p. 1.

143. A. S. Rabson, L. Kilham, and R. L. Kirschstein, J. Natl. Cancer Inst., 27, 1217 (1961).
144. W. Bernhard, F. H. Kasten, and C. Chang, Compt. Rend. Herb. Sea. Acad. Crevas, Paris, 257, 1566 (1963).
145. L. Cheong, J. Fogh, and R. Barclay, Fed. Proc., 24, 596 (1965).
146. R. L. Sinsheimer, J. Mol. Biol., 1, 43 (1959).
147. H. D. Mayor, R. M. Jamison, L. E. Jordan, and J. L. Melnick, J. Bacteriol., 90, 235 (1965).
148. E. Bereczky and I. Archetti, Arch. Ges. Virusforsch., 22, 426 (1968).
149. H. D. Mayor, L. Jordan, and M. Ito, J. Virol., 4, 191 (1969).
150. J. A. Rose, K. I. Berns, M. D. Hoggan, and F. J. Koczot, Proc. Natl. Acad. Sci. USA, 64, 863 (1969).
151. H. D. Mayor, K. Torkiai, J. L. Melnick, and M. Mandel, Science, 166, 1280 (1969).
152. K. I. Berns and J. A. Rose, J. Virol., 5, 693 (1970).
153. K. I. Berns and S. Adler, J. Virol., 9, 394 (1972).
154. A. H. Barwise and I. O. Walker, FEBS Lett., 6, 13 (1970).
155. G. Siegl, Arch. Ges. Virusforsch., 43, 334 (1973).
156. S. K. Vernon, J. T. Stansy, A. R. Neurath, and B. A. Rubin, J. Gen. Virol., 10, 267 (1971).
157. H. W. Gerry, T. J. Kelly, Jr., and K. I. Berns, J. Mol. Biol., 79, 207 (1973).
158. C. F. Garon, K. W. Berry, and J. A. Rose, Proc. Natl. Acad. Sci. USA, 69, 2391 (1972).
159. J. Wolfson and D. Dressler, Proc. Natl. Acad. Sci. USA, 69, 3054 (1972).
160. K. I. Berns and T. J. Kelly, Jr., J. Mol. Biol., 82, 267 (1974).
161. F. W. Studier, J. Mol. Biol., 11, 373 (1965).
162. D. Freifelder, A. Kleinschmidt, and R. L. Sinsheimer, Science, 146, 254 (1964).
163. J. A. Rose, M. D. Hoggan, F. Koczot, and A. J. Shatkin, J. Virol., 2, 999 (1968).
164. B. J. Carter and G. Khoury, Virology, 63, 523 (1975).
165. P. May and E. May, J. Gen. Virol., 6, 437 (1970).
166. B. J. Carter, G. Khoury, and D. T. Denhardt, J. Virol., 16, 559 (1975).
167. K. I. Berns, J. Kort, K. H. Fife, E. W. Grogan, and I. Spear, J. Virol., 16, 712 (1975).
168. C. Chany and C. Brailovsky, Proc. Natl. Acad. Sci. USA, 57, 87 (1967).
169. N. Ledinko and H. W. Toolan, J. Virol., 2, 155 (1968).
170. N. Ledinko, S. Hopkins, and H. W. Toolan, J. Gen. Virol., 5, 19 (1969).
171. H. Toolan and N. Ledinko, Virology, 35, 475 (1968).
172. J. L. Melnick, and W. P. Parks, Identification of multiple defective and noncytopathic viruses in tissue culture, Relazione al V1 Congresso Internazionale di Patologia Clinico, Roma, October 1966, pp. 237-262.
173. R. W. Atchison, Virology, 42, 155 (1970).
174. N. R. Blacklow, M. D. Hoggan, and M. S. McClanahan, Proc. Soc. Exp. Biol. Med., 134, 952 (1970).

175. N. R. Blacklow, R. Dolin, M. D. Hoggan, J. Gen. Virol., 10, 29 (1971).
176. R. Dolin and A. S. Rabson, J. Natl. Cancer Inst., 50, 205 (1973).
177. B. J. Carter and J. A. Rose, J. Virol., 10, 9 (1972).
178. C. J. Henry, L. P. Merkow, M. Pardo, and C. McCabe, Virology, 49, 618 (1972).
179. M. Ishibashi and M. Ito, Virology, 45, 317 (1971).
180. H. D. Mayor and J. D. Ratner, Nature New Biol., 239, 20 (1972).
181. M. Ito and E. Suzuki, J. Gen. Virol., 9, 243 (1970).
182. H. D. Mayor and J. D. Ratner, Biochim. Biophys. Acta, 299, 189 (1973).
183. R. W. Tennant, K. R. Layman, and R. E. Hand, Jr., J. Virol., 4, 872 (1969).
184. R. W. Tennant and R. E. Hand, Jr., Virology, 42, 1054 (1970).
185. P. Tattersall, J. Virol., 10, 586 (1972).
186. G. Siegl and M. Gautschi, Arch. Ges Virusforsch, 40, 105 (1973).
187. G. E. Hampton, Can. J. Microbiol., 16, 266 (1970).
188. R. W. Tennant, J. Virol., 8, 402 (1971).
189. P. Tattersall, L. V. Crawford, and A. J. Shatkin, J. Virol., 12, 1446 (1973).
190. G. Siegl and M. Gautschi, Arch. Ges Virusforsch., 40, 119 (1973).
191. G. A. Cole and N. Nathanson, Acta Virol., 13, 515 (1969).
192. H. A. Fields and B. L. Nicholson, Can. J. Microbiol., 18, 103 (1972).
193. L. A. Salzman, W. L. White, and L. McKerlie, J. Virol., 10, 573 (1972).
194. B. J. Carter, G. Khoury, and J. A. Rose, J. Virol., 10, 1118 (1972).
195. J. A. Rose and F. Koczot, J. Virol., 10, 1 (1972).
196. S. L. Rhode, III, J. Virol., 11, 856 (1973).
197. F. Al-Lami, J. Gen. Virol., 5, 485 (1969).
198. R. E. Harris, P. H. Coleman, and D. S. Morahan, Appl. Microbiol., 28(3), 351 (1974).
199. J. C. Parker, S. S. Cross, M. J. Collins, Jr., and W. P. Rowe, J. Natl. Cancer Inst., 45, 297 (1970).
200. D. W. Boucher, J. L. Melnick, and H. D. Mayor, Science, 173, 1243 (1971).
201. D. W. Boucher, W. P. Parks, and J. L. Melnick, Virology, 39, 932 (1971).
202. A. M. El Mishad, V. J. Yates, K. J. McCormick, and J. J. Trentin, in Abstract 73rd Annual Meeting of the American Society of Microbiologists, Vol. 217, p. 230 (1973).
203. J. R. Gorham, G. R. Hartsough, N. Sato, and S. Lust, Vet. Med., 61 35 (1966).
204. I. I. Singer, Exp. Cell Res., 99, 346 (1976).
205. I. I. Singer and H. W. Toolan, Virology, 65, 40 (1975).
206. I. I. Singer in Viral Immunediagnosis (E. Kurstak and K. Maramorosch eds.), Academic, New York, 1974, pp. 101-124.
207. M. D. Hoggan, A. J. Shatkin, N. R. Blacklow, F. Koczot, and J. A. Rose, J. Virol., 2, 850 (1968).

208. G. Torpier, J. D'Halluin, and P. Boulanger, J. Microsc., 11, 259 (1971).
209. R. W. Atchison, B. C. Casto, W. McD. Hammon, Virology, 29, 353 (1966).
210. N. R. Blacklow, M. D. Hoggan, and W. P. Rowe, J. Exp. Med., 125, 755 (1967).
211. H. D. Mayor and L. E. Jordan, Exp. Mol. Pathol., 5, 580 (1966).
212. G. Cocuzza, G. Ricceri, D. Duscio, and G. Nicoletti, Boll. Ist. Sieroter Milan, 46, 9, (1967).
213. C. K. Y. Fong, N. Ledinko, and H. W. Toolan, Proc. Soc. Exp. Biol. Med., 134, 1199 (1970).
214. J. T. Parsons, J. T. Gardner, and M. Green, Proc. Natl. Acad. Sci. USA, 68, 557 (1971).
215. M. Ito, J. L. Melnick, and H. D. Mayor, J. Gen. Virol., 1, 199 (1967).
216. S. L. Rhode, III, J. Virol., 17, 659 (1976).
217. H. D. Mayor, S. Drake, and L. E. Jordan, J. Ultrastruct. Res., 52, 52 (1975).
218. R. L. Sinsheimer, R. Knippers, and T. Komano, Cold Spring Harbor Symp. Quant. Biol., 33, 443 (1968).
219. P. R. Dobson and C. W. Helleiner, Can. J. Microbiol., 19, 35 (1973).
220. P. Tattersall, Fed. Proc., 31, 3973 (1972).
221. S. L. Rhode, III, J. Virol., 14, 791 (1974).
222. B. J. Carter, F. J. Koczot, J. Garrison, R. Dolin, and J. A. Rose, Nature New Biol., 244, 71 (1973).
223. S. E. Strauss, H. S. Ginsberg, and J. A. Rose, J. Virol., 17, 140 (1976).
224. S. E. Strauss, E. Sebring, H. S. Ginsberg, and J. A. Rose, Proc. Natl. Acad. Sci. USA, 73, 742 (1976).
225. K. I. Berns, T. C. Pinkerton, G. F. Thomas, and M. D. Hoggan, Virology, 68, 556 (1975).
226. C. K. Y. Fong, H. W. Toolan, and M. S. Hopkins, Proc. Soc. Exp. Biol. Med., 135, 585 (1970).
227. L. A. Salzman and B. H. Redler, J. Virol., 14, 434 (1975).
228. B. J. Carter and J. A. Rose, Virology, 61, 182 (1974).
229. B. C. Casto, J. A. Armstrong, R. W. Atchison, and W. McD. Hammon, Virology, 33, 452 (1967).
230. W. P. Parks, A. M. Casazza, and J. L. Melnick, J. Exp. Med., 127, 91 (1968).
231. V. V. Berg, J. Natl. Cancer Inst., 38, 481 (1967).
232. V. V. Berg, Cancer Res., 29, 1669 (1969).
233. R. L. Kirschstein, K. O. Smith, and E. A. Peters, Proc. Soc. Exp. Biol. Med., 128, 670 (1968).
234. R. V. Gilden, J. Kern, T. G. Beddow, and R. J. Huebner, Nature, 219, 80 (1968).
235. R. V. Gilden, J. Kern, T. G. Beddow, and R. J. Huebner, Nature, 220, 1139 (1968).
236. H. D. Mayor, G. S. Houlditch, and D. M. Mumford, Nature New Biol., 241, 44 (1973).

237. G. Cukor, N. R. Blacklow, S. Kibrick, and I. C. Swan, J. Natl. Cancer Inst., 55, 957 (1975).
238. M. Horzinek, M. Mussgay, J. Maess, and K. Petzoldt, Arch. Ges. Virusforsch., 21, 98 (1967).
239. E. Luchsinger and G. Wellemans, Arch. Ges. Virusforsch., 35, 203 (1971).
240. M. Gautschi and G. Siegl, Arch. Ges. Virusforsch., 43, 326 (1973).
241. A. J. Dalton, L. Kilham, and R. E. Zeigler, Virology, 20, 391 (1963).

Chapter 11

The Molecular Biology of the Papovaviruses

Joseph Sambrook

Cold Spring Harbor Laboratory
Cold Spring Harbor, New York

I. Introduction 590

II. Subgroup A: The Papilloma Viruses 591

III. Subgroup B: Polyoma Virus and Simian Virus 40 593

A. Discovery of the Viruses 593
B. Polyoma Virus and Simian Virus Particles 596
C. Interactions of Polyoma Virus and Simian Virus 40 with Cultured Cells 612
D. The Lytic Cycle 614
E. Genetics of Polyoma Virus and Simian Virus 40 636
F. Transformation 641
G. Conclusion 651

Acknowledgments 652

References 652

I. Introduction

Of all animal viruses those of the papova group are the most fiercely discussed and intensively investigated. At first blush, this may seem surprising because they apparently offer little to excite the imagination: their particles are constructed according to standard icosahedral principles and contain as genetic information only a single small molecule of duplex DNA. In the case of polyoma virus and simian virus 40 (SV40), this genetic information is expressed as 3 or 4 proteins, 2 of which are used to form the coat of the virus particles. Why are objects of such straightforward constitution the cause of such sedulous industry and why should they arouse in their adherents such powerful feelings?

The answer to this paradox lies in the rich variety of interactions that occur between these simple viruses and their hosts.. All of the papovaviruses are tumor viruses. Some of them (the papilloma viruses) regularly cause benign growths in their natural hosts, but do not produce tumors after injection into exotic animals. Others (for example, polyoma virus and SV40) do not seem to induce malignancies of any sort in populations of wild animals, but do cause tumors after inoculation into newborn rodents. The papilloma viruses rarely have any visible effect on cells in tissue culture. Polyoma virus and SV40, however, interact with cells in vitro in a variety of ways, the exact outcome determined solely by the species of the host cell. At one pole is a productive or lytic response in which there is an ordered appearance of virus-specific functions: the infection proceeds through a series of well-defined episodes which culminate in cell death and the concomitant release of a new crop of virus particles. The whole affair lasts no longer than 48 hr. At the other pole is an incomplete infection in which very little virus is produced and the cells survive. Between these two extremes are mixed responses in which some cells of the culture make new virus and die, while others continue to divide. Most of the cells that live through their encounter with the virus show no permanent change in their phenotype. However, a small proportion assumes a new set of stable properties which closely resemble the properties of cells derived directly from virus-induced tumors. No longer are they sensitive to controls which regulate the growth of cells in culture and often their division becomes independent of the presence of a solid support. They are said to be transformed. After injection into susceptible animals, transformed cells usually grow to form tumors, and it is clear that the act of transformation provides a model system in which at least some of the events that precede tumor formation in vivo can be duplicated in vitro. It provides a means to study in a quantitative way the interaction between DNA tumor viruses and well characterized populations of cells. Events that normally take place deep inside an animal at sites which may be shielded by an immune system can now be arranged in time and space to suit the experimenter. It is this heady combination of a virus of simple structure with a very small number of genes, acting on defined populations of cells in culture to produce urgent and permanent changes in phenotype that fascinates and excites molecular biologists.

All viruses of the papovavirus group share a common architecture. Each consists of a circular double-stranded DNA molecule that is associated with histones and is coated with protein capsomeres arranged in the form of a skew icosahedron [1-3]. The family encompasses 2 genera, called subgroup A (the papilloma viruses) and subgroup B (which includes SV40 and polyoma virus) [4]. The member viruses of these subgroups differ substantially in size and nucleic acid content, and their modes of intercourse with animals or cells in culture are so dissimilar that it will be easier for us to consider the molecular biology of the 2 subgroups separately.

II. Subgroup A: The Papilloma Viruses

Several species of mammals commonly carry persistent tumors called warts or papillomas, which appear after infection of the basal epidermal cells of the skin by a papilloma virus. The infected cells are stimulated to divide and daughter cells are shed to the outside where they gradually keratinize and form the crusty surface of the tumor. At least in humans, each wart seems to be the consequence of an infection of a single basal cell with human papilloma virus. Individual warts of women heterozygous for electrophoretic variants of the enzyme glucose-6-phosphate dehydrogenase, always contain only one of the isozymes [5]. The gene for this enzyme is located on the X chromosome so that every cell of heterozygous females contains the information for both isozymes; however, only one of the X chromosomes (chosen at random) functions in any given cell [6]. The fact that individual warts are not mosaic for the forms of the enzyme proves that there is no spread of virus with recruitment of adjacent basal cells into the tumor and indicates that each wart is probably a clonal descendant of a single infected cell. These observations are in accord with the finding that there is no infectious virus detectable in the proliferating basal cells. Virus multiplication seems to occur only in the nuclei of older cells which have differentiated and lost the capacity for division. It is believed that the viral genome is present in the dividing cells in a latent form and is activated to produce infectious mature virus during the migration of the daughter cells to the surface of the wart, at about the time that they start to keratinize [7-9]. Why virus fails to replicate in dividing cells is unknown, as is the nature of the induction event.

In its natural host, the cottontail rabbit, Shope papilloma virus causes large persistent tumors which contain large quantities of virus particles arranged in crystalline arrays in the nuclei of keratinized cells [10, 11]. Similar lesions develop after inoculation of the virus into domestic rabbits but they rarely contain infectious virus [10]. However, considerable quantities of infectious viral DNA are present [12] and it therefore seems likely that virus multiplication in the keratinized cells of papillomas of domestic rabbits is blocked at a stage after viral DNA replication, perhaps during capsid synthesis or assembly of virus particles.

The papillomas of most species of animals persist for a few weeks and then regress. However in both cottontail and domestic rabbits, squamous cell carcinomas develop after 4-9 months at the site of the papilloma [13, 15]. Although no virus can be recovered from the malignant tumors (a result which suggests that they may arise directly from the basal cells of papillomas), non-virion antigens are present [15] and it is tempting to believe that these are analogous to the T-antigen found in tumors induced by SV40 and polyoma virus (see Section III.F.3).

At least 5 different papilloma viruses have been isolated from various species of mammals. While morphologically identical, they do not show detectable serological cross-reactivity and are usually highly species-specific. The purified virus particles, 55 nm in diameter, are skew icosahedral structures with 72 capsomeres and a triangulation number of 7. Rabbit papilloma virus has a left-hand surface lattice [2] and human papilloma virus a right-hand lattice [1]; the capsomeres of the virus seem to be squat hollow cylinders about 10 nm in length [16]. The virus particles contain at least 6 polypeptides, the major component having a molecular weight of 50,000-60,000 [17, 18], with a blocked N-terminal amino acid and a C-terminal threonine [19]. At least 5 minor components are present, some of which appear to bind tightly to the viral DNA and to have the properties of histones [20].

The DNAs of papilloma viruses make up 12% of the weight of the particles and range in molecular weight from 4-5.3 $\times$ 10^6 daltons [21-23], sufficient to code for 300,000 daltons of protein. The DNA is double-stranded, superhelical, and circular, and therefore sediments through density gradients in ways which originally were thought to be anomalous [21, 22, 24], but now are satisfactorily explained by the topology of the molecules (see review [25]). The DNA of Shope papilloma virus has a dGMP + dCMP content of 49% [21] whereas those of bovine, canine, and human papilloma viruses have 45, 43, and 41%, respectively. No homology can be detected by hybridization between papilloma viruses or between polyoma virus DNA and rabbit papilloma virus DNA (see review [26]). The nearest neighbor patterns of the DNA of rabbit and human papilloma viruses have been determined [27, 28] and are similar to each other and to that of mammalian cell DNA. Analysis of partially denatured molecules shows that human papilloma virus DNA has 7 regions rich in adenine and thymidine [29].

Papilloma viruses do not replicate in cells grown in vitro. Human papilloma virus has been reported occasionally to produce cytopathic changes in cultures of monkey kidney, human, and murine fetal cells [30, 31]. However, the effects are too sporadic to form the basis of an infectivity assay.

Human papilloma virus transforms human cells with low efficiency [32] and bovine papilloma virus and its isolated DNA transform bovine and murine cells [33-35], again with extremely low efficiency.

It is this lack of adequate tissue culture systems that is responsible for our ignorance of the molecular biology of the papilloma viruses. Were such systems available, the viruses, which are of great biological interest, would no doubt have received as much attention as their better known relatives, polyoma virus and SV40. Things being as they are, the papilloma viruses remain as distant objects of curiosity.

III. Subgroup B: Polyoma Virus and Simian Virus 40

Both polyoma virus and SV40 were isolated as unwanted contaminants of other virus preparations. From such inauspicious beginnings has grown a literature so massive and perceptive that our understanding of the molecular biology of the two viruses is now good, and within a year or two will be total.

A. Discovery of the Viruses

1. Poloyma Virus

In November 1951, Ludwig Gross, in the course of carrying out experiments on murine leukemia virus, observed that a single C3H mouse that had been injected with a cell-free Ak leukemic extract developed small, bilateral tumors in its neck, while remaining free of all histological signs of leukemia. Gross felt that the occurrence of the tumor was unusual enough to justify a full description in his notes [36]. His curiosity was further aroused when similar tumors appeared, several weeks later, in the additional 2 mice of the same litter that had been injected with the same extract. Gross began to search for the agent that had caused the parotid tumors and by early 1953 he had obtained unequivocal evidence that the Ak leukemic extracts contain 2 oncogenic viruses: one that induced leukemia and one that caused the parotid tumors. He was able to separate the two by filtration and by ultracentrifugation and he could free the parotid tumor agent from any leukogenic activity by heating the cell-free extracts to 63°C for 30 min [37].

The existence of an agent in leukemic mouse extracts which could cause neck tumors when injected into newborn mice was confirmed during the next few years by a large number of investigators (see review [36]), but the major breakthrough came when the virus which induced the neck tumors was grown in vitro in cultures of mouse embryo cells [38, 39]. For the first time purified virions became available in adequate amounts and it was possible to assay the infectivity of the virus efficiently. When large quantities of the virus that had been grown in vitro were inoculated into susceptible mice, it was quickly discovered that parotid tumors were not the only kind of malignancy to develop. They were the most frequent, but a variety of other solid tumors (including epithelial thymic tumors, breast and renal carcinomas, adrenal tumors, and fibrosarcomas) also appeared [40]. Because of its capacity to induce tumors in such diverse types of cells, the parotid agent was renamed "polyoma virus."

We now recognize that infection of mice by polyoma virus is a common event, for the virus can be found in a high proportion of healthy mice, both wild and laboratory bred, of several strains [41, 42]. It is clear that Gross' original isolation of the virus from a leukemic mouse was fortuitous: as far as we know murine leukemia and infection with polyoma virus are in no way related. But we now are faced with a paradox. If polyoma virus can regularly be isolated from mice, and if it efficiently induces tumors, why is it that parotid malignancies in untreated mice are extremely rare [41]? A possible answer comes from epidemiological studies of Rowe and his colleagues [42-44] who found that polyoma virus is most frequently isolated from adult mice: younger animals generally show no evidence of any encounter with the virus. Gross [37] had previously found that polyoma virus causes tumors only when it is inoculated into newborn mice. In adult mice the virus multiplies but does not produce a detectable disease, presumably because of the strong immunological response raised by the animal against new, virus-specific antigens [45]. So it seems likely that in its natural host, the adult mouse, the virus is a harmless passenger, a situation, we shall see, which is quite common with other papova viruses of subgroup B.

2. *Simian Virus 40 and Human Papovaviruses*

In 1960, Sweet and Hilleman were conducting a systematic study of indigenous viruses of cultures of rhesus monkey kidney cells which at the time were used for growth of poliovirus vaccine [46]. Already a considerable number of simian viruses had been discovered in such cells [47] and had been assigned serial numbers SV1, SV2, etc. Most of them had been detected by their ability to cause cytopathic changes in cultured cells derived from rhesus monkeys. SV40, however, had remained undiscovered because it did not cause any significant change in the cells of its natural host. It was only when Sweet and Hilleman used kidney cells derived from African green monkeys as an indicator system that the intense vacuolization, so typical of SV40, was first observed. At the time, the existence of another monkey virus to add to the plethora did not seem to be very significant and Sweet and Hilleman [46] described SV40 as "just one more of the troublesome simian agents to be screened and then eliminated from virus stocks and then from live virus vaccines." But within the next few years SV40 had been shown to induce tumors when injected into hamsters [48, 49] and to transform human cells in tissue culture [50, 51]. A short time later it became clear that such transformed cells could grow to form nodules when injected subcutaneously into humans [52]. These results, coupled to the discovery that SV40 caused persistent but apparently harmless infections in the kidneys of virtually all adult rhesus monkeys [53, 54] led to the painful awareness that a virus with proven oncogenic potential had been a common contaminant of inactivated poliovirus vaccines that had been injected into 10-20 million people in the United States alone. Epidemiological studies were immediately begun to determine whether the exposed population developed neoplasms at an increased rate. From the results of these studies [55, 56], it is now reassuringly clear that there is no urgent or large-scale consequence experienced by the infected population. However, none of the epidemiological surveys have been sufficiently sensitive to exclude completely the possibility that SV40 induces neoplasms or other diseases of humans with low frequency. And in view of the following recent developments it seems important to examine the question of involvement of SV40 in human disease more thoroughly. When humans were exposed to the virus on a large scale in the 1950s, there was no evidence that SV40 was already endemic in human populations. Recently, however, isolations of the virus have been made from humans, twice from patients suffering from the rare demyelinating disease, progressive multifocal leukoencephalopathy (PML [57]) and once from a tumor of a person with metastatic melanoma [58]. In other studies a nonstructural antigen characteristic of papovaviruses, T-antigen, has been detected in the nuclei of cells cultured from 2 meningiomas, while another SV40-specific antigen, U-antigen, has been found in the cells of a third tumor of the same type [59]. Furthermore, 3 new papovaviruses have been isolated from the brains of patients with PML (JC virus [60]), from the urine of a patient carrying a renal allograft (BK virus [61]) and from a reticulum cell sarcoma and the urine of a patient with the sex-linked recessive disorder, Wiskott-Aldrich syndrome, (MM virus [62]). All these viruses share antigenic and biological properties with SV40. The virus particles are identical in size and architecture [63]; the nonstructural intracellular T-antigen, which appears to be coded by the A gene of SV40 (see Section III.D.4a) cross-reacts extensively with antigens found in cells infected or transformed by BK and JC viruses [64, 65]; both JC and BK viruses induce tumors in newborn hamsters [64, 65]; and BK virus causes transformation of hamster cells in culture [66, 67]. However, the viruses also differ in some major respects. The antigenic cross-reaction between their virion antigens is not strong and the viruses can easily be distinguished by straightforward immunological tests [68, 69]: the virions of JC and BK viruses hemagglutinate human group O erythrocytes, whereas those of SV40 do not. Finally, the DNA fragments obtained by digesting the genomes of the 3 viruses with restricting endonucleases are significantly different from one another [70].

Clearly, papovaviruses circulate widely in human populations. Both JC and BK appear to be endemic viruses of humans [71, 72] and they seem to have reached the same sort of accomodation with their host species as has polyoma virus with the mouse and SV40 with the rhesus monkey. Generally they cause asymptomatic infections that do not inconvenience their hosts in any significant way. Only under certain rare conditions, particularly when the immune system has been suppressed, are JC and BK reactivated and allowed to multiply extensively in their tissue of choice. Under these circumstances it is difficult to discern whether the association of the viruses with diseases such as PML is causal or inconsequential. There is no doubt that the rhesus monkey is the natural host of SV40, nor that there was a massive exposure of humans to the virus when contaminated vaccines were used between 1954 and 1963. However the population seems to have suffered remarkably few ill effects as a consequence of the encounter. Only 3 bona fide isolations of SV40 have been made from humans, and antibody surveys reveal remarkably little evidence of ongoing SV40 activity in the human population [73]. It therefore seems reasonable to conclude that SV40 may not be able to be maintained in a nonsimian species, even after introduction to that species on the grandest scale.

3. *Other Papovaviruses*

Papovaviruses of subgroup B are not restricted to primates and isolation of viruses with typical papovavirus morphology have been reported from rabbits (rabbit vacuolating virus [74]) and mice (K virus [75]). Almost nothing is known of the molecular biology of these viruses and they will not be mentioned again in this chapter.

B. Polyoma Virus and Simian Virus Particles

1. *The Proteins*

Polyoma virus and SV40 are the smallest viruses that contain double-stranded DNA. Their particles are about 45 nm in diameter, with a sedimentation coefficient s20.W of 240, and their genome (3.5×10^6 daltons) can potentially encode only 2.0×10^5 daltons of protein.

Estimates of the number of polypeptides in purified particles of SV40 and polyoma virus have varied widely over the years, ranging from 1 [76] to as many as 12 [77]. A possible reason for this wide discrepancy is the difficulty of determining exactly what a virus particle is. By the standards of most animal viruses, polyoma virus and SV40 are easy to purify. Even so, the best preparations contain 10-100 times more noninfectious than infectious particles [26]. At least in theory, this bias could be due to the presence of badly packed virus particles containing some species of proteins that have no right to be present at all, others that are present in abnormally high or low amounts, or still others that may not have undergone complete cleavage from their precursors or are aggregates or break-down products. One or other of these possibilities coupled to the many variations in purification procedure that were in use a few years ago may be the cause of the diverse estimates of the number of polypeptides reported in early papers. Nowadays, however, the methods of purification are more or less standardized and there is general agreement that polyoma virus and SV40 particles contain 6 polypeptides detectable by sodium dodecylsulfate (SDS) polyacrylamide gel electrophoresis that are always present in stoichiometric amounts. The molecular weights of these polypeptides are shown in Table 11.1.

Table 11.1 Molecular Weights of Polypeptides of Polyoma Virus and SV40 and SV40 Particles[a]

Polyoma virus[b]			SV40		
Equivalent host protein	Polypeptide number	Molecular weight	Equivalent host protein	Polypeptide number	Molecular weight
	P1	86.000			
	P2	48.000		VP1	45.000
	P3	35.000		VP2	32.000
	P4	23.000		VP3	23.000
F3	P5	19.000	F3	VP4	14.000
F2b	P6	17.000	F2b	VP5	13.000
F2a1	P7	15.000	F2a1	VP6	12.000

[a]Molecular weights were determined by electrophoresis through SDS-polyacrylamide gels.
[b]The data on polyoma virus polypeptides are taken from references [78-82]; those on SV40 from references [83-89].

The major protein component of particles of SV40 is VP1; that of polyoma virus particles is P2. In each case a single polypeptide accounts for over 70% of the total protein mass of the virions and seems to be the sole component of the 72 morphological capsomeres. Crawford [90] suggested that VP1 and P2 are represented a total of 420 times in each virus particle, 60 of the capsomeres (those in the interfaces) containing 6 copies of the polypeptide, and 12 of the capsomeres (those at the vertices) containing 5 polypeptide chains. The total mass of protein in the virions would therefore amount to about 20×10^6 daltons ($45{,}000 \times 420$), a figure that is in good agreement with the observed size and buoyant density (1.34 g/cc) of the particles. Recently Walter and Deppert [91] have presented evidence in favor of this model. By using SDS they were able to isolate from intact polyoma virus particles large polymeric structures that sedimented at the same rate as shells of empty capsids and consisted only of P2 monomers. The structures were resistant to heat but were rapidly broken down to monomeric units of P2 by the addition of dithiothreitol. Thus it seems that the outer shell of polyoma virus consists of a polymeric network of P2 (or VP1) polypeptides, which are attached to each other by disulfide bridges. The other virus polypeptides are released from the polymeric structures with SDS. In mature virions they probably lie to the inside of the capsid shell and are bound to it by hydrophobic bonds.

Although VP1 and P2 are of similar size and perform similar functions in SV40 and polyoma virus, respectively, they are unrelated immunologically and yield different peptides after digestion with trypsin [85]. The amino acid sequence at the N-terminal end of VP1 of SV40 has been determined to be ala-pro-thr-lys-arg-lys-gly- [87]. It contains at least 1 phosphoryl group bound to either seryl or threonyl residues, or both [92]. P2 of polyoma virus also has alanine at its N-terminal end and has been shown to be responsible for the viral-mediated agglutination of guinea pig red blood cells [90].

In addition to the major capsid protein, particles of SV40 and polyoma virus contain a polypeptide of molecular weight 23,000, called VP3 in the case of SV40 and P4 in the case of polyoma virus. It comprises 10% of the protein mass of the virions and very probably lines the inside of the capsid shell. At least in the case of SV40, VP3 has a blocked N-terminal residue [87] and is phosphorylated [92].

The remaining polypeptides of the virion are minor components. P1 of polyoma virus is thought to be a dimer of P2 [78]. VP2 of SV40 and P3 of polyoma virus yield tryptic peptides that are also obtained from VP3 and P4, respectively [81, 93], and there is evidence that P4 is derived from P3 by a proteolytic cleavage at a unique site [81]. Thus we are left with three alternative explanations for the existence of P4 and VP3. Either they are artifacts produced by the casual action of proteases on the virus particle, or they are produced by a proteolytic cleavage that is essential, perhaps for assembly of virus particles; or they are synthesized from a subset of the nucleotides which code for VP2 and P3.

The 3 polypeptides of lowest molecular weight (VP4, VP5, and VP6 in the case of SV40, and P5, P6, and P7 in the case of polyoma virus) are a group of basic proteins that are specified by the host cell. During lytic infection they become complexed with viral DNA and are packed along with it into mature virions. Thus if cells are grown in the presence of radioactive amino acids only before they are infected with polyoma virus or SV40, the 3 proteins of lowest molecular weight in the progeny virus particles are found to be labeled to a disproportionately high degree [80, 88]. Furthermore, the peptide patterns obtained from these proteins are identical to those obtained directly from uninfected host cells [80]. The major viral histone of SV40, VP4, corresponds to host histone F3, VP5 corresponds to F2b, and VP6 to $F2a_1$ [88, 89]. Neither polyoma virus nor SV40 particles contain any detectable F2 histone [80, 88, 93]. "Full" particles of SV40 and polyoma virus contain about 40-50 molecules of each of the low molecular weight peptides, that is, a total weight of histone approximately equal to the weight of the DNA [84], a situation that is typical of chromatin extracted from mammalian cells [94]. "Empty" particles, which are devoid of DNA, lack the 3 histone components entirely [80].

Why the DNAs of polyoma virus and SV40 should form complexes with cell histones is an interesting question. On the one hand, it has been argued that histones attach to any sort of DNA, that their binding to SV40 and polyoma virus DNA could be unrelated to any functional role they may have in animal cell chromatin, and that the association occurs only because viral DNA and histones are both present in the nucleus of an infected cell. However, the DNAs of other viruses that are synthesized in the cell nucleus, for example adenoviruses, become bound to virus-specified basic polypeptides and not to cell histones [95]. Perhaps the reason for the difference in behavior between adenoviruses on the one hand and SV40 and polyoma virus on the other is that the former inhibits cellular DNA and histone synthesis during lytic infection [96] whereas the latter stimulate them [97-99]. Thus in cells infected with polyoma virus or SV40, large quantities of histones are available to the viral DNAs. On the other hand, it is possible that the binding of cell histones to SV40 and polyoma virus DNAs is not a casual attachment. Perhaps the histones play an essential role in regulating the expression of viral DNA or perhaps they cause it to fold into a suitable configuration for packing into particles. Whatever their function or functions, the structures of the nucleoprotein complexes of polyoma virus and SV40 seem to be quite similar to that of chromatin extracted from uninfected cells. There is a flexible DNA filament to which are attached globular particles called nucleosomes [100, 101] which contain about 200 base-pairs of DNA associated with the 4 histones $F2a_1$, $F2a_2$, $F2_b$ and F3 [102-104]. The DNA in the filament exists in 2 states, as shown by electron microscopy. In the nucleosomes it is highly condensed and is protected from the action of nucleases [105]; the bridges between the nucleosomes contain about 40 base-pairs of DNA that is accessible to nuclease. Nucleoprotein complexes of SV40 and polyoma virus can be extracted from virions [84, 93], from infected cells [106-109], or they can be reconstituted in vitro from naked viral DNA and histones [110]. The complexes extracted from purified virions and infected cells are very similar in chemical composition [111, 112]. The DNA in the bridge region is extended and apparently not under strain [109, 110]. However the length of the total filament is but one-seventh to one-fifth of the length of naked viral DNA. Only after the histones have been removed from it by phenol extraction does the DNA expand to its full length and assume its characteristic superhelical form with about 23 negative superhelical turns per viral genome. Therefore in the relaxed complexes the DNA must be under a torsional constraint which is confined to the nucleosomes. After the histones have been removed this constraint spreads along the entire length of the DNA molecule and induces superhelical turns. In all probability the torsional deformation caused by each nucleosome is equivalent to unwinding of the double-helix of DNA by 1 turn, so that after removal of histones, the naked DNA contains as many superhelical turns as the original number of nucleosomes [110].

The morphology of the nucleoprotein complexes that are made in vitro by reconstitution depends on the relative proportions of histones and DNA in the reaction. As the ratio of histone/DNA increases, more nucleosomes form on the viral DNA; the number of superhelical turns in the bridge regions of DNA decreases as the nucleoprotein complexes appear progressively more relaxed and the circumference of the DNA decreases [110]. As far as is known, nucleosomes do not have any fixed or preferred sites on SV40 DNA. McCarthy and Polisky [113] have examined the location of histone molecules on SV40 DNA by digesting nucleoprotein complexes obtained from virions with a restriction endonuclease (Hind III) that cleaves naked SV40 DNA at 6 well-defined sites. It seems that in the nucleoprotein complex, none of the recognition sites were protected against cleavage by the enzyme. From this result, it seems that the histones either are fixed at positions that are randomly located on viral DNA or they are mobile; therefore, we would not expect them to have any role in controlling the expression of viral genes. However Huang et al. [114] have shown that the activity of *Escherichia coli* DNA-dependent RNA polymerase on the virion-derived nucleoprotein complex of SV40 is only about 30% of that obtained using an equivalent amount of naked viral DNA. Furthermore, competition hybridization experiments showed that RNA synthesized from the complex was complementary only to a portion of the SV40 genome [114]. By contrast, the entire sequence of 1 of the 2 strands is transcribed from naked SV40 DNA [115]. After the histones have been stripped by centrifugation of the complex in cesium chloride, the efficiency of transcription of viral sequence increases, but not to a level equivalent to that of naked DNA.

Huang et al. [114] were unable completely to characterize the small amount of protein that remains attached to the viral DNA after dissociation of the histones. However, when the same series of manipulations is carried out with viral nucleoprotein complexes obtained from infected cells, the only protein that can be detected after the equilibrium centrifugation step is VP1 (Botchan, M. and Maniatis, T., personal communication). This result is highly surprising: it suggests that a few molecules of VP1 are bound tightly to the intracellular viral genome, presumably at a specific site. If they are involved in the control of expression of viral functions they act perhaps by preventing free movement of histones around the viral genome or perhaps by diverting polymerase to the segment of the viral DNA that is needed to be expressed. Clearly this area of research with polyoma and SV40 is particularly active and exciting. It is everyone's hope to use the viral nucleoprotein complex as a paradigm for cell chromatin. In this way it may be possible to find out not only about the way that expression of virus genes is controlled, but also something of the ways used by mammalian cells to control their affairs. However, a cautionary note is necessary: in a strict sense none of the proteins present in virions are required for expression of virus functions in infected cells, since naked viral DNA carries both plaque forming and transforming activities [116]. However, the specific infectivity of purified DNA is very low and even in the most sensitive assay systems only about 1 molecule in 10^4 or 10^5 successfully establishes a productive infection. On the other hand, at least 1 virus particle in every 100 is infectious. Therefore, while the proteins of the virions are dispensable, they improve the efficiency of infection by at least 2 or 3 orders of magnitude.

There are no enzymic activities known to be associated with any of the viral polypeptides. However, preparations of SV40 and polyoma virus particles commonly contain a nuclease that has a high preference for single-stranded DNA [117-119]. The claims that this enzyme cleaves polyoma virus DNA at a specific site [120] are unconvincing and there is no reason to suppose that the enzyme is necessary for infectivity of virus particles nor that it is any more than a particularly tenacious contaminating host cell protein.

2. *The DNA*

a. Topology Polyoma virus and SV40 are similar in structure and biological activity. Their genomes are identical in topology and size and are organized and expressed in the same way. Given this synonymy, it is surprising to find that their DNAs have different base compositions (polyoma virus 49% guanosine plus cytosine [G + C] [121-123]); SV40 41% G + C [124] and are nonhomologous to one another over much of their lengths. Only in 1 small region, which encompasses 15% of the genome and maps in the segment of the viral DNAs that codes for the major coat protein, is there detectable homology [125].

In any preparation of polyoma virus or SV40 DNA most of the molecules are circular, double-stranded, and superhelical with a molecular weight of about 3.5×10^6 daltons (see review [96]). These molecules are known as component I. Each of them consists of 2 covalently closed circular DNA strands, with the bases of 1 strand paired with the bases of the other to form a Watson-Crick helix. Thus the 2 strands are topologically joined and cannot be dissociated from one another by conditions such as high pH or elevated temperature which destroy hydrogen bonds [123, 126]. In neutral conditions component I molecules have a sedimentation coefficient of about 21S; at pH 12.5 or higher, all the base pairs are disrupted, then because the DNA strands cannot separate, the molecules collapse into dense supercoils which sediment at 53S (see reviews [25, 26]).

Purified component I DNA contains at least 20-24 right-handed superhelical turns per molecule [127, 128]. The origin of these turns has been mentioned previously. At the moment of completion of its last phosphodiester bond, the viral DNA is most likely organized into a number of nucleosomes [99, 129] each of which distorts the DNA by an amount equivalent to unwinding the helix by 1 Watson-Crick turn [110]. When the proteins are removed from the DNA during purification, the component I molecules are placed under strain which is relieved partially by building superhelical turns into the DNA.

A number of techniques have been used to determine the number of superhelical turns in SV40 and polyoma virus DNAs. Among these are alkaline buoyant titration [130] and buoyant separation in the presence of intercalating dyes [131, 132]. However, these tedious and indirect methods have all been rendered obsolete by the discovery [128] that electrophoresis through agarose gels can be used to examine the structure of superhelical DNA. Molecules that differ by as little as 1 superhelical turn can be separated by this technique. Using the electrophoretic mobility of SV40 DNA through agarose gels as an assay, Keller and Wendel were able to purify from mammalian cells an enzyme, originally discovered by Champoux and Dulbecco, that acts on component I DNA, gradually removing its superhelical turns [128, 133]. By using the purified DNA relaxing protein to generate reaction intermediates that could be separated on agarose gels, Keller and Wendel [128] were able directly to determine that component I SV40 DNA contains at least 20-24 superhelical turns.

In theory the superhelical turns in closed circular SV40 and polyoma virus DNAs could result from either unwinding or overwinding of the DNAs by nucleosomes. That the former possibility is correct was shown by Bauer and Vinograd [127] using sedimentation analysis of SV40 DNA in the presence of the intercalating dye ethidium bromide. Upon intercalation, ethidium bromide causes the DNA duplex to unwind [127, 134, 135]. As the amount of dye bound per molecule of SV40 DNA increases, the sedimentation velocity of the molecule decreases to a point where it approaches that of nonsuperhelical DNA. Because the binding of ethidium bromide unwinds helical DNA and simultaneously reduces the number of superhelical turns, the superhelical turns in SV40 DNA must result from a deficiency of Watson-Crick turns which could occur only if nucleosomes unwind DNA helices.

When a single phosphodiester bond in component I DNA is broken, the superhelical turns immediately disappear, because free rotation around the phosphodiester bond opposite the break is possible and the deficiency of Watson-Crick turns can be rapidly corrected [122, 126, 136, 137].

Preparations of viral DNA always contain some "nicked" (component II) molecules, which can be separated from component I DNA by 3 methods. First, component II, being less compact than component I, sediments more slowly than the superhelical form in neutral conditions. Second, components I and II reach to different extents with intercalating dyes such as ethidium bromide and propidium diiodide, which unwind DNA and alter its buoyant density. Because of its closed circular structure, component I DNA cannot bind as many molecules of the dyes per base-pair as component II. Consequently the 2 forms of viral DNA have different buoyant densities when centrifuged to equilibrium in gradients of cesium chloride-ethidium bromide [138] or cesium chloride-propidium diiodide [139]. Third, components I and II migrate through agarose or starch gels at different rates [128, 140, 141].

Component II, like component I carries both infectivity and transforming ability [142]. A third form (component III) of the viral DNA consists of linear molecules which result from a double-stranded break in component I DNA and it too is infectious [26].

b. Mapping. Because the genomes of polyoma virus and SV40 are circular, they offer no obvious natural coordinates with which one might fix the positions of genetic loci or template functions. However, in recent years a large number of restriction enzymes has been isolated from a variety of bacteria. These enzymes are highly specific endonucleases which produce double-strand cleavages of native, unmodified DNA (see reviews [143-147]). They recognize base sequences that possess 2-fold rotational symmetry about an axis perpendicular to the axis of the DNA duplex, in other words, palindromic sequences of the following type.

$$
\begin{array}{c} - \text{A B C C}' \text{B}' \text{A}' - \\ - \text{A}' \text{B}' \text{C}' \text{C B A} - \end{array}
$$

Because different enzymes recognize different palindromes, each of them generates a characteristic set of cleavage products when reacted with DNA. The specific fragments of DNA can be isolated in pure form by electrophoresis through polyacrylamide and/or agarose gels, and the order of fragments along the parental DNA can be established by a variety of techniques, including heteroduplex and denaturation loop mapping, and analysis of partial digestion products.

The enzymes that are most commonly used to cleave the DNA of polyoma virus and SV40 are shown in Table 11.2 and maps of the resulting DNA fragments are shown in Fig. 11.1 and Fig. 11.2. Restriction endonucleases, by making the isolation of any desired segment of polyoma virus and SV40 DNAs a routine matter, have been invaluable in elucidating the structure and function of the 2 viral genomes. However, they are not the only enzymes that cleave polyoma and SV40 DNAs at specific sites. Surprisingly, single-strand specific nucleases isolated from *Aspergillus* [165, 166] and *Neurospora* [167] cut both strands of superhelical SV40 and polyoma DNAs at one of a limited number

Table 11.2 Restriction Enzymes Commonly Used to Generate Specific Fragments of Polyoma Virus and SV40 DNAs[a]

		Number of sites of cleavage on the DNAs of	
Name	Source	Polyoma virus	SV40
Eco R1	*E. coli*	1	1
Bam I	*B. amyloliquefaciens*	1	1
Hae II	*H. aegyptius*	1	1
Hae III	*H. aegyptius*	17	18
Hpa I	*H. parainfluenzae*	0	4
Hpa II	*H. parainfluenzae*	8	1
Hha I	*H. hemolyticus*	3	2
Bum I	*B. umbra*	?	3
Hind II	*H. influenzae d*	2	7
Hind III	*H. influenzae d*	2	6
Hin f	*H. influenzae f*	?	10
Eco R2	*E. coli*	?	16
Alu I	*A. luteus*	?	33

[a]The data for this table were taken from references [148-164].

of sites on the viral genomes to form unit-length, linear duplex molecules [167-170]. The fact that these enzymes are active on superhelical SV40 and polyoma virus DNAs indicates that unpaired bases are present in the viral genomes, a conclusion that is confirmed by the observations that superhelical DNAs react with formaldehyde [171], methylmercury hydroxide [172], the protein coded by gene 32 of bacteriophage T4 [154, 173-175], and the water-soluble carbodiimide, N-cyclohexyl-N′-B(methylmorpholinium)ethylcarbodiimide-p-toluene sulfonate [176], all reagents that are specific for single-strand nucleic acids. The positions of the unpaired regions of superhelical SV40 and polyoma virus DNAs have been mapped in relation to sites of cleavage by restriction endonucleases. On polyoma DNA there seem to be at least 3 sites which are cut by S1 and 5 sites at which T4 gene 32 protein binds preferentially. These are located at 0.09, 0.22, 0.47, 0.56, and 0.80 genome lengths from the single Eco R1 cleavage site, [174, 175]. On superhelical SV40 DNA there are 2 sites that are preferentially attacked by S1, at about 0.15-0.25 and 0.45-0.55 genome lengths from the site of cleavage by Eco R1 [169]. Gene 32 protein binds at only 1 site located 0.46 genome lengths from the Eco R1 cleavage position [173].

It seems likely that the regions in superhelical molecules that have the greatest tendency to contain unpaired bases are rich in adenine and thymidine because they are the first to become denatured when linear viral DNA molecules are exposed to high pH or high concentrations of formamide [155, 177-179]. Whether the existence of such easily-denatured regions in the DNAs of polyoma virus and SV40 has any functional significance is unknown.

Finally, about 30% of the nucleotide sequence of SV40 DNA has been determined. Total viral DNA or specific restriction enzyme fragments are isolated and transcribed in vitro by means of *E. coli* RNA polymerase in the presence of ribonucleotide triphosphates, 1 of which is labeled in the α position with ^{32}P. The RNA product is purified and, if necessary, selected by hybridization on a specific segment of viral DNA; its sequence is then determined [180-185].

c. Substitutions, Deletions, and Rearrangements. When SV40 and polyoma virus are grown in cells infected at low multiplicity (less than 0.1 plaque forming units [PFU] per cell), the viral DNA extracted from the progeny particles consists of molecules of defined length and sequence: very few aberrant molecules are detectable. However, when the viruses are grown for several successive generations at high multiplicities of infection, the yield of physical particles remains more or less unchanged, while the yield of infectious particles is drastically reduced [186-189]. The defective viruses produced in this way are often heterogeneous in density and in most cases are lighter than infectious particles [190]. Their genomes contain extensive alterations that are detectable by hybridization [191], restriction enzyme analysis [192-194], and by electron

Polyoma DNA – A2 Strain

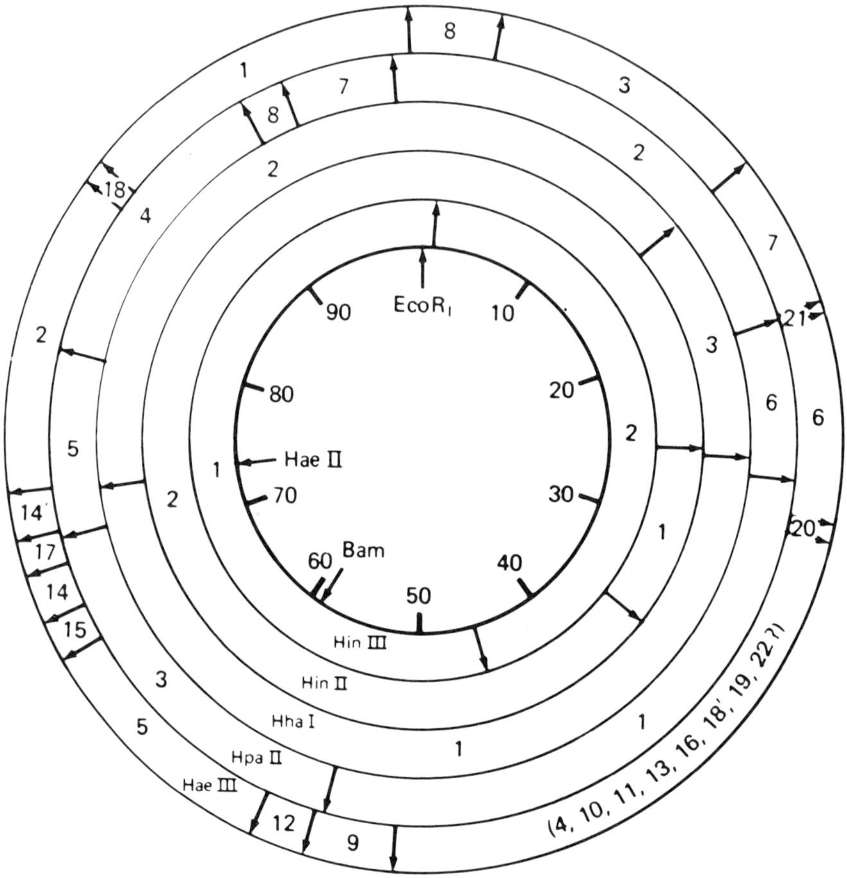
EcoR1
90
10
80
20
Hae II
70
30
60
Bam
40
50
Hin III
Hin II
Hha I
Hpa II
Hae III
(4, 10, 11, 13, 16, 18', 19, 22?)

Fig. 11.1 Sites of cleavage by restriction in polyoma virus DNA. The positions of the cleavage sites were taken from the following sources:

Enzyme	Reference
Hpa II	148
Eco R1	148
Hind III	148
Hind II	150-152
Bam 1	152
Hha I	152
Hae II	149
Hae III	149, 153

This figure was kindly supplied by Beverly Griffin and Mike Fried [149].

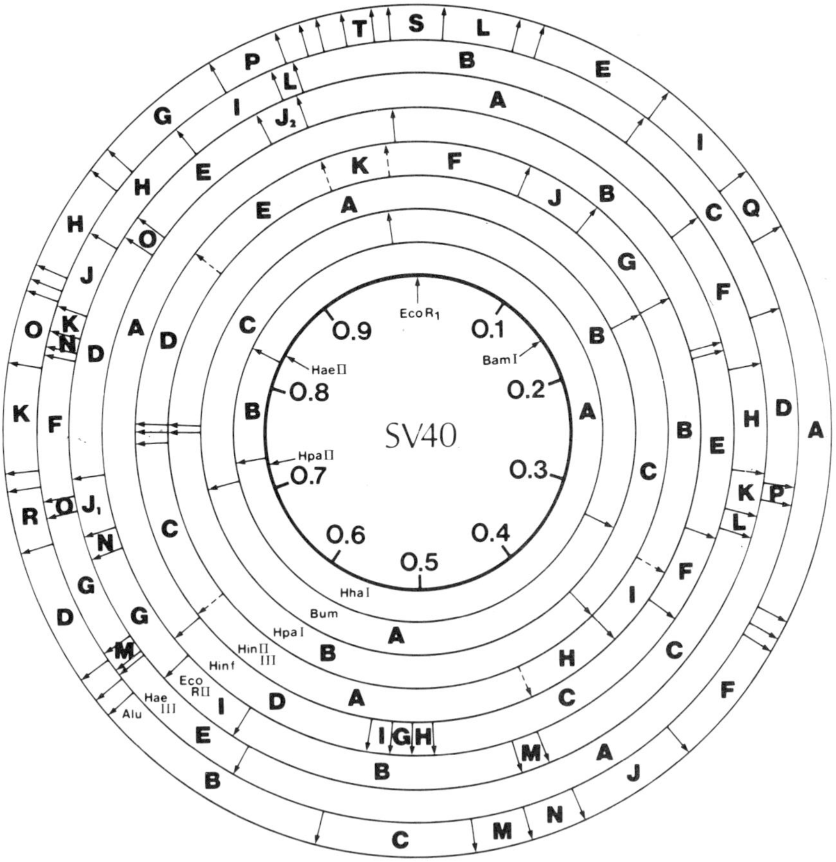
SV40
Eco R1
Bam I
Hae II
Hpa II
0.1
0.2
0.3
0.4
0.5
0.6
0.7
0.8
0.9
Hha I
Bum
Hpa I
Hin II III
Hinf
Eco RII
Hae III
Alu

Fig. 11.2 Sites of cleavage by restriction enzymes in simian virus 40 DNA. The positions of the cleavage sites were taken from the following sources.

Enzyme	Reference
Eco RI	154, 155
Bam H1	156
Hae II	158
Hae III	163
Hpa I	157, 161
Hpa II	157
Hha II	159
Bum I	160
Hind II	161, 162
Hind III	161, 162
Hin f	159
Eco R2	163
Alu I	162

microscopic analysis of heteroduplex molecules [195]. The majority of defective particles are unable to replicate unassisted and they can be propagated only in cells that are coinfected with a helper particle.

Several workers have devised methods to clone and propagate individual viruses from heterogeneous populations containing defective particles [193, 194, 196-199]. Although the details vary slightly, the principle of the methods used in different laboratories is the same. Cells are infected simultaneously with a low multiplicity of a population containing defective particles, or DNA extracted from them, and with a higher multiplicity of a helper virus, either wild-type polyoma virus or SV40, or a temperature-sensitive (*ts*) mutant. Ideally, only 1 defective particle is present in a cell that also receives a helper virus. The yield obtained from such cells consists of helper particles and a single type of defective virus which can be propagated indefinitely in the constant presence of helper. In principle, any virus that contains a defect in a region that can be complemented in trans can be isolated and grown in this fashion. The DNA of the cloned defective virus can be separated from that of its helper either by virtue of differences in the sizes of the 2 genomes or because of differential sensitivity to cleavage by restriction endonucleases. In this way it is possible to isolate from populations of SV40 and polyoma virus grown at high multiplicity, defective genomes that have deletions, substitutions, or duplications. Further, by enzymically removing the DNA sequences around or between sites of cleavage by restriction enzymes [198, 200-202], viral genomes can be constructed in vitro that contain specific alterations in designated sites. Thus the opportunity now exists to use polyoma virus and SV40 as vectors, by replacing segments of their genomes with "foreign" DNA, a development which has implications for the future extending far beyond the parochial world of SV40 and polyoma virus genetics. For the moment, however, the major contribution of the ability to clone defective genomes has been the accurate physical characterization of molecules of polyoma virus and SV40 DNA that contain substitutions or duplications or from which a segment of viral DNA has been deleted.

(1) Substituted molecules. The first formal demonstration of cellular DNA sequences in superhelical DNA extracted from SV40 virions was reported in 1969 by Aloni and his coworkers [203] who showed that there was a significant amount of homology between highly purified viral DNA and DNA extracted from uninfected cells. Previously, however, several workers had noticed an unexpectedly high level of hybridization between RNA transcribed in vitro from superhelical polyoma virus and SV40 DNAs and DNA from uninfected, untransformed cells (see for example [204, 205]). Despite this corroborative circumstantial evidence, Aloni's findings remained in doubt for several years. In some laboratories his results were repeated consistently, while in others it was extremely difficult to find any significant amount of hybridization between viral and cellular DNAs. These differences were finally resolved when

it was shown that host sequences were present only in stocks of viruses that had been grown for several cycles at high multiplicities of infection, and that the degree of substitution of viral DNA by host sequences was related to the number of times that the particular virus stock under study had been passaged at high multiplicity. During such serial passage a class of defective particles appears with closed circular genomes in which portions of the viral DNA have been replaced by covalently attached host sequences. The phenomenon occurs with SV40 [191, 192, 206-210] and poloyma virus [211].

Once substituted molecules have appeared in a population of viral DNA they accumulate at a rapid rate, presumably because of an advantage in replication [206, 209]. Perhaps the attached host sequences provide additional sites for initiation of DNA synthesis and thereby enable the substituted molecules to be replicated more efficiently than molecules of unit-length nondefective DNA, which contains a single site at which DNA synthesis is initiated (see section III. D.4b). Under the crowded conditions imposed by growth of polyoma virus and SV40 at high multiplicity there is heavy competition between different classes of substituted molecules [192, 206, 209]. Consequently, it is common to find that after many serial passages, one or a few species of substituted viral DNA, presumably because those which replicate or are packed into virus particles most effectively, come to dominate the population of defective molecules. Often, these evolutionarily successful molecules consist of tandemly repeated segments each of which consists of a segment of host DNA covalently attached to a subset of SV40 sequences that includes the initiation site for DNA replication. Different cellular DNA sequences are present in different defective molecules; the number of tandem repeats per molecule and the length of the SV40 segment included in each repeat vary from one defective to another [199]. But because of the ability of some substituted molecules to outgrow their siblings, it is difficult to know with certainty whether there is any portion of the host genome that is taken into substituted molecules more efficiently than any other. However, the fact that substituted defectives can contain cellular DNA sequences that are reiterated many times within the host genome as well as sequences that appear only once [192, 199, 206, 207, 209] suggests that there are multiple sites in the host genome that can recombine with viral DNA. As for the segment of viral DNA that is deleted from substituted molecules, there is clear indication from electron microscopic and restriction enzyme analysis that during early passages at high multiplicity certain segments of the SV40 genome are replaced more frequently than others. Chow et al. [210] have shown that in such defective viral stocks, parts of the viral sequences between 0.31 and 0.59 on the conventional map are commonly deleted and replaced by a piece of DNA that is equivalent in length to about 10-20% of the viral genome. However longer substitutions are also possible and the point of insertion into the viral genome is by no means constant.

Substituted viral genomes must originate as a consequence of recombination events between host and viral DNA. The heterogeneity of cellular sequences in early passage, substituted molecules and the failure to identify in viral DNA a region of mandatory replacement together suggest that substitution occurs by a rather nonspecific mechanism with neither the cell nor the virus showing absolute preference for the site of recombination. The details of the process remain obscure. Evidence has been reported for massive integration of viral sequences into cellular DNA during lytic growth of polyoma virus [212] and SV40 [213-215]. If these accounts are correct, it seems likely that the substituted molecules arise by inaccurate excision of viral DNA from an integrated state. However, the whole idea of integration of viral sequences into cellular DNA during lytic infection has recently been called into question [216]. Much of the hybridization that previously was thought to represent virus sequences that had become covalently attached to host DNA now can be explained by the discovery in lytically infected cells of giant oligomers of viral DNA. Given these uncertainties it remains possible that substituted molecules arise by recombination between free viral genomes and fragments of host cell DNA that are often produced during productive infections by polyoma virus and SV40 [217-219].

(2) Deleted and rearranged molecules. The first detectable change that occurs in the genomes of polyoma virus and SV40 during passage at high multiplicity is deletion of segments of viral DNA. Thus, preparations of DNA isolated from early passages contain a significant number of molecules that are shorter than unit-length viral DNA and are predominantly noninfectious [186-190, 195-199, 220-223]. The size and location of the deletions relative to the known sites of cleavage by restriction enzymes have been determined by heteroduplex analysis [198, 222, 223]. Defective genomes obtained from early passaged virus range from 0.70-0.98 fractional lengths of viral DNA. The deletions seem to be distributed essentially at random throughout the viral genome. Virus particles of late passage stocks generally contain a much more homogenous set of deleted molecules, presumably as a consequence of selection. Several deleted variants have been cloned and propagated in the presence of transcomplementing helper particles [193, 194, 196-201], and the structure of their genomes has been examined in detail. There seems to be no clear indication that any regions of the viral DNA are particularly vulnerable to deletion. However there is one segment, the initiation site for DNA synthesis, that must be retained if the deleted viral genome is to survive and replicate. Presumably if this region of the viral genome is lost, replication occurs either not at all or at such a low frequency that the deleted molecule is placed at a tremendous competitive disadvantage. In fact, many of the molecules that have suffered deletions also contain a duplication which in every case so far examined spans the region of the viral genome that contains the site of initiation of

DNA synthesis [192, 195-199]. In the most extreme examples the cloned defective molecules consist of nothing but a reiterated series of initiation sites and a few of their flanking sequences [152, 197, 207, 224-229].

Four ways have been used to obtain deletions in designated portions of the SV40 genome. The least precise method involves arranging a system of reciprocal complementation, in which defective DNA molecules are complemented by a temperature-sensitive mutant of the virus; the mutant is in turn complemented by the defective DNA. By choosing appropriate temperature-sensitive mutants it is possible to select and propagate DNA molecules that have suffered alterations either in the early or the late regions of the virus [195, 196, 198, 199].

Variants can also be selected that are resistant to the action of Eco R1 and Hpa II, restriction enzymes that cleave SV40 DNA at a single site (see Table 11.2). The genomes of these resistant variants range from 0.71-0.96 fractional lengths of wild-type SV40 DNA and the deleted region always spans the appropriate restriction site. The viral DNA molecules that are resistant to Eco R1 are invariably noninfectious and can be grown only by complementation [200, 202]. In addition to the mandatory deletion, they often contain a duplication of the region spanning the initiation site for viral DNA synthesis. Some molecules that are resistant to Hpa II have suffered extensive deletions and are noninfectious. However, it is also possible to isolate DNA that is resistant to digestion by Hpa II and yet does not require a helper in order to replicate. Each of these viable deletion mutants lacks a small segment of viral DNA (80-190 base-pairs) and contains no other detectable rearrangements. Thus the region in SV40 DNA around the cleavage site for Hpa II is not essential for the lytic growth of the virus, at least in vitro [198, 202]. This conclusion is in agreement with the observation that insertion of a short stretch (< 50 base-pairs) of poly(dA:dT) at the site of cleavage of SV40 by Hpa II results in the production of a virus that is nondefective and can replicate without assistance, albeit at a perceptibly slower rate than wild-type virus [198].

The 2 remaining methods for obtaining deletions are highly specific and involve treating wild-type viral DNA with various enzymes in vitro. (1) Restriction endonucleases have been used to excise specific segments of viral DNA and the resulting deleted molecules have been cloned and propagated by complementation with temperature-sensitive mutants [198, 200, 201]. Two types of defective molecules are generated by this procedure: "exisional" and "extended" deletions. In excisional deletions the boundaries of the deletion correspond to restriction enzyme cleavage sites [198, 200, 201]. Presumably such molecules arise as a consequence of perfect cyclization at the cohesive termini generated by cleavage of DNA with the restriction enzymes used (Hind III and Eco R2). In molecules with extended deletions, on the other hand, the limits of the deletion extend beyond the restriction sites [198, 200, 201] and it seems likely that they must have been generated by a kind of "illegitimate" recombination. (2) Linear molecules obtained by cleavage of component I SV40 DNA with Eco R1 or Hpa II have been digested with 5′-exonuclease to remove 10-20 nucleotides from the 5′ end of each of the DNA strands [202]. The infectivity of linear molecules generated by Hpa II is not destroyed by this treatment, although the genomes of the progeny particles lack a small segment of DNA that spans the Hpa II cleavage site. Like naturally occurring Hpa II-resistant variants, they grow more slowly and form smaller plaques than wild-type virus. The linear molecules produced by cleavage of SV40 DNA with Eco R1 lost their infectivity during digestion with exonuclease and require a complementing helper in order to grow. The genomes of the defective progeny viruses manifest small deletions of the DNA sequences (15-50 base-pairs) around the Eco R1 cleavage site. How "stripped" linear DNA molecules recircularize is unknown. However the fact that the resulting deletions do not seem to extend beyond the regions of viral DNA that are digested by exonuclease strongly suggests a mechanism involving base-pairing between the exposed single-stranded tails of DNA.

Clearly, all this recently found ability to manipulate the genomes of polyoma virus and SV40 in precise and delicate ways is an immense advance. Given the large number of restriction endonucleases now available and the efficient complementation of defective viruses by helper particles, it now seems possible to generate a complete set of overlapping deletion mutants. As we shall see, these are sorely needed in order to establish precise and perfect maps of the genomes of polyoma and SV40 with identification of the template regions for regulatory as well as structural elements.

C. Interactions of Polyoma Virus and Simian Virus 40 with Cultured Cells

Whether poloyma virus and SV40 evoke productive or nonproductive responses depends on the species of the cells being infected (see Table 11.3). During infection of cells of the virus' native hosts (mouse cells for polyoma virus and monkey cells for SV40), the overwhelming response is productive: new virus is synthesized by cells that are fated to die. Exposure to the virus of cells of other animal species results in a nonproductive infection in which very little or no progeny virus is produced and the cells survive. We do not know why some sorts of cells support lytic growth while others allow only incomplete infections. In abortively infected cells [230], early virus-specific RNA [231, 232] and T-antigen [233-235] are synthesized. The activity of some of the enzymes associated with DNA metabolism increases after infection [236, 237] and cellular DNA synthesis is often induced [231, 237-244]. By contrast to lytic infection,

Table 11.3 Complementing Groups of SV40 *ts* Mutants[a]

Complementation group	A	B	C	BC	D
A	–				
B	+	–			
C	+	+	–		
BC	+	–	–	–	
D	–	–	–	–	–

[a]Complementation occurs between mutants (+); complementation cannot be detected (–). Data for this table were taken from references [345, 413, 414, 417, 488, 489].

however, neither viral DNA replication nor expression of late viral genes can be detected. The failure of polyoma virus and SV40 to discharge their late functions in nonpermissive cells is not understood. The restricted expression of viral genes could be due either to absence of essential host products or to suppressive effects present in nonpermissive cells. Most of the available evidence favors the first hypothesis. Basilico et al. [245] have shown that the amount of polyoma virus replication in a series of hybrid clones formed by fusing permissive mouse cells with nonpermissive hamster cells is proportional to the number of mouse chromosomes present in the hybrid cells. This result strongly suggests that the ability to support virus replication is a consequence of the production by permissive cells of a factor or factors necessary for full expression of virus genes. Because susceptibility to virus growth is dominant over nonsusceptibility, the result also implies that hamster cells are nonpermissive because they fail to synthesize the putative factor or factors and rules out the possibility that nonpermissive cells produce a substance which inhibits virus replication.

The inability of the virus to replicate in nonpermissive cells does not seem to be due to the inactivity of the host RNA polymerase on the viral DNA, for Amati et al. [246] have shown that hybrids formed between a hamster cell line resistant to α-amanitin (a toxin extracted from the mushroom *Amanita phalloides*) and α-amanitin-sensitive mouse cells support efficient growth of polyoma virus both in the presence and absence of the drug.

Recently, Graessman and Graessman [247] have reported that after free SV40 DNA molecules have been introduced into nonpermissive mouse cells by microinjection, rather than by conventional means, late viral functions are expressed. Taken at its face value this result does not support the idea of

Basilico et al. [245] that cell-coded factors are necessary for synthesis of late virus-coded products. But there is one way to resolve the apparent disagreement, although it seems to me to be more a concoction to accommodate the data than a reasoned scientific hypothesis. One has to argue that in nonpermissive cells, the virus embarks upon a series of stochastic events. Should it fail to complete any one of them the infection will abort. After entering the cell by the regular route, the chances that the virus will execute all the necessary steps are very low. After microinjection, when the virus genome enters the cell by an abnormal route in great numbers, it may be that its chances of success are commensurately higher. In this scheme, the function of the factor or factors produced by permissive cells would be to increase greatly the probability that the virus would successfully negotiate the stochastic barriers. Clearly this explanation is contrived and rather than taking it or one of its variations with any seriousness perhaps it would be best to admit that our present knowledge of the molecular basis for permissivity is too poor to allow any firm conclusions to be drawn.

D. The Lytic Cycle

1. Absorption and Penetration

At least in the case of polyoma virus, penetration of cultured cells by virions occurs in 3 steps. First, the virus particles attach to cell receptors that contain sialic acid [248]. This initial binding must be loose and electrostatic in nature for it can be reversed by alterations in ionic strength or pH of the medium [249]. In the second stage, the infectivity of the virus becomes resistant to changes in pH, but it remains sensitive to inactivation by antiserum. Finally, about 50% of the particles enter the cell [249, 250-253], are ferried across the cytoplasm in pinocytotic vesicles [254], and enter the nuclei, probably through nuclear pores [255]. For the first hour or 2 after infection virus particles can be detected by electron microscopy in the cell nuclei, but they disappear within the next 2 hr, presumably as a consequence of uncoating.

2. The Time Course of Viral Infection

The replication cycle of polyoma virus and SV40 occurs in 2 stages (see review [96]. The first occurs early after infection and continues until the onset of viral DNA synthesis, some 10-12 hr later (see reviews [256-259]). In this period the only change detectable in the infected cells is the appearance of virus-specific antigens which are presumably translated from the small fraction of RNA that is found to hybridize to viral DNA.

Synthesis of polyoma virus and SV40 DNA begins about 10-12 hr after infection and its rate increases constantly during the succeeding 10-12 hr until a maximum is reached and held for about 16-20 hr. During this phase there is continued production of early viral mRNA and protein and specifically late mRNA and viral proteins appear in ever-increasing quantities. Synthesis of host DNA and histones is induced and there is a progressive increase in specific activity of many of the cell-coded enzymes concerned with metabolism of DNA. Three to six hr after the onset of viral DNA synthesis virus capsid protein can be detected in the nuclei and the first virus progeny particles appear about 24 hr after infection. Because of the high degree of asychrony within the infected cell population, virus continues to be produced by the infected cell culture for at least a further 24 hr.

After this period, there is a perceptible lessening of metabolic activity: viral DNA and RNA synthesis stop and in some types of host cells, chromosomal DNA is degraded. By 50 hr after infection many of the cells have detached from the surface of the Petri dish and are dead.

3. *Transcription*

Neither polyoma virus nor SV40 particles contain DNA-dependent RNA polymerase, so that the uncoated viral DNA must be transcribed into RNA by host enzymes. There are 3 principal forms of DNA-dependent RNA polymerase (I, II, and III) in mammalian cells [260], which differ in their chromatographic properties, intranuclear location [261], salt optima and sensitivity to α-amanitin [262]. Whereas enzyme II is totally inhibited by low concentrations of α-amanitin, enzyme III is moderately resistant and enzyme I is completely insensitive. Two kinds of experiments show that form II DNA-dependent RNA polymerase, the nucleoplasmic activity responsible for synthesis of heterogeneous nuclear RNA in uninfected cells, transcribes SV40 and polyoma virus DNA at late times during lytic infection. First, replication of polyoma virus in mouse cells is inhibited by the presence of α-amanitin, yet the virus grows with equal efficiency in the presence or absence of the toxin in cells that contain a polymerase that is resistant to α-amanitin [246]. Second, it has been shown that the synthesis of virus-specific RNA in vitro by nuclei isolated from permissive cells at late times after infection by SV40 or polyoma virus is totally inhibited by α-amanitin [263, 264]. Both types of experiment leave open the possibility that a polymerase other than enzyme II is responsible for *early* transcription of the viral genomes. In fact there is morphological and autoradiographic evidence [265] that there is an association between SV40 DNA and the nucleoli of permissive cells at all stages of the lytic cycle, and it still seems possible that RNA polymerase I, an enzyme that is present exclusively in nucleoli and is responsible for the synthesis of ribosomal RNA in uninfected cells [266], may play a role in early transcription or replication of the viral genome.

No way has yet been found to isolate viral genomes during the act of transcription. Consequently neither the topology of the viral DNA template, nor the proteins that are associated with it during synthesis of RNA are known. But it seems clear that the major form of intracellular viral DNA, the 55S nucleoprotein complex, is not the template for RNA synthesis. Schmookler et al. [264] lysed infected cells with gentle detergent treatment and separated the resulting components by velocity sedimentation through sucrose gradients. When each fraction was assayed for its ability to direct the synthesis of viral RNA in vitro, it was found that the 55S nucleoprotein complex contained only low levels of endogenous RNA polymerase activity. Over 80% of the viral RNA synthesized in vitro was generated by a fast sedimenting component that pelleted during the period of centrifugation.

While the difficulties posed by the system are formidable, the failure to characterize and visualize viral transcription complexes is perhaps slightly surprising. After all, about 10% of the RNA synthesized by infected cells at late times after infection is virus-specific. And the properties of viral DNA are so idiosyncratic that separation of viral and cell transcription complexes should be fairly straightforward. It may be that too little attention has been paid to the possibility that single-stranded, rather than duplex viral DNA is the template for transcription. Purified DNA-dependent RNA polymerase II has a high preference for single-stranded SV40 DNA [267], and there are large quantities of single-stranded SV40 DNA present in cells at late times after infection (Sambrook, unpublished results). Together these observations make a circumstantial case that seems plausible enough to warrant further enquiry. By contrast to many animal viruses, SV40 and polyoma virus stimulate the synthesis of host cell RNAs of all kinds during lytic infection [268] and the only way to detect viral RNA sequences amongst the ongoing clamor of cellular metabolism is by nucleic acid hybridization. By 2-4 hr after infection there is present in infected cells RNA which anneals to purified viral DNA immobilized on nitrocellulose filters. For the succeeding 8-10 hr there is a gradual accumulation in the cytoplasm of a species of viral RNA that sediments at 19S (equivalent to a molecular weight of 9×10^5 daltons) [258, 269, 270], but at no time before the onset of viral DNA synthesis are large quantities of viral RNA detectable. Only 0.01-0.001% of the total RNA extracted from infected cells during the early phase of the virus growth cycle hybridizes to viral DNA [271, 272]. As viral DNA replication proceeds, however, there is a vast increase in the rate of synthesis of viral transcripts so that by late times after infection 10% of the pulse-labeled RNA hybridizes to viral DNA [273-275] and 0.1-1.0% of the total weight of RNA in the cell is virus-specific [271, 272]. Competition hybridization experiments between the RNA species isolated at different times after infection show that there are different sets of sequences transcribed in the early and late phases of the viral replication cycle. Early after infection only about

30-40% of the sequences of viral DNA is transcribed into stable cytoplasmic RNA [274, 276-283]; at late times the equivalent of 1 whole strand of the DNA is copied into stable transcripts [284, 285]. Sedimentation velocity analysis shows that at late times after infection there are species of polyoma virus and SV40 RNA present on the polysomes of infected cells with molecular weights of 9×10^5 (19S) and 7×10^5 (16S) [258, 269, 270]. Preliminary evidence suggests that the 19S peak may contain 2 species of RNA, 1 that is transcribed at both early and late times after infection and 1 that is synthesized only at late times [275]. This second species, a true late RNA, shares nucleotide sequences with some of the viral RNA that sediments at 16S [286] and is thought to be a precursor of at least a part of it [287]. However, there seems to be a portion of the 16S fraction, unrelated to 19S RNA [288], that may well be a separate viral transcript. All species of late cytoplasmic viral RNA are thought to contain methylated base groups at their 5′ ends [289] and tracts of poly(A) 150-200 nucleotides in length at their 3′ ends [290, 291].

From these results, it is clear that the start of DNA synthesis is a watershed. It is coincident not only with a sizeable increase in the rate of RNA synthesis but also with notable alterations in the pattern of transcription. The inescapable conclusion is that at least some control over virus gene expression is exerted at the level of RNA. Thus during the last few years considerable effort has been expended to find out more about the arrangement of template regions in the viral genomes: which segments of polyoma virus and SV40 DNA code for early and late viral functions, which strand of the viral DNA is transcribed in any given region, and in which direction transcription proceeds around the viral genomes. Finding the answers to these problems depended on the development of 2 techniques: first, the ability to obtain defined segments of polyoma virus and SV40 DNAs by cleavage with restriction endonucleases, and second, the knack of separating each of these subviral fragments into their component strands.

The 4 enzymes that have proved to be the most useful in determining the patterns of transcription of polyoma virus and SV40 are Hpa I and Hpa II isolated from *Hemophilus parainfluenza* [157, 292], EcoR1 isolated from strains of *E. coli* that harbor $f1^+$ drug resistance factors [293], and Hind II/ Hind III isolated from *H. influenzae* serotype d [294]. The locations on the viral genomes of the fragments that have been used in the analysis of viral transcription are shown in Fig. 11.3 and Fig. 11.4.

The complementary strands of polyoma virus and SV40 DNAs cannot be separated by any of the standard methods, such as equilibrium centrifugation at alkaline pH or unequal binding to polynucleotides. However when superhelical DNA of either virus is transcribed in vitro with *E. coli* RNA polymerase, the product is a species of highly asymmetric RNA (cRNA) that is complementary to only 1 of the 2 viral DNA strands [115, 295, 296]. Thus when denatured

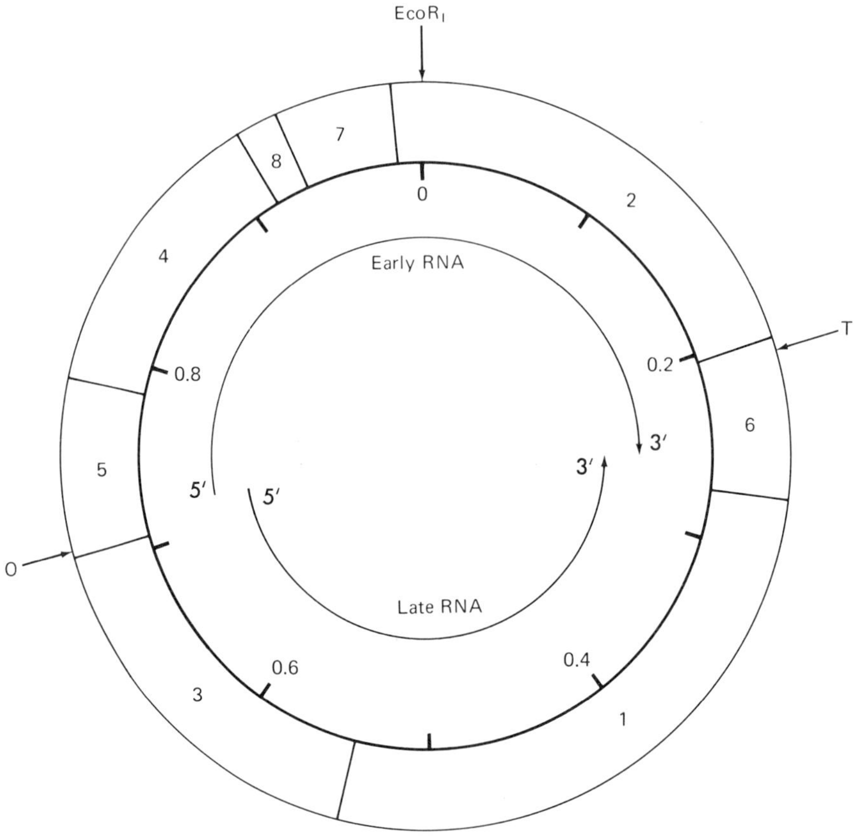
EcoR$_I$
7
8
0
2
4
Early RNA
T
0.8
0.2
6
3′
5
5′
5′
3′
O
Late RNA
0.6
0.4
3
1

Fig. 11.3 Template map of polyoma virus. The positions of the sites that are cleaved by the restricting endonuclease Hpa II are shown as reference points as well as the sites at which DNA synthesis originates (0) and terminates (T) [362]. The data for the transcription map were obtained from reference [296].

DNA is allowed to reanneal in the presence of a vast excess (20-50-fold) by weight of cRNA, one of the strands of DNA very rapidly forms hybrid with the RNA while the other remains in a single-stranded form. These two forms of DNA can easily be separated by chromatography on hydroxylapatite and the isolated strands can then be purified to homogeneity [271, 272, 296]. The strand of SV40 that is complementary to cRNA is called the E or (–) strand because it is transcribed into stable species of RNA at early times after infection; the other is called the L strand. For polyoma virus DNA, this nomenclature is reversed because *E. coli* RNA polymerase happens to transcribe in vitro the strand

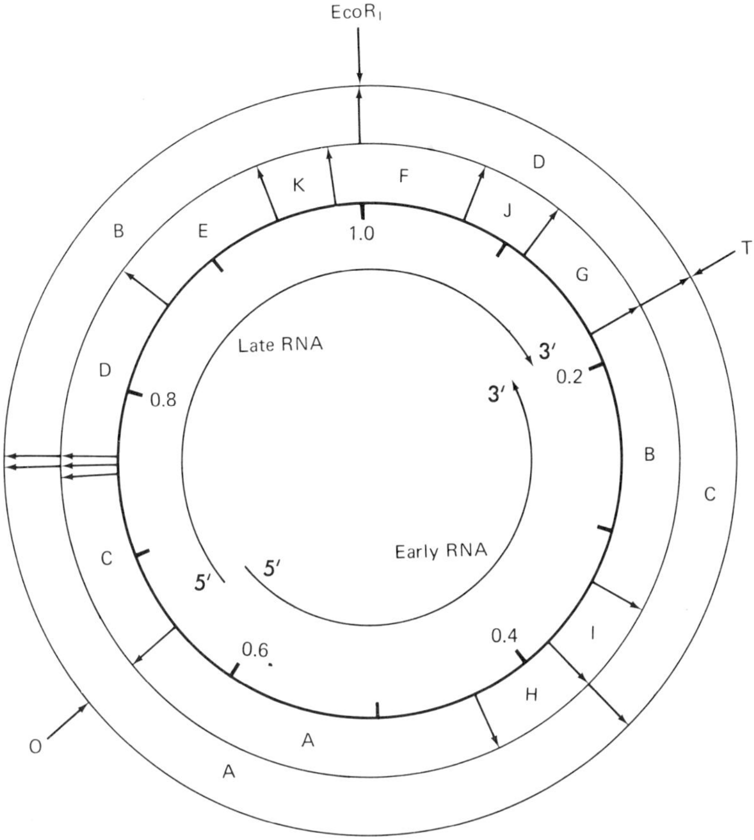
EcoR$_I$
B
E
K
F
D
J
G
T
1.0
D
0.8
Late RNA
3′
3′
0.2
B
C
C
5′
5′
Early RNA
0.6
0.4
I
H
A
A
O

Fig. 11.4 Template map of SV40. The positions of the sites that are cleaved by restricting endonucleases Eco R1, Hpa I, and Hind II/III are shown as reference points, as well as the sites at which DNA synthesis originates (O) and terminates (T) [363]. The data for the transcription map were obtained from references [180, 232, 271, 272, 297, 298, 300].

of DNA that is copied into stable species of RNA only at late times after infection. Because *E. coli* RNA polymerase copies all of the sequences of one strand of each of the two viral DNAs at approximately equal frequencies, it is possible to use cRNA to separate not only unit-length viral genomes but also the strands of subviral fragments [272, 296-298]. Most of the work on mapping transcription products of polyoma virus and SV40 has been carried out using single separated strands of DNA obtained in the manner just described. Recently, however, another method, a modification of a technique first described by Hayward [299], has been used. When fragments of SV40 DNA are denatured and subjected to electrophoresis through agarose gel, one strand of DNA is found to migrate at a faster rate than its partner [300]. Because it is possible to recover the DNA strands in high yield and because the method is technically less demanding than the hybridization technique, it seems likely that electrophoretic separation of strands will become the technique of choice.

When the appropriate reagents are available, it is a comparatively simple exercise in hybridization to assign each strand of each DNA segment to the early or late sector of the viral genome. The transcription maps of polyoma virus and SV40 that have been obtained in this way are shown in Figs. 11.3 and 11.4. About 55% of the sequences of polyoma virus DNA are appropriated by early functions transcribed into early cytoplasmic RNA [296]. In the case of SV40 most estimates were obtained by saturation hybridization experiments [276-279] and from studies in which asymmetric cRNA was annealed to RNAs extracted from lytically infected cells at early and late stages after infection [301]. About 40% of the coding capacity of the virus is set aside for early genes [271, 272, 297, 298]. Recently, however, there have been suggestions that a slightly larger segment of the viral DNA, perhaps 45-50% of the genome, is expressed at early times after infection [180, 232, 298]. This higher estimate fits in well with the apparent size of the early SV40 messenger RNA (9×10^5 daltons), calculated on the basis of sedimentation velocity. But in view of the weight of earlier data, assignment of an exact length to the early and late regions probably should be deferred until the nucleotide sequences of early and late messenger RNAs have been completed.

The early and late functions of both viruses are located on different strands of DNA. Thus the early genes must stretch along the viral genome in one direction, the late genes in the other, a conclusion that is in agreement with early findings that some but not all virus-specific RNA obtained from lytically infected cells can hybridize with asymmetric cRNA [301, 302].

The results described so far describe the topology of the early and late template regions of the genomes of polyoma virus and SV40, but they do not indicate the direction of transcription around the circular DNAs. Several kinds of experiments have been carried out to secure this knowledge. All of them are similar in design and involve determining the direction of synthesis of cRNA on viral DNA. An example, taken from Sambrook et al. [297], is shown in Fig. 11.5. SV40 DNA was treated with endonuclease Eco R1 which cleaves the viral genome at a unique site by making 2 staggered, single-strand scissions, 4 bases apart [303]. The resulting linear molecules contain at each of their ends a 3′-hydroxyl group and a protruding 5′-phosphoryl tail, and they serve as excellent primer templates for DNA synthesis. It was therefore possible to incorporate [^{3}H]dTMP at the ends of the molecules using RNA-dependent DNA polymerase isolated from avian myeloblastosis virus. The DNA was then cleaved

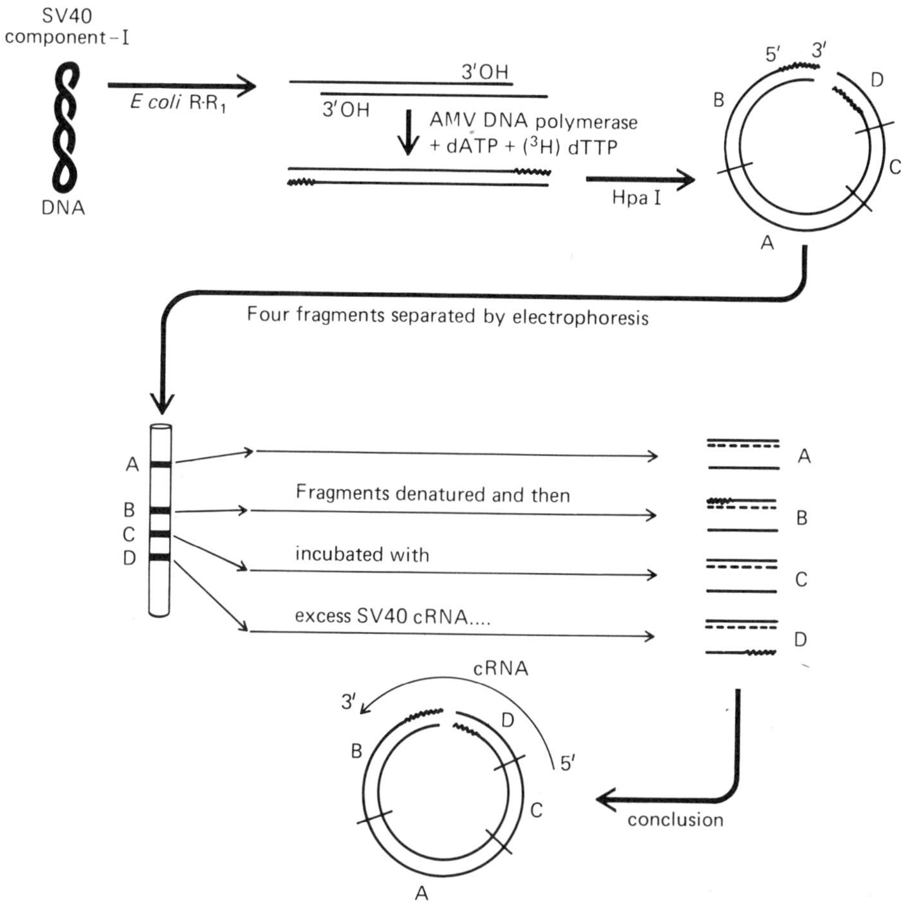
SV40
component-I
DNA
E coli R·R1
3'OH
3'OH
AMV DNA polymerase
+ dATP + (3H) dTTP
Hpa I
5'
3'
B
D
C
A
Four fragments separated by electrophoresis
A
B
C
D
Fragments denatured and then
incubated with
excess SV40 cRNA....
A
B
C
D
cRNA
3'
5'
B
D
C
A
conclusion

Fig. 11.5 Determination of the direction of transcription of SV40 DNA. From reference [297]. For explanation, see text.

with endonuclease Hpa I and the 4 resulting DNA fragments were separated by electrophoresis, denatured, and hybridized to asymmetric cRNA (see Fig. 11.5). Only the 2 fragments (B and D) that flank the Eco R1 site contained radioactivity. Because of the nature of the staggered ends produced by Eco R1, the radioactivity must be present in one of the two strands of the DNA of fragment B and in the opposite strand of fragment D. Thus the radioactivity in only one of the two fragments should hybridize to cRNA. The results showed that over 90% of the ^{3}H counts in fragment B annealed to cRNA whereas less than 6% of the ^{3}H counts in fragment D were complementary to cRNA [297]. Knowing that

RNA-dependent DNA polymerase catalyzes the incorporation of nucleotides in a $5' \rightarrow 3'$ direction, and because the position of fragment B is in the viral DNA counterclockwise from the Eco R1 site, then RNA-dependent DNA polymerase must synthesize DNA in a clockwise direction on fragment B and a counterclockwise direction on fragment D. Because the radioactivity incorporated into fragment B is complementary to cRNA, RNA must be synthesized in vitro from SV40 by *E. coli* RNA polymerase in a counterclockwise direction and because cRNA and early RNA have the same sense, the early gene of SV40 must also be transcribed in a counterclockwise direction as drawn on the conventional SV40 map. The direction of synthesis of cRNA has been determined for polyoma virus [296] and, by other workers [298], for SV40, using procedures similar in principle to that described above. In the case of polyoma virus, however, it is important to remember that cRNA is synthesized in a clockwise direction on the conventional map and is complementary to the L strand of viral DNA.

This rather minor distinction apart, it is very clear from these results that polyoma virus and SV40 possess arresting similarities. In both cases the 5′ ends of the stable species of RNA map close to the point at which DNA synthesis originates, while the 3′ ends map near the point of termination of DNA synthesis. In view of the dependence of transcription of the late genes of the viruses upon viral DNA synthesis, it is difficult to believe that this close geographical relationship is accidental. However, the exact mechanism by which the viruses regulate their transcription is still unknown. It is clear that synthesis of stable late RNA does not occur when initiation of viral DNA synthesis is blocked by the presence of an inhibitor such as cytosine arabinoside [279, 280, 283]. However, studies with temperature-sensitive mutants of SV40 [303] and inhibitors [305-307] have shown that the shift from the early to the late phase transcription depends only on initiation of viral DNA synthesis and not upon its continuation. Once transcription of late sequences has begun, it cannot be stopped merely by interrupting viral DNA synthesis.

The following 3 possible mechanisms have been proposed to explain these data.

(1) The template model proposes that an alteration in the physical structure of the viral genome is responsible for the change in transcription pattern. Such an alteration could occur by removal of histones or other proteins from the viral DNA during replication. Or perhaps late RNA can be synthesized only from viral genomes that are in the act of replication, or from progeny molecules that have some special structural feature, such as a specific nick in 1 or other of the DNA strands.

(2) Alterations in RNA polymerase: there could be synthesis of a new polymerase or modification of an existing host polymerase so that it can recognize and transcribe late viral DNA sequences. The only constraint upon

this model is that the hypothetical polymerase has to resemble host RNA polymerase II in its properties, because as we have seen, transcription of polyoma virus and SV40 DNAs, at least at late times, is carried out by an enzyme indistinguishable in its properties from host RNA polymerase.

(3) Changes in processing: perhaps there is no control of virus gene expression exerted at the level of transcription. It is possible, for instance, that late viral RNA is made at all stages of infection but is rapidly degraded within the nucleus at early times. Its appearance in the cytoplasm at late times during infection would then reflect a change in the RNA processing machinery. This suggestion merits serious consideration because of results obtained from 2 sorts of experiments. First, it is clear that the viral RNA sequences present in the cytoplasm are often only a subset of those found in the nucleus. At late times after infection with polyoma virus and SV40, the nuclei of cells contain short-lived species of RNA which seem to be transcribed from the entire sequences of viral DNA [275, 291, 296, 297, 308, 309]. Second, Aloni [275] has reported the isolation from cells infected with SV40 of double-stranded RNA molecules which are equal in length to 1 strand of the viral DNA, and a number of other workers have concluded that the nascent transcripts of the virus consist, at least in part, of molecules whose size is at least the length of 1 DNA strand [258, 270, 274, 310-314]. All these data have led to a hypothesis, proposed by Aloni [308] first for SV40 and later [309] for polyoma virus, that at late times after infection all or nearly all of the sequences of each of the 2 viral DNA strands are transcribed. Considerable portions of the RNA molecules are then degraded and it is the surviving noncomplementary sequences that are transported to the cytoplasm for decoding.

There are at least 2 key experiments that should be capable of distinguishing between the alternative hypothesis of transcriptional control mediated by changes in template or polymerase, and posttranscriptional control mediated by processing. The first experiment would ask whether at *early* times after infection there is any transcription of *late* strand DNA at all. The second, whether both early and late strands of viral DNA are transcribed *at equal frequencies* at late times after infection. If the answers to both these questions are yes, then we are forced to accept that posttranscriptional processing of RNA plays a major role in controlling viral gene expression. If the answers are no, control exerted at the level of transcription will be proven. In the latter case we would be left to find an explanation for the presence of symmetric viral transcripts in the nuclei of infected cells. Perhaps they are aberrant transcripts, arising not from free unit-length viral DNA molecules but from the putative integrated viral DNA sequences [312, 314], or from defective viral DNA molecules that have lost the region of the viral genome that contains the termination site for viral RNA synthesis; or perhaps they are produced by only a small fraction of the infected cells in which viral transcription has become totally disordered.

Given all these uncertainties it seems impossible to make a choice between the 2 kinds of models. Clearly this area of research on polyoma virus and SV40 is highly active and it should not be too long before definitive results are available.

Perhaps it would be best to end this section with a brief discussion of a topic that is understood with some clarity: the transcription of SV40 DNA in vitro. The superhelical form of the viral DNA is a very efficient template for *E. coli* RNA polymerase and as has been discussed above, the entire sequence of only 1 strand of the duplex DNA is transcribed in the in vitro reaction [115, 295, 315, 316]. The product is heterogeneous in size and contains some molecules whose length is at least equivalent to unit-length SV40 DNA [115, 295, 315, 316]. Thus *E. coli* DNA-dependent RNA polymerase can, during the course of transcription, pass through and transcribe the DNA sequence at which RNA synthesis was initiated.

Several molecules of *E. coli* RNA polymerase, the actual number depending upon the ionic strength of the reaction mixture [317, 318], can interact simultaneously with a single molecule of superhelical SV40 DNA [319, 320]. All of them initiate synthesis of RNA at the same time and travel around the circular DNA in the same direction [319]. After completion of the reaction a fraction of the product remains complexed with the DNA [321], presumably because the template, being topologically underwound, contains regions of denatured DNA with which RNA can form stable hybrids.

When SV40 DNA that has been converted to a linear form by Eco R1 is used as template, there is one strong promoter located at 0.16 fractional genome lengths from the Eco R1 cleavage site. From this site transcription proceeds towards the short arm of the DNA [319]. The same promoter is preferentially used by *E. coli* polymerase when the transcription reaction is carried out at reduced temperature with superhelical SV40 as template [184]. Because cRNA is synthesized in a counterclockwise direction on SV40 DNA [297], the *E. coli* RNA polymerase begins its traverse of SV40 DNA by heading into the late region of the viral genome, transcribing the E strand.

Thus the major promoter for *E. coli* RNA polymerase maps on the SV40 genome at a position almost diametrically opposite to that of the 5′ end of stable, early messenger RNA [297, 298] and we are given no simple indication of the possible location of any promoter or promoters that may function in vivo.

Under the same ionic conditions that allow asymmetric transcription of superhelical SV40 DNA, component I polyoma virus DNA is transcribed symmetrically (J. P. Monjardino, quoted in [296]). However, asymmetric RNA can be obtained either by transcription of certain classes of defective polyoma virus DNA molecules (M. Vogt, quoted in [296]) or by transcription of nondefective polyoma virus DNA at high ionic strength [296].

4. *DNA Synthesis in Infected Cells*

Soon after infection of cultures of resting cells with SV40 or polyoma virus, the activity of many of the enzymes concerned with DNA synthesis rises [236, 242, 243, 322-330], probably due to net synthesis of new enzyme molecules [323, 325, 330]. By about 12-15 hr after infection with high multiplicities of virus, these enzymes reach their greatest levels of activity, and soon afterwards, synthesis of viral DNA, cellular DNA (both nuclear [238, 239, 322, 331-335] and mitochondrial [336]), and of histones [97-99, 337, 338] begins. The mechanism of the induction is not understood, but it is clear from inactivation studies [238, 339, 340] and from examination of the physiology of temperature-sensitive mutants of polyoma virus [341-343] and SV40 [343, 345] that a virus-coded protein, probably the early gene product [345], is involved.

Studies of synchronized cultures of permissive cells have shown that the length of time that elapses between infection and the onset of viral DNA synthesis is related to the period of the cell cycle in which the cells are infected [346, 347]. For example, when monkey cells in the G1 phase of the growth cycle are infected with SV40, the first progeny DNA molecules are synthesized during the subsequent S period. If the cells are infected during the S or G2 phases of the cell cycle, no progeny viral DNA molecules are detectable until the S period of the next cell cycle. Thus it seems that some event occurs late during G1 or early in S that is required for replication of viral DNA. The virus has the ability to stimulate cells that are arrested early in the G1 phase to resume their progression through the growth cycle. However, when cultures of randomly growing cells are infected with polyoma virus or SV40, there is no apparent interruption to the movement of cells through their division cycles. Because the virus is unable to influence the behavior of growing cells, it must wait for the critical event to happen at the appointed time in the cell cycle before it can replicate its DNA. The molecular nature of the event is unknown.

a. Structure of Replicating Viral DNA. Without methods that selectively extract viral DNA molecules from permissively infected cells, it would be very difficult indeed to study the process of replication of polyoma and SV40 DNA. The technique that is most commonly used was devised by Hirt [348]. Cells are lysed by the addition of SDS and the high molecular weight cellular DNA is removed from the preparation by precipitation with high concentrations of sodium chloride. Many, if not all of the viral DNA molecules (both replicating and mature) remain in the supernatant and can easily be purified for further analysis.

When permissive cells at late times after infection with polyoma virus or SV40 are exposed to [^{3}H]thymidine for short periods of time (less than 10 min) the radioactive DNA molecules in the Hirt supernatant are found to sediment through neutral sucrose gradients slightly more rapidly than do mature molecules of superhelical viral DNA (22-26S compared with 21S [349-352]). If the brief labeling period is followed by progressively longer chases, the radioactivity gradually disappears from the heterogeneous 22-26S material and emerges in component I DNA, sedimenting at 21S. Newly labeled, synthesized, mature superhelical DNA is not detected until the sum of the pulse plus chase periods exceeds 10 min [349-352]. These results mean that the labeled material sedimenting between 22 and 26S must consist of replicating molecules of viral DNA and that the time taken to synthesize a complete copy of the genome of polyoma virus or SV40 is about 10 min. In the electron microscope, the majority of the replicating molecules are seen to contain 3 branches, 2 branch points, and no free ends, in other words, classical Cairns-type structures [353]. The Cairns model of DNA replication predicts first, that no strands of DNA in replicating molecules will be longer than the parental genome, and second, that the 2 replicating progeny strands are of equal length. Both these conditions are fulfilled by replicating polyoma virus and SV40 DNA. The fact that newly synthesized DNA in replicative intermediates always sediments more slowly through alkaline gradients than does unit-length viral DNA [351, 352, 354, 355] is enough to eliminate the possibility that SV40 and polyoma virus DNA replicate as rolling circles [356].

In early studies [349, 350, 358, 359] the purified replicating molecules generally appeared in an extended configuration with no apparent superhelices. More recently, however, evidence has been accumulating that most of the replicating molecules of SV40 and polyoma virus DNAs contain a superhelical region in the unreplicated portion of the molecules [351, 352, 355, 357]. Thus the replicating molecules consist of 2 circles of equal length (the replicating regions) attached to a common region of DNA that contains superhelical turns. One circle plus the superhelical region DNA is equal in length to mature SV40 DNA [351, 355]. The superhelical structure of the replicating intermediates is due to the presence of covalently closed template strands [351, 352, 360] to which the newly synthesized daughter strands are hydrogen-bonded.

The fact that the unreplicated portion of purified replicating molecules is superhelical while the replicated portions are not, can be explained as follows. We have seen (see Section III.B.2a) that superhelical turns in SV40 and polyoma virus DNAs are, in a sense artifacts. In the cells, the molecules, complexed with histones, are probably flat; only after removal of the histones during purification of the DNA do superhelices form. Thus one possible explanation for the failure to find superhelical turns in replicated regions of replicative intermediates of viral DNA is that these segments are not bound to histones in vivo. However, it seems more likely that histones are indeed bound evenly over the entire length of the replicated and unreplicated portions of the replicative intermediates. When the histones are removed from the unreplicated segments of the molecules, superhelices form. However, the replicated portions of the molecules contain discontinuities in the daughter DNA strands near the replicating forks. These nicks or gaps could act as "swivels" through which the superhelical twists can be dissipated after the histones have been removed.

Because of their structural features, it is possible to fractionate replicating molecules on the basis of their buoyancy in cesium chloride gradients containing an intercalating dye such as ethidium bromide [351, 352, 355, 357]. Because superhelical turns are present only in the unreplicating portion of the molecules, "young" replicative molecules contain a great proportion of their sequences in the superhelical form. They therefore band in cesium chloride-ethidium bromide equilibrium gradients at a position close to that of component I DNA. As replication proceeds, more and more of the sequences of the molecules are transferred to the nonsuperhelical region. There, topologically unconstrained, they are free to bind greater quantities of ethidium bromide per nucleotide pair. As replication proceeds the number of superhelical turns in the molecules becomes smaller and smaller and the replicative intermediates band in cesium chloride-ethidium bromide gradients at positions that are progressively closer to that of component II DNA.

b. The Origin and Termination of DNA Replication. Two kinds of experiments prove that DNA replication begins at a specific site on the genomes of polyoma virus and SV40.

First, when young replicating molecules are treated with a restricting endonuclease such as Eco R1 that cleaves the viral DNA at only 1 site, linear structures containing a small "bubble" are seen by electron microscopy. The midpoint of the bubble, which consists of the DNA sequences that have been replicated is always located at the same position in the viral genome. Molecules in a more advanced state of replication contain larger bubbles and the lengths of *both* arms of the flanking unreplicated DNA are concomitantly and equally reduced. These results mean that the genomes of polyoma virus and SV40 each contain a single site at which DNA synthesis is initiated, that replication is bidirectional, and that the replication forks move in opposite directions at about the same rate [361, 362]. The origin has been located on the SV40 physical map at a distance 0.33 fractional lengths from the Eco R1 site (i.e., map position 0.33 or 0.67). On polyoma virus DNA, the origin is located at position 0.71, near the junction of Hpa II DNA fragments 3 and 5, as judged by electron microscopic analysis of replicative forms cleaved with Hind III, an enzyme that cleaves polyoma virus DNA in 2 places [148].

Second, Danna and Nathans [363] have devised a technique, based on the kinetic analysis of hemoglobin synthesis [364], that involves restriction enzyme analysis of pulse-labeled, replicating forms of viral DNA. In molecules which complete replication during the pulse, regions of the DNA near the origin will be least radioactive and those near the terminus most radioactive. Conversely, molecules which begin replication during the pulse will be more radioactive near the origin and less radioactive near the terminus. In analyses of SV40 DNA replication, Danna and Nathans [363] concluded that replication begins in Hind II/III fragment C (near position 67). By the same technique Crawford et al. [365, 366] showed that the origin of replication of polyoma virus DNA is located near the junction between Hpa II DNA fragments 3 and 5 (near position 71). The replicating forks move at about the same rate so that termination occurs 0.5 fractional lengths away from the origin (i.e., at 0.17 in SV40 DNA and position 0.21 in polyoma virus DNA). It seems that replication terminates merely when the replication forks meet opposite the initiation site, rather than at any specific DNA sequence, since some defective DNAs of both polyoma virus [152] and SV40 [227, 228] appear to lack that portion of the viral genome where termination normally occurs. Furthermore Brockman et al. [199] have shown that in certain SV40 defective molecules the replication fork appears to pass through the region of Hin G which contains the normal termination site for DNA synthesis.

c. Polynucleotide Chain Elongation. If replicative intermediates are labeled for very brief times (15-60 sec) and analyzed by sedimentation through alkaline gradients, 2 populations of radioactive DNA strands are found. One consists of the growing daughter chains of viral DNA, is heterogeneous, and ranges in size from 6-16S; the other, homogeneous in size, sediments at 4S. These latter fragments are about 150 nucleotides long and can be chased into the growing daughter chains and are thus likely to be true intermediates in the replication process [367, 368]. About 20-30 ribonucleotides have been detected covalently linked to the 5′ end of 4S DNA pieces in polyoma virus DNA synthesized in isolated nuclei [368-370]; presumably this short stretch of RNA acts as a primer for the synthesis of 4S fragments of DNA. From these results it seems likely that elongation of the progeny strands of polyoma virus or SV40 DNA occurs by a discontinuous process much like that proposed by Okazaki et al. [371, 372] for the replication of *E. coli* DNA. However, the 4S fragments of polyoma virus and SV40 DNAs differ from Okazaki pieces in two major respects. First they are much smaller (4S compared to 10S); second, they contain at the RNA-DNA link all 4 common ribo- and deoxyribonucleotides [368-370]. This result, which indicates that there is no specific sequence at the junction between RNA and DNA, is in contrast to the report of a specific RNA-DNA link in replicating *E. coli* DNA [373].

Finally, there is disagreement over the origin of the 4S fragments. On the one hand, the extensive self-annealing exhibited by isolated fragments suggests that they are synthesized on both strands of DNA at both of the replicating forks [367, 369, 374, 375]; on the other, the fact that about 50% of the radioactivity is incorporated into long strands of DNA during short pulse [370, 376] argues strongly that synthesis of 4S DNA occurs from only 1 strand of DNA, and that the second strand of DNA is extended in a continuous fashion. In my opinion the evidence favors the second scheme, for there is a possible alternative way to explain the self-annealing characteristics of 4S DNA. Robbersson et al. [377] have shown that, in addition to bidirectionally replicating forms, a small minority (10%) of polyoma virus and SV40 DNAs replicate unidirectionally from an additional origin. In such a situation, a fraction of the 4S chains of DNA would be expected to self-anneal even though they are synthesized in a strictly asymmetric fashion.

Whatever the origin of 4S fragments, discontinuous synthesis of polyoma virus and SV40 DNA is enhanced in the presence of drugs such as hydroxyurea or fluorodeoxyuridine [368, 375, 378, 379]. These inhibitors lower the concentration of deoxynucleotide triphosphate pools in mammalian cells, and by slowing down the rate of polynucleotide propagation cause an accumulation of young replicative intermediates [368, 380]. The 4S fragments in these molecules are separated from each other by single-stranded gaps, for they cannot be "joined" in vitro to growing chains by *E. coli* DNA ligase, but only by a combination of T4 DNA polymerase and ligase [378]. These results suggest that there may be 2 different DNA polymerases involved in the replication of polyoma and SV40 DNA: one responsible for the synthesis of 4S pieces and the other responsible for "gap-filling." While the enzymes that carry out these functions have not yet been identified with certainty, it has been argued by Laipis and Levine [378] that the most likely candidate for synthesis of 4S fragments is DNA polymerase α (see review by Weissbach [381]: Of the 3 known mammalian DNA polymerases only α can synthesize DNA from a ribonucleotide primer and it is least sensitive to low concentrations of deoxynucleotide triphosphates. Either the β or γ polymerase could be responsible for filling the gaps between 4S DNA fragments and attaching them to the growing DNA chain.

In addition to polymerases several other proteins, presumably all of them host-coded, are required for efficient growth of DNA chains. Among these are the following.

A protein analogous to gene 32 protein of bacteriophage T4 that denatures the DNA strands ahead of the replication fork.

An RNA polymerase that synthesizes the oligoribonucleotide primer used by polymerase α.

An enzyme that removes the RNA and creates a gap that is filled by the action of a second polymerase; an enzyme (RNAse H) whose properties in vitro are consistent with this role has been purified from mammalian cells [382].

Polynucleotide ligase, an enzyme that is present in large quantities in mammalian cells [330].

An unwinding protein. So that denaturation of DNA can continue ahead of the replication forks, the superhelical turns that are introduced into the DNA by circular parental denaturation must be removed. An enzyme that acts as this sort of "swivel" is present in mammalian cells [133]; it acts by breaking one of the parental DNA strands, allowing the superhelical twists to dissipate by free rotation of one DNA strand around the other. Finally, the enzyme reseals the gap and restores the DNA to a closed circular structure [128]. The enzyme is remarkable since it requires no exogenous energy source such as ATP [383].

d. Termination of DNA Synthesis and Segregation of Daughter Molecules. When the 2 replicating forks meet in the terminal region of SV40 DNA, they fuse and 2 interlocked circles of viral DNA are formed. These segregate to give rise to 2 component II DNA molecules each of which consists of a strand of closed circular parental DNA and a daughter strand that contains a specific nick or gap at position 0.17, the terminal region for DNA synthesis [384]. After a short delay, the interruption in the daughter strand is repaired and the process of replication is complete. Occasionally (about once in every 75 rounds of replication), repair is carried out before segregation of the interlocked circles can occur and a catenated dimer is formed [385]. The frequency of formation of such forms is increased about 4-fold in the presence of cycloheximide [385, 386] a result which suggests that the act of segregation may depend on a labile protein. Catenated dimers of viral DNA are unstable and they are rapidly converted by recombination into circular dimers or monomers [385].

5. *Virus-coded Proteins in Lytically Infected Cells*

Infection of permissive cells with polyoma virus or SV40 does not result in a substantial diminution in the rate of cellular protein synthesis [77, 82, 387-389]. In fact, as we have already seen (see Section III.D.4), lytic infection leads to a comprehensive stimulation of cell-coded enzymes concerned with synthesis of DNA. The major problem, therefore, is to identify viral proteins within the large background of host proteins. Three general approaches have been used, often in combination: (1) detection of novel species of antigens; (2) analysis of polypeptides from virus-infected cells by electrophoresis through SDS polyacrylamide gels; and (3) translation of virus-specific RNA in cell-free systems.

a. T-Antigen. Several virus-induced early antigens have been identified in infected cells; some of these are also found in cells transformed by SV40 or polyoma virus. By far the best known is T-antigen, a protein present in the nucleus of infected and transformed cells. For many years it successfully eluded definitive characterization. Indeed, following its initial detection by complement fixation [390, 391] and immunofluorescent techniques [392-395], an extended debate arose, at times quite lively, concerning whether the antigen was really a bona fide virus gene product, or was merely a novel, derepressed, cellular protein. As we shall see, recent evidence obtained from studies of *ts* mutants of polyoma virus and SV40 and from translation of viral RNA in cell-free systems, heavily favors the idea that T-antigen is at least in part a virus-coded substance.

Although the origin of T-antigen was at one time in question, there was never any doubt that it was virus-specific; polyoma virus and SV40 induce immunologically distinct T-antigens and yet each virus induces the same T-antigen no matter what species of cells are used for infection [390, 395-401]. The appearance of the antigen after infection depends on the expression of part of the viral genome, since mutants of polyoma virus [402] and SV40 [403] that are defective at any early stage in lytic infection, either do not induce the antigen, or manufacture it in reduced quantities or in altered form. Furthermore, inactivation of the viruses by UV-irradiation or chemicals reduces their ability to induce T-antigen [339, 340, 404-409], and a deletion mutant of SV40 whose genome is 13% shorter than that of wild-type virus fails to elicit the synthesis of the antigen [179, 223]. None of these arguments, however, provides rigorous proof that T-antigens are virus-coded.

However, in recent years several important advances have been made in the purification and analysis of T-antigen and there is now strong evidence that it is at least in part the product of the single early gene of polyoma virus and SV40. Direct proof of the best sort has come from synthesis in vitro of polypeptides that are specifically precipitated by antiserum to T-antigen. Two groups of workers [410, 411] have shown that SV40 cRNA is translated in a wheat germ system into a number of polypeptides with molecular weights of 92,000, 82,000, 68,000, 59,000, and 25,000. Because of the asymmetric nature of cRNA [115] some of these products are probably translated from anti-late RNA sequences. However, the 59,000 and 25,000 components are specifically precipitated by hamster antiserum to SV40 T-antigen.

Cells lytically infected or transformed by SV40 produce a protein whose molecular weight is variously estimated to be 87,000-100,000 [389, 402, 410] that is precipitated by antiserum to T-antigen. Polyadenylated viral RNA extracted from these cells has been shown to direct the synthesis in vitro of 2 polypeptides (of molecular weights 59,000 and 87,000) that also react with antiserum to T-antigen [410]. Purified T-antigen is known to be extremely sensitive to proteolytic attack [402] and it seems possible that the 59,000 and 25,000 components are degradation products that have retained the capacity to react with antiserum. Direct evidence that the 87,000-100,000 protein is encoded by the early gene of SV40 comes from 2 sources. First, messenger RNA extracted from infected cells and purified by hybridization to DNA of Hae III fragment A (located entirely within the early region of the SV40 genome) directs the synthesis in vitro of T-antigen [410]. Second, a mutant of SV40 that carries a deletion within the A gene produces a protein that reacts with antiserum to T-antigen and is smaller than that induced by wild-type virus [412]. This information together with the fact that the 87,000 protein is large enough to account for the entire coding capacity of the early region of the viruses argues strongly that T-antigen is the sole early product of SV40 and polyoma virus. This hypothesis is also consistent with genetic evidence: more than 50 early mutants of SV40 have been isolated [344, 413-419] and all of them fall into 1 complementation group.

Estimates of the molecular weight of T-antigen that has been purified by the conventional techniques of protein chemistry [420-422] range from 70,000-100,000. However, sedimentation values obtained in nondenaturing solvents indicate that the molecular weight of the antigen is much larger [423-427]; presumably the native immunoreactive material consists of either a series of T-antigen subunits or a mixture of T-antigen and host cell components. What these putative cellular components might be is an open question. However, purified T-antigen has been shown to bind in vitro to both double-stranded calf thymus DNA [426] and SV40 DNA [428, 429] and it is reasonable to believe that the antigen may be associated with viral and/or host DNA during infection.

If we accept that T-antigen is the product of the single early gene of SV40 and polyoma virus, then it is possible to make reasonable conjectures about its function. First, T-antigen must be involved in the initiation of DNA synthesis. Mutants of the early gene (group A mutants) are inhibited in their ability to induce the synthesis of either cellular [345, 430] or viral [431] DNA at nonpermissive temperatures. If viral DNA synthesis is allowed to start at permissive temperature and the infected cells are then warmed to nonpermissive temperature, no new rounds of viral DNA synthesis are initiated. Thus the early gene product is required to function only when a viral DNA molecule begins to replicate. Second, the early gene product is required to establish the late phase of viral transcription. Mutants of the A group synthesize early viral RNA at the nonpermissive temperature [304] but none of the late sequences of the viral genome are transcribed into stable RNA. Furthermore, the early gene product seems to regulate its own synthesis, since under restrictive conditions, group A mutants synthesize large quantities of an unstable form of T-antigen [403]. This result implies that cells infected by wild-type SV40 contain a specific quantity of T-antigen. Presumably once a prescribed amount of the active protein is available, no more of it is synthesized. Thus directly or indirectly, the early gene product controls the expression of all of the known virus genes. Third, the early gene is involved in the establishment, and probably the maintenance, of the transformed state. Group A mutants can transform cells only at the nonpermissive temperature. When such cells are incubated at the nonpermissive temperature they often manifest a change of phenotype and assume many of the properties characteristic of untransformed cells (see Section III.E.1b).

Taking all this evidence together, it is clear that the early protein is a pivotal molecule that controls many aspects of viral and cellular metabolism. The mechanism by which this regulation is achieved is unknown but given the affinity of T-antigen for DNA [426, 428, 429] it seems highly likely to involve binding of the antigen to specific nucleotide sequences. Indeed there have been claims that T-antigen binds preferentially to the site on the SV40 genome at which replication begins [428]. The present evidence indicates that viral DNA synthesis and transcription of the early and late regions of the viral genome begin in the same segment of the viral DNA, and may be that T-antigen controls these processes by binding to that self-same site. How the antigen exerts regulation over cellular functions is an open question.

b. U-Antigen. Little is known of the structure and function of this antigen which is detected by complement fixation and immunofluorescence at the nuclear membrane (but not in the nucleus) of cells infected and transformed by SV40 [432]. It is produced early during lytic infection, a result that immediately raises a paradox. If T-antigen is the sole product of the 1 early gene of SV40 how can 2 different SV40-specific antigens be present at early times in infected cells? There are 2 possible explanations. Either U-antigen is not a virus-coded substance, or it is a degradation product of T-antigen. While definitive studies, such as cell-free synthesis of U-antigen and comparison of the peptides of U- and T-antigens are lacking, the available evidence favors the hypothesis that U-antigen is a degradation product of T-antigen. There is a nondefective hybrid virus, $Ad2^{+}ND1$, whose genome consists of a molecule of adenovirus 2 DNA from which a small segment of nonessential sequences has been deleted [433]. In its place a piece of SV40 DNA is integrated that includes the region of the genome coding for the C-terminal end of the early protein [433-436]. $Ad2^{+}ND1$ induces the synthesis of SV40-specific U-antigen, but not T-antigen [435]. Because host protein and RNA synthesis is efficiently suppressed by adenovirus infection [437, 438], it seems likely that U-antigen is coded by the small fragment of SV40 DNA that is integrated into the genome of $Ad2^{+}ND1$. The size of U-antigen is unknown; in theory, any fragment of protein derived from the C-terminal end of the early SV40 protein might carry

U-antigen activity, and it may be that U-antigen is a mixture of molecular species, each of which consists of different subsets of the sequences of T-antigen.

c. Surface Changes. During the course of lytic infection with SV40 or polyoma virus, permissive cells show the following changes in their surface properties.

Infection with polyoma virus causes the rate at which cells take up 2-deoxyglucose to increase by about 10-fold [439]. This stimulation, which occurs during the early phase of virus infection, is a consequence of an accelerated transport of hexoses across the cell membrane.

In those cells in which host DNA synthesis is stimulated after infection, there is a change in the outer membrane, so that the cells agglutinate efficiently in the presence of plant lectins such as concavalin A or wheat agglutinin [341, 440].

Cells infected with SV40 or polyoma virus display on their surfaces novel antigens that can be detected by immunofluorescence [441-443] or the ability to confer resistance to SV40 enhanced tumors upon hamsters [444].

All of these alterations also occur on the surfaces of cells transformed by polyoma virus or SV40. Whether any of them are due to incorporation of virus-coded or virus-induced material into cell membranes or whether they merely represent a nonspecific response by the cell to the presence of active viral genomes is unresolved.

d. Helper Function of Simian Virus 40. Monkey kidney cells are semi-permissive for the replication of adenoviruses: However, coinfection of the cells with SV40 greatly enhances the yield of adenovirus from such cells [445, 446]. $Ad2^{+}ND1$, along with other nondefective Ad2-SV40 hybrids [434], replicates efficiently in monkey cells and it seems that the integrated segment of SV40 provides a helper function for replication of the adenovirus. Mutants of $Ad2^{+}ND1$, selected for their inability to grow on monkey cells [447], fail to induce normal quantities of U-antigen, and it therefore seems likely that helper activity might be provided by, or might depend upon the activity of, U-antigen. How helper function acts is not understood. In the absence of SV40, infection of monkey cells by adenoviruses results in the production of normal quantities of adenovirus T-antigen [448, 449] and DNA [450, 451]. Viral RNA synthesis seems to proceed normally [452, 453], and the concentrations of adenovirus messenger RNA capable of directing protein synthesis in vitro are equivalent in enhanced and unenhanced infections [454]. The defect in growth of adenovirus 2 in monkey cells is thought to reside in the production of structural components of the adenovirus virion (notably hexon and fiber), which are found in greatly reduced quantities in unenhanced cells [447, 454-458]. Thus, it seems that the block to adenovirus multiplication and the site of action of the SV40-coded helper function may both be found in the level of translation. In both the presence and absence of SV40 coinfection, adenovirus messenger in monkey cells becomes attached to ribosomes and it can be released by treatment with EDTA. However, recent observations [452,458-460] suggest that certain species of late adenovirus RNA fail to become incorporated into functional polysomes.

Several temperature-sensitive mutants of SV40 fail to aid the replication of adenoviruses in monkey cells incubated at nonpermissive temperatures [461, 462]. Many of these belong to complementation group D, whose members seem to be defective at an extremely early stage of infection. The virions absorb and penetrate normally but they do not uncoat at the restrictive temperature [344, 345, 463]. Thus no viral functions are expressed, nor is RNA synthesized, so that it is not surprising that D group mutants cannot enhance adenovirus infection. Only 1 mutant is known that fails to assist adenoviruses to grow in monkey cells and possibly belongs to a complementation group other than group D [461]. Called *ts* 640 (414) it differs from classical group D mutants [345] in that it efficiently complements late temperature-sensitive mutants of SV40 [416]. However, no reports have been made of the physical location on the SV40 genome of the mutation of *ts* 640, nor of complementation studies between *ts* 640 and mutants of SV40 known to be located in the early gene. Whether *ts* 640 synthesizes U- or T-antigens at the restrictive temperature is unclear. So for the moment it seems prudent to conclude that none of the known *ts* mutants of SV40 have been proven to affect specifically the helper function of the virus.

The molecular mechanism of helper function is unknown, but its effect can be simulated by treating monkey cells with iododeoxyuridine before infection [464] with adenoviruses, and suppressed by using a line of monkey cells that fails to show induction of cell DNA synthesis after infection with SV40 [465]. These results suggest that helper substance is cell-coded and is synthesized when the cell's pattern of DNA synthesis is altered, either by infection with SV40 or by treatment with base analogs. It would be interesting to know whether the subset of early SV40 mutants that are inhibited in their ability to induce cellular DNA synthesis at elevated temperature [345, 430] also fails to elicit helper activity.

e. Polypeptides of the Virion. Late during infection with polyoma virus or SV40, the cytoplasm of permissive cells contains large quantities of RNA that code for the 2 major polypeptides of the virions [466-468]. The manufacture of these proteins accounts for 10-20% of the total protein synthesis in cells in the terminal stages of lytic infection [77, 82, 387, 388, 469, 470]. They are synthesized in the cytoplasm but are transported rapidly to the nucleus for assembly into progeny virus particles. The mechanism by which mature virions are formed is largely unknown, but because SV40 and polyoma are viruses of simple structure, it has often been suggested that self-assembly processes may play a large role in the production of progeny particles [256,

471, 472]. Empty shells consisting of virus capsid proteins appear to form readily, and although the relationship between empty capsids and infectious virus has not been firmly established, the available data [469, 473, 474] indicate clearly that empty capsids are not breakdown products of infectious virus particles but are precursors of them. In mature DNA, presumably associated with host histones, it is withdrawn at random from the pool of viral genomes and becomes encapsidated within preassembled shells of capsid proteins [475]. Eventually large crystalline arrays of progeny virions form, arranged on various cellular membranes as monolayers of tightly-packed particles [476-478]. At about this stage in the virus growth cycle the nucleus of the cell stains very brightly with fluorescent antiserum raised against intact virus particles [479] and, in the case of SV40, the cytoplasm begins to show signs of vacuolization. Within 12 hr the cells round up and detach from the substrate and the lytic cycle is complete.

E. Genetics of Polyoma Virus and Simian Virus 40

For such small viruses there is a surprising wealth of material available for genetic studies. There are host-range mutants of polyoma virus, which are able to replicate on certain sorts of transformed mouse cells, but not on contact-inhibited 3T3 cells [480, 481]; mutants that fail to absorb to some kinds of cells [482]; mutants that fail to grow at low temperature [483]; and a variety of mutants with defective, substituted, and rearranged genomes of both polyoma virus and SV40 that has already been described (see Section III.B.2c). However, the most extensively studied mutants are those whose growth at high temperature is significantly less than at low temperature (*ts* mutants). Over 200 temperature-sensitive mutants of SV40 [344, 345, 413-416, 418, 427, 461, 484, 485], and about 100 of polyoma virus [342, 486, 487] have been isolated.

The genetic lesions in these mutants have been classified on the basis of complementation tests and, in some cases have been assigned to physical locations on the viral genome. The effects of the mutations have been analyzed by comparing the events during infection of permissive cells at restrictive temperature with those that occur in cells infected either with wild-type viruses at the same temperature or with mutant viruses at permissive temperature.

1. *Simian Virus 40 ts Mutants*

a. Complementation Tests and Physiological Mapping. The temperature-sensitive mutants of SV40 have been classified into 5 groups (A, B, C, BC, and D) on the basis of their behavior in complementation tests ([345, 413, 414,

417, 488, 489]; see Table 11.3). Mutants of group A are early mutants that fail to synthesize viral DNA at nonpermissive temperature [345, 416, 417, 431]; those of groups B, C, and BC are late [345, 414, 415, 419]. The virus particles of mutants belonging to group D contain a defective structural protein which seems to interfere with uncoating of the viral genome. No virus functions are detectable in cells infected at restrictive temperature with these mutants [488]. Two methods are available to assign physical locations to *ts* mutants on the viral genome. Cells are infected at the restrictive temperature with specific restriction enzyme fragments of wild-type DNA that have been annealed to a complete single-strand of DNA extracted from a *ts* mutant. When a fragment of DNA is used that is homologous to the mutated sequence, marker rescue occurs and wild-type virus is produced by the infected cell [200, 490]. Because the map positions of the specific fragments on the genome of SV40 are known (see Fig. 11.2), this type of experiment allows the mutation to be assigned unambiguously to defined segments of the viral DNA. The second technique depends on the fact that heteroduplex molecules formed between the complementary DNA strands of a mutant and its wild-type parent or a revertant must contain a mismatched base-pair at the site of mutation. These unnatural base-pairs are slightly more susceptible than are Watson-Crick pairs to attack by nucleases such as S1, which degrade single-stranded but not double-stranded DNA [491, 492]. By measuring the distance of the S1 cleavage site from a known position of cleavage by a restriction enzyme, it is possible to map the mismatched base-pair. Mutants of group A map in the early region of the viral genome between 0.43 and 0.32 map units. The mutants of all other groups map in the late region, those of groups B, C, and BC between 0.94 and 0.17 map units, those of group D between 0.85 and 0.94 map units. For unknown reasons, there are no mutants that map between 0.43 and 0.85 map units [493]. Mutants of the B, C, and BC groups map in the late region of the viral genome [200, 490, 491, 493]. The virions of these mutants are thermolabile and are defective at a late stage of the lytic cycle. Probably all of them are located in a section of the viral DNA that codes for a single polypeptide, the major coat protein VP1; they are divided into 3 groups as a consequence of intracistronic complementation [345, 413-414, 416, 418]. All of them transform efficiently at nonpermissive temperature.

Mutants of the D group, like those of groups B, C, and BC, lie in the late region of the viral genome [200, 490, 491, 493]. In all probability they are located in the gene coding for the second major capsid protein, VP3. While the virus particles of group D mutants are not thermally labile [344]; they cannot uncoat in permissive cells incubated at restrictive temperatures [344, 463, 488] so that virus infection aborts at a very early stage. However, the thermosensitive event can be bypassed when viral DNA rather than intact virions is used to initiate infection [488]. After infection with naked DNA at the restrictive temperature, progeny virions are synthesized that are infectious at permissive temperature but which are again unable to uncoat in cells incubated at the restrictive temperature. Thus, there is no requirement for the D gene product at any stage during lytic infection between uncoating and assembly of new particles. Obviously, mutants that are defective in uncoating, can neither complement, nor themselves be complemented by other temperature-sensitive mutants [345].

The role of the D gene in transformation is unclear. As expected, D group mutants cannot efficiently initiate transformation of cells at the non-permissive temperature [494] and Martin et al. [345] report that cells transformed by D-group mutants at the permissive temperature maintain their transformed phenotype during incubation at the restrictive temperature. However, Robb [495, 496] has reported that the transformation frequency by *ts* 101, a member of complementation group D, is extremely low at both permissive and restrictive temperatures. Furthermore, when clones of mouse cells transformed by *ts* 101 at the permissive temperature are incubated at 39°C (restrictive temperature), T-antigen, as assayed by immunofluorescence, is lost. There is no easy way to explain these results. However, I believe that they have little to do with the mechanism of virus-mediated transformation. First, it is possible to transform cells with subgenomic fragments of SV40 obtained by cleavage of the viral genome with restricting endonucleases [497]. The only region that carries transforming activity is the early region [497, 498]. Second, many lines of cells transformed by SV40 do not contain any detectable sequences of RNA complementary to the late segment of the viral genome [270, 499, 500]. In the absence of any evidence for transcription of the D gene in transformed cells it is difficult to sustain for long any argument that assigns a positive function to the D gene product in the maintenance of the transformed state.

b. Physiological Studies. Mutants of complementation group A map in the early region of the viral genome [200, 490, 491, 492, 493]. As we have seen, it is highly likely that the only early viral product is T-antigen and it follows that the phenotype exhibited by mutants in the A gene are due to alterations in T-antigen [403]. During infection of permissive cells at restrictive temperature, the group A mutants synthesize early viral RNA [304] but fail to produce viral DNA or capsid antigens [414]. The A gene product is continuously required for initiation but not for propagation of each round of viral DNA synthesis, nor for segregation of the progeny molecules [345, 431]. However, it is required at least once in the cycle to allow the establishment of the late phase of viral transcription [304]. At the nonpermissive temperature group A mutants are unable to establish stable transformation of cells of several species, including rat [501, 502], Swiss mice [389, 503], Balb/C mice [504, 505], Syrian hamster [389, 503-505], Chinese hamster [345, 506], rabbit [389, 503-505], and human [504, 505]. Thus, there seems to be no doubt that the A gene product is needed for the initiation of transformation. If cells are transformed at the permissive temperature by mutants of the A group and are then tested at the nonpermissive temperature, at least some of the phenotypic properties of at least some of the transformed cell lines revert [389, 502, 503, 505, 506]. Thus, it seems that continuous production of the A gene product may be necessary to maintain some of the phenotypic markers of transformation. Unfortunately, however, the experiments are not clear-cut, and there is disagreement between the results obtained by different groups even when the same combinations of mutants and host cells are used. For example, Tegtmeyer [389, 503] and Brugge and Butel [504, 505] do not see any disappearance of T-antigen from cells shifted from low to high temperature, while Osborn and Weber [425, 501] report that by 3 days after the shift, 85% of the cells have become T-antigen negative. Tegtmeyer [389, 503] finds that all of his lines of mouse cells transformed by A gene mutants can form colonies when grown at high temperature in the presence of high concentrations of serum; similar cell lines derived by Brugge and Butel [504, 505] do not.

These papers show that induction of transformation by A gene mutants is much more sensitive to high temperature than is maintenance of the transformed state, and taken together they also indicate a dependence by the cell on the A gene product to maintain a full-blown transformed phenotype. But the dependence is not total for reasons that, at the moment, are not entirely clear.

2. *Polyoma Virus ts Mutants*

Temperature-sensitive mutants of polyoma virus have been divided into 4 groups on the basis of complementation tests. Two of these (groups I and IV) are composed of late mutants [507] which are analogous in their properties to SV40 *ts* mutants of the B and C complementation groups.

The early mutants fall into a single complementation group, and define a function that is required for the initiation of each round of viral DNA replication [508] for synthesis of T-antigen detectable by immunofluorescence [402], and for the establishment of transformation [343, 507, 509, 510]. However, it is not required for initiation of cellular DNA synthesis [511, 512] nor for the changes that are detected by lectin-mediated agglutinability on the surfaces of infected cells [341]. By sharp contrast to the early mutants of SV40, none of the early mutants of polyoma virus have been shown to affect the maintenance of the properties of transformed cells [343, 507, 509].

The single remaining *ts* mutant of polyoma virus (*ts* 3) resembles the SV40 mutants of complementation group D. Cells infected with intact virions at the nonpermissive temperature are unable to carry out early viral functions [342, 515]; naked viral DNA is, however, fully infectious under these conditions

[513]. These properties are consistent with a defect in a virion-associated protein which prevents uncoating of the viral genome. It is puzzling then to find that hamster cells transformed by *ts* 3 exhibit temperature-sensitive growth properties as well as temperature-sensitive surface characteristics [341, 514]. Because of the weight of evidence supporting the idea that the product of the A gene of SV40 is necessary and sufficient to maintain cells in the transformed state, and in view of the cogent reasons for believing that the D gene product cannot be involved in the process of transformation, there is no easy explanation of these results. Perhaps the group D mutants of polyoma virus and SV40 are double mutants or perhaps they have a predisposition to transform cells which already carry mutations that prevent full expression of the transformed phenotype at restrictive temperatures.

3. Host-Range Mutants of Polyoma Virus

Host-range mutants of polyoma virus have been selected [480] which will not grow on untransformed mouse cells of the 3T3 line, but will replicate in a variety of other cell types such as polyoma virus-transformed cells, RNA virus-transformed cells, primary baby mouse kidney epithelial cells, and primary mouse embryo fibroblasts [481]. While the nature of the permissive factors that are present in cells capable of supporting lytic infection is unknown, they are certainly host-coded because the permissive state does not require a functioning integrated polyoma virus genome [481].

In nonpermissive cells, the host range mutants synthesize viral DNA [515] and are therefore late mutants in a classical sense. However, they fail to induce lectin-mediated agglutinability in 3T3 cells [440] and they are reportedly not able to transform rat or hamster fibroblasts, functions that are ascribed to the action of early viral gene products. Clearly this paradox can be resolved by physical mapping of the genetic lesions in the host-range mutants.

4. Recombination Studies

While recombination between *ts* mutants of SV40 [418] and of polyoma virus [516] has been observed, the frequency is too low to allow the construction of meaningful genetic maps. In the most extensive study reported to date, Dubbs et al. [418] showed that *ts* mutants of complementation groups B and C recombine to give wild-type recombinants at a frequency of about 1 in 10^4. They were also able to devise several techniques such as treatment of infected cells with inhibitors of DNA synthesis and irradiation of the input virus that led to apparently enhanced recombination frequencies. However, the advent of restriction enzymes and the development of ways to map the physical positions of *ts* mutations directly, has removed much of the need to develop sensitive and reproducible recombination assays. These new techniques are so powerful that no one nowadays would consider using conventional genetic methods to locate a mutation in the SV40 genome.

F. Transformation

The main feature that distinguishes cells transformed by polyoma virus or SV40 from untransformed cells is their insensitivity to the controls that regulate their multiplication in vitro. Thus, they will divide when suspended in medium containing agar or methylcellulose [517] or in medium depleted of serum growth factors [518], while untransformed cells will not. It is this sort of differential growth that provides the basis for assays that are commonly used to select colonies of transformed cells from a background of untransformed cells [519]. By picking and subculturing transformed clones it is possible to establish lines of transformed cells and to ask in what ways they differ from their untransformed parents. Cell biologists have not been reluctant to carry out this type of work and there has been no dearth of papers which contrast one aspect or another of transformed and normal cell behavior. The list of accumulated differences is much too long to be discussed here, but includes changes in the cell surface (see review [520]) intracellular organization [521], and requirement for growth factors and cell contact, (see review [522], and the proceedings of 2 recent symposia [523, 524]). From all these studies it is abundantly clear that a number of physiological and social characteristics of cells are drastically and permanently altered by transformation. The gene products of polyoma virus and SV40 cannot be directly and separately responsible for each of these changes, for there are considerably more of them than there are virus-coded products. Thus one or a few viral proteins must act in a highly pleiotropic manner, or the complicated transformed phenotype is the sum of many secondary and tertiary events that are the consequence of a small number of primary events. Whatever its mechanism, the hope of those working in the field is to relate the phenomenon of transformation to the action of specific molecules which regulate cellular metabolism.

There are 2 lines of evidence which prove that virus-coded functions are necessary to cause stable transformation. First, the transforming activity is inactivated by irradiation [340, 405-407] and, second, conditional lethal mutants of SV40 and polyoma virus exist which are unable to transform cells under nonpermissive conditions. However, not all the virus functions that are required for lytic growth are essential for transformation.

1. Some types of defective virus genomes which cannot replicate in the absence of helper virus have an unimpaired capacity to transform cells [525].
2. Certain groups of temperature-sensitive mutants of polyoma virus and SV40 retain full transforming ability at nonpermissive temperature (see Section III.E).
3. Fragments of DNA obtained by cleavage of intact viral genomes with restricting endonucleases can cause transformation [497, 498].

Transformation is an inefficient process in that about 10^4-10^5 infectious units (that is, 10^6-10^7 particles) are required per transforming event [526-528]. However, at low virus concentrations, the number of cells transformed is proportional to the multiplicity of infection, a result which suggests that a single virus particle is sufficient to initiate transformation, albeit with very low probability. There is no evidence to suggest that this poor efficiency of transformation is due to genetic inhomogeneity of either the cell or the virus populations [517, 529]. Consequently, it is the belief of most people that stable transformation is a stochastic event which occurs with low probability in virus-infected cells.

1. Virus Genetic Material in Transformed Cells

After lines of nonpermissive cells transformed by polyoma virus or SV40 have been established, it is usually impossible to find any evidence of infectious virus or infectious viral DNA [530] and all attempts to induce virus replication in such cells by physical and chemical methods have failed. However, the presence of the viral genome or a part of it can be detected both by direct and indirect means.

1. Hybridization Studies. The amount of viral DNA in cells transformed by polyoma virus and SV40 has been measured by 2 techniques. The first method, worked out by Westphal and Dulbecco, 1968 [205] involves synthesis in vitro of highly radioactive RNA from purified viral DNA using *E. coli* DNA-dependent RNA polymerase. The RNA is annealed to transformed or control cell DNAs immobilized on nitrocellulose filters, and the number of copies of viral DNA per transformed cell is normally calculated from reconstruction experiments using mixtures of untransformed cell DNA and known amounts of viral DNA. The method is extremely sensitive and is capable of detecting as little as 1 part of SV40 DNA in 10 million parts of cell DNA. Westphal and Dulbecco [205] found that different lines of SV40-transformed cells contained different numbers of viral genome equivalents, ranging from about 10-30 copies

per diploid quantity (3.9×10^{12} daltons) of mammalian cell DNA. Polyoma virus-transformed cells contain rather less viral DNA, from 4-10 copies per diploid quantity of cell DNA. Westphal and Dulbecco's experiment was the first direct demonstration that viral DNA sequences could be detected in the genome of transformed cells. Stimulating though the technique was, it was never entirely free of criticism. A rather polemic discussion arose which centered on 3 issues. First, Westphal and Dulbecco had shown that there was always some hybridization of the probe RNA to DNA extracted from untransformed cells. For several years it was unclear why this hybridization should occur, and during this period, the value of the technique remained unproved. In 1972, however, Lavi and Winocour [191] showed that growth of SV40 at high multiplicities resulted in the appearance in the virus stocks of particles which contained covalently linked sequences of host and viral DNA. We now recognize that Westphal and Dulbecco used as probe RNA synthesized from preparations of SV40 DNA of doubtful pedigree and, in retrospect, it is in no way surprising that the RNA manifested some ability to hybridize to untransformed cell DNA. Clearly the problem can be circumvented by using as template DNA isolated from stocks of virus particles that have been synthesized in cells infected at low multiplicity.

The second criticism of the filter hybridization system stemmed from skepticism concerning the fidelity of the reconstruction experiment. Haas et al. [531] showed that SV40 DNA may be preferentially lost from nitrocellulose filters during the hybridization procedure, a result which suggests that the reconstruction experiment may suffer from a systematic error that could lead to an inflated assessment of the number of copies of viral DNA per transformed cell. Recently this problem has been solved by using as probe saturating quantities of RNA whose specific activity is accurately known [532]. From the number of counts bound, it is possible to calculate directly the amount of viral DNA present in the genome of transformed cells without resorting to reconstruction experiments.

Clearly, despite the important results that were obtained using it, the filter hybridization technique has had its share of problems. With time most of these have been overcome, and the strengths and limitations of the method are now in plain view. But because of the twilight that surrounded the technique for several years, it became increasingly desirable to have available another way to identify and quantitate viral DNA in transformed cells. This second method became available in 1971, when Gelb et al. [533] showed that the number of copies of viral DNA in the genome of transformed cells could be determined by measuring the rate of reannealing of small amounts of highly radioactive DNA in the presence of large quantities of unlabled transformed cell DNA. This technique depends on the fact that the rate of reannealing of a given DNA sequence in solution is proportional to its initial concentration

[534, 535]. The number of viral genomes per transformed cell is generally calculated from the magnitude of the increase in the rate of hybridization of labeled probe DNA in the presence of transformed-cell DNA compared with control DNA. By using this type of analysis, it has been shown that different lines of mouse cells transformed by SV40 contain viral sequences which range in amount from about 1-2 copies to as many as 8-10 copies of SV40 DNA per diploid quantity of cell DNA [500, 533]. Cells transformed by wild-type polyoma virus contain between 0.6 and 2.9 copies of the viral genome per diploid equivalent of cell DNA [296]. Thus, the values obtained from studies of reassociation kinetics are for both viruses considerably lower than those obtained by the filter hybridization technique. In 1 sense, this is hardly surprising, for the 2 methods measure 2 quite different things. The filter hybridization technique simply assays the total weight of viral sequences that are complementary to the probe; the parameters that determine the kinetics of reassociation of viral DNA in the presence of transformed cell DNA are much more complicated. If transformed cells contain only a partial copy of the viral genome, or if some parts of the viral genome are present at a much higher frequency than others, the observed rate of renaturation of the labeled probe DNA in the presence of transformed cell DNA will be an average of the rates of reannealing of each independent segment of the DNA. However, until recently it has been conventional to assume in calculating the number of copies of viral DNA per transformed cell that all sequences of the labeled probe are present at equal frequencies. This assumption may lead to estimates of the number of complete genomes which are lower than the total quantity of viral DNA present in the transformed cells [536-539]. The best way to resolve this problem is to use as probes in kinetic hybridization experiments defined segments of viral DNAs produced by cleavage of the intact viral genomes with restricting endonucleases. So far this sort of experiment has been carried out with only 1 SV40 transformed cell line, SVT2, which by conventional analysis has been shown to contain between 1.56 and 2.2 genome equivalents of viral DNA per diploid quantity of cell DNA [500, 533]. It was found [538, 539] that the early region of the SV40 genome was present in the transformed cells at a frequency of about 6.0 copies per diploid quantity of cell DNA and that the late region was represented only once. Clearly it will be important to determine whether this sort of bias commonly applies to other lines of transformed cells and if so, when the selective amplification of certain viral sequences occurs in the life of the transformed cell.

The physical state of viral DNA in nonpermissive cells transformed by SV40 has been examined using the DNA/RNA hybridization technique [540]. The ability of cellular DNA to hybridize with SV40-specific RNA remains associated with the high molecular weight fraction after zonal centrifugation and equilibrium sedimentation in alkaline gradients, in cesium chloride gradients containing ethidium bromide, and during Hirt [348] extractions of DNA from transformed cells. From these results, it was concluded that the viral DNA sequences are integrated into the cellular genome in SV40-transformed cells. A similar conclusion was later reached for polyoma virus DNA sequences in transformed cells [541-543].

The details of the arrangement of the integrated sequences are unknown. Several possible structures have been proposed [540], but so far there is no conclusive experiment to favor any 1 of them over the others.

At least in the case of transformed human cells, there is evidence for an integration site for SV40 DNA in chromosome C-7. By making hybrids between mouse cells and 2 SV40 transformants of cells that originally had been obtained from patients with Lesch-Nyan disease, Croce and his coworkers have been able to correlate the expression of virus functions with the presence of specific human chromosomes [544-547]. Only chromosome C-7 segregates coordinately with SV40-T antigen and tumor-specific transplantation antigen [544-547]. Fusion of hybrid clones with permissive African green monkey kidney cells results in the rescue of infectious SV40 only from those clones that contain human chromosome 7 *and* show the presence of T-antigen [544, 546, 547]. Finally, only those hybrids that contain both chromosome 7 and T-antigen have the tumorigenic capacity and growth properties of transformed cells [548, 549]. This series of experiments conclusively demonstrates that at least 1 copy of SV40 DNA that is active in transformed cells resides on chromosome 7. But it does not rule out the possibility that the virus genome had also integrated into other chromosomes which were lost from the hybrid cells at an earlier stage.

2. *Transcription Studies*

Virus-specific RNA is invariably present in cells that have been transformed by SV40 and polyoma virus and from saturation hybridization data it appears that 0.001-0.002% of the total cellular RNA is complementary to viral DNA [550].

The map position of the segment of viral DNA that is transcribed into stable RNA in transformed cells has been determined in saturation hybridization experiments using the separated strands of each of the specific fragments of the 2 viral genomes [296, 538]. Always present in mouse cells transformed by SV40 are RNA sequences which originate from the segment of viral DNA that codes for early functions, a finding that agrees with the results of earlier competition hybridization experiments [276-278] and provides further support for the idea that the viral function that is responsible for maintenance of transformation is an early gene product. However, from these data it is impossible to be certain whether there is an exact correspondence between the viral sequences that are expressed early during lytic infection and those that are transcribed in transformed cells. Virus-specific RNA is present in such low concentrations that it is impossible to saturate the ^{32}P-labeled probe DNA. Consequently the exact proportion of the early sequences of SV40 that is transcribed in transformed cells and the precise positions of the 5′ and 3′ ends of the RNA cannot be mapped. While some lines of SV40-transformed cells contain only those RNA sequences which are derived from the early segment of the viral genome, others contain very complicated sets of transcripts. Commonly, more RNA sequences complementary to the E strand of viral DNA are present than are ever expressed during lytic infection. In productively infected cells, close to half of the sequences of the E strand of SV40 DNA are detectable in stable species of RNA. Many transformants, however, contain RNA complementary to between 50% and 80% of the E strand sequences [499, 500]. Therefore, at least part of the viral RNA present in transformed cells must consist of anti-late sequences. From the pattern of hybridization of RNA extracted from transformed cells to the separated strands of the specific fragments of SV40, it appears that the anti-late sequences originate from a region of the E strand DNA that is close to, or contiguous with, the 3′ end of the early genes.

Finally, those lines of transformed cells that contain many copies of SV40 DNA per diploid quantity of cell DNA often contain low levels of RNA that hybridize to the L strand DNA of SV40 DNA [500]. However, it has never been possible to add enough RNA into the hybridization mixtures to saturate the ^{32}P-labeled probe, and it is not known what proportion of the L strand sequences are transcribed. We are sure, though, that this RNA plays no role in maintaining the transformed state because it is not present in all SV40-transformed cell lines. Perhaps it is transcribed viral DNA which is carried along in the genome of transformed cells but which plays no role in maintaining the transformed state.

All cell lines transformed by polyoma virus examined to date contain transcripts of a segment of viral DNA which spans about two-thirds of the early region of the viral genome and which begins close to the 5′ end of the early genes [296]. Because the cells contain no RNA which hybridizes to the distal part of the early region it seems, perhaps surprisingly, that full expression of the early functions of polyoma virus is not required to maintain cells in their transformed state.

The model that has been proposed [271, 500] to account for the patterns of viral RNA synthesis in cells transformed by papovaviruses is based on the following 3 premises.

1. That the transcribed copy of viral DNA integrates into host DNA by means of a break somewhere in its late genes, thereby preserving intact those viral functions which are required to maintain cells in the transformed state

2. That integration does not necessarily occur at the same site within the viral genome in different cell lines
3. That integration places viral DNA under the control of host promoters.

Given these assumptions, we can generate a simple model which is consistent with all the data so far accumulated. After initiating transcription at a host promoter, the RNA polymerase enters the integrated viral DNA on the E strand somewhere in the late region of the genome. It continues along viral sequences synthesizing first anti-late and then early RNA. Assuming that, in different cell lines, viral DNA is integrated with breaks at different positions in the viral genome, then the percentage of the viral genome that is transcribed will vary, depending on the distance between the integration site and the beginning of the early genes. Because the transcription of the integrated genome is under the control of a host promoter, RNA molecules should be synthesized that contain covalently linked host and viral sequences and are considerably larger than a single strand of viral DNA. Such molecules are in fact present in the nuclei of at least 1 line of SV40-transformed cells [551, 552]. To account for the observation that the viral sequences present in the cytoplasm of these cells are not attached to host sequences [552], it has been proposed that the primary transcription product undergoes processing during which polyadenylic acid residues are attached to the 3′ end and the host sequences are removed from the 5′ end.

Clearly, this model is tentative and several of its major predictions could turn out to be wrong. For instance, we have no evidence that the 5′ end of the early genes really is a terminator for RNA synthesis; it could equally well be a processing point. And we cannot be certain that the only way to transcribe viral DNA is by synthesis of hybrid host-virus RNA molecules; we could explain the presence of viral RNA sequences in varying abundances if there is an early viral promoter which is active at least some of the time.

Whether or not the site of integration within 1 viral genome turns out to be important in regulating expression of viral genes, evidence is already available showing that control of viral transcription in transformed cells is mediated at least in part by the structure of chromatin. When chromatin extracted from SV40-transformed cells is used as a template for *E. coli* RNA polymerase, the same region of the viral DNA is transcribed in vitro that is transcribed in vivo [553-555]. By contrast, transcription of purified transformed-cell DNA by the polymerase yields transcripts from all regions of the SV40 genome with roughly equal frequencies. There is also evidence for temporal control of viral transcription. Swetly and Watanabe [556] have shown that in a line of SV40-transformed rat liver cells, the viral DNA sequences are transcribed only during the S phase; furthermore, only chromatin isolated from cells during S phase can serve as a template for viral RNA synthesis in vitro. Thus, there are both spatial and temporal constraints applied to the expression of integrated viral genomes by chromosomal proteins. While this result is intriguing, the available data do not distinguish whether these constraints are the consequence of transcriptional control of viral sequences or a cause of it.

3. *Virus Proteins in Transformed Cells*

Cells transformed by polyoma virus and SV40 contain T-antigen in their nuclei [390, 400, 557]. All SV40-transformed cells yield an antigen with an S value of 22S; in some cell lines an additional form of the antigen is detected, which sediments at 15S [425]. The functional significance of these 2 forms of T-antigen is unclear but may be related to the observation that the antigen elutes from columns of phosphocellulose or double-stranded DNA cellulose in 2 peaks [422, 426].

U-antigen has also been demonstrated at the nuclear membrane of certain cells transformed by SV40 [432], but no attempts have been made to purify the antigen.

Finally, 2 new classes of antigens can be detected on the surface of transformed cells. The first class comprises the tumor-specific transplantation antigens (TSTA) which are demonstrated by showing that animals which have been immunized with infectious polyoma virus or SV40 or with virus-transformed cells from another animal species, more readily reject isologous cells transformed by the same virus than do unimmunized animals [558-562]. The antigens are virus-specific and their structure is unknown although they do not appear to be related to virus capsid or T-antigens. Whether they are virus- or cell-coded is uncertain.

The second class are the surface (S) antigens. Cells transformed by polyoma virus or SV40 often but not invariably display [563, 564] on their surfaces antigens that can be detected by any one of a variety of in vitro immunological tests (see review [565]). Some of these antigens may be identical to TSTA but others certainly are not. At least some of the antigens are cell-coded since they can be found on the surfaces of embryonic cells [566], on unfertilized mouse eggs [567], and can be exposed on the surfaces of normal cells by mild treatment with protease [568].

4. *Rescue of Virus from Transformed Cells*

When cells transformed by SV40 are fused with permissive monkey cells, the heterokaryons commonly produce infectious SV40 [569-579]. Virus replication occurs first in the nucleus of the transformed cell, later passing to the nucleus of the permissive cell [580-582]. The system is very inefficient, with at most 5% of the heterokaryons producing virus. The reason for this poor response is not known, although it does not seem to be due to genetic inhomogeneity among the transformed cells, because virus can be rescued from every one of several subclones of transformed cells, albeit at different frequencies [571, 572, 579]. The percentage of heterokaryons that show SV40-specific cytopathic changes is dependent upon the ratio of nonpermissive to permissive nuclei in the heterokaryons. Yet it is known that all heterokaryons between mouse and monkey cells, whatever their nuclear constitution, are able to support the growth of SV40 [574]. Thus, differential permissivity does not seem to be a factor in the observed low frequency of SV40 production in heterokaryons. Possibly the rate-limiting step is the release of SV40 from its integrated state. Watkins [575, 578] has shown that the yield of SV40 is substantially increased if the transformed cells are treated with base analogs before fusion. It is easy to believe that such substances may act directly or indirectly to disturb the structure of the integrated genome and thereby allow its more efficient release. The mechanism of release is unknown: it could involve either direct excision of viral sequences from chromosomal DNA or induction of a type of viral DNA synthesis with production of a small number of progeny genomes [539].

The contribution of the permissive cell to the process of rescue is unknown. However, infectious SV40 can be rescued by fusion of SV3T3 cells with enucleate cytoplasm [583, 584]. Thus the nucleus of the permissive cell is not required for the initiation and maintenance of SV40 replication and the production of infectious virions.

Up to now, we have had a rather narrow view of the kinds of SV40 produced in heterokaryons of transformed and permissive cells. More extensive cataloging of the released viral genomes is required and the helper lawn systems developed for isolating defective SV40 [194] and polyoma virus particles [193] should provide a useful tool in examining partial genomes in the fusion lysates.

Two other methods have been used to recover SV40 from transformed cells. When high molecular weight DNA extracted from transformed hamster, mouse, or monkey cells is introduced into permissive monkey cells, infectious SV40 is produced by the recipient cells [588]. The technique is suprisingly sensitive. The specific infectivity of the integrated viral sequences is close to that of naked purified viral DNA and infectious SV40 can be rescued from lines of transformed cells which do not yield SV40 after heterokaryon formation with permissive cells. Second, when metaphase chromosomes isolated from a line of SV40-transformed Chinese hamster cells are transferred into permissive monkey cells, infectious virus is produced [589]. The method is quite inefficient. T-antigen can be detected in 1 recipient cell in 10^4 and infectious virus in 1 in 10^5. In no case is the frequency of rescue of SV40 from transformed cells sufficiently high to allow examination of the event at a molecular level. Hopefully, it will be possible to at least eliminate some of the alternative mechanisms by careful examination of the structure of the rescued virus genomes.

Until now, we have been concerned almost entirely with the properties of transformed, virus-free cells. In general, these have been derived by infection of nonpermissive cells. However, it is also possible to isolate transformants from cultures of semipermissive and permissive cells. Such transformants, especially those that contain a complete set of viral DNA sequences, are not easy to obtain without manipulating the experimental conditions in ways that prevent the spread of progeny wild-type virus through the culture [525, 590, 591]. In the absence of such procedures the most commonly isolated transformants of permissive and semipermissive cells are those that contain only partial copies of the viral genome. Most lines of polyoma virus-transformed cells fall into this category, for the sorts of cells that are most often used for transformation studies with polyoma virus are partly permissive. The transformants derived from them do not yield virus after fusion with mouse cells [571, 592] and are thought to have been transformed by defective virus mutants that cannot grow and kill the cells. However, there are lines of semipermissive transformed cells that contain at least 1 complete copy of the viral DNA. Good examples of these are lines of BHK or 3T3 cells that have been transformed by temperature-sensitive mutants of polyoma virus [542, 591, 592]. Such cell lines are obtained by allowing infection with the virus mutant to proceed at nonrestrictive temperature for a few days and then growing the cells only at restrictive temperature. Presumably, under these conditions, production of infectious virus and its propagation through the culture are severely reduced. Very little or no infectious virus is detectable in the culture when the cells are grown at the restrictive temperature, but virus replication can be induced in a small proportion of the transformed cell population merely by returning the cells to permissive temperature [542, 591]. In the case of transformed BHK cells, however, the efficiency of virus rescue can be improved by fusing the transformed cells with permissive mouse cells [542]. Other lines of transformed cells have been isolated from cultures of semipermissive cells that have been exposed to wild-type polyoma virus or SV40. Often, these yield infectious virus at a low spontaneous rate and at a higher level after treatments that are known to affect DNA metabolism [593-603]. The molecular mechanisms involved in these processes have not been studied.

Finally, there exists a class of transformants which by rights should be forbidden. These are transformed permissive cells which contain a rescuable viral genome and are susceptible to infection by wild-type virus [570, 588, 604]. Why the resident genome fails to initiate a lytic cycle is unknown.

G. Conclusion

To the question, "What, and how much, does the virus contribute to the maintenance of the transformed state?" we still have no satisfactory answer. Clearly the viral DNA becomes associated with the cellular genome and virus-coded products influence the phenotype of the cell. Equally clearly, cellular genes also affect the transformed phenotype. It is possible to obtain from populations of transformed cells variants that have regained 1 or more properties typical of untransformed cells [605-615]. Most of these variants, which show fundamental changes in their phenotype, have no alteration in their content or expression of viral DNA sequences. In the best-studied case, the "revertants" of SV40-transformed mouse cells, the cells still contain SV40 specific T-antigen [605-609] and SV40-specific RNA sequences [500] and, as judged by reassociation kinetics, the amount of SV40 DNA contained in a diploid quantity of revertant cell DNA is identical to that present in the genome of the parental cells [500]. Finally, SV40 rescued from most types of revertant cells by heterokaryon formation is not detectably different in its infectivity and shows only minor alterations in transforming ability compared with virus rescued from the homologous transformant [572, 605, 607]. Comparisons of the chromosome constitutions of revertants and transformants, however, have invariably shown that revertants contain more chromosomes than the cells from which they were derived [606, 608, 609, 616, 617]. Taken together, the changes in chromosome composition and the failure to uncover any defect in the integrated viral genome suggest that in most of the cases so far examined, reversion has occurred as a consequence of alterations to the cell genome [500, 616]. It appears then that the transformed cell phenotype is firmly under the control of cellular genes. A similar conclusion was reached by Kelly and Sambrook [618] who isolated variants of SV40-transformed cells that contained less viral DNA than the parental cells, and in which no T-antigen was detectable. Apparently no viral genes were active in these cells, yet they retained their transformed morphology and growth characteristics. So, at least in well established transformed cell lines, the virus may have little to do with maintaining the transformed phenotype. However, there is overwhelming evidence from many sources that viral functions are required to establish the transformed phenotype. In addition, the close correlation between the expression of transformed characteristics and the presence in hybrid cells of a specific chromosome that carries the active viral DNA, as well as studies of cells transformed by temperature-sensitive mutants indicates that functions supplied by integrated viral sequences can be necessary to maintain the transformed phenotype.

Clearly, both of these conclusions cannot apply to any individual line at the same time. But it seems possible that each is correct at different stages in the cell's history. Soon after infection, perhaps, the virus is responsible for maintaining the cell in the transformed state. As the generations pass, the cell may adapt or accumulate mutations that allow it to accomodate to its new way of life. Conceivably, the virus directs the cell into a state where it no longer depends on virus gene products and eventually comes to ignore them.

It seems that the only viral gene product required to maintain transformation is the A protein, which carries the antigenic determinants for T-antigen. How this one substance can interact with the cell to provide the radical changes in metabolism that are collectively called transformation is totally unclear. Many models have been proposed (see for example [258, 345, 496]), all of which involve a pleiotropic virus gene product that interacts with both cellular and viral DNA. While many of these models have attractive features, none can be thoroughly tested given the current lack of adequate cellular mutants. However, it is widely accepted that the molecular biology of polyoma virus and SV40, at least in permissive cells, will be completely elucidated within a few years [619]. It seems reasonable to hope that knowledge of the way that virus coded products act during the lytic cycle might also apply to the cognate problem of transformation.

Acknowledgments

I am very greatly indebted to Terri Grodzicker and Mike Fried for their critical and helpful reading of the manuscript, and for their assistance in organizing the references. Without their support and friendship, this paper would not exist.

My thanks also to Beverly Griffin for permission to use the restriction enzyme map of polyoma virus DNA that appears in Fig. 11.1.

This chapter was written during the summer of 1975 and was completed in August of that year.

References

1. A. Klug and J. T. Finch, J. Mol. Biol., 11, 403-423 (1965).
2. J. T. Finch and A. Klug, J. Mol. Biol., 13, 1-12 (1965).
3. A. Klug, J. Mol. Biol., 11, 424-431 (1965).
4. J. L. Melnick, A. C. Allison, J. S. Butel, W. Eckhart, B. E. Eddy, S. Kit, A. J. Levine, J. A. R. Miles, J. S. Pagano, L. Sachs and V. Vonka, Intervirology, 3, 106-120 (1974).
5. R. F. Murray, J. Hubbs, and P. Payne, Nature, 232, 51 (1971).
6. M. Lyon, Nature New Biol., 232, 229-232 (1971).
7. W. F. Noyes, J. Exp. Med., 109, 423-429 (1959).
8. R. S. Stone, R. E. Shope, and D. H. Moore, J. Exp. Med., 110, 543-546 (1959).

9. J. D. Almeida, A. F. Howatson, and M. G. Williams, J. Invest. Dermatol., 38, 337-345 (1959).
10. R. W. Shope and E. W. Hurst, J. Exp. Med., 58, 607-610 (1933).
11. W. F. Noyes and R. C. Mellors, J. Exp. Med., 106, 555-561 (1957).
12. Y. Ito, Cold Spring Harbor Symp. Quant. Biol., 27, 387-394 (1962).
13. P. Rous and J. W. Beard, J. Exp. Med., 62, 523-548 (1935).
14. J. T. Syverton, Ann. N.Y. Acad. Sci., 54, 1126-1140 (1952).
15. C. A. Evans, R. S. Weiser, and Y. Ito, Cold Spring Harbor Symp. Quant. Biol., 27, 453-462 (1962).
16. A. F. Howatson and L. V. Crawford, Virology, 21, 1-6 (1963).
17. F. Pass and J. V. Maizel, J. Invest. Dermatol. 60, 307-311 (1973).
18. G. Spira, M. K. Estes, G. R. Dressman, J. S. Butel, and W. E. Rawls, Intervirology, 3, 220-231 (1974).
19. S. J. Kass and C. A. Knight, Virology, 27, 273-281 (1965).
20. M. Favre, F. Breitburd, O. Croissant, and G. Orth, J. Virol., 15, 1239-1247 (1975).
21. J. D. Watson and J. W. Littlefield, J. Mol. Biol., 2, 161-165 (1960).
22. L. V. Crawford, J. Mol. Biol., 13, 362-372 (1965).
23. A. K. Kleinschmidt, S. J. Kass, R. C. Williams, and C. A. Knight, J. Mol. Biol., 13, 749-756 (1965).
24. L. V. Crawford, J. Mol. Biol., 8, 489-495 (1964).
25. J. Vinograd and J. Lebowitz, J. Gen. Physiol., 49, 103-125 (1966).
26. L. V. Crawford, Adv. Virus Res., 14, 89 (1969).
27. J. Subak-Sharpe, R. R. Bürk, L. V. Crawford, J. M. Morrison, J. Hay, and H. M. Keir, Cold Spring Harbor Symp. Quant. Biol., 31, 737-748 (1966).
28. J. M. Morrison, H. M. Keir, H. Subak-Sharpe, and L. V. Crawford, J. Gen. Virol., 1, 101-108 (1967).
29. E. A. C. Follett and L. V. Crawford, J. Mol. Biol., 28, 461-467 (1967).
30. C. G. Mendelson and A. M. Kligman, Arch. Dermatol., 83, 559-562 (1961).
31. S. Oroszlan and M. A. Rich, Science, 146, 531-533 (1964).
32. W. F. Noyes, Virology, 25, 358-363 (1965).
33. P. H. Black, J. W. Hartley, W. P. Rowe, and R. J. Huebner, Nature, 199, 1016-1018 (1963).
34. M. Thomas, J. P. Levy, J. Tanzer, M. Boiron, and J. Bernard, Comp. Rend. Acad. Sci., Paris, 257, 2155-2158 (1963).
35. M. Thomas, M. Boiron, J. Tanzer, J. P. Levy, and J. Bernard, Nature, 202, 709-710 (1964).
36. L. Gross, Oncogenic Viruses, 2nd Ed., Pergamon, New York, 1970.
37. L. Gross, Proc. Soc. Exp. Biol., 83, 414-421 (1953).
38. S. E. Stewart, B. E. Eddy, A. M. Gochenour, N. G. Borgese, and G. Grubbs, Virology, 3, 380-400 (1957).
39. B. E. Eddy, S. E. Stewart, and W. Berkeley, Proc. Soc. Exp. Biol., 98, 848-851 (1958).
40. S. E. Stewart, B. E. Eddy, and N. G. Borgese, J. Natl. Cancer Inst., 20, 1223-1243 (1958).
41. L. Gross, Proc. Soc. Exp. Biol., 88, 362-368 (1955).

42. W. P. Rowe, Bact. Rev., 25, 18-31 (1961).
43. W. P. Rowe, J. W. Hartley, L. W. Law, and R. J. Huebner, J. Exp. Med., 109, 449-462 (1959).
44. W. P. Rowe, R. J. Huebner, and J. W. Hartley, Perspectives in Virology, Vol. 2, Burgess, Minneapolis. 1961, pp. 177-194.
45. K. Habel, Adv. Immunol., 10, 229-250 (1969).
46. B. H. Sweet and M. R. Hilleman, Proc. Soc. Exp. Biol. Med., 105, 420-427 (1960).
47. R. N. Hull, J. R. Minner, and J. W. Smith, Am. J. Hyg., 63, 204-215 (1956).
48. E. Borman and M. R. Ross, Am. Soc. Micro. Annual Meeting, 62, 142 (1961).
49. A. J. Girardi, B. H. Sweet, V. B. Slotnick, and M. R. Hilleman, Proc. Soc. Exp. Biol. Med., 109, 649-653 (1964).
50. H. Shein and J. T. Enders, Proc. Natl. Acad. Sci. USA, 48, 1164-1172 (1962).
51. H. Koprowski, J. A. Ponten, F. Jensen, R. G. Ravdin, P. Moorhead, and E. Saksela, J. Cell. Comp. Physiol., 59, 281-292 (1962).
52. F. C. Jensen, H. Koprowski, J. S. Pagano, J. Ponten, and R. G. Ravdin, J. Natl. Cancer Inst., 32, 917-932 (1964).
53. G. D. Hsiung, Bact. Rev., 32, 185-205 (1968).
54. G. D. Hsiung, T. Atoyantan, and C. W. Lee, Am. J. Epidemiol., 89, 464-471 (1969).
55. J. F. Fraumeni, Jr., C. R. Stark, E. Gold, and M. L. Lepow, Science, 167, 59-60 (1970).
56. A. M. Stewart and D. Hewitt, Lancet, ii, 789 (1965).
57. L. Weiner, R. Herndon, O. Narayan, R. T. Johnson, K. Shah, L. G. Rubinstein, T. J. Prezozisi, and F. K. Conley, N. Engl. J. Med., 286, 385-390 (1972).
58. F. Soriano, C. E. Shelburne, and M. Gökcen, Nature, 249, 421-424 (1974).
59. A. F. Weiss, R. Portmann, H. Fischer, J. Simon, and K. D. Zang, Proc. Natl. Acad. Sci. USA, 72, 609-613 (1975).
60. B. L. Padgett, D. L. Walker, G. M. zuRhein, R. I. Eckroade, and B. H. Dessel, Lancet, i, 1257-1260 (1971).
61. S. D. Gardner, A. M. Field, D. V. Coleman, and B. Hulme, Lancet, i, 1253-1257 (1971).
62. K. K. Takemoto, A. S. Rabson, M. F. Mullarkey, R. M. Blaese, C. F. Garon, and D. Nelson, J. Natl. Cancer Inst., 53, 1205-1207 (1974).
63. C. R. Madeley, in Virus Morphology, Churchill-Livingstone, London. 1972, pp. 134-135.
64. D. L. Walker, B. L. Padgett, G. M. zuRhein, A. E. Albert, and R. F. Marsh, Science, 181, 674-676 (1973).
65. K. V. Shah, R. W. Daniel, and J. Strandberg, J. Natl. Cancer Inst., 54, 945-950 (1975).
66. E. D. Major and G. DiMayorca, Proc. Natl. Acad. Sci. USA, 70, 3210-3212 (1973).
67. M. Portolani, A. Barbanti, G. Brodano, and M. LaPlaca, J. Virol., 15, 420-422 (1975).

68. J. B. Penney and O. Narajani, Infect. Immun., 8, 299-300 (1973).
69. K. K. Takemoto and M. F. Mullarkey, J. Virol., 12, 625-631 (1973).
70. J. E. Osborn, S. M. Robertson, B. L. Padgett, G. M. zuRhein, D. L. Walker, and B. Weisblum, J. Virol., 13, 614-622 (1974).
71. S. D. Gardner, Br. Med. J., 1, 77-78 (1973).
72. B. L. Padgett and D. L. Walker, J. Infect. Dis., 127, 467-470 (1973).
73. K. V. Shah, Am. J. Epidemiol., 95, 199-206 (1972).
74. J. T. Hartley and W. P. Rowe, Science, 143, 258-260 (1964).
75. L. Kilham, Science, 116, 391-392 (1952).
76. H. V. Thorne and D. Warden, J. Gen. Virol., 1, 135-137 (1967).
77. G. Walter, R. Roblin, and R. Dulbecoo, Proc. Natl. Acad. Sci. USA, 69, 921-924 (1972).
78. R. Roblin, E. Härle, and R. Dulbecco, Virology, 45, 555-566 (1971).
79. T. Friedman and D. David, J. Virol., 10, 776-782 (1972).
80. P. M. Frearson and L. V. Crawford, J. Gen. Virol., 14, 141-155 (1972).
81. W. Gibson, Virology, 62, 319-336 (1974).
82. J. G. Seehafer and R. Weil, Virology, 58, 75-85 (1974).
83. M. Girard, L. Marty, and F. Suarez, Biochem. Biophys. Res. Commun., 40, 97-102 (1970).
84. M. K. Estes, E. S. Huang, and J. S. Pagano, J. Virol., 7, 635-641 (1971).
85. B. Hirt and R. Gesteland, Experimentia, 27, 736 (1971).
86. S. Barban and R. S. Goor, J. Virol., 7, 198-203 (1971).
87. E. Lazarides, J. G. Files, and K. Weber, Virology, 60, 584-587 (1974).
88. D. M. Pett, M. K. Estes, and J. S. Pagano, J. Virol., 15, 379-385 (1975).
89. R. S. Lake, S. Barban, and N. P. Salzman, Biochem. Biophys. Res. Commun., 54, 640-647 (1975).
90. L. V. Crawford, Br. Med. Bull, 29, 253-258 (1973).
91. G. Walter and W. Deppert, Cold Spring Harbor Symp. Quant. Biol., 39, 225-257 (1974).
92. K. B. Tan and F. Sokol, J. Virol., 10, 985-994 (1972).
93. G. Fey and B. Hirt, Cold Spring Harbor Symp. Quant. Biol., 39, 235-241 (1974).
94. J. Bonner, M. Dahmus, D. Fambrough, R. C. C. Huang, K. Marushige, and D. Y. Tuan, Science, 159, 47-58 (1968).
95. E. Everitt and L. Philipson, Virology, 62, 253-269 (1974).
96. J. Tooze, The Molecular Biology of Tumor Viruses, (J. Tooze, ed.), Cold Spring Harbor, New York. 1973.
97. E. Winocour and E. Robbins, Virology, 40, 307-317 (1970).
98. R. Hancock and R. Weil, Proc. Natl. Acad. Sci. USA, 63, 1144-1150 (1969).
99. T. Seebeck and R. Weil, J. Virol., 13, 567-576 (1974).
100. A. L. Olins and D. E. Olins, Science, 183, 330-332 (1974).
101. P. Oudet, M. Gross-Bellard, and P. Chambon, Cell, 4, 281-300 (1975).
102. R. D. Kornberg and J. O. Thomas, Science, 184, 865-868 (1974).
103. M. Noll, Nature, 251, 249-251 (1974).
104. L. A. Burgoyne, D. R. Hewish, and J. Mobbs, Biochem. J., 143, 67-72 (1974).

105. D. R. Hewish and L. A. Burgoyne, Biochem. Biophys. Res. Commun., 52, 475-479 (1973).
106. M. H. Green, H. I. Miller, and S. Hendler, Proc. Natl. Acad. Sci. USA, 68, 1032-1036 (1971).
107. M. White and R. Eason, J. Virol., 8, 363-371 (1971).
108. A. J. Louie, Cold Spring Harbor Symp. Quant. Biol., 39, 259-266 (1974).
109. J. D. Griffith, Science, 187, 1202-1203 (1975).
110. J. E. Germond, B. Hirt, P. Oudet, M. Gross-Bellard, and P. Chambon, Proc. Natl. Acad. Sci. USA, 72, 1843-1847 (1975).
111. W. Meinke, M. R. Hall, and D. A. Goldstein, J. Virol., 15, 439-448 (1975).
112. A. Sen, R. Hancock, and A. J. Levine, Virology, 61, 11-21 (1974).
113. B. McCarthy and T. Polisky, unpublished observations.
114. E. S. Huang, M. Nonoyama, and J. S. Pagano, J. Virol., 9, 930-937 (1972).
115. H. Westphal, J. Mol. Biol., 50, 407-420 (1970).
116. L. Crawford, R. Dulbecco, M. Fried, L. Montagnier, and M. Stoker, Proc. Natl. Acad. Sci. USA, 52, 148-152 (1964).
117. F. Cuzin, D. Blangy, and R. Rouget, Comp. Rend. Acad. Sci., Paris, Series D, 273, 2650-2653 (1971).
118. J. C. Kaplan, S. M. Wilbert, and P. H. Black, J. Virol., 9, 800-803 (1972).
119. W. R. Kidwell, R. Saral, R. G. Martin, and H. L. Ozer, J. Virol., 10, 410-416 (1972).
120. A. Parodi, P. Rouget, O. Croissant, D. Blangy, and F. Cuzin, Cold Spring Harbor Symp. Quant. Biol., 39, 247-257 (1974).
121. J. D. Smith, G. Freeman, M. Vogt, and R. Dulbecco, Virology, 12, 185-196 (1960).
122. L. V. Crawford, Virology, 19, 279-282 (1963).
123. R. Weil, Proc. Natl. Acad. Sci. USA, 49, 480-486 (1963).
124. L. V. Crawford and P. H. Black, Virology, 24, 388-392 (1964).
125. J. Ferguson and R. W. Davis, J. Mol. Biol., 94, 135-150 (1975).
126. J. Vinograd, J. Lebowitz, R. Radloff, R. Watson, and P. Laipis, Proc. Natl. Acad. Sci. USA, 53, 1104-1111 (1965).
127. W. Bauer and J. Vinograd, J. Mol. Biol., 33, 141-171 (1968).
128. W. Keller and I. Wendel, Cold Spring Harbor Symp. Quant. Biol., 39, 199-208 (1974).
129. D. A. Goldstein, M. R. Hall, and W. Meinke, J. Virol., 12, 887-900 (1973).
130. J. Vinograd, J. Lebowitz, and R. Watson, J. Mol. Biol., 33, 173-197 (1968).
131. W. Bauer and J. Vinograd, J. Mol. Biol., 54, 281-298 (1970).
132. H. B. Gray, Jr., W. B. Upholt, and J. Vinograd, J. Mol. Biol., 62, 1-19 (1971).
133. J. J. Champoux and R. Dulbecco, Proc. Natl. Acad. Sci. USA, 69, 143-146 (1972).
134. L. V. Crawford and M. J. Waring, J. Mol. Biol., 25, 23-30 (1967).
135. M. Schmir, B. M. J. Revet, and J. Vinograd, J. Mol. Biol., 83, 35-45 (1974).
136. R. Dulbecco and M. Vogt, Proc. Natl. Acad. Sci. USA, 50, 236-243 (1963).
137. R. Weil and J. Vinograd, Proc. Natl. Acad. Sci. USA, 50, 730-738 (1963).
138. R. Radloff, W. Bauer, and J. Vinograd, Proc. Natl. Acad. Sci. USA, 57, 1514-1521 (1967).

139. B. Hudson, W. B. Upholt, T. Devinny, and J. Vinograd, Proc. Natl. Acad. Sci. USA, 63, 813-820 (1969).
140. H. V. Thorne, J. Mol. Biol., 24, 203-211 (1967).
141. C. Aaij and P. Borst, Biochim. Biophys. Acta, 269, 192-200 (1972).
142. J. Pagano and C. A. Hutchinson, III, in Methods in Virology, Vol. V, (K. Maramorosch and H. Koprowski, eds.), Academic, New York. 1971, p. 79.
143. M. Meselson, Ann. Rev. Biochem., 41, 447-466 (1972).
144. H. W. Boyer, Fed. Proc., 33, 1125-1127 (1974).
145. S. Linn, J. A. Lantenberger, B. Eskin, and D. Lackey, Fed. Proc., 33, 1128-1134 (1974).
146. W. Arber, Proc. Nucl. Acid Res., 2, 751 (1972).
147. K. Murray and R. W. Old, Prog. Nucl. Acid Res. Mol. Biol., 14, 101-155 (1974).
148. B. Griffin, M. Fried, and A. Cowie, Proc. Natl. Acad. Sci. USA, 71, 2077-2081 (1974).
149. M. Fried and B. E. Griffin, Adv. Cancer Res., 24, 67-107 (1977).
150. M. C. Y. Chen, K. S. S. Chang, and N. P. Salzman, J. Virol., 15, 191-198 (1975).
151. W. Folk, B. R. Fishel, and D. M. Anderson, Virology, 64, 277-280 (1975).
152. B. E. Griffin and M. Fried, Nature, 256, 157-179 (1975).
153. J. Summers, J. Virol., 15, 946-953 (1975).
154. J. F. Morrow and P. Berg, Proc. Natl. Acad. Sci. USA, 69, 3365-3369 (1972).
155. C. Mulder and H. Delius, Proc. Natl. Acad. Sci. USA, 69, 3215-3219 (1972).
156. J. Sambrook and M. Mathews, unpublished results, 1974.
157. P. A. Sharp, W. Sugden, and J. Sambrook, Biochemistry, 12, 3055-3063 (1973).
158. R. J. Roberts, J. Breitmeyer, N. Tabachnik, and P. Myers, J. Mol. Biol., 91, 121-123 (1975).
159. K. N. Subramanian, S. Zain, R. J. Roberts, and S. Weissman, unpublished results, 1974.
160. M. Mathews, unpublished results, 1975.
161. K. J. Danna, G. H. Sack, Jr., and D. Nathans, J. Mol. Biol., 78, 363-376 (1973).
163. K. N. Subramanian, J. Pan, S. Zain, and S. Weissman, Nucl. Acid Res., 1, 727-752 (1974).
164. R. Yang, K. J. Danna, A. Van der Voorde, and W. Fiers, Virology, 68, 260-265 (1975).
165. T. Ando, Biochim. Biophys. Acta, 114, 158-168 (1966).
166. V. Vogt, Eur. J. Biochem., 33, 192-200 (1973).
167. A. C. Kato, K. Bartok, M. J. Fraser, and D. T. Denhardt, Biochim. Biophys. Acta, 308, 68-78 (1973).
168. M. Mechali, A. M. de Recondo, and M. Girard, Biochem. Biophys. Res. Commun., 54, 1306-1320 (1973).

169. P. Beard, J. F. Morrow, and P. Berg, J. Virol., 12, 1303-1313 (1973).
170. J. Germond, V. Vogt, and B. Hirt, Eur. J. Biochem., 43, 591-600 (1974).
171. W. W. Dean and J. Lebowitz, Nature New Biol., 231, 5-8 (1971).
172. T. A. Beerman and J. Lebowitz, J. Mol. Biol., 79, 451-470 (1973).
173. J. F. Morrow and P. Berg, J. Virol., 12, 1361-1362 (1973).
174. M. Yaniv, O. Croissant, and F. Cuzin, Biochem. Biophys. Res. Commun., 57, 1074-1077 (1974).
175. J. Montjardino and A. W. James, Nature, 255, 249-251 (1975).
176. N. P. Salzman, J. Lebowitz, M. Chen, E. Sebring, and C. F. Garon, Cold Spring Harbor Symp. Quant. Biol., 39, 209-218 (1974).
177. M. F. Bourgignon, Biochim. Biophys. Acta, 166, 242-245 (1968).
178. E. A. C. Follett and L. V. Crawford, J. Mol. Biol., 34, 565-573 (1968).
179. K. Yoshiike, A. Furano, and K. Suzuki, J. Mol. Biol., 70, 415-423 (1972).
180. R. Dhar, K. Subramanian, B. S. Zain, J. Pan, and S. Weissmann, Cold Spring Harbor Symp. Quant. Biol., 39, 153-160 (1974).
181. R. Dhar, S. M. Weissman, B. S. Zain, J. Pan, and A. M. Lewis, Jr., Nucl. Acid Res., 1, 595-613 (1974).
182. R. Dhar, B. S. Zain, S. M. Weissman, J. Pan, and K. Subramanian, Proc. Natl. Acad. Sci. USA, 71, 371-375 (1974).
183. B. S. Zain, R. Dhar, S. M. Weissman, P. Lebowitz, and A. M. Lewis, Jr., J. Virol., 11, 682-693 (1973).
184. B. S. Zain, S. M. Weissman, R. Dhar, and J. Pan, Nucl. Acid Res., 1, 595-613 (1974).
185. W. Fiers, K. Danna, R. Rogiers, A. Van der Voorde, J. Van Herreweghe, H. Van Heuverswyn, G. Volkaert, and R. Yang, Cold Spring Harbor Symp. Quant. Biol., 39, 179-186 (1974).
186. S. Uchida, S. Watanabe, and M. Kato, Virology, 28, 135-141 (1966).
187. M. E. Blackstein, C. P. Stanners, and A. J. Farmilo, J. Mol. Biol., 42, 301-313 (1969).
188. H. V. Thorne, J. Mol. Biol., 35, 215-226 (1968).
189. H. V. Thorne, J. Evans, and D. Warden, Nature, 219, 728-730 (1968).
190. S. Uchida, K. Yoshiike, S. Watanabe, and A. Furuno, Virology, 34, 1-8 (1968).
191. S. Lavi and E. Winocour, J. Virol., 9, 309-316 (1972).
192. W. W. Brockman, T. N. H. Lee, and D. Nathans, Virology, 54, 384-397 (1973).
193. M. Fried, J. Virol., 13, 939-946 (1974).
194. W. W. Brockman and D. Nathans, Proc. Natl. Acad. Sci. USA, 71, 942-946 (1974).
195. H. Tai, C. A. Smith, P. A. Sharp, and J. Vinograd, J. Virol., 9, 317-325 (1972).
196. J. E. Mertz and P. Berg, Proc. Natl. Acad. Sci. USA, 71, 4879-4883 (1974).
197. M. Fried, B. E. Griffin, E. Lund, and D. L. Robberson, Cold Spring Harbor Symp. Quant. Biol., 39, 45-52 (1974).
198. J. E. Mertz, J. Carbon, M. Herzberg, R. W. Davis, and P. Berg, Cold Spring Harbor Symp. Quant. Biol., 39, 69-84 (1974).

199. W. W. Brockman, T. N. H. Lee, and D. Nathans, Cold Spring Harbor Symp. Quant. Biol., 39, 119-127 (1974).
200. C. J. Lai and D. Nathans, Cold Spring Harbor Symp. Quant. Biol., 39, 53-60 (1974).
201. C. J. Lai and D. Nathans, J. Mol. Biol., 89, 179-193 (1974b).
202. J. Carbon, T. E. Shenk, and P. Berg, Proc. Natl. Acad. Sci. USA, 72, 1392-1396 (1975).
203. Y. Aloni, E. Winocour, L. Sachs, and J. Torten, J. Mol. Biol., 44, 333-345 (1969).
204. E. Winocour, Virology, 31, 15-28 (1967).
205. H. Westphal and R. Dulbecco, Proc. Natl. Acad. Sci. USA, 59, 1158-1165 (1968).
206. E. Winocour, N. Frenkel, S. Lavi, M. Osenholts, and S. Rozenblatt, Cold Spring Harbor Symp. Quant. Biol., 39, 101-108 (1974).
207. M. A. Martin, L. D. Gelb, G. C. Fareed, and J. B. Milstein, J. Virol., 12, 748-757 (1973).
208. S. Rosenblatt, S. Lavi, M. F. Singer, and E. Winocour, J. Virol., 12, 501-510 (1973).
209. N. Frenkel, S. Lavi, and E. Winocour, Virology, 60, 9-20 (1974).
210. L. T. Chow, H. W. Boyer, E. G. Tischer, and H. M. Goodman, Cold Spring Harbor Symp. Quant. Biol., 39, 109-117 (1974).
211. S. Lavi and E. Winocour, Virology, 57, 296-297 (1974).
212. R. K. Ralph and J. S. Colter, Virology, 48, 49-58 (1972).
213. K. Hirai and V. Defendi, J. Virol., 9, 705-707 (1972).
214. W. Waldeck, K. Kammer, and G. Sauer, Virology, 54, 452-464 (1973).
215. F. Holzel and F. Sokol, J. Mol. Biol., 84, 423-444 (1974).
216. M. A. Martin, personal communication, 1975.
217. A. J. Levine and A. J. Teresky, J. Virol., 5, 451-457 (1970).
218. D. M. Trilling and D. Axelrod, Science, 168, 268-271 (1970).
219. M. R. Michel, B. Hirt, and R. Weil, Proc. Natl. Acad. Sci. USA, 58, 1381-1388 (1967).
220. K. Yoshiike, Virology, 34, 391-401 (1968a).
221. K. Yoshiike, Virology, 34, 402-409 (1968b).
222. R. Risser and C. Mulder, Virology, 58, 424-438 (1972).
223. K. Yoshiike, A. Furano, S. Watanabe, S. Uchida, K. Matsubara, and Y. Takagi, Cold Spring Harbor Symp. Quant. Biol., 39, 85-93 (1974).
224. G. C. Fareed, J. C. Byrne, and M. A. Martin, J. Mol. Biol., 87, 275-288 (1974).
225. G. Khoury, G. C. Fareed, K. Berry, M. A. Martin, T. N. H. Lee, and D. Nathans, J. Mol. Biol., 87, 289-301 (1974).
226. D. Davoli and G. C. Fareed, Cold Spring Harbor Symp. Quant. Biol., 39, 137-146 (1974).
227. M. A. Martin, L. D. Gelb, C. Garon, K. K. Takemoto, T. N. H. Lee, G. H. Sack, Jr., and D. Nathans, Virology, 59, 179-189 (1974).
228. M. A. Martin, G. Khoury, and G. C. Fareed, Cold Spring Harbor Symp. Quant. Biol., 39, 129-136 (1974).

229. D. L. Robbersson and M. Fried, Proc. Natl. Acad. Sci. USA, 71, 3497-3501 (1974).
230. M. Stoker, Nature, 218, 234-237 (1968).
231. E. May, P. May, and R. Weil, Proc. Natl. Acad. Sci. USA, 70, 1654-1658 (1973).
232. G. Khoury, P. Howley, M. Brown, and M. Martin, Cold Spring Harbor Symp. Quant. Biol., 39, 147-152 (1974).
233. V. Defendi, B. Ephrussi, and H. Koprowski, Nature, 203, 495-496 (1964).
234. M. Fogel and V. Defendi, Proc. Natl. Acad. Sci. USA, 58, 967-973 (1967).
235. P. H. Black, Virology, 28, 760-763 (1966).
236. S. Kit, J. L. Melnick, D. R. Dubbs, L. Piekarski, and R. A. DeTorres, Persp. Virol., 5, 63-86 (1967).
237. S. Kit, Adv. Cancer Res., 11, 73-221 (1968).
238. D. Gershon, P. Hausen, L. Sachs, and E. Winocour, Proc. Natl. Acad. Sci. USA, 54, 1584-1592 (1965).
239. D. Gershon, L. Sachs, and E. Winocour, Proc. Natl. Acad. Sci. USA, 56, 918-925 (1966).
240. G. Sauer and V. Defendi, Proc. Natl. Acad. Sci. USA, 56, 452-457 (1966).
241. P. Henry, P. H. Black, M. Oxman, and S. M. Weissman, Proc. Natl. Acad. Sci. USA, 56, 1170-1176 (1966).
242. R. Sheinin, Virology, 28, 47-55 (1966).
243. S. Kit, J. L. Melnick, M. Anken, D. R. Dubbs, R. A. DeTorres, and T. Kitahara, J. Virol., 1, 684-692 (1967).
244. E. May, P. May, and R. Weil, Proc. Natl. Acad. Sci. USA, 68, 1208-1211 (1971).
245. C. Basilico, Y. Matsuya, and H. Green, Virology, 41, 295-305 (1970).
246. P. Amati, F. Blasi, U. diPorzio, A. Riccio, and C. Traboni, Proc. Natl. Acad. Sci. USA, 72, 753-757 (1975).
247. M. Graessman and A. Graessman, Virology, 65, 591-594 (1975).
248. R. Mori, J. H. Schieble, and W. W. Ackerman, Proc. Soc. Exp. Biol. Med., 109, 685-690 (1962).
249. L. V. Crawford, Virology, 18, 177-181 (1962).
250. P. Bourgaux, Virology, 23, 46-55 (1964).
251. H. Ozer and K. K. Takemoto, J. Virol., 4, 408-415 (1969).
252. G. Barbanti-Brodano, P. Swetly, and H. Koprowski, J. Virol., 6, 78-86 (1970).
253. C. C. Howe, K. B. Tan, and F. Sokol, J. Gen. Virol., 27, 11-24 (1975).
254. K. Hummeler, N. Tomassini, and F. Sokol, J. Virol., 6, 87-93 (1970).
255. N. M. Maraldi, G. Barbanti-Brodano, M. Portolani, and M. LaPlaca, J. Gen. Virol., 27, 71-80 (1975).
256. M. Green, Ann. Rev. Biochem., 39, 701-756 (1970).
257. A. S. Kaplan and T. Ben-Porat, Ann. Rev. Microbiol., 22, 427-450 (1968).
258. R. Weil, C. Salomon, E. May, and P. May, Cold Spring Harbor Symp. Quant. Biol., 39, 381-395 (1974).
259. S. Manteuil, J. Pages, D. Stehelin, and M. Girard, J. Virol., 11, 98-106 (1973).

260. R. G. Roeder and W. J. Rutter, Nature, 224, 234-237 (1969).
261. R. G. Roeder and W. J. Rutter, Proc. Natl. Acad. Sci. USA, 65, 675-682 (1970).
262. F. Stirpe and L. Fiume, Biochem. J., 105, 779-782 (1967).
263. A. H. Jackson and W. Sugden, J. Virol., 10, 1086-1091 (1972).
264. R. J. Schmookler, J. Buss, and M. H. Green, Virology, 57, 122-127 (1974).
265. M. Geuskens and E. May, Exp. Cell Res., 87, 175-188 (1974).
266. R. H. Reeder and R. G. Roeder, J. Mol. Biol., 67, 433-441 (1972).
267. W. Sugden and W. Keller, J. Biol. Chem., 248, 3777-3788 (1973).
268. K. Oda and R. Dulbecco, Virology, 35, 439-444 (1968).
269. R. A. Weinberg, S. O. Warnaar, and E. Winocour, J. Virol., 10, 193-201 (1972).
270. E. Buetti, J. Virol., 14, 249-260 (1974).
271. J. Sambrook, P. A. Sharp, and W. Keller, J. Mol. Biol., 70, 57-71 (1972).
272. G. Khoury, J. C. Byrne, and M. A. Martin, Proc. Natl. Acad. Sci. USA, 69, 1925-1928 (1972).
273. T. L. Benjamin, J. Mol. Biol., 16, 359-373 (1966).
274. S. Tonegawa, G. Walter, A. Bernardini, and R. Dulbecco, Cold Spring Harbor Symp. Quant. Biol., 35, 823-831 (1970).
275. Y. Aloni, Cold Spring Harbor Symp. Quant. Biol., 39, 165-178 (1974).
276. Y. Aloni, E. Winocour, and L. Sachs, J. Mol. Biol., 31, 415-429 (1968).
277. K. Oda and R. Dulbecco, Proc. Natl. Acad. Sci. USA, 60, 525-532 (1968).
278. G. Sauer and J. R. Kidwai, Proc. Natl. Acad. Sci. USA, 61, 1256-1263 (1968).
279. R. I. Carp, F. Sokol, and G. Sauer, Virology, 37, 214-226 (1969).
280. J. Hudson, D. Goldstein, and R. Weil, Proc. Natl. Acad. Sci. USA, 65, 226-233 (1970).
281. S. Tonegawa, G. Walter, and R. Dulbecco, in The Biology of Oncogenic Viruses, North Holland, Amsterdam. (L. C. Sylvestri, ed.), pp. 64-76.
282. W. P. Cheevers and R. Sheinin, Can. J. Biochem., 48, 1104-1112 (1970).
283. G. Sauer, Nature New Biol., 231, 135-138 (1971).
284. M. A. Martin and D. Axelrod, Science, 164, 68-70 (1969).
285. M. A. Martin and D. Axelrod, Proc. Natl. Acad. Sci. USA, 64, 1203-1210 (1969).
286. R. A. Weinberg, Z. Ben-Ishai, and J. E. Newbold, J. Virol., 13, 1263-1273 (1974).
287. Y. Aloni, M. Shani, and Y. Reuveni, Proc. Natl. Acad. Sci. USA, 72, 2587-2591 (1975).
288. S. O. Warnaar and A. W. DeMol, J. Virol., 12, 124-129 (1973).
289. S. Lavi and A. S. Shatkin, Proc. Natl. Acad. Sci. USA, 72, 2012-2016 (1975).
290. R. A. Weinberg, Z. Ben-Ishai, and J. E. Newbold, Nature New Biol., 238, 111-113 (1972).
291. Y. Aloni, Nature New Biol., 243, 2-6 (1973).
292. R. Gromkova and S. H. Goodgal, J. Bacteriol., 109, 987-992 (1972).

293. R. N. Yoshimori, Ph.D. Thesis, University of California at San Francisco, Microbiology Dept., 1971.
294. H. O. Smith and K. Wilcox, J. Mol. Biol., 51, 379-391 (1970).
295. H. Westphal and E. D. Kiehn, Cold Spring Harbor Symp. Quant. Biol., 35, 819-821 (1970).
296. R. Kamen, D. M. Lindstrom, H. Shure, and R. W. Old, Cold Spring Harbor Symp. Quant. Biol., 39, 187-198 (1974).
297. J. Sambrook, W. Sugden, W. Keller, and P. A. Sharp, Proc. Natl. Acad. Sci. USA, 70, 3711-3715 (1973).
298. G. Khoury, M. A. Martin, T. N. H. Lee, K. J. Danna, and D. Nathans, J. Mol. Biol., 78, 377-389 (1973).
299. G. Hayward, Virology, 49, 342-344 (1972).
300. J. Flint, J. Wewerka-Lutz, A. S. Levine, J. Sambrook, and P. A. Sharp, J. Virol., 16, 662-673 (1975).
301. D. M. Lindstrom and R. Dulbecco, Proc. Natl. Acad. Sci. USA, 65, 1089-1096 (1972).
302. N. Mueller, J. Zemla, and G. Brandner, FEBS Lett., 31, 222-224 (1973).
303. J. Hedgpeth, H. M. Goodman, and H. Boyer, Proc. Natl. Acad. Sci. USA, 69, 3448-3452 (1972).
304. K. Cowan, P. Tegtmeyer, and D. D. Anthony, Proc. Natl. Acad. Sci. USA, 70, 1927-1930 (1973).
305. G. Brandner and N. Mueller, FEBS Lett., 42, 124-126 (1974).
306. S. Manteuil and M. Girard, Virology, 60, 438-454 (1974).
307. D. M. Glover, Biochem. Biophys. Res. Commun., 57, 1137-1143 (1974).
308. Y. Aloni, Proc. Natl. Acad. Sci. USA, 69, 2404-2409 (1972).
309. Y. Aloni and H. Locker, Virology, 54, 495-505 (1973).
310. M. A. Martin, Cold Spring Harbor Symp. Quant. Biol., 35, 833-841 (1970).
311. N. H. Acheson, E. Buetti, K. Scherrer, and R. Weil, Proc. Natl. Acad. Sci. USA, 68, 2231-2235 (1971).
312. R. Jaenisch, Nature New Biol., 235, 46-47 (1972).
313. R. Kajioka, Virology, 48, 284-287 (1972).
314. S. Rosenblatt and E. Winocour, Virology, 50, 558-566 (1972).
315. H. Westphal and E. D. Kiehn, Cold Spring Harbor Symp. Quant. Biol., 35, 819-821 (1970).
316. A. H. Fried and F. Sokol, J. Gen. Virol., 17, 69-79 (1972).
317. L. V. Crawford, E. M. Crawford, J. P. Richardson, and H. S. Slayter, J. Mol. Biol., 14, 593-597 (1965).
318. D. Pettijohn and T. Kamiya, J. Mol. Biol., 29, 275 (1967).
319. H. Delius, H. Westphal, and N. Axelrod, J. Mol. Biol., 74, 677-687 (1973).
320. M. Herzberg and E. Winocour, J. Virol., 6, 667-676 (1970).
321. J. J. Champoux and B. C. McConnaughy, Biochemistry, 14, 307-346 (1975).
322. R. Dulbecco, L. H. Hartwell, and M. Vogt, Proc. Natl. Acad. Sci. USA, 53, 403-410 (1965).
323. P. M. Frearson, S. Kit, and D. R. Dubbs, Cancer Res., 25, 737-744 (1965).
324. P. M. Frearson, S. Kit, and D. R. Dubbs, Cancer Res., 26, 1653-1660 (1966).

325. L. H. Hartwell, M. Vogt, and R. Dulbecco, Virology, 27, 262-272 (1965).
326. S. Kit, P. M. Frearson, and D. R. Dubbs, Proc. Soc. Exp. Biol., 24, 596-599 (1965).
327. S. Kit, D. R. Dubbs, and P. M. Frearson, Cancer Res., 26, 638-646 (1966).
328. S. Kit, D. R. Dubbs, P. M. Frearson, and J. L. Melnick, Virology, 29, 69-83 (1966).
329. J. Kara and R. Weil, Proc. Natl. Acad. Sci. USA, 57, 63-70 (1967).
330. J. Sambrook and A. Shatkin, J. Virol., 4, 719-726 (1969).
331. R. Weil, M. R. Michel, and G. K. Ruschmann, Proc. Natl. Acad. Sci. USA, 53, 1468-1475 (1965).
332. M. Hatanaka and R. Dulbecco, Proc. Natl. Acad. Sci. USA, 56, 736-740 (1966).
333. P. Molteni, V. deSimone, E. Grosso, P. A. Bianchi, and E. Polli, Biochem. J., 98, 78-81 (1966).
334. H. Werchau, H. Westphal, G. Maass, and R. Hass, Arch. Ges. Virusforsch., 19, 351-360 (1966).
335. E. Ritzi and A. J. Levine, J. Virol., 5, 686-692 (1970).
336. A. J. Levine, Proc. Natl. Acad. Sci. USA, 68, 717-720 (1971).
337. H. Shimono and A. S. Kaplan, Virology, 37, 690-694 (1969).
338. G. Rovera, R. Baserga, and V. Defendi, Nature New Biol., 237, 240-241 (1972).
339. V. Defendi and F. Jensen, Science, 157, 703-705 (1967).
340. A. Brown, R. A. Consigli, J. Zabielski, and R. Weil, J. Virol., 14, 840-845 (1974).
341. W. Eckhart, R. Dulbecco, and M. Burger, Proc. Natl. Acad. Sci. USA, 68, 283-286 (1971).
342. W. Eckhart and R. Dulbecco, Virology, 60, 359-369 (1974).
343. W. Eckhart, Cold Spring Harbor Symp. Quant. Biol., 39, 37-40 (1974).
344. J. A. Robb and R. G. Martin, J. Virol., 9, 956-968 (1972).
345. R. G. Martin, J. Y. Chou, J. Avila, and R. Saral, Cold Spring Harbor Symp. Quant. Biol., 39, 17-24 (1974).
346. J. Pages, S. Manteuil, D. Stehelin, M. Fizman, M. Marx, and M. Girard, J. Virol., 12, 99-107 (1973).
347. H. V. Thorne, J. Gen. Virol., 18, 163-169 (1973).
348. B. Hirt, J. Mol. Biol., 26, 365-369 (1967).
349. A. J. Levine, H. S. Kang, and F. Billheimer, J. Mol. Biol., 50, 549-568 (1970).
350. P. Bourgaux, D. Bourgaux-Ramoisy, and P. Seiler, J. Mol. Biol., 59, 195-206 (1971).
351. E. D. Sebring, T. J. Kelly, Jr., M. M. Thoren, and N. P. Salzman, J. Virol., 8, 478-490 (1971).
352. A. Roman, J. J. Champoux, and R. Dulbecco, Virology, 57, 147-160 (1974).
353. J. Cairns, Cold Spring Harbor Symp. Quant. Biol., 28, 43-46 (1963).
354. P. Bourgaux, D. Bourgaux-Ramoisy, and R. Dulbecco, Proc. Natl. Acad. Sci. USA, 701-708 (1969).

355. R. Jaenisch, A. Mayer, and A. Levine, Nature New Biol., 233, 72-75 (1971).
356. W. Gilbert and D. Dressler, Cold Spring Harbor Symp. Quant. Biol., 33, 473-484 (1968).
357. B. Hirt, J. Mol. Biol., 40, 141-144 (1969).
358. W. Meinke and D. A. Goldstein, J. Mol. Biol., 61, 543-563 (1971).
359. P. Bourgaux and D. Bourgaux-Ramoisy, Nature, 235, 105-107 (1972).
360. A. Mayer and A. J. Levine, Virology, 50, 549-568 (1972).
361. G. C. Fareed, G. F. Garon, and N. P. Salzman, J. Virol., 10, 484-491 (1972).
362. L. V. Crawford, C. Syrett, and A. Wilde, J. Gen. Virol., 21, 515-521 (1973).
363. K. J. Danna and D. Nathans, Proc. Natl. Acad. Sci. USA, 69, 3097-3100 (1972).
364. H. M. Dintzis, Proc. Natl. Acad. Sci. USA, 47, 247-261 (1961).
365. L. V. Crawford, A. K. Robbins, and P. M. Nicklin, J. Gen. Virol., 25, 133-142 (1974).
366. L. V. Crawford, A. K. Robbins, P. M. Nicklin, and K. Osborn, Cold Spring Harbor Symp. Quant. Biol., 39, 219-225 (1974).
367. G. C. Fareed and N. P. Salzman, Nature New Biol., 233, 277-279 (1972).
368. G. Magnusson, V. Pigiet, E. L. Winnacker, R. Abrams, and P. Reichard, Proc. Natl. Acad. Sci. USA, 70, 412-415 (1973).
369. V. Pigiet, R. Eliasson, and P. Reichard, J. Mol. Biol., 84, 197-216 (1974).
370. T. Hunter and B. Franke, J. Mol. Biol., 83, 123-130 (1974).
371. A. Sugino, S. Hirose, and R. Okazaki, Proc. Natl. Acad. Sci. USA, 69, 1863-1867 (1972).
372. R. Okazaki, T. Okazaki, K. Sakabe, K. Sugimoto, R. Kainuma, A. Sugino, and N. Iwatsuki, Cold Spring Harbor Symp. Quant. Biol., 33, 129-144 (1968).
373. S. Hirose, R. Okazaki, and F. Tamanoi, J. Mol. Biol., 77, 501-517 (1973).
374. G. C. Fareed, G. Khoury, and N. P. Salzman, J. Mol. Biol., 77, 457-462 (1973).
375. G. Magnusson, J. Virol., 12, 600-608 (1973).
376. B. Franke and M. Vogt, Cell, 5, 205-211 (1973).
377. D. L. Robberson, L. V. Crawford, C. Syrett, and E. W. James, J. Gen. Virol., 26, 59-69 (1975).
378. P. Laipis and A. J. Levine, Virology, 56, 580-594 (1973).
379. N. P. Salzman and M. Thoren, J. Virol., 11, 721-729 (1973).
380. G. Magnusson, R. Craig, M. Markhammer, P. Reichard, M. Staub, and H. Warner, Cold Spring Harbor Symp. Quant. Biol., 39, 227-234 (1974).
381. A. Weissbach, Cell, 5, 101-108 (1975).
382. W. Keller and R. Crouch, Proc. Natl. Acad. Sci. USA, 69, 3360-3364 (1972).
383. W. Keller, Proc. Natl. Acad. Sci. USA, 72, 2550-2554 (1975).
384. G. C. Fareed, M. L. McKerlie, and N. P. Salzman, J. Mol. Biol., 74, 95-111 (1973).
385. R. Jaenisch and A. S. Levine, J. Mol. Biol., 73, 199-212 (1973).
386. R. Jaenisch and A. S. Levine, Virology, 48, 373-379 (1972).

387. C. W. Anderson and R. F. Gesteland, J. Virol., 9, 758-765 (1972).
388. E. D. Kiehn, Virology, 56, 313-333 (1973).
389. P. Tegtmeyer, Cold Spring Harbor Symp. Quant. Biol., 39, 9-15 (1974).
390. P. H. Black, W. P. Rowe, H. C. Turner, and R. J. Huebner, Proc. Natl. Acad. Sci. USA, 50, 1148-1156 (1963).
391. K. K. Takemoto and K. Hable, Proc. Soc. Exp. Biol. Med., 120, 124-127 (1965).
392. J. H. Pope and W. P. Rowe, J. Exp. Med., 120, 121-127 (1964).
393. F. Rapp, T. Kitahara, J. S. Butel, and J. L. Melnick, Proc. Natl. Acad. Sci. USA, 52, 1138-1142 (1964).
394. R. V. Gilden, R. I. Carp, F. Taguchi, and V. Defendi, Proc. Natl. Acad. Sci. USA, 53, 684-692 (1965).
395. M. D. Hoggan, W. P. Rowe, P. H. Black, and R. J. Huebner, Proc. Natl. Acad. Sci. USA, 52, 12-19 (1965).
396. F. Rapp, J. S. Butel, L. A. Feldman, T. Kitahara, and J. L. Melnick, J. Exp. Med., 121, 935-944 (1965).
397. P. H. Black and W. P. Rowe, Proc. Natl. Acad. Sci. USA, 60, 606-613 (1963).
398. A. B. Sabin, H. M. Shein, M. A. Koch, and J. F. Enders, Proc. Natl. Acad. Sci. USA, 52, 1316-1318 (1964).
399. F. Rapp, J. S. Butel, and J. L. Melnick, Proc. Soc. Exp. Biol. Med., 116, 1131-1135 (1964).
400. K. Habel, Virology, 25, 55-61 (1965).
401. K. Habel, F. C. Jensen, J. Pagano, and H. Koprowski, Proc. Soc. Exp. Biol. Med., 118, 4-9 (1965).
402. M. N. Oxman, K. K. Takemoto, and W. Eckhart, Virology, 49, 675-682 (1972).
403. P. Tegtmeyer, M. Schwartz, J. K. Collins, and K. Rundell, J. Virol., 16, 168-178 (1975).
404. R. I. Carp and R. V. Gilden, Virology, 27, 639-641 (1965).
405. C. Basilico and G. DiMayorca, Proc. Nat. Acad. Sci. USA, 54, 125-127 (1965).
406. T. L. Benjamin, Proc. Natl. Acad. Sci. USA, 54, 121-124 (1965).
407. R. Latarjet, R. Cramer, and L. Montagnier, Virology, 33, 104-111 (1967).
408. A. Coppey and R. Wicker, Ann. Inst. Pasteur, 115, 478-485 (1968).
409. H. Yamamoto and H. Shimojo, J. Virol., 7, 419-425 (1971).
410. W. F. Mangel, S. T. Bayley, T. Wheeler, and A. E. Smith, unpublished observations.
411. B. E. Roberts, M. Gorecki, R. C. Mulligan, K. J. Danna, S. Rosenblatt, and A. Rich, Proc. Natl. Acad. Sci. USA, 72, 1922-1926 (1975).
412. K. Rundell, D. Tegtmeyer, C. J. Lai, and D. Nathans, personal communication, 1975.
413. P. Tegtmeyer, C. Dohan, and C. Reznikoff, Proc. Natl. Acad. Sci. USA, 66, 745-752 (1970).
414. P. Tegtmeyer and H. L. Ozer, J. Virol., 8, 516-524 (1971).
415. G. Kimura and R. Dulbecco, Virology, 49, 394-403 (1972).
416. G. Kimura and R. Dulbecco, Virology, 52, 529-534 (1973).

417. J. Y. Chou and R. G. Martin, J. Virol., 13, 1101-1109 (1974).
418. D. R. Dubbs, M. Rachmeler, and S. Kit, Virology, 57, 161-174 (1974).
419. N. Yamaguchi and T. Kuchino, J. Virol., 15, 1297-1301 (1975).
420. B. C. DelVillano and V. Defendi, Virology, 51, 34-46 (1973).
421. D. M. Livingstone and I. C. Henderson, Cold Spring Harbor Symp. Quant. Biol., 39, 283-289 (1974).
422. I. C. Henderson and D. M. Livingstone, Cell, 3, 65-70 (1975).
423. H. M. Lazarus, M. B. Sporn, J. B. Smith, and W. R. Henderson, J. Virol., 1, 1093-1095 (1967).
424. C. W. Potter, B. C. McLaughlin, and J. S. Oxford, J. Virol., 4, 574-579 (1969).
425. M. Osborn and K. Weber, Cold Spring Harbor Symp. Quant. Biol., 39, 267-276 (1974).
426. R. B. Carroll, L. Hager, and R. Dulbecco, Proc. Natl. Acad. Sci. USA, 71, 3754-3757 (1974).
427. T. Kuchino and N. Yamaguchi, J. Virol., 15, 1302-1307 (1975).
428. S. I. Reed, J. Ferguson, R. W. Davis, and G. Stark, Proc. Natl. Acad. Sci. USA, 72, 1605-1609 (1975).
429. T. Spillman, B. Spomer, and L. Hager, unpublished observations.
430. J. Y. Chou and R. G. Martin, J. Virol., 15, 145-150 (1975).
431. P. Tegtmeyer, J. Virol., 10, 591-598 (1972).
432. A. M. Lewis and W. P. Rowe, J. Virol., 7, 189-197 (1971).
433. T. J. Kelly, Jr., and A. M. Lewis, J. Virol., 12, 643-652 (1973).
434. M. J. Levin, C. S. Crumpacker, A. M. Lewis, M. H. Oxman, P. H. Henry, and W. P. Rowe, J. Virol., 7, 343-351 (1971).
435. A. M. Lewis, A. S. Levine, C. S. Crumpacker, M. J. Levin, R. J. Samaha, and P. H. Henry, J. Virol., 11, 655-664 (1973).
436. J. F. Morrow, P. Berg, T. J. Kelly, Jr., and A. M. Lewis, J. Virol., 12, 653-658 (1973).
437. L. J. Bello and H. S. Ginsberg, J. Virol., 3, 106-113 (1969).
438. D. O. White, M. D. Scharff, and J. V. Maizel, Jr., Virology, 38, 395-406 (1969).
439. W. Eckhart and M. Weber, Virology, 61, 223-228 (1974).
440. T. L. Benjamin and M. Burger, Proc. Natl. Acad. Sci. USA, 67, 929-934 (1970).
441. T. E. Kluchareva, K. L. Schachanina, S. Belova, V. Chibisova, and G. I. Deichman, J. Natl. Cancer Inst., 39, 825-832 (1967).
442. I. S. Irlin, Virology, 32, 725-728 (1967).
443. R. A. Malmgren, K. K. Takemoto, and P. G. Carney, J. Natl. Cancer Inst., 40, 263-268 (1968).
444. A. J. Girardi and V. Defendi, Virology, 42, 688-698 (1970).
445. A. S. Rabson, G. T. O'Conor, I. K. Berezesky, and F. J. Paul, Proc. Soc. Exp. Med., 116, 187-190 (1964).
446. G. T. O'Conor, A. S. Rabson, I. K. Berezesky, and F. J. Paul, J. Natl. Cancer Inst., 31, 903-917 (1963).

447. T. Grodzicker, C. Anderson, P. A. Sharp, and J. Sambrook, J. Virol., 13, 1237-1244 (1974).
448. L. A. Feldman, J. S. Butel, and F. Rapp, J. Bacteriol., 91, 813-818 (1966).
449. R. A. Malmgren, A. S. Rabson, P. G. Carney, and F. J. Paul, J. Bacteriol., 91, 262-265 (1966).
450. P. R. Reich, S. G. Baum, J. A. Rose, W. P. Rowe, and S. M. Weissman, Proc. Natl. Acad. Sci. USA, 55, 336-341 (1966).
451. S. G. Baum, W. H. Wiese, and P. R. Reich, Virology, 34, 373-376 (1968).
452. R. Fox and S. Baum, J. Virol., 10, 220-227 (1972).
453. J. J. Lucas and H. S. Ginsberg, J. Virol., 10, 1109-1118 (1972).
454. L. Eron, H. Westphal, and G. Khoury, J. Virol., 15, 1256-1261 (1975).
455. M. P. Friedman, M. J. Lyons, and H. S. Ginsberg, J. Virol., 5, 586-597 (1970).
456. C. J. Henry, M. Slifkin, and L. Merkow, Nature New Biol., 233, 39-41 (1971).
457. S. Baum, M. Horwitz, and J. Maizel, Jr., J. Virol., 10, 211-219 (1972).
458. S. Baum and R. I. Fox, Cold Spring Harbor Symp. Quant. Biol., 39, 567-575 (1974).
459. K. Hashimoto, K. Nakajima, K. Oda, and H. Shimojo, J. Mol. Biol., 81, 207-223 (1973).
460. K. Nakajima, H. Ishitsuka, and K. Oda, Nature, 252, 652-653 (1974).
461. G. Kimura, Nature, 248, 590-592 (1974).
462. M. Jerkofsky, Virology, 65, 579-582 (1975).
463. J. Y. Chou, J. Avila, and R. G. Martin, J. Virol., 14, 116-124 (1974).
464. M. Jerkofsky and F. Rapp, J. Virol., 15, 253-258 (1975).
465. M. Jerkofsky andG JRapp, Virology, 51, 466-473 (1973).
466. H. L. Lodish, R. L. Weinberg, and H. L. Ozer, J. Virol., 13, 590-595 (1974).
467. C. L. Prives, A. Aviv, E. Gilboa, M. Revel, and E. Winocour, Cold Spring Harbor Symp. Quant. Biol., 39, 309-316 (1974).
468. C. L. Prives, H. Aviv, B. M. Paterson, B. E. Roberts, S. Rozenblatt, M. Revel, and E. Winocour, Proc. Natl. Acad. Sci. USA, 71, 302-306 (1974).
469. H. L. Ozer, J. Virol., 9, 41-51 (1972).
470. H. Fischer and G. Sauer, J. Virol., 9, 1-9 (1972).
471. F. Fenner, B. R. McAuslan, C. A. Mims, J. F. Sambrook, and D. O. White, The Biology of Animal Viruses, 2nd Ed., Academic, New York. 1974.
472. R. Leberman in The Molecular Biology of Viruses (L. V. Crawford and M. Stoker, eds.). Cambridge, Cambridge, England. 1968.
473. H. L. Ozer and P. Tegtmeyer, J. Virol., 9, 52-60 (1972).
474. K. B. Tan and F. Sokol, J. Gen. Virol., 25, 37-51 (1974).
475. M. Girard, D. Stehelin, S. Manteuil, and J. Pages, J. Virol., 11, 107-115 (1973).
476. C. F. T. Mattern and A. M. DeLeva, Virology, 36, 683-685 (1968).
477. C. F. T. Mattern, K. K. Takemoto, and W. A. Daniel, Virology, 30, 242-256 (1966).
478. N. Granboulan, P. Tournier, R. Wicker, and W. Bernhard, J. Cell. Biol., 17, 423-441 (1963).

479. H. D. Mayor, S. E. Stinebaugh, R. M. Jamison, L. E. Jordan, and J. L. Melnick, Exp. Mol. Pathol., 1, 397 (1962).
480. T. Benjamin, Proc. Natl. Acad. Sci. USA, 67, 394-399 (1970).
481. T. Benjamin and E. Goldman, Cold Spring Harbor Symp. Quant. Biol., 39, 41-44 (1974).
482. C. Basilico and G. diMayorca, J. Virol., 13, 931-934 (1974).
483. G. Kimura, Jap. J. Med., 17, 537-539 (1973).
484. K. K. Takemoto and M. A. Martin, Virology, 42, 938-945 (1970).
485. S. Kit, S. Tokuno, K. Nakajima, D. Trkula, and D. R. Dubbs, J. Virol., 6, 286-294 (1970).
486. M. Fried, Virology, 25, 669-671 (1965).
487. G. diMayorca, J. Callender, G. Marin, and R. Giordano, Virology, 38, 126-133 (1969).
488. J. A. Robb, P. Tegtmeyer, A. Ishikawa, and H. L. Ozer, J. Virol., 13, 662-665 (1974).
489. M. J. Tevethia, L. W. Ripper, and S. S. Tevethia, Intervirology, 3, 245-255 (1974).
490. C. J. Lai and D. Nathans, Virology, 60, 466-475 (1974).
491. T. E. Shenk, C. Rhodes, P. W. J. Rigby, and P. Berg, Cold Spring Harbor Symp. Quant. Biol., 39, 61-67 (1974).
492. T. E. Shenk, C. Rhodes, P. Rigby, and P. Berg, Proc. Natl. Acad. Sci. USA, 72, 989-993 (1975).
493. C. J. Lai and D. Nathans, Virology, 66, 70-81 (1975).
494. J. A. Robb, H. S. Smith, and C. D. Scher, J. Virol., 9, 969-972 (1972).
495. J. A. Robb, J. Virol., 12, 1187-1190 (1973).
496. J. A. Robb, Cold Spring Harbor Symp. Quant. Biol., 39, 277-281 (1974).
497. F. L. Graham, P. J. Abrahams, C. Mulder, H. L. Heijneker, S. O. Warnaar, F. A. J. deVries, W. Fiers, and A. J. van der Eb, Cold Spring Harbor Symp. Quant. Biol., 39, 637-650 (1974).
498. J. Arrand and J. Sambrook, unpublished results, 1974.
499. G. Khoury, J. C. Byrne, K. K. Takemoto, and M. A. Martin, J. Virol., 11, 54-60 (1973).
500. B. Ozanne, P. A. Sharp, and J. Sambrook, J. Virol., 12, 90-98 (1973).
501. M. Osborn and K. Weber, J. Virol., 15, 636-644 (1975).
502. G. Kimura and A. Hagaki, Proc. Natl. Acad. Sci. USA, 72, 673-677 (1975).
503. P. Tegtmeyer, J. Virol., 15, 613-618 (1975).
504. J. Brugge and J. Butel, J. Virol., 15, 619-635 (1975).
505. J. Butel, J. Brugge, and C. Noonan, Cold Spring Harbor Symp. Quant. Biol., 39, 25-36 (1974).
506. R. G. Martin and J. Y. Chou, J. Virol., 15, 599-612 (1975).
507. G. diMayorca and J. Callender, Prog. Med. Virol., 12, 284-301 (1970).
508. B. Francke and W. Eckhart, Virology, 55, 127-135 (1973).
509. M. Fried, Proc. Natl. Acad. Sci. USA, 53, 486-491 (1965).
510. W. Eckhart, Virology, 38, 120-125 (1969).
511. M. Fried, Virology, 40, 605-617 (1970).
512. W. Eckhart, Proc. Roy. Soc., London, Series B, 177, 59-63 (1971).

513. R. Dulbecco and W. Eckhart, Proc. Natl. Acad. Sci. USA, 67, 1775-1781 (1970).
514. B. Ozanne and J. Sambrook, Nature New Biol., 232, 156-160 (1971).
515. T. L. Benjamin, personal communication, 1975.
516. A. Ishikawa and G. diMayorca, Second Lepetit Symp. The Biology of Oncogenic Viruses (L. G. Sylvestri, ed.), North Holland, Amsterdam. 1971, pp. 294-301.
517. I. A. Macpherson and M. G. P. Stoker, Virology, 16, 147-151 (1962).
518. H. S. Smith, C. D. Scher, and G. J. Todaro, Virology, 44, 359-370 (1970).
519. J. Sambrook and R. Pollack, Meth. Enzymol., 32, 583-592 (1974).
520. R. O. Hynes, Cell, 1, 147-156 (1974).
521. R. Pollack, M. Osborn, and M. Weber, Proc. Natl. Acad. Sci. USA, 72, 994-998 (1975).
522. R. Dulbecco, Proc. Roy. Soc., London, Series B, 189, 1-14 (1975).
523. Cold Spring Harbor Symp. Quant. Biol., 39, (1974).
524. B. Clarkson and R. Baserga, Control of Proliferation in Animal Cells, Cold Spring Harbor, New York. 1974.
525. K. Shiroki and H. Shimojo, Virology, 45, 163-171 (1971).
526. M. G. P. Stoker and M. Abel, Cold Spring Harbor Symp. Quant. Biol., 27, 375-386 (1962).
527. I. Macpherson and L. Montagnier, Virology, 23, 291-294 (1964).
528. G. Todaro and H. Green, Virology, 28, 756-759 (1966).
529. C. Basilico and G. Marin, Virology, 28, 429-437 (1966).
530. R. Dulbecco and M. Vogt, Proc. Natl. Acad. Sci. USA, 67, 1775-1781 (1970).
531. M. Haas, M. Vogt, and R. Dulbecco, Proc. Natl. Acad. Sci. USA, 69, 2160-2164 (1972).
532. M. Botchan and G. McKenna, Cold Spring Harbor Symp. Quant. Biol., 38, 391-395 (1973).
533. L. Gelb, D. E. Kohne, and M. Martin, J. Mol. Biol., 57, 129-145 (1971).
534. J. Wetmur and N. Davidson, J. Mol. Biol., 31, 349-370 (1968).
535. R. Britten and D. E. Kohne, Science, 168, 529-540 (1968).
536. P. A. Sharp, U. Pettersson, and J. Sambrook, J. Mol. Biol., 86, 709-726 (1974).
537. P. H. Gallimore, P. A. Sharp, and J. Sambrook, J. Mol. Biol., 89, 49-72 (1974).
538. M. Botchan, B. Ozanne, W. Sugden, P. Sharp, and J. Sambrook, Proc. Natl. Acad. Sci. USA, 71, 4183-4187 (1974).
539. J. Sambrook, M. Botchan, P. H. Gallimore, B. Ozanne, U. Pettersson, J. Williams, and P. Sharp, Cold Spring Harbor Symp. Quant. Biol., 39, 615-632 (1974).
540. J. Sambrook, H. Westphal, P. R. Srinivasan, and R. Dulbecco, Proc. Natl. Acad. Sci. USA, 60, 1288-1295 (1968).
541. M. Shani, Z. Rabinowitz, and L. Sachs, J. Virol., 10, 456-461 (1972).
542. W. Folk, J. Virol., 11, 424-431 (1973).
543. H. Manor, M. Fogel, and L. Sachs, Virology, 53, 174-185 (1973).

544. C. M. Croce, A. J. Girardi, and H. Koprowski, Proc. Natl. Acad. Sci. USA, 70, 3617-3620 (1973).
545. C. M. Croce and H. Koprowski, J. Exp. Med., 139, 1350-1353 (1973).
546. C. M. Croce, K. Huebner, A. J. Girardi, and H. Koprowski, Virology, 60, 276-281 (1974).
547. C. M. Croce, K. Huebner, A. J. Girardi, and H. Koprowski, Cold Spring Harbor Symp. Quant. Biol., 39, 335-343 (1974).
548. C. Croce and H. Koprowski, Proc. Natl. Acad. Sci. USA, 72, 1658-1660 (1975).
549. C. M. Croce, D. Aden, and H. Koprowski, Proc. Natl. Acad. Sci. USA, 72, 1397-1400 (1975).
550. J. Sambrook, P. A. Sharp, B. Ozanne, and U. Pettersson, in Control of Transcription (B. B. Biswas, R. K. Mandal, A. Stevens, and W. E. Cohn, eds.), Plenum, New York. 1973, pp. 167-179.
551. U. Lindberg and J. E. Darnell, Proc. Natl. Acad. Sci. USA, 65, 1089-1096 (1970).
552. R. Wall and J. E. Darnell, Nature New Biol., 232, 73-77 (1971).
553. S. M. Astrin, Proc. Natl. Acad. Sci. USA, 70, 2304-2308 (1973).
554. S. M. Astrin, Biochemistry, 14, 2700-2704 (1975).
555. T. Y. Shih, G. Khoury, and M. A. Martin, Proc. Natl. Acad. Sci. USA, 70, 3506-3510 (1973).
556. P. Swetly and Y. Watanabe, Biochemistry, 13, 4122-4126 (1974).
557. K. Habel, Cancer Res., 26, 2018-2024 (1966).
558. K. Habel, Proc. Soc. Exp. Biol. Med., 106, 722-725 (1961).
559. H. O. Sjogren, I. Hellstrom, and G. Klein, Cancer Res., 21, 329-337 (1961).
560. V. Defendi, Proc. Soc. Exp. Biol. Med., 113, 12-16 (1963).
561. V. Defendi, Transplantation, 6, 642-643 (1968).
562. M. A. Koch and A. S. Sabin, Proc. Soc. Exp. Biol. Med., 113, 4-12 (1963).
563. M. J. Levin, M. N. Oxman, G. T. Diamandopoulos, A. S. Levine, P. H. Henry, and J. F. Enders, Proc. Soc. Exp. Biol. Med., 62, 589-596 (1969).
564. A. S. Levine, M. N. Oxman, P. H. Henry, M. J. Levin, G. T. Diamandopoulos, and J. F. Enders, J. Virol., 6, 199-207 (1970).
565. J. S. Butel, S. S. Tevethia, and J. L. Melnick, Adv. Cancer Res., 15, 1-55 (1972).
566. R. Duff and F. Rapp, J. Immunol., 105, 521-523 (1970).
567. W. Baranska, P. Koldovsky, and H. Koprowski, Proc. Natl. Acad. Sci. USA, 67, 193-199 (1970).
568. P. Hayry and V. Defendi, Virology, 41, 22-29 (1970).
569. P. Gerber, Virology, 28, 501-509 (1966).
570. H. Koprowski, F. C. Jensen, and Z. Steplewski, Proc. Natl. Acad. Sci. USA, 58, 127-133 (1967).
571. J. Watkins and R. Dulbecco, Proc. Natl. Acad. Sci. USA, 58, 1396-1403 (1967).
572. M. Botchan, B. Ozanne, and J. Sambrook, Cold Spring Harbor Symp. Quant. Biol., 39, 95-100 (1974).
573. S. Kit and M. Brown, J. Virol., 4, 226-230 (1969).

574. K. Huebner and H. Koprowski, Virology, 58, 609-611 (1974).
575. J. F. Watkins, J. Cell Sci., 6, 721-737 (1970).
576. B. B. Knowles, F. C. Jensen, Z. Steplewski, and H. Koprowski, Proc. Natl. Acad. Sci. USA, 58, 127-133 (1968).
577. F. C. Jensen and H. Koprowski, Virology, 37, 687-690 (1969).
578. J. F. Watkins, Cold Spring Harbor Symp. Quant. Biol., 39, 355-362 (1974).
579. D. R. Dubbs and S. Kit, J. Virol., 2, 1272-1282 (1968).
580. G. H. Wever, S. Kit and D. R. Dubbs, J. Virol., 5, 578-585 (1970).
581. S. Kit, D. R. Dubbs, and K. Somers, in CIBA Symp. The Strategy of the Viral Genome (Wolstenholme and Knight, eds.), Churchill-Livingstone, London. 1971, pp. 229-243.
582. R. Glaser and R. Farringia, Intervirology, 1, 135-140 (1973).
583. C. M. Croce and H. Koprowski, Virology, 51, 227-229 (1973).
584. G. Poste, B. Schaeffer, P. Reeve, and D. J. Alexander, Virology, 50, 85-95 (1974).
585. K. Huebner, C. M. Croce, and H. Koprowski, Virology, 59, 570-573 (1974).
586. K. Huebner, D. Santoli, C. M. Croce, and H. Koprowski, Virology, 63, 512-522 (1975).
587. G. J. Todaro and K. K. Takemoto, Proc. Natl. Acad. Sci. USA, 62, 1031-1037 (1969).
588. V. A. L. Boyd and J. S. Butel, J. Virol., 10, 399-409 (1972).
589. M. Shani, E. Huberman, Y. Aloni, and L. Sachs, Virology, 61, 303-305 (1974).
590. M. V. Fernandes and P. S. Moorhead, Texas Rep. Biol. Med., 23, 242-258 (1965).
591. M. Vogt, J. Mol. Biol., 47, 307-316 (1970).
592. J. Summers and M. Vogt, in Second Lepetit Symposium: The Biology of Oncogenic Viruses (L. G. Sylvestri, ed.), North Holland, Amsterdam. 1971, pp. 306-311.
593. M. Fogel and L. Sachs, Virology, 37, 327-334 (1969).
594. M. Fogel and L. Sachs, Virology, 40, 174-177 (1970).
595. M. Saito, F. Taguchi, K. Hasegawa, Y. Yoshida, and D. Nagaki, Jap. J. Microbiol., 14, 512-515 (1970).
596. M. Fogel, Virology, 49, 12-22 (1972).
597. M. Fogel, Nature New Biol., 241, 182-184 (1973).
598. M. Fogel, Virology, 65, 446-454 (1975).
599. W. H. Burns and P. H. Black, Virology, 39, 625-634 (1969).
600. W. H. Burns and P. H. Black, J. Virol., 2, 606-609 (1968).
601. H. Rothschild and P. H. Black, Virology, 42, 251-256 (1970).
602. S. Kit and S. Dubbs, in Second Lepetit Symposium: The Biology of Oncogenic Viruses (L. G. Sylvestri, ed.), North Holland, Amsterdam. 1971, pp. 28-41.
603. J. C. Kaplan, S. M. Wilbert, and P. H. Black, J. Virol., 9, 448-453 (1972).
604. P. Swetly, G. Brodano, B. Knowles, and H. Koprowski, J. Virol., 4, 348-355 (1969).

605. R. E. Pollack, H. Green, and G. J. Todaro, Proc. Natl. Acad. Sci. USA, 60, 126-133 (1968).
606. B. Ozanne and J. Sambrook, Second Lepetit Symp.: The Biology of Oncogenic Viruses, (L. G. Sylvestri, ed.), North Holland, Amsterdam. 1971, pp. 248-256.
607. B. Ozanne, J. Virol., 12, 90-99 (1973).
608. L. A. Culp and P. H. Black, J. Virol., 9, 611-620 (1972).
609. A. Vogel, R. Risser, and R. Pollack, J. Cell Physiol., 82, 181-188 (1973).
610. Z. Rabinowitz and L. Sachs, Nature, 220, 1203-1206 (1968).
611. Z. Rabinowitz and L. Sachs, Virology, 38, 343-346 (1969).
612. Z. Rabinowitz and L. Sachs, Virology, 38, 336-342 (1969).
613. Z. Rabinowitz and L. Sachs, Nature, 225, 136-144 (1970).
614. J. A. Wyke, Exp. Cell Res., 66, 203-208 (1971).
615. J. Wyke, Exp. Cell Res., 66, 209-223 (1971).
616. R. E. Pollack, S. Wollman, and A. Vogel, Nature, 228, 938-970 (1970).
617. S. Hitotsumachi, Z. Rabinowitz, and L. Sachs, Nature, 231, 511-514 (1971).
618. F. Kelly and J. Sambrook, Cold Spring Harbor Symp. Quant. Biol., 39, 345-353 (1974).
619. R. Dulbecco, Cold Spring Harbor Symp. Quant. Biol., 39, 1-7 (1974).

Chapter 12

Adenoviruses

William S. M. Wold, Maurice Green, and Werner Büttner

Institute for Molecular Virology
St. Louis University Medical School
St. Louis, Missouri

I.	Introduction	675
	A. Model for Investigating the Molecular Biology of Human and Transformed Cells	675
	B. Adenoviruses: Occurrence and Pathology	676
	C. Experimental Systems for Studying Productive Infection and Cell Transformation	677
II.	Oncogenicity and Classification of the Human Adenoviruses	678
III.	The Adenovirus DNA Genome	681
	A. Molecular Weight, Conformation, and Intramolecular Structure	681
	B. DNA-DNA Homology Relationships among Human Adenoviruses	685
	C. Specific Fragments of the Adenovirus Genome and Localization of Transforming Genes	686
	D. Biological Activity of Viral DNA and Viral DNA Fragments	690
IV.	Adenovirus: Structure and Protein Constituents	690
	A. Introduction	690
	B. Morphology	691

C. Polypeptide Components 691
D. The Hexon 697
E. The Fiber 698
F. The Penton Base 698
G. The Core 698

V. Adenovirus Genetics 699

A. *Cyt* Mutants 699
B. Ad 5 *ts* and *hr* Mutants 700
C. Ad 12 *ts* Mutants 710
D. Ad 31 *ts* Mutants 711
E. CELO *ts* Mutants 712
F. Ad 2 *ts* Mutants 712
G. Host-Range Mutants of Ad_2-SV40 Hybrid Viruses 713

VI. Adenovirus Replicative Cycle (Productive Infection) 714

A. Early Events 714
B. The Molecular Events of Productive Infection 714
C. Viral Gene Transcription 717
D. Virus-Associated RNA 718
E. Biogenesis of Adenovirus mRNA 718
F. The Switch from Early to Late Viral Gene Expression 720
G. Biosynthesis of Adenovirus Proteins 721

VII. Adenovirus DNA Replication 726

A. A Model for DNA Replication in Mammalian Cells 726
B. Viral-Coded Proteins Involved in Adenovirus DNA Replication 726
C. Replicative Forms and Viral DNA Intermediates 727
D. Enzymes Involved in Adenovirus DNA Replication 728
E. Current Model for Adenovirus DNA Replication 729

VIII. Cell Transformation 730

A. Viral DNA Sequences in Adenovirus-Transformed Cells 732
B. Transcription of Viral RNA in Adenovirus-Transformed Cells 733
C. Adenovirus-Induced Proteins in Transformed Cells 738

IX. Adenovirus-Simian Virus 40 (SV40) Hybrids 738

A. Simian Virus 40 Enhancement of Adenovirus Replication in Monkey Cells 738
B. Ad-SV40 Hybrids: Isolation and Properties 739
C. Defective Ad-SV40 Hybrids 744
D. Nondefective Ad_2-SV40 Hybrids 746

X. Mapping the Adenovirus Genome 749

A. Mapping DNA Regions Encoding mRNA 750
B. Mapping Size-Fractionated mRNA 751
C. Mapping Viral-Coded Proteins by Cell-Free Translation of Restriction Fragment-Specific mRNA 752
D. Physical Mapping of Adenovirus *ts* Mutants 754

Acknowledgments 756

References 757

Recent Developments 891

I. Introduction

A. Model for Investigating the Molecular Biology of Human and Transformed Cells

The adenoviruses (Ads) are nonenveloped, icosahedral (an icosahedron is a polyhedron with 20 faces) viruses containing linear, double-stranded DNA of molecular weight 20-25 $\times$ 10^6 daltons [83]. The human Ads provide some of the best experimental systems for investigating DNA tumor virus replication, cell transformation, and the molecular biology of human and transformed cells. And of course they also are human pathogens, and may even be involved in human cancer. Ads replicate in the nuclei of human cells, with 2 stages of gene expression, "early" (before the initiation of viral DNA replication), and "late." Early genes encode functions for viral DNA replication, cell transformation, the "switch" to late stages of gene expression, possibly the turn off of host DNA synthesis, and other unknown requisites for replication. Late genes primarily encode viral structural proteins, and other functions that probably include inhibition of host protein synthesis and formation of cytoplasmic rRNA. The mature virion is apparently devoid of enzymes, except possibly DNA endonucleases, so host cell mechanisms must be used to express early, and to a large extent late genes. Because of this, analysis of viral replication, especially at early stages, provides keen insights into many aspects of host cell molecular biology, since the synthesis of Ad macromolecules reflects that of the host cell. The viral genome is expressed in 2 stages, so the productive infection can be viewed as a very simple "differentiating" system, and therefore may be especially instructive with regard to regulation of eukaryotic gene expression. The late stage of infection is interesting because viral macromolecules are synthesized in prodigious amounts while those of the host cell are repressed. Thus the late stage of infection is an ideal model to study viral DNA replication, and viral mRNA transcription, processing, and translation. From a practical viewpoint, Ads are easily grown, and large quantities of viral-specific DNA, RNA, and proteins are readily obtained. Finally, the viral genome is relatively small, encoding some 20-40 proteins, thus facilitating the combined genetic and biochemical analyses essential for elucidating function. Given the present rate of progress, it is possible that the next decade will see a relatively complete description of Ad replication.

Other groups of oncogenic DNA viruses are the papovaviruses and herpesviruses. Polyoma and simian virus 40 (SV40) viruses do not block host macromolecular synthesis, so the Ad system is superior in this respect. The herpesviruses suffer the experimental disadvantage that their genomes are 4-5 times more complex than the Ads. The wart-producing papilloma viruses are very difficult to grow in cultured cells, so little is known about their replication.

Ads are of special interest because the functions of only 1 or 2 early viral genes can transform a cell to malignancy. For several human Ads, viral DNA segments containing the "transforming genes" have been identified [78]. This offers the exciting opportunity to study the protein products of transforming genes, and how their interaction with the mammalian cell induces and maintains the neoplastic state. Human Ads fall into 3 oncogenic groups that express 3 unrelated transforming gene sequences [56-59, 83]. This implies that 3 different classes of viral-coded transforming proteins can initiate and/or maintain the growth properties of Ad-transformed cells. Thus studies on cell transformation by members of different Ad groups should permit functional comparisons of cellular and viral factors that regulate growth in mammalian cells.

The past 5 years have seen the dawning of a new era in adenovirology. Much has been learned, especially with regard to the early events of productive infection, mRNA biogenesis, Ad genetics, and the mapping of viral DNA and mRNA sequences in infected and transformed cells. This chapter will concentrate on these more recent developments, building upon our previous knowledge, and present a comprehensive but focused description of the molecular biology of Ad replication and cell transformation. The most recent reviews on Ads are [208, 209, 347, 352].

B. Adenoviruses: Occurrence and Pathology

Over 80 serologically distinct Ads have been isolated from a variety of animal species, including 31 human (referred to as Ad 1-31), 23 simian, 10 bovine, 8 avian, 4 porcine, 2 canine, 2 murine, 1 ovine, and 1 opposum (reviewed in [73, 83, 208, 242, 287]). Ads were first isolated by Rowe and co-workers [229] from adenoids and by Hilleman and Werner [107] from patients with acute respiratory disease. As pointed out by Schlesinger [242], this was the year of the double helix [309] and the plaque assay for animal viruses [42].

Thus the births of adenovirology and modern molecular biology coincide, and the study of Ads encompasses an era of phenomenal growth in animal molecular virology.

The human Ads are associated mainly with mild respiratory illness, often reaching epidemic proportions (especially types 4 and 7) in military recruits. Several Ad serotypes infect most children before the age of 5, can remain latent in the gastrointestinal tract, persist in lymphoid tissues, and can be cultured from 50-90% of surgically removed adenoids and tonsils. Ads have been associated with or isolated from patients with a variety of disorders, including ocular and gastrointestinal infections, rashes, and renal lesions (reviewed in [257, 358]).

Ads 12 and 18 were the first human viruses shown to possess oncogenic properties, i.e., they induced tumors when inoculated into newborn hamsters [115, 288]. Oncogenicity in newborn rodents and the ability to transform cultured cells of a foreign species is a property of many human and animal Ads [83, 84, 355]. The ubiquity of the human Ads and their ability to integrate into susceptible cells viral DNA segments containing transorming genes [63, 66] or most of the viral genome [94], raises the possibility that Ad genes may be integrated into human tissue cells during infection. If so, then possibly Ads play a role in human cancer and other chronic human diseases, by affecting the properties of cells through expression of integrated transforming or other genes. To investigate this possibility, a large number of human tumors and normal tissues are being analyzed for the presence of Ad DNA and mRNA by molecular hybridization techniques [353, 355]; to date, no evidence has been obtained for integrated Ad 2 and 12 transforming genes in human gastrointestinal and lung tissues [354].

C. Experimental Systems for Studying Productive Infection and Cell Transformation

Human Ads can productively infect permissive human cells, culminating in cell death and the replication of up to 10^5 viral progeny particles per cell [83]. In contrast, infection of nonpermissive or semipermissive rodent cells results in abortive infection, where many of the cells may be killed, but viral replication is very limited. However, after abortive infection a small fraction of the cells can become transformed to malignancy. Although Ad-transformed cells fail to synthesize viral particles, they retain viral DNA sequences [88], and synthesize virus-specific RNA [56] and tumor (T) antigens [116]. During productive infection, the virus penetrates the cell and uses both host enzymes and viral-coded proteins to replicate viral DNA and convert viral genetic information into proteins. Molecular studies with Ads have utilized most often Ad 2-infected suspension cultures of human KB cells [85], one of the best understood virus-mammalian cell systems; analysis of human HeLa cells infected with Ad 2 have yielded comparable results. Our current understanding of Ad replication is derived mainly from studies with the human Ads. Recent studies with the CELO virus, an oncogenic avian Ad, have provided complementary data, concerning mainly the structure and replication of viral DNA.

Studies on cell transformation by Ads have analyzed mainly transformed rat or hamster cells. Progeny virions are not produced in Ad-transformed cells. Viral DNA segments containing transforming genes of Ad 2, Ad 5 [66], and Ad 7 [63], and most if not all the viral genome of the highly oncogenic Ad 12 [94], become stably integrated into host cell chromosomes. It is widely believed, although not proven, that the expression of integrated Ad-transforming genes as virus-specific proteins induces and maintains the neoplastic state. The elucidation of the functions of viral transforming proteins presents one of the most difficult and exciting challenges in biology.

Tremendous advances have been made recently in the understanding of Ad-transformed cells. These include: (1) the number of viral DNA copies per transformed cell; (2) segments of the viral genome integrated into host DNA; (3) regions of integrated viral DNA expressed as mRNA; and (4) the number and sizes of viral proteins encoded by each segment of viral DNA in transformed cells. Because DNA tumor viruses in general do not efficiently transform cultured cells, studies on the molecular events during cell transformation are not possible. Instead, our understanding of cell transformation is extrapolated from the properties of the transformed cell, or from the limited molecular analysis of early and abortive infection. The early events of cell transformation are more opportunely studied with RNA tumor viruses where rapid and synchronous cell transformation can be achieved with mammalian [235] and avian [241] systems. Since the genome of an RNA tumor virus is converted to a DNA replica by the viral reverse transcriptase and then integrated into the cellular genome (reviewed in [86]), the initiation of cell transformation and the regulation of growth may involve similar mechanisms for both RNA and DNA tumor viruses.

II. Oncogenicity and Classification of the Human Adenoviruses

The 31 human Ads have been purified, analyzed chemically, and their DNAs characterized with regard to base composition and molecular weight [83, 84, 212]. The human Ads consist of protein and 12-14% DNA, possess a particle weight of 175×10^6 daltons, and contain a single DNA molecule of 20-25 $\times$ 10^6 daltons [92]. Of the 31 human Ads, 13 form 3 well defined subgroups, A, B, and C (Table 12.1), based on oncogenicity, T-antigen cross-reactivity, guanine plus cytosine (G + C) content, hybridization of viral DNA to transformed cell virus-specific RNA, but most decisively by reciprocal DNA-DNA homology measurements among viral DNAs (see Section III). Inoculation of purified preparations of the 31 serotypes into newborn hamsters has confirmed the existence of 3 subgroups [84]: highly oncogenic group A (Ad 12, 18, and 31) induced tumors in most animals in weeks to months; weakly oncogenic group B (Ad 3, 7, 11, 14, 16, and 21) induced tumors (all but type 11) in a small fraction of animals in months to years; and "nononcogenic" group C (Ad 1, 2, 5, and 6) did not induce tumors but transformed cultured rat cells in vitro [55]. The oncogenicity of Ad 14, 16, and 21 were demonstrated only with purified Ad [84]. Several simian Ads as well as avian, bovine, and canine Ads have also been shown to induce tumors [83, 209]. The prime factor determining the oncogenicity of different Ads could be the type of surface antigens elicited by virus-infected cells: although Ad 12 and Ad 2 have similar transformation efficiencies in vitro, Ad 12-transformed cells are highly tumorigenic but Ad 2-transformed rat cells vary in their tumorigenicity in syngeneic or immunosuppressed rats, or nude mice.

The 18 unclassified human Ads have base compositions similar to those of group C viruses ([212] and Green, unpublished data). Of these, all but Ad 4 have been tentatively placed in group D [172] based on transformation of rat and hamster embryo cells by Ad 9, 10, 13, 15, 17, 19, and 26, and detection (by complement fixation with sera from hamsters with tumors induced by Ad 19- and Ad 26-transformed cells) of cross-reacting antigens (presumably a common T-antigen) in KB cells infected with the 17 group D viruses. However, virus-specific RNA was not detected in Ad 19-transformed cells, and further attempts to detect viral DNA and RNA in these transformed cells have not been reported. Whether Ads outside groups A, B, and C can transform cells will require confirmation. DNA-DNA homology measurements have identified at least 4 Ad groups (Section III) but the transforming ability of all Ad groups is not known.

Group A, B, and C Ads have been well characterized. Virus-specific RNA has been isolated from cells transformed by 10 different Ads and was characterized with regard to base composition and homology to different Ad DNAs [83]. In addition, cells transformed by Ad 2, 5, 7, and 12 [66, 88] have been shown to possess virus-specific DNA sequences.

Base composition analysis of the human Ad DNAs revealed a correlation between oncogenicity in newborn hamsters and a low G + C content [212]: 48-49% G + C for highly oncogenic group A; 50-52% for weakly oncogenic group B; and 55-61% for nononcogenic human Ads including transforming group C. But this correlation does not hold for simian Ads, since highly oncogenic SA-7 has 58-60% G + C whereas nononcogenic SV15 has 56% G + C [213]. As concluded by Pina and Green [213], there is no reason to expect

Table 12.1 Properties of Oncogenic and Transforming Human Adenoviruses[a]

Group	Members	Oncogenicity	DNA (%)	DNA[b] (% G + C)	DNA/mRNA[c] (% Homology)
A	Ad 12, 18, 31	Highly oncogenic[d]	11.6-12.5	48-49	30-60
B	Ad 3, 7, 11, 14, 16, 21	Weakly oncogenic[d] (except Ad 11)	12.5-13.7	49-52	40-100
C	Ad 1, 2, 5, 6	Nononcogenic in newborn hamsters but transforms rat embryo cells[e]	12.5-13.7	57-59	90-100

[a] Adapted from Green [83].
[b] Pina and Green [212] and Green (unpublished data).
[c] Hybridization of virus-specific RNA from transformed cells with viral DNA from other group members, given as percent of homologous hybridization; Fujinaga and Green [56-58]; Fujinaga et al. [62].
[d] Highly oncogenic adenoviruses induced tumors in a large proportion of newborn hamsters within 2 months after injection with a purified virus; wekaly oncogenic adenoviruses induced tumors in a small proportion of animals after 4 to 18 months [84].
[e] Freeman et al. [55].

that the overall base composition of the viral genome would reflect the base composition of the viral genes essential for transformation.

III. The Adenovirus DNA Genome

A. Molecular Weight, Conformation, and Intramolecular Structure

The DNAs of representative human Ads of groups A, B, and C (Ad 2, 5, 7, 12, 18, 21, and 31) are uninterrupted, linear, duplex molecules of molecular weights ranging from 20-25 $\times 10^6$, as determined by electron microscopy (see Fig. 12.1) [92] and by sedimentation velocity analysis of native and alkali-denatured DNA [92, 293]. Human Ad DNAs lack both the complementary single-stranded terminal repetition of the type found in bacteriophage λ (cohesive ends), as well as the duplex terminal repetition of the type ABCDE-----ABCDE found in the T-even bacteriophage [92]. The DNA genomes of several well studied bacteriophages consist of populations of circularly permuted molecules which contain the same linear sequence of nucleotides but differ in the sequence at each terminus [282]. The following 5 lines of evidence prove that the Ad genome is not circularly permuted.

1. Circular double-stranded molecules are not obtained after denaturing and reannealing Ad 2 DNA [133].
2. Ad 2 DNA is sheared into half molecules that are separable by equilibrium density centrifugation [39, 133].
3. Partial denaturation maps of Ad 2 DNA (but not Ad 12 DNA) are unique, as revealed by electron microscopy [39].
4. The end fragments isolated from Ad 2 and SA-7 DNA after digestion with *Escherichia coli* exonuclease III represent only a fraction of the DNA sequences of the viral genome [179].
5. Digestion with restriction endonucleases generates a limited number of unique fragments from Ad 2, 3, 5, 7, and 12 DNAs [177, 178].

CELO DNA, of molecular weight 28 $\times 10^6$, has structural properties similar to those of Ad 2 [225, 335, 336]. Therefore, all Ad genomes thus far studied are nonpermuted linear DNA molecules, containing neither cohesive ends nor duplex terminal repetitions, suggesting that their mode of replication differs from that of bacteriophage DNA molecules, and furthermore that a circular DNA structure may not be essential for integration of the viral genome during cell transformation.

Ad DNA molecules have 2 unique structural features: "inverted terminal

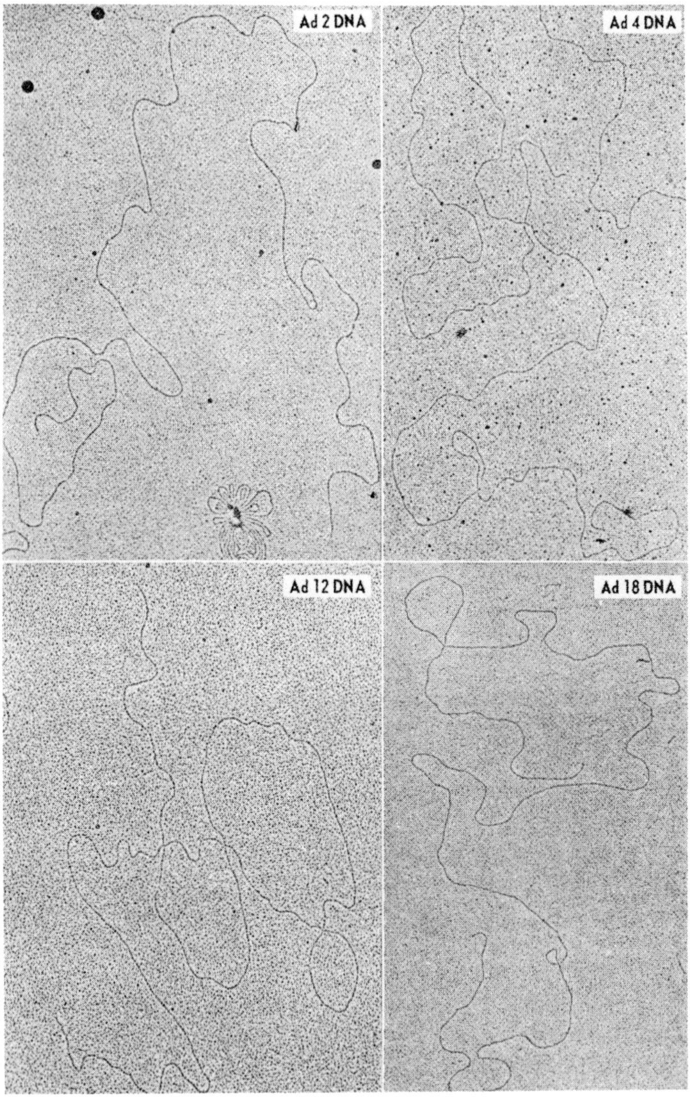
Ad 2 DNA
Ad 4 DNA
Ad 12 DNA
Ad 18 DNA

Fig. 12.1 Electron micrograph of linear molecules of Ad 2, 4, 12, and 18 DNAs. Magnification of photographic plate X7000 (from Green et al. [92]).

repetitions," and palindromic sequences near the termini. Ad 1, 2, 3, 7, 18, and 31 DNA molecules form single-stranded circles when denatured and renatured at low DNA concentrations [67, 329] due to the presence of complementary sequences at the end of each DNA strand (see Fig. 12.2). Inverted terminal repetitions, ABCD · · · · · D′C′B′A′ (A′-D′ are the complementary nucleotides), are present also in the DNA of Ad-associated viruses [14, 137]. The Ad 2 repetition has been estimated at 100-140 nucleotides with the use of specific endonucleases to cleave DNA termini [222], and by electron microscopy [381]. Ad 18 DNA molecules have unusually long terminal repetitions, ranging from 300-6000 nucleotides, that form visible panhandles upon denaturation and reannealing [68]. The 3′ terminal sequences of Ad 2 DNA and Ad 5 DNA are the same, ···pCpC···pGpApTpG$^{3'}$ [261].

The second unusual structural feature of Ad 2 DNA molecules is the presence of palindromes with the structure ab·t·a′b′ (t is the turnaround sequence) as described by Padmanabhan et al. [193]. Either the entire palindrome or 1 arm (e.g., ab) is located approximately 180 nucleotides from each terminus, and was detected by digestion of the 3′ end of Ad 2 DNA by exonuclease III. The exposed single-stranded DNA segment folds back on itself to form a hairpin structure, as indicated by resistance to DNA polymerase I-catalyzed repair at low temperature. Either of 2 structures are compatible with these data [193]; the entire palindrome is located internally (Fig. 12.2, structure I) or a sequence complementary to the terminus is present internally (structure II). The hairpin structure was isolated by digestion with S1 nuclease (a single-strand specific random endonuclease) and was shown by electrophoresis on polyacrylamide gels to comprise about 50 base-pairs. The function of the inverted terminal repetition and palindromes in the Ad 2 DNA molecule is unknown. Perhaps they bind the protein or proteins reported to circularize viral DNA [224, 225] and/or play a role in viral DNA replication or integration during cell transformation.

Green and Pina [89] reported that viral DNA could not be isolated from purified adenovirions by standard phenol extraction procedures without treatment with a proteolytic enzyme, and consequently they concluded that it was necessary to remove a protein component from viral DNA. The standard procedure used subsequently by all workers for isolation of Ad DNA utilized treatment with either papain or pronase. Recently, however, Robinson et al. [225] isolated circularized viral DNA molecules by treatment of virions with 4 M guanidine hydrochloride. Proteolytic enzymes were not used. They concluded that the DNA termini were held together by a tightly bound protein or proteins [224], and proposed that the Ad chromosome may be a DNA-protein complex. Recent studies on this protein are presented in the Recent Developments section at the end of this volume.

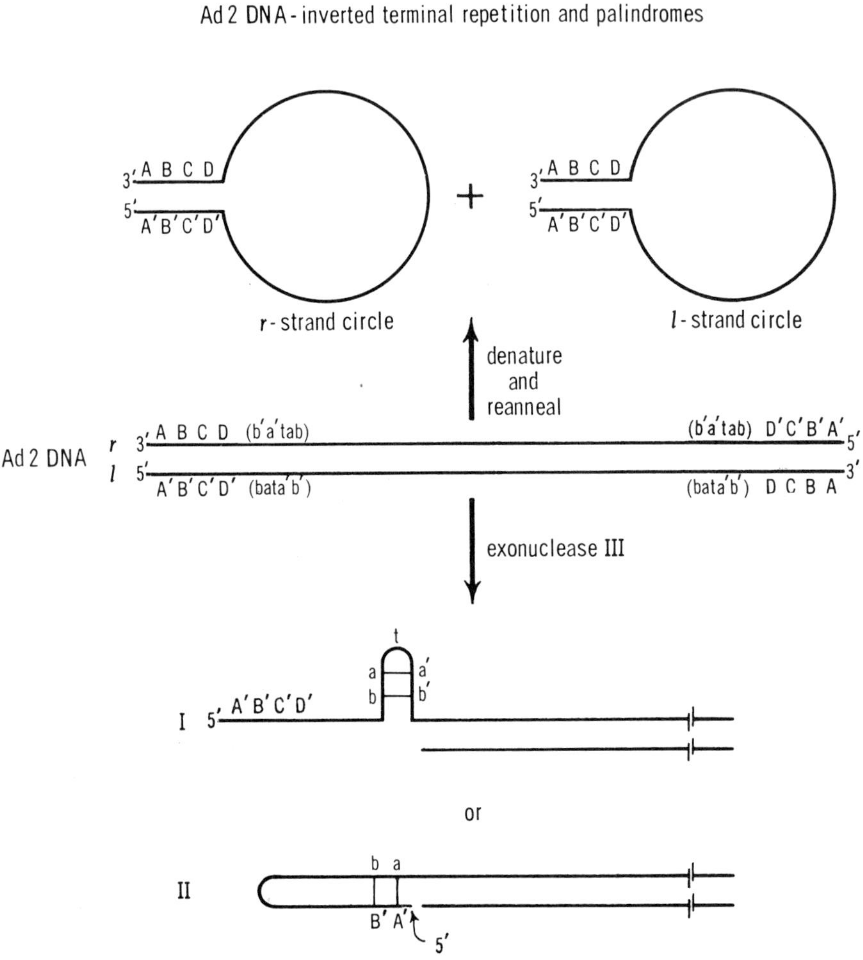
Ad 2 DNA - inverted terminal repetition and palindromes
3' A B C D
5' A' B' C' D'
r - strand circle
+
3' A B C D
5' A' B' C' D'
l - strand circle
denature
and
reanneal
Ad 2 DNA
r 3' A B C D (b'a'tab)
l 5' A' B' C' D' (bata'b')
(b'a'tab) D' C' B' A' 5'
3'
(bata'b') D C B A
exonuclease III
t
a a'
b b'
I 5' A' B' C' D'
or
b a
II
B' A'
5'

Fig. 12.2 Some intramolecular features of Ad 2 DNA-inverted terminal repetition and palindrome. Denaturation of Ad 2 DNA results in the formation of single-stranded circles from *r*- and *l*-strands due to the inverted terminal repetition, represented by ABCD ··· D′C′B′A′ (A′-D′ are the complementary nucleotides). The palindrome abta′b′ (a′b′ are the complementary sequences and t is the turnabout region) is located about 180 nucleotides from each terminus. DNA sequences in the region a b may be the same as in the region A′B′ (to form structure II) or may be different (structure I).

B. DNA-DNA Homology Relationships among Human Adenoviruses

The relationships among various Ad has been investigated by DNA-DNA hybridization, as summarized in Table 12.2. Viruses within each group (A-D) are closely related and share 50-100% of their base sequences, while those of different groups share only 5-28% of their sequences ([62, 138-140, 213] and Green, Mackey, and Wold, unpublished data). Viruses within group B,

Table 12.2 DNA-DNA Homology Relationships Among Adenovirus DNAs

	Percent homology	
DNA Pair hybridized	Immobilized[a] viral DNA	DNA in[b] solution
Group A X group A (Ad 12, 18, 31)	80-85	55-68
Group B X group B (Ad 3, 7, 11, 14, 16, 21)	70-100	80-95
Group C X group C (Ad 1, 2, 5, 6)	85-95	
Group D X group D (Ad 8-10, 13, 15, 19, 20, 22-24, 26-30). The group status of Ad 4, 17, and 25 is unknown at present.		92
Group A-D interrelationships	10-25	5-28
Ad 2 Hind III-G transforming fragment		
X group C		98-100
X 27 nongroup C, SA7, SV40		0
Ad 12 (Huie) Eco RI-C transforming fragment		
X Ad 18, 31, and 5 Ad 12 isolates		53-68
X 28 nongroup A serotypes		0
Simian SA7 and SV15 X Ad 12	9-12	
Simian SA7 and SV15 X Ad 7	16-19	
Simian SA7 and SV15 X Ad 2	19-23	
Bovine adenovirus 3 X Ad 5		25[c]

[a]Adapted from [83]; data from hybridization using DNA in agar, from [138-140]; data from hybridization with DNA on filters, from [62, 213].
[b]Green, Mackey, and Wold, unpublished data.
[c]Niiyama et al. [365].

C, and D are closely related, some indistinguishable by hybridization, while those in highly oncogenic group A show considerable, interesting heterogeneity in sequence. The Eco RI-C fragment from the left-hand 16% (G + C rich end) of Ad 12 (Huie) DNA, which was shown by transfection experiments to contain "the transforming genes" [78], hybridized only 53-68% with the DNAs of 6 different Ad 12 isolates, Ad 18, and Ad 31 (Table 12.2). Since the Ad 12 Eco RI-C fragment harbors group A Ad-transforming genes, this result raises the possibility that some variability may exist in the region of the viral genome harboring the transforming genes of group A viruses. Large differences were also detected in the virus-specific RNA sequences expressed in cells transformed by group A viruses [57], although this result may not deal directly with the question of heterogeneity of group A Ad-transforming genes, because Ad-transformed cells may contain and express Ad genes not involved in cell transformation. In contrast to group A viruses, the transforming genes in group C viruses are identical, since radioactive transforming Ad 2 Hind III-G fragment, which contains information for both initiation and maintenance of transformation, hybridizes 98-100% with the DNA of other group C viruses (Table 12.2).

An interesting finding is that neither the transforming Ad 12 Eco RI-C fragment, nor the transforming Ad 2 Hind III-G fragment, hybridizes at all to DNA from Ad serotypes outside the oncogenic groups A and C, respectively. Therefore, it is clear that the proteins specified by transforming genes of viruses from groups A, B, and C differ. Whether these differences account for the oncogenic differences between the virus groups remains to be established.

C. Specific Fragments of the Adenovirus Genome and Localization of Transforming Genes

The first attempt to subdivide physically the Ad genome was based on the observation by Green and Pina [90] of the broad thermal melting curves of Ad 2, 4, 12, and 18 DNA molecules, which they interpreted as indicating intramolecular heterogeneity of base composition. To prove this point, Kimes and Green [133] fractionated Ad 2 DNA into half molecules by controlled shearing and equilibrium centrifugation in CsC1 gradients; the half molecules differed in G + C content by 7%. Similarly, Doerfler and Kleinschmidt [39] separated Ad 2 DNA molecules in CsC1 gradients as Hg complexes.

Two further technological advances have allowed the detailed structural and functional analysis of the Ad genome. The first was the separation and purification by Landgraf-Leurs and Green [142] of the complementary strands of Ad 2 DNA. Figure 12.3 shows the preparative separation of the *r* (light)

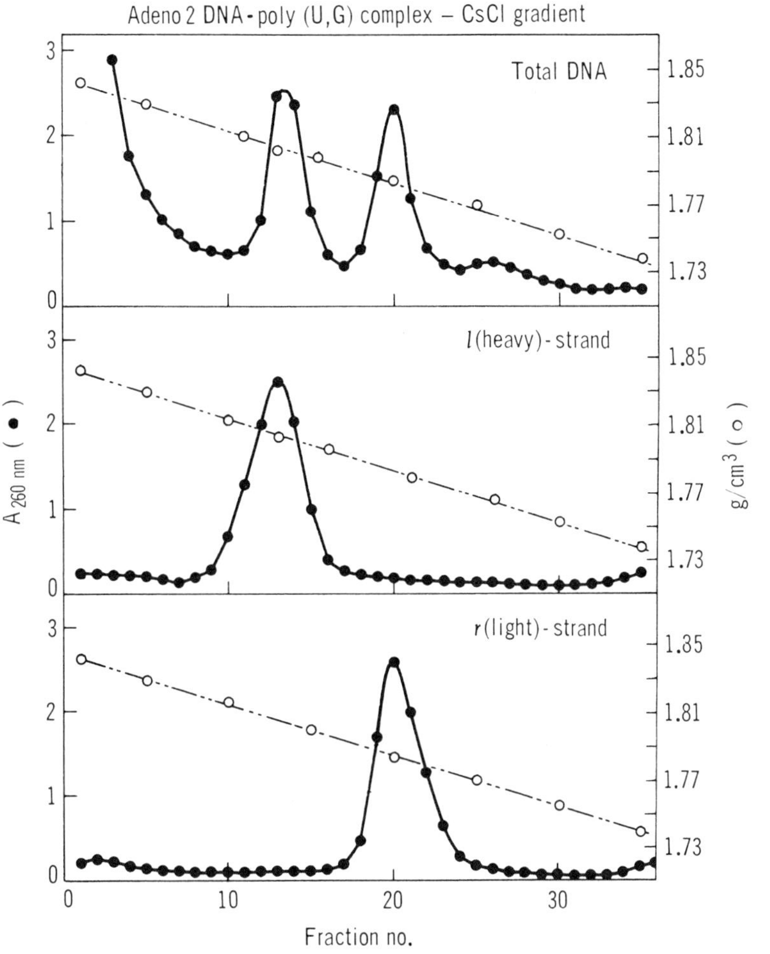
Adeno 2 DNA-poly (U,G) complex – CsCl gradient
Total DNA
l(heavy)-strand
r(light)-strand
A 260 nm (●)
g/cm³ (○)
Fraction no.
0
10
20
30
1.85
1.81
1.77
1.73
3
2
1
0

Fig. 12.3 Separation of unlabeled Ad 2 heavy (H) and light (L) DNA strands. Unlabeled Ad 2 DNA was denatured, annealed to a poly(U:G) ribocopolymer, and centrifuged to equilibrium in CsC1 (Total DNA). Fractions with H- and L-strands were pooled and further purified by rebanding in CsC1. The light and heavy strands, previously designated L and H, are now designated *r* and *l*, to indicate rightward and leftward transcription [337].

and *l* (heavy) strands of Ad 2 DNA as complexes with poly(UG) by CsC1 equilibrium density gradient centrifugation. The second advance, pioneered by scientists at the Cold Spring Harbor Laboratories, was the cleavage of the duplex Ad DNA molecule into specific fragments by several bacterial restriction enzymes, and the linear ordering of each restriction fragment along the genome [177, 203, 244].

The endonuclease Eco R1 cleavage map (6 fragments) and the endonuclease Hind III cleavage map (12 fragments) of the Ad 2 genome, and the Eco R1 map (6 fragments) of the Ad 12 genome are shown in Fig. 12.4. The G + C rich region is positioned on the left side of the map. The left end fragment of Ad 2 and Ad 5 produced by endonuclease Hind III, the G-fragment of molecular weight 1.6×10^6, has been shown to transform rat embryo cells in vitro [78]. The Ad 12 DNA Eco R1-C fragment (3.4×10^6 daltons) also has transforming activity [78; F. Graham, S. Mak, and A. van der Eb, personal communication]. Strands of each Eco R1 fragment of Ad 2 DNA were separated from each other by electrophoresis on agarose gels after denaturation, as shown in Fig. 12.5 [53]. The availability of separated strands, fragments, and fragment strands representing defined portions of the viral genome have allowed a number of studies, including the following.

1. Estimates of the fraction of heavy (*l*) and light (*r*) DNA strands transcribed and expressed as mRNA early and late during productive infection and in transformed cells, and quantitative estimates of different abundance classes of viral RNA species [88, 143, 283, 284, 326, 327].

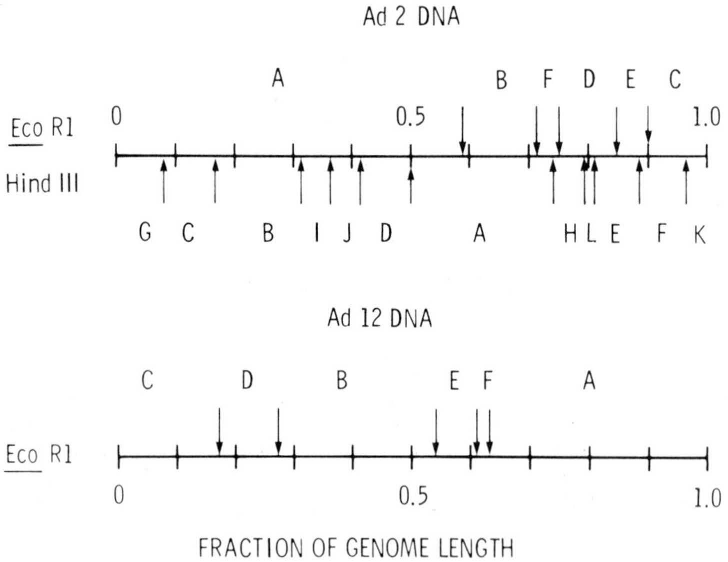
Ad 2 DNA
A
B F D E C
Eco R1
0
0.5
1.0
Hind III
G C B I J D A H L E F K
Ad 12 DNA
C D B E F A
Eco R1
0
0.5
1.0
FRACTION OF GENOME LENGTH

Fig. 12.4 Enconuclease Eco R1 and Hind III cleavage maps of Ad 2 DNA ([177]; R. J. Roberts, J. Sambrook, and P. Sharp, personal communication) and endonuclease Eco R1 cleavage map of Ad 12 DNA (Delius and Mulder, personal communication). Arrows refer to points of cleavage.

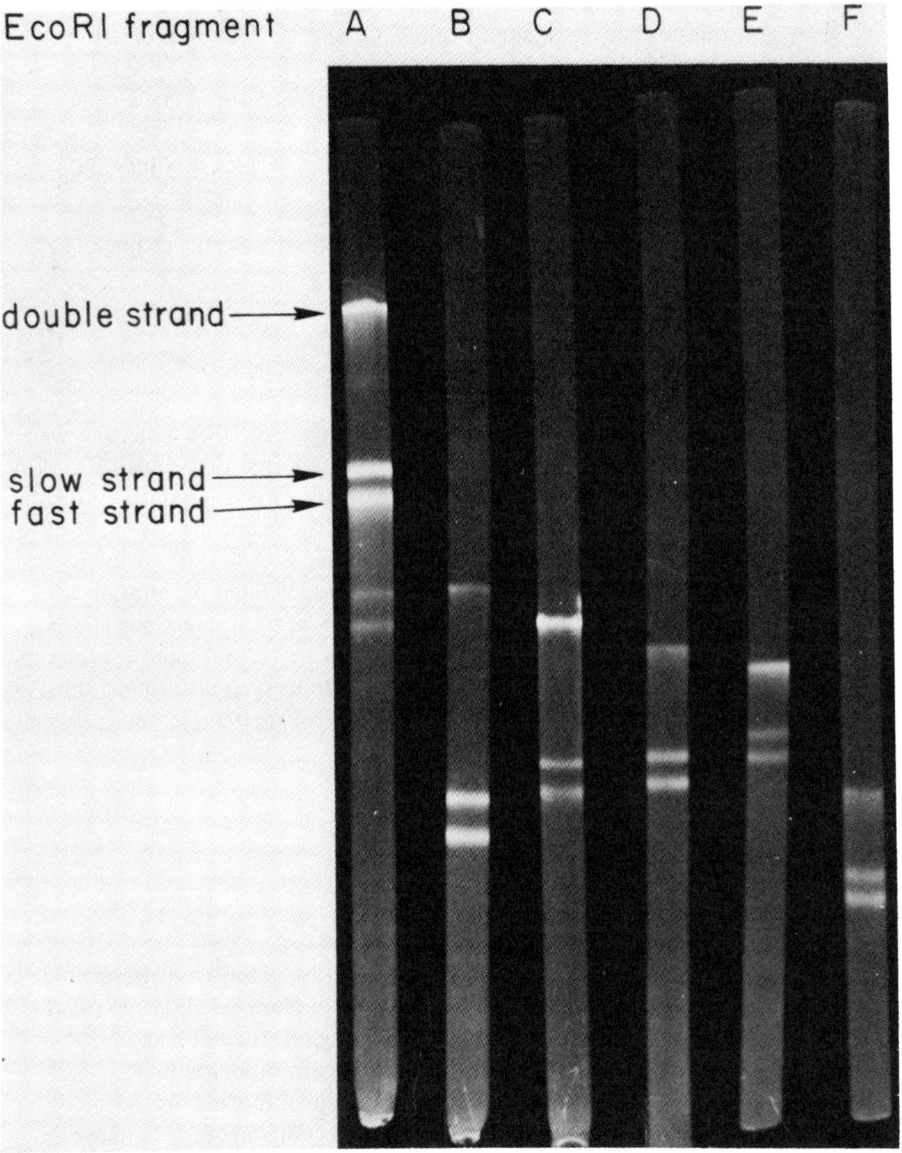
EcoRI fragment
A
B
C
D
E
F
double strand
slow strand
fast strand

Fig. 12.5 Separation of the Eco R1 fragments strands of Ad 2 DNA by electrophoresis in agarose gels [53].

2. The identification of the portions of the viral genome incorporated into transformed cells, thus distinguishing between essential and non-essential viral information for cell transformation [66, 236, 246].
3. The determination of viral DNA segments transcribed into mRNA in transformed cells, and in cells early and late during productive infection [29, 53, 245, 341].
4. The location of origin and termination sites for viral DNA replication [240, 286, 356, 376, 379].
5. The identification of large Ad 2 RNA molecules that may be further processed [8].

The results of these exciting studies are presented later in this chapter.

D. Biological Activity of Viral DNA and Viral DNA Fragments

Early attempts to demonstrate infectivity of human Ad DNA molecules were unsuccessful. Burnett and Harrington [19, 20] reported that simian SA-7 DNA was infectious and tumorigenic. Ad 1 DNA was shown to produce cytopathic effect (CPE) on infected cells [183]. However, the major improvement in the technology was developed by Graham and van der Eb [79, 80], who coprecipitated viral DNA with calcium phosphate, which facilitated penetration of DNA into the cell, and showed that Ad 5 DNA was both infectious and possessed transforming activity. However, the efficiency is still quite low and requires 1 μg of DNA to produce from 1-10 infectious or transforming units. DNA protein complexes are 100-fold more efficient in transfection.

IV. Adenovirus: Structure and Protein Constituents

A. Introduction

The architecture of the adenovirion (reviewed in [208, 209]) has attracted much attention since the classical electron micrographs of Horne et al. [109] (see Fig. 12.6). Ads are particularly suitable for morphological studies because of their large size (diameter 65-80 nm), paracrystalline accumulations within the cell nucleus, ease of purification in abundant quantity, and disruption by a variety of conditions into characteristic subviral components. Moreover, most virion proteins are soluble under nondenaturing conditions and therefore are suitable for further biochemical studies.

B. Morphology

When adenovirions are examined by electron microscopy using the negative staining technique, the individual morphological subunits, the capsomers which comprise the viral capsid shell, are revealed (see Fig. 12.6). The 252 capsomers are arranged in characteristic cubic symmetry, as an icosahedron with 20 triangular facets and 12 vertices. Thin section electron microscopy disclosed a nucleoid or core structure containing tightly packed viral DNA which is associated with 2 polypeptides, the core proteins. The 240 capsomeres that face 6 neighbors are called hexons whereas the 12 capsomeres at the vertices that face 5 neighbors are termed pentons [73]. Each penton consists of a penton base with a projection called a "fiber", which is of variable length depending upon the serotype, and terminates in a knoblike structure [185, 291]. The 60 hexons encircling the 12 pentons are called peripentonal hexons and differ topologically from the 180 hexons that form the triangular facets of the icosahedron. Several types of treatment can remove the pentons in association with the peripentonal hexons. The resulting "cornerless structure" [231] may be dissociated into 20 interlocking groups of 9 hexons [256], called "ninemers" [198], corresponding to the 20 facets of the icosahedron.

C. Polypeptide Components

The polypeptide composition of whole virions and subviral components has been studied primarily by SDS polyacrylamide gel electrophoresis. Maizel et al. [165, 166] described 9 different polypeptides for Ad 2. More recent studies revealed up to 15 structural polypeptides [5, 49] (Fig. 12.7). Polypeptides II, III, and IV are the major polypeptides of the capsid. Polypeptide II is the subunit of the hexon with a molecular weight of 120,000; 3 polypeptide chains constitute 1 hexon protein. Polypeptide III (85,000) is the subunit of the penton base; probably 5 subunits assemble into 1 penton base. Polypeptide IV (62,000) is the subunit of the fiber; 3 subunits form the fiber protein. Polypeptides V (48,500) and VII (18,500) are found in association with the core structure.

Several of the remaining polypeptides have been assigned tentative positions in the virion by a variety of techniques. See Fig. 12.8 for the current model described by Everitt et al. [49]. Polypeptide IX (12,000) may serve to bind hexons together into ninemers, and appears to be partially exposed at the virion surface [50, 165, 166]. Polypeptide VIII (13,000) is found in association with individual hexons and appears to be located on the inner surface of the capsid. Polypeptide VI (24,000) is also localized internally, firmly

A

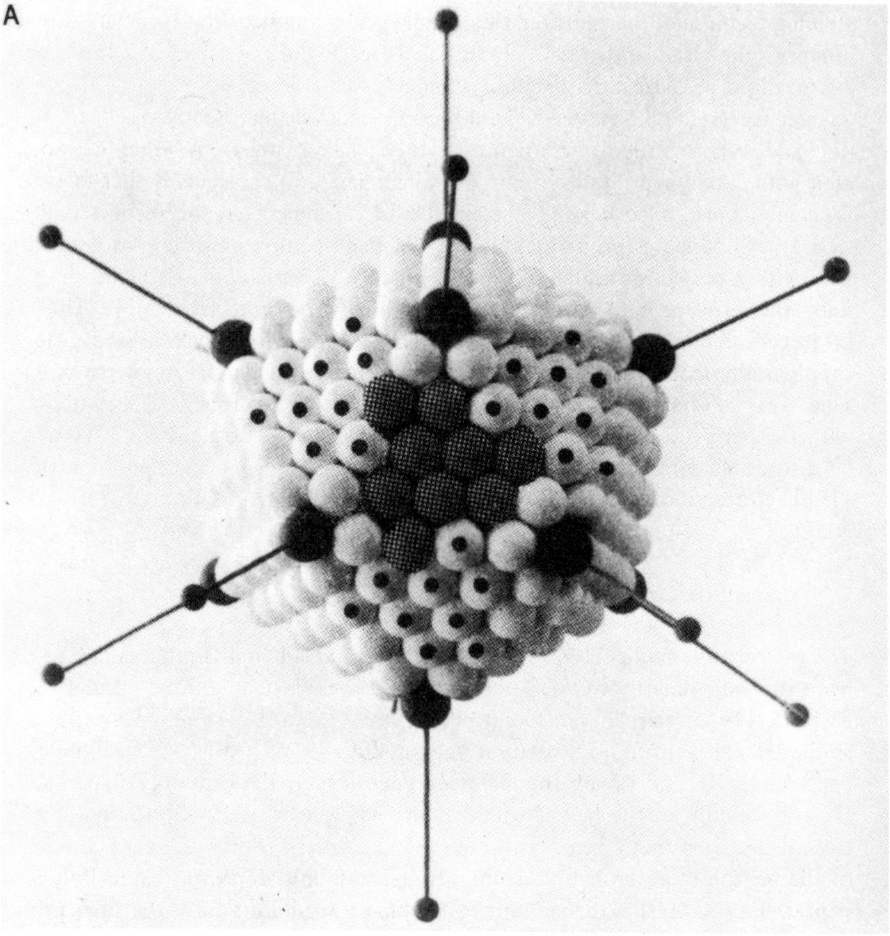

Fig. 12.6 Structure of the Ad capsid. (A) Model of Ad, taken from Pereira and Wrigley [198]. The pentons (penton base plus projecting fiber) are indicated in black. One group of 9 hexons (hexon ninemer) is shaded and its 3 neighboring ninemers are pointed out by dots. (B) Electron micrograph of an Ad 5 virion, negatively stained by sodium silicotungstate [291]). Note the antenna-like fibers. (C) Electron micrograph of Ad 5, negatively stained, showing the icosahedral array of capsomers [291].

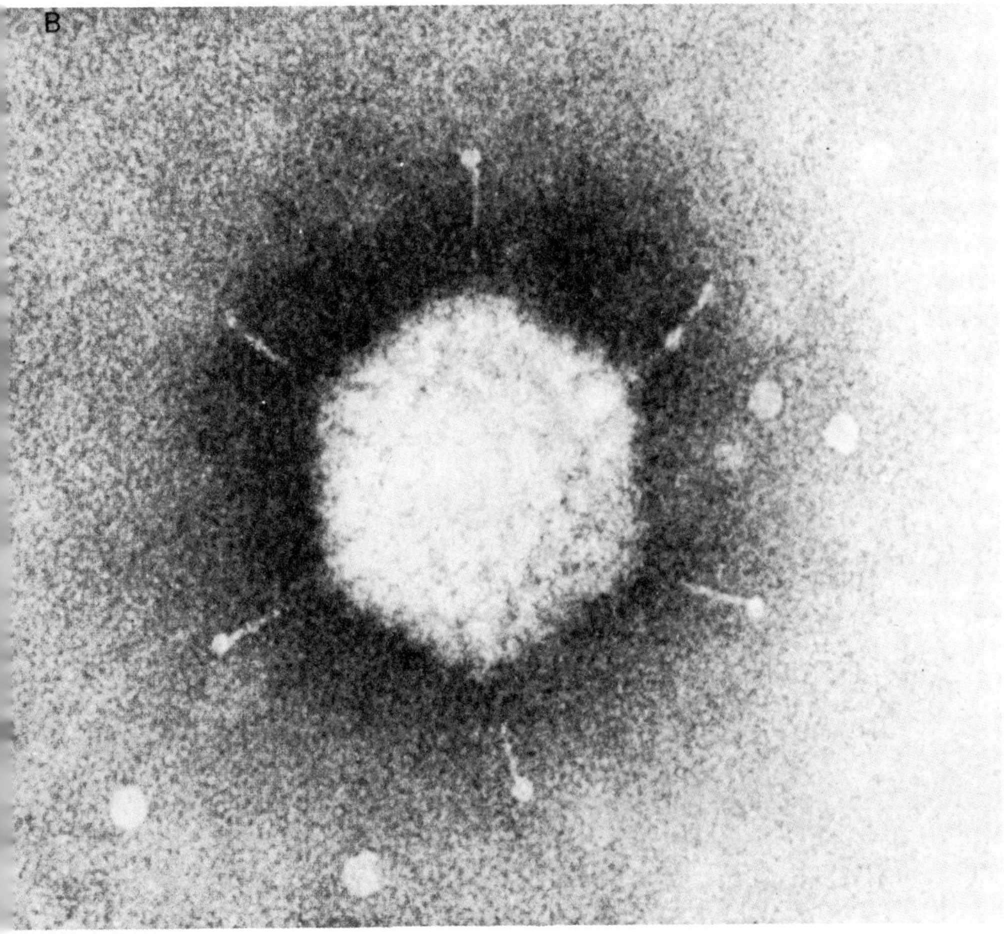
B

Fig. 12.6

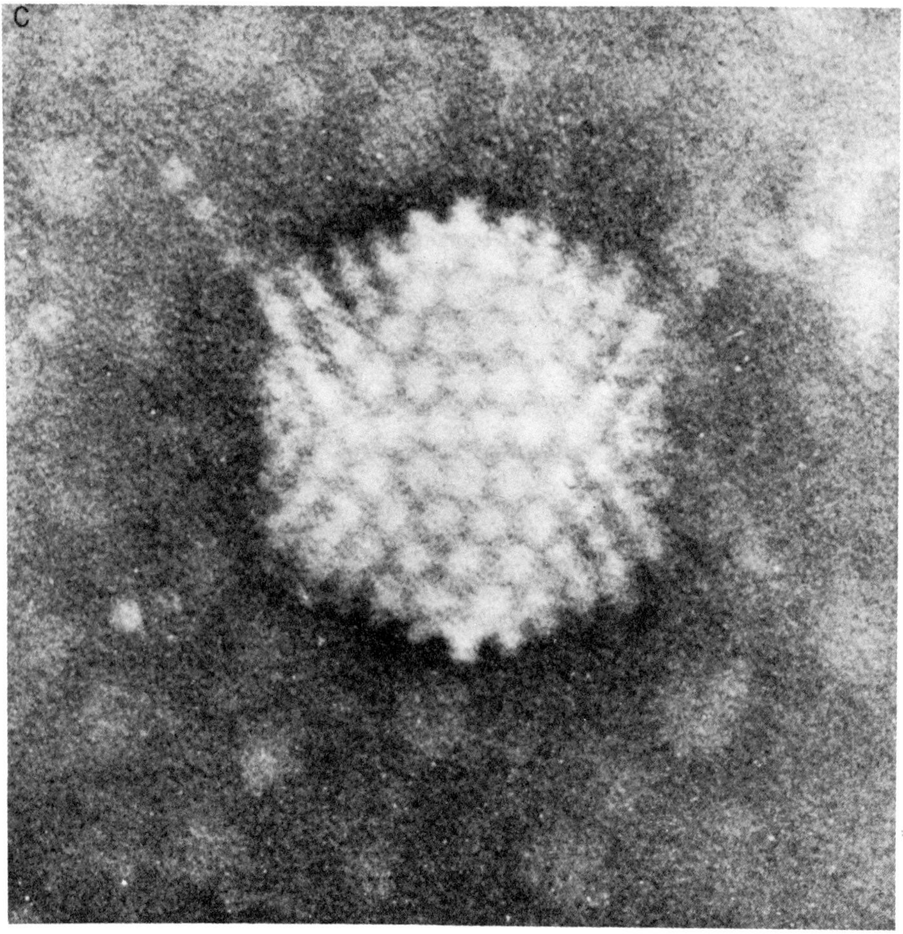
C

Fig. 12.6

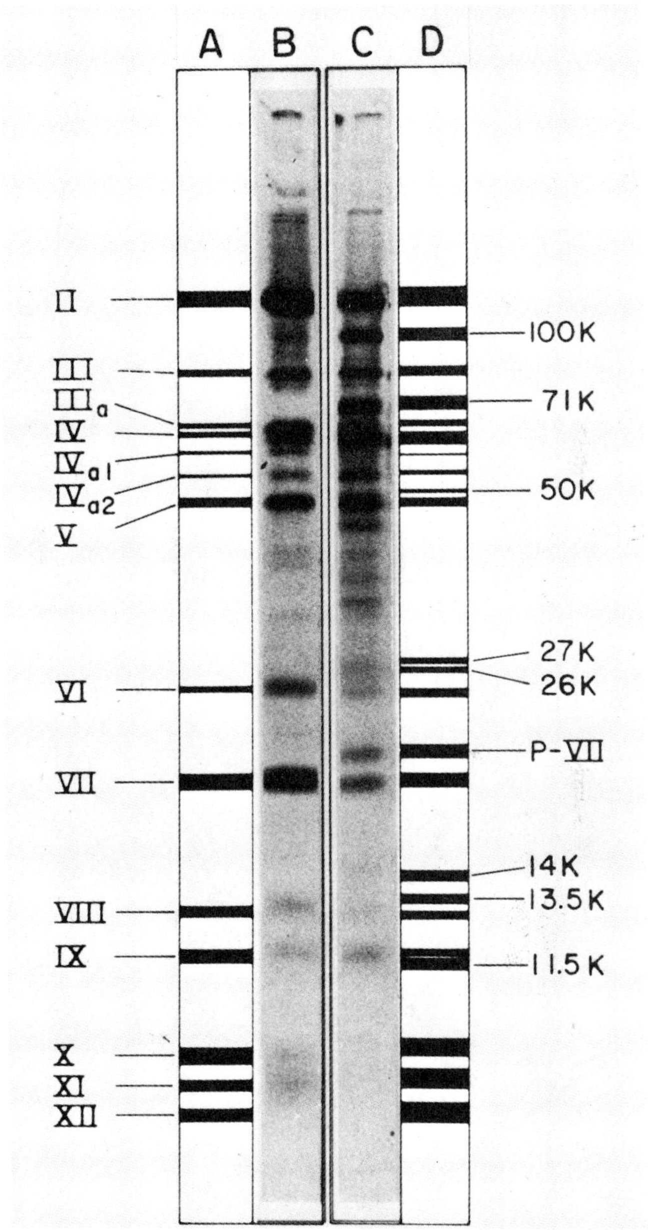
A
B
C
D
II
III
IIIa
IV
IVa1
IVa2
V
VI
VII
VIII
IX
X
XI
XII
100K
71K
50K
27K
26K
P-VII
14K
13.5K
11.5K

Fig. 12.7 Polypeptide composition of Ad 2 virions and infected cells. SDS polyacrylamide gel autoradiograms of [^{35}S] methionine-labeled purified virus (B), and an infected cell extract (C). Parts (A) and (D) are idealized drawings of virion proteins and polypeptides, respectively, found in the infected cell. Reprinted from Anderson et al. [5].

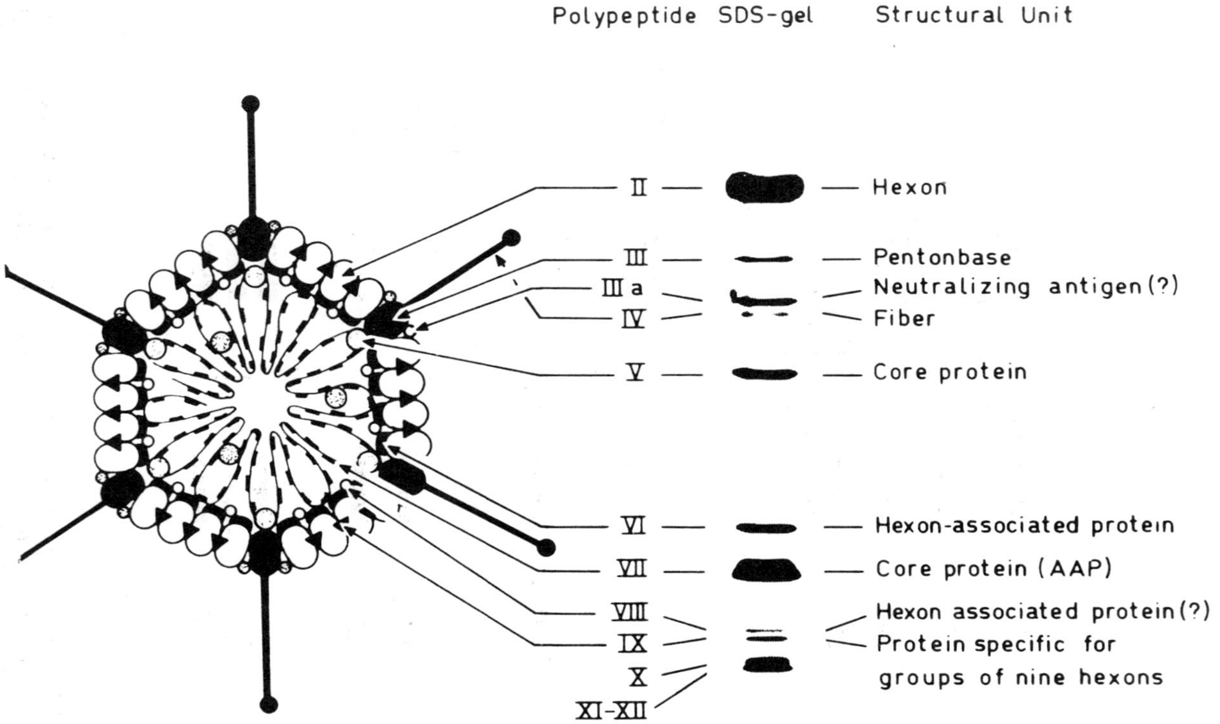
Polypeptide SDS-gel Structural Unit
II — Hexon
III — Pentonbase
IIIa — Neutralizing antigen(?)
IV — Fiber
V — Core protein
VI — Hexon-associated protein
VII — Core protein (AAP)
VIII — Hexon associated protein(?)
IX — Protein specific for groups of nine hexons
X
XI-XII

Fig. 12.8 Model of the tentative location of structural polypeptides in the Ad 2 virion. Taken from Everitt et al. [49].

associated with hexons, and is isolated with ninemers. Polypeptide IIIa (66,000) is localized externally in a position close to the vertex regions of the virion, and may extend into the interior of the virion, since it can be chemically cross-linked to polypeptide VII, which is internal. IIIa may be a phosphoprotein [230].

The localization of the smallest polypeptides X, XI, and XII is not yet established [17, 49]. In addition to the structural polypeptides II to XII, a number of constituents representing a minor portion of the viral protein mass are in purified virions [5, 50, 120]. At least 3 polypeptides may be precursors to mature viral proteins. For example, VI, VII, and very likely VIII, are derived from larger precursors, P-VI, P-VII, and P-VIII, respectively. Since cleavage of the precursors may be one of the last steps in virion assembly, the precursors can be incorporated into virions.

D. The Hexon

Nermut [182] has recently provided refined measurements of several structural features of Ad hexons by electron microscopic freeze-drying and freeze-etching techniques. The edge of the Ad 5 icosahedron is 43 nm while the calculated diameter of the virions in the 5-fold symmetry orientation is 73 nm. The hexon was visualized as a "conical triangular prism" with a base 8.6 × 9.8 nm, ending in a narrow "top" about 7.4 nm in diameter. The average height of the hexon is 11 nm. The hexon is composed of 3 subunits which are in close contact at the bottom but are comparatively free and mobile at the top. The ninemers released from disrupted virions display a left-handed orientation on the virus capsid [198]. The mutual orientation of hexon within the ninemers is always from corner to edge, but since the triangular facets of the icosahedron are inclined at an angle of 60°, the peripheral hexons of 2 adjacent ninemers in the capsid are not bonded from corner to edge but from corner to corner. Thus the binding of hexons within a ninemer is different from that between 2 ninemers. The corner bonding between 2 ninemers is weaker, apparently explaining why capsids break into the characteristic ninemers and not at random. Pereira and Wriggley [198] have reconstituted empty Ad capsid shells from isolated ninemers in vitro.

Gruetter and Franklin [99] reexamined the molecular weight of hexon and subunits by sedimentation equilibrium, light scattering, and diffusion measurements. They concluded that the hexon has a molecular weight from 355,000-363,000 and consists of 3 identical subunits of 120,000.

Analysis of recently isolated Ad 5 temperature-sensitive (*ts*) mutants suggested that several gene products may be required for the assembly and transport of hexon capsomers (see Section V). Leibowitz and Horwitz [148] have examined 2 Ad 5 *ts* mutants from different complementation groups that are defective in hexon assembly. Tryptic and chymotryptic peptide maps did not reveal any differences between the hexon polypeptides derived from wild-type and *ts* virions. This indicates that viral genes other than those coding for the primary structure of the hexon polypeptide are required for assembly.

E. The Fiber

The Ad fiber was initially described as a shaft (2 × 16 nm) terminated on 1 end by a sphere (4 nm in diameter) [291, 317]. Depending upon the serotype, the length of the fiber varies from 9-31 nm [186]. The molecular weight of purified intact Ad 2 and Ad 5 fiber has been estimated to be 200,000 [264] and 183,000 [40], and that of fiber subunit to be 60,000-65,000 [264]. Peptide maps and electrophoretic separation of polypeptide chains indicated that each fiber molecule consists of 3 polypeptides of molecular weight about 61,000, 2 identical and 1 unique [40]. A model is suggested in which 2 identical chains form the shaft and 1 the knob. Fiber contains a sugar residue, probably N-acetyl-glucosamine [357].

F. The Penton Base

The penton base is much more difficult to purify than hexon and fiber and therefore has not been as thoroughly studied. Valentine and Pereira [291] described the penton base as a sphere of average diameter 8 nm. Pettersson and Hoglund [202] observed a globular structure which sometimes exhibited a pentagonal outline after negative staining. The molecular weight of the Ad 2 penton base was estimated at 400,000 [202]. The polypeptide subunits of the penton base have been reported as 70,000 [166] and 85,000 [5] by sodium dodecylsulfate (SDS) gel electrophoresis, suggesting that the penton base consists of about 5 subunits.

G. The Core

Ad cores have been released by treatment of virions with heat [231], formamide [260], pyridine [215], and sarcosyl [17]. Cores have been visualized by electron microscopy of thin sections of virions and more satisfactorily by freeze-cleaving intranuclear virus [17]. The core structures prepared by various procedures differ. "Sarkosyl cores" are tightly packed and consist of discrete clusters of approximately ten (21.6 nm) spheres. The predominant polypeptide associated with the sarkosyl cores is the arginine-alanine rich major core polypeptide VII. The cores exposed by freeze-cleavage also consist of several closely packed spheres. Visualization of the substructure of "pyridine cores" is hampered by a halo of associated material. Pyridine cores contain polypeptides VII and V, and it is assumed that polypeptide V coats the viral DNA-protein VII complex and thus masks the internal core structure. Two functions have been suggested for the core proteins: packaging DNA (polypeptide VII), and binding the DNA-polypeptide VII complex of the nucleoid to the penton [49] (see Fig. 12.8). Brown et al. [17], have suggested that the viral DNA is folded into 12 spheres with icosahedral symmetry; protein V could link the 12 spheres to the 12 pentons in the capsid. In addition, an interaction between the hexon capsomer and the core structure could be mediated by internally localized proteins VI and VIII.

V. Adenovirus Genetics

It is clear from studies on microbial systems that a complete understanding of Ad replication will depend upon combined biochemical and genetic analyses, for it is an overwhelming task for biochemical studies alone to characterize a series of undiscovered proteins, coded by both the virus and the cell, that carry out unknown functions essential for virus replication within the uncharted enzymatic milieu of the host cell. However, viral mutants unable to replicate, and host cell mutants unable to support viral replication, provide a relatively easy means to establish both viral and host cell functions involved in virus replication. Comparative studies on wild-type and mutant viruses and host cells are extremely instructive in understanding the biochemistry of the mutated gene function. Not much is known about Ad genetics [reviewed in refs. 209 and 352]. Because of their importance, in this section we will examine in detail some of the properties of the known Ad mutants.

A. *Cyt* Mutants

Takemori et al. [274] isolated the first Ad mutants, cytocidal (*cyt*) mutants of Ad 12. These mutants arose spontaneously from Ad 12 stocks at a frequency of 0.001%; UV-irradiation increased the frequency 5-6-fold. The *cyt* mutants cause increased cellular destruction, i.e., large clear plaques, instead of the small, fuzzy plaques induced by parental Ad 12. The *cyt* mutants are less tumorigenic than cyt^+ (i.e., wild-type phenotype) in newborn hamsters, and fail to transform newborn hamster kidney cells in vitro. The cyt^+ phenotype is dominant over *cyt* in terms of the CPE produced. Some mutants (*cyt kb*) fail to grow in a line of KB cells (KB-1), but grow in another line (KB-2);

other mutants (*cyt* kb^+) grow in both KB-1 and KB-2 lines [275]. The kb^+ phenotype is dominant over *kb*. Although *cyt kb* do not grow or synthesize hexon antigen in KB-1 cells, they produce CPE and T-antigen. Different *cyt* isolates (both *kb* and kb^+) fail to complement in plaque morphology, CPE, or tumorigenicity, and different *cyt kb* mutants fail to complement growth in KB-1 cells.

Recombinants were isolated from crosses of different *cyt kb* isolates by exploiting the nonpermissiveness of KB-1 cells [273]. All recombinants have cyt^+ kb^+ properties, and resemble the parental virus in plaque morphology, CPE, tumorigenicity, and growth on KB-1 cells. Because phenotypic characters (plaque morphology, CPE, tumorigenicity, growth on KB-1 cells) are linked in recombinants, and because the different *cyt* mutants do not complement, it was suggested [273, 274] that all *cyt* mutations reside in a single gene. Recombination was considered to be intragenic.

Mutants with *cyt* phenotype have also been isolated by Yamamoto et al. [330]. These mutants, named *lt*, form large clear plaques and are less tumorigenic than the parental strain. It is not known whether *cyt* and *lt* mutants are identical.

Unfortunately, the physical and biochemical properties of mutants with *cyt* phenotype have not been investigated extensively. The *cyt* mutants are similar to wild-type in density and DNA size, but wild-type stocks contain more defective virions (capable of cell killing, but unable to induce T- or V-antigen) [51]. The *lt* mutants [330] resemble wild-type in virion structure (antigenicity, polypeptides, and heat stability), viral replication, T-antigen induction, induction of cellular DNA synthesis in nongrowing hamster cells, and integration of viral DNA into cellular DNA. However, unlike wild-type, *lt* mutants do not induce cell surface alterations in HEK cells, and are relatively defective in inducing transplantation immunity in hamsters. Based on these observations, Yamamoto et al. [330] pointed out that cell surface changes may be related to cell transformation by Ad 12. In view of the high oncogenicity of Ad 12, an interesting possibility is that all *cyt* mutations are in a single gene, and that 1 gene may harbor information for both cell lysis and tumorigenicity [273]. Clearly, further study is warranted on the role of the *cyt* gene product in Ad 12 oncogenicity.

B. Ad 5 *ts* and *hr* Mutants

Conditional-lethal temperature-sensitive (*ts*) mutants of several Ad serotypes have recently become available. These *ts* mutants replicate at low permissive temperatures (PT, 32-33°C), but not at high nonpermissive temperatures (NPT, 38-41°C). It is usually held that the *ts* phenotype results from a single amino acid substitution in a protein, causing a thermolabile protein that can only function at low temperatures. Such *ts* mutants are valuable in biochemical studies because the function of the thermolabile protein can be established by examining the effect of shifting the mutant cultures from PT to NPT and vice versa.

Mutagenic agents used to generate these *ts* mutants were UV-irradiation, nitrosoguanidine, nitrous acid, hydroxylamine, and 5-bromodeoxyuridine. Most mutants are genetically stable, not unreasonably "leaky", have low reversion rates, and have proven suitable for genetic and biochemical analysis. The nomenclature proposed for Ad mutants by Ginsberg et al. [77] will be used in this discussion.

The most extensive genetics has been done with Ad 5 *ts* mutants [46, 272, 320]. Table 12.3 and Table 12.4 summarize the properties of mutants in the known complementation groups. Workers in Williams' laboratory have isolated some 80 *ts* mutants, and have classified 51 into 16 complementation groups, A to P [323]. Ginsberg's group [75] classified 15 *ts* mutants into 6 complementation groups, I to VI. Complementation groups N (*ts* 36 and 37) and VI (*ts*149) are the same, while Ginsberg's complementation group I (*ts*125) is absent from Williams' mutants (Williams' group Q). Thus a total of 17 complementation groups of *ts* mutants have been discovered. Groups N and Q are early DNA-negative mutants, while the remaining 15 groups are apparently late mutants with defects in proteins synthesized at late stages of infection, following the initiation of viral DNA replication.

Harrison et al. [101] isolated 2 groups (represented by H5*hr*1 and H5*hr*6) of Ad 5 host range (*hr*) mutants. H5*hr*1 grows only on Ad 5-transformed HEK cells; H5*hr*6 grows on these cells as well as untransformed HEK cells. H5*hr*1 and *hr*6 complement each other and all Ad 5 *ts* mutants, so at least 19 Ad 5 complementation groups exist.

A linear genetic recombination map has been constructed [334, 352], based on recombination frequencies from 2-factor crosses (see Fig. 12.9D). Preliminary recombination experiments have placed H5*hr*1 and H5*hr*6 towards the left end of the map [101]. A heat-stable variant of Ad 5 has been selected; 3-factor crosses, using this mutant in conjunction with *ts* mutants, suggest that the heat-stable mutation is in a single gene located near the left end of the genetic map [334]. Recombination studies have also been done between Ad 5 and $Ad_2{}^+ND_1$ *ts* mutants (see below), and based on these some of the *ts* lesions have been located on the physical map of the viral genome (Fig. 12.9C), thus correlating the genetic and physical maps [238]. Takahashi [272] has also isolated Ad 5 conditional-lethal *hr* mutants that grow in HEK cells but not in hamster kidney cells, but further studies with these mutants have not been reported.

Table 12.3 Characteristics of Williams' Ad 5 *ts* Mutants

Serological class	Complementation group	Mutant designation	Complementation by Ad 12	Characteristics at nonpermissive temperature[a]
I.				
All capsid antigens made	II(2)[b]	*ts*18,12	+	*ts*18 is heat labile
	I(1)	*ts*19		*ts*18 and 19 do not induce interferon in chick embryo cells
	L(1)	*ts*24	+	
	M(3)	*ts*31	–	
II.				
No capsid antigens made,	N(3)[c]	*ts*36,37	–	Defective in initiation but not maintenance of cell transformation, possibly defective in initiation of viral DNA synthesis, no capsid proteins, not induce intranuclear crystals, potentiates growth of AAV.
	Q[d]	*ts*125	–	See Table 12.4
III.				
Fiber antigen negative	G(2)	*ts*13,9	–	Not induce intranuclear crystals
	J(1)	*ts*5	–	Defective in fiber structural gene
	K(3)	*ts*22	–	

IV.				
Hexon antigen negative	D(1)	*ts*17	+	*ts*17 induces intranuclear crystals, both *ts*17 and *ts*20 are defective in assembly of hexon polypeptides into hexon capsomeres
	E(1)	*ts*20		
V.				
Abnormal hexon transport (hexon antigen detected in cytoplasm by immunofluorescence)	A(20)	*ts*1,7,16,30	+	*ts*1 has same heat stability as wild-type
	B(1)	*ts*3	+	
	C(1)	*ts*4	+	
	F(7)	*ts*2,10,14	+	Defective in hexon structural gene
Unclassified	O(3)			
	P(1)			Not determined

[a]At NPT all mutants block host DNA synthesis, and all late mutants, except possibly *ts*22, block host protein synthesis. SV40 enhances growth of *ts*3, 4, 19, 31, and 125 in monkey cells. The genetic and physical maps of these mutants are shown in Fig. 12.9.
[b]Numbers in parenthesis indicate the number of *ts* mutants isolated in the complementation group [323].
[c]Same complementation group as H5*ts*149 (Table 12.4).
[d]H5*ts*125 was isolated by Ensinger and Ginsberg [46].

Table 12.4 Characteristics of Ginsberg's Ad 5 *ts* Mutants[a]

Complementation group	Prototype mutant designation	Characteristics at nonpermissive temperature
I (1)[b]	*ts*125[c]	Defective in structural gene for 72K single-stranded DNA binding protein that is required for initiation of viral DNA synthesis; transforms cells with 3-8 times greater frequency than wild type; not synthesize capsid antigens; potentiates growth of AAV
II (6)	*ts*116	Synthesizes hexon polypeptides but not hexon antigen
III (5)[d]	*ts*135	Defective in assembly of intact virions, although all capsid proteins and arginine-rich core protein are made and are transported into the nucleus
IV (1)	*ts*142	Not make fiber antigen
V	*ts*147	Defective in transport of hexon polypeptides into the nucleus; same complementation group as H5*ts*F2; mutation may be in hexon structural gene
VI[e]	*ts*149	DNA-negative, transforms cells with same frequency as wild-type, potentiates growth of AAV

[a]Adapted from [75]. At NPT all mutants block host DNA synthesis.
[b]Numbers in parentheses indicate number of isolates in the complementation group.
[c]Williams' complementation group Q. Levine et al. [149] have reported that SV40 complements growth of *ts*125 in monkey cells, but SV40 *ts*A mutants do not. The *ts*A gene is believed to encode an early SV40 protein, the T-antigen, which binds to double-stranded DNA.
[d]There may be from 1-5 complementation groups defective in virion assembly.
[e]Same complementation group as Williams' group N, which contains mutants defective in cell transformation.

1. *Early Mutants*

Biochemical characterizations of Ad 5 *ts* mutants are in their beginning stages (summarized in Table 12.3 and 12.4). There are 2 complementation groups of *ts* mutants with defects in early genes. Since there probably are 6-11 early genes (see below), presumably more groups will be discovered. The 2 early groups are group I or Q, represented by H5*ts*125 [46], and group VI or N, represented by H5*ts*149 [75] and H5*ts*36 and 37 [323]. At NPT, H5*ts*125 fails to synthesize viral DNA [46, 75, 149, 295, 297], reduces host DNA synthesis to 25% of control levels and does not make immunologically reactive capsid antigens [46]. Similarily, H5*ts*36 is viral DNA-negative, blocks host DNA synthesis [318], and at NPT does not produce capsid proteins, and possibly some other viral proteins [232, 234]. These results confirm that major late proteins are not synthesized without initiation of viral DNA replication. Mutants in both groups resemble wild-type virus in inhibition of host cell DNA synthesis [46, 318].

Most interesting, at NPT rat embryo cells are transformed at much reduced frequencies by H5*ts*36 [323], but by as much as 8-fold greater frequency by H5*ts*125 [75]. Both mutants transform like wild-type at PT. Unexplicably, H5*ts*149 (same complementation group as H5*ts*36) has been reported to transform normally at NPT [75]. The H5*ts*36 gene is apparently not involved in maintenance of cell transformation. First, it transforms with frequencies similar to wild-type when infected rat cells are shifted up from PT to NPT 48 hr postinfection, and second, cells transformed at PT retain their transformed phenotype at NPT [323]. This contrasts with SV40 *ts*A mutants which are also transformation-defective, but which seem to lose their transformed phenotype at NPT, suggesting that the A gene product is necessary for maintenance of transformation [18, 168, 190, 279]. No cell lines transformed by Ad *ts* mutants have been isolated that lose their transformed phenotype after shift-up to NPT [75, 323]; this is necessary to prove that an Ad gene product is directly responsible for the maintenance of transformation.

H5*ts*125 and 36 replicate viral DNA normally at PT. After shift-up to NPT, the rate of DNA replication of H5*ts*125 declines rapidly (within 30 min), but that of H5*ts*36 does not decline until 4-6 hr [149, 297]. DNA synthesis by H5*ts*149 declines 1-2 hr after shift-up [75]. Since H5*ts*36 and 149 are in the same complementation groups (groups N or VI), it is not clear whether the relatively slow response to shift-up merely reflects peculiarities in the particular mutants under study (H5*ts*36 is somewhat leaky [323]). Experiments where viral DNA was pulse labeled with [^{3}H]thymidine at PT, followed by a chase of the label at NPT, suggested that the products of H5*ts*125, and possibly H5*ts*36, are required for initiation (as opposed to completion) of new rounds of viral DNA replication [297].

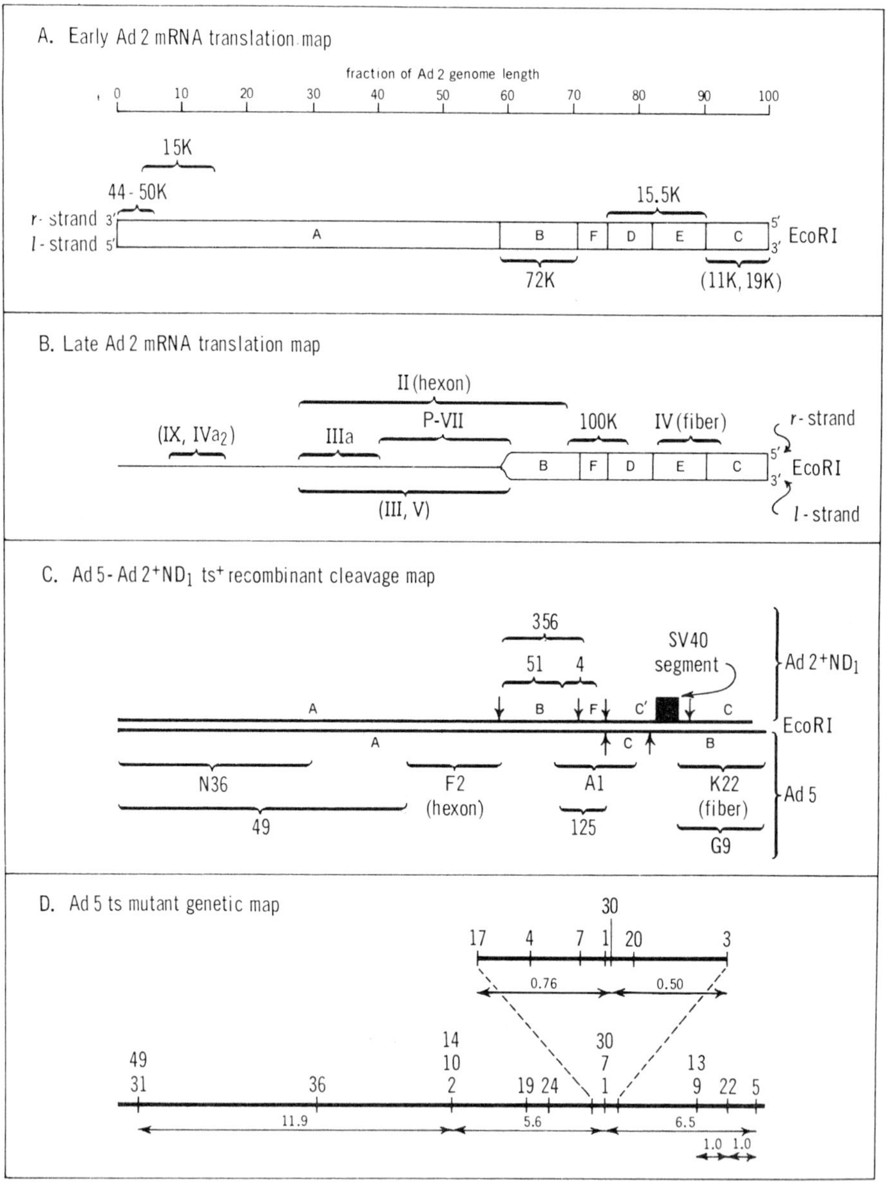
A. Early Ad 2 mRNA translation map
fraction of Ad 2 genome length
0 10 20 30 40 50 60 70 80 90 100
15K
44-50K
15.5K
r-strand 3'
l-strand 5'
A
B
F
D
E
C
5'
3'
EcoRI
72K
(11K, 19K)
B. Late Ad 2 mRNA translation map
II (hexon)
P-VII
100K
IV (fiber)
(IX, IVa2)
IIIa
r-strand
B
F
D
E
C
EcoRI
(III, V)
l-strand
C. Ad 5-Ad 2+ND1 ts+ recombinant cleavage map
356
51
4
SV40 segment
Ad 2+ND1
A
B
F
C'
C
EcoRI
A
C
B
N36
F2 (hexon)
A1
K22 (fiber)
Ad 5
49
125
G9
D. Ad 5 ts mutant genetic map
30
17 4 7 1 20 3
0.76
0.50
49 31
36
14 10 2
19 24
30 7 1
13 9 22 5
11.9
5.6
6.5
1.0 1.0

Fig. 12.9 Maps of early and late Ad 2-coded polypeptides, and physical and genetic maps of Ad 5 *ts* mutants. (A) and (B) Map obtained by cell-free translation of early and late Ad 2-specific mRNA, isolated by hybridization to and elution from Ad 2 Eco RI, Hind III, Hpa I, and Bam DNA restriction endonuclease fragments [158, 363]. The genes encoding these proteins could lie

Ad 5-infected cells contain 2 infected cell specific proteins of 72,000 and 46,000-50,000, as determined by SDS polyacrylamide gel electrophoresis, that bind to single-stranded DNA [294]. These are early proteins. Tryptic fingerprints [370] suggest that the 46,000-50,000 proteins are degradation products of the 72,000 protein. The level of these proteins is markedly reduced in H5*ts*125 following shift-up to NPT, and DNA-binding proteins prepared from H5*ts*125 grown at PT elute from DNA-cellulose at lower temperatures than wild-type or H5*ts*36 proteins [149, 297]. These results indicate that the 72,000 protein from H5*ts*125 is thermolabile and are strong evidence that the lesion of H5*ts*125 resides in the gene of the DNA-binding protein. This protein initiates and elongates Ad DNA chain synthesis [378].

It is interesting that H5*ts*125 and H5*ts*36, as well as Ad 12 DNA-negative *ts* mutants (see later), all appear able to shut off host DNA synthesis like wild-type virus. Since DNA-negative early mutants apparently do not express late genes, this may mean that an early Ad gene functions in blocking host DNA synthesis.

H5*hr*1 grows only on Ad 5-transformed cells, and so may be defective in a gene function provided by the transformed cells. H5*hr*1 does not synthesize virus DNA and late proteins. Most interesting, *hr*1 does not transform rat cells, does not produce Ad 5 T-antigen, and may map within the left 4% of the Ad 5 genome (E. Frost and J. Williams, personal communication), the probable location of Ad 5 transforming gene(s). Therefore, H5*hr*1 may be defective in a protein required for transformation. Further studies on this, and other early *hr* and *ts* mutants, should prove rewarding.

Fig. 12.9 (continued)

anywhere within the braces, and do not necessarily span the whole area enclosed by the braces. The strand specificity of the polypeptides was deduced from the early and late mRNA maps [210, 245] as shown in Fig. 12.12(B). In section B, the strand specificity of the polypeptides to the left of Eco RI-B fragment is not known, although most probably arise from *r*-strand. (C) Map of Ad 5-$Ad_2^+ND_1$ *ts* recombinants, based upon differing cleavage patterns of the Ad 5 and $Ad_2^+ND_1$ sections of the recombinant genome by restriction and endonucleases [98, 170, 238, 321]. The numbers refer to the mutant designation. Ad 5 *ts* mutants were isolated by Williams et al. [323], and their properties are summarized in Table 12.3. H5*ts*125 was isolated by Ensinger and Ginsberg [46] (Table 12.4), and is defective in the structural gene for the early single-stranded 72,000 DNA binding protein [295]. The $Ad_2^+ND_1$ *ts* mutants were isolated by Grodzicker et al. [98]. The braces represent the maximum areas in the genome that could harbor the *ts* lesion. (D) Ad 5 *ts* mutant genetic map, based upon recombination frequencies of Ad 5 *ts* mutants in 2-factor crosses. Adapted from Williams et al. [323].

2. *Late Mutants*

There are 15 complementation groups of late mutants known. All late mutants tested synthesize viral DNA and switch off host DNA at NPT [46, 318]. All of Williams' late mutants tested, possibly excepting H5*ts*22, inhibit host polypeptide synthesis [234]. Williams' mutants fall into 5 serological categories (I-V, not to be confused with Ensinger and Ginsberg's complementation groups) based on a variety of immunological tests using antisera to specific virion components [232, 323]. Group I includes mutants with no major differences from wild-type in the production of capsid antigens; group II, no production of any capsid antigens (these are the early mutants discussed above); group III, no production of fiber antigen; group IV, no production of hexon antigen; and group V, abnormal hexon transport. Mutants, in complementation groups 0 and P, have not yet been serologically characterized. Mutants from 12 complementation groups, representing the 5 serological groups, have been further examined by SDS polyacrylamide gel electrophoresis, followed by autoradiography [234]. After a short labeling period (30 min) at NPT, no differences were observed in the mobilities or relative quantity of any of the polypeptides discernible, indicating that the appropriate polypeptides are synthesized by all late mutants. On the other hand, long labeling periods revealed 3 types of responses that correlated with the serological groupings: (1) H5*ts*18, 19, no differences compared to wild type (serological group I); (2) H5*ts*5, 9, 13, and 22, reduced labeling of fiber relative to hexon and penton base (serological group III); and (3) H5*ts*1, 2, 3, 14, 17, and 20, reduced labeling of hexon relative to fiber and penton base (serological groups IV and V). It was suggested that the reduced quantities of these polypeptides after a long labeling period resulted because of enhanced proteolytic degradation of the defective polypeptides.

A number of different gene functions seem to be involved in the assembly of hexon monomer polypeptide into mature hexon trimer capsomere, and the transport of hexon capsomere from the cell cytoplasm to the nucleus where virus assembly takes place (Tables 12.3, 12.4). Williams' H5*ts*17, and 20 (complementation groups D and E), which produce hexon monomer but not hexon capsomere antigen (antibody against hexon capsomere will not react well with hexon monomer), may be defective in assembly of hexon monomeric subunits into the trimer capsomeres [148]. Hexon monomer polypeptides are identical to each other and to wild-type in size and in trypsin and chymotrysin peptide maps, suggesting that neither mutant is defective in the hexon structural gene. Hexon polypeptides from both mutants assemble after shift-down from NPT to PT. Assembly of *ts*17 is increased by cycloheximide and is unaffected by cyanide, while assembly of *ts*20 is unaffected by cycloheximide and is blocked by cyanide. These results support the genetic data that these mutants are defective in different genes, and imply that at least 2 genes are required for hexon assembly. Mutants in Ginsberg's complementation group II seem similar in that they do not make hexon antigen, but do make hexon polypeptide [45, 75].

Mutants in Williams' serological group V, comprising complementation groups A, B, C, and F, synthesize hexon capsomere antigen (fluorescent antibody tests) [232], and hexon monomer polypeptide [234], but the hexon antigen is not transported to the nucleus and accumulates in the cytoplasm at the nuclear membrane. Other capsid antigens are synthesized and transported normally. The group A, B, and C mutants are located close to each other and to group D and E mutants on the genetic map (Fig. 12.9D). H5*ts*A1 is located in the same coordinates as the gene for 100K polypeptide on the physical map (Fig. 12.9B,C). H5*ts*F2 maps to the left of group A-E mutants on the genetic map, and within the coordinates of the hexon structural gene on the physical map (Fig. 12.9C,D) [170]. Thus, group F mutants may be in the hexon structural gene; if so, then at NPT the defective hexon monomeres apparently are assembled into capsomeres, but the capsomeres are not transported into the nucleus. Ginsberg's H5*ts*147, defective in hexon transport [76], is in the same complementation group as group F (H. Ginsberg, personal communication).

Mutants in 3 complementation groups are defective in fiber antigen synthesis [76, 323]. From physical mapping results, Mautner et al. [170] suggested that the H5*ts*K22 mutation is in the structural gene for fiber. H5*ts*K22, and H5*ts*G9 (fiber-defective) both map to the extreme right of the viral genome [170, 238]; it is not known why mutants in complementation groups G and J are deficient in fiber antigen. No fiber transport mutants are known.

Comparatively little else is known about the late Ad 5 mutants. H5*ts*H18 and H5*ts*I19 fail to induce interferon in chick embryo cells [290]. Neither the DNA-negative H5*ts*N36 nor the fiber antigen-negative H5*ts*G9 and H5*ts*G13 induce Ad-specific intranuclear crystals, whereas the hexon-negative H5*ts*D17 does [324]. Wild-type Ad 5 and Ad 12, and DNA-negative H5*ts*N36, H5*ts*149, H5*ts*125, (complementation groups Q or I), H12*ts*A275, H12*ts*B221, and H12*ts*C295 (see later) all potentiate the growth of Ad-associated virus type 1 in HEK cells at 40°C [100].

Intertypic complementation studies between Ad 5 and Ad 12 have been done. Williams et al. [321] found that wild-type Ad 12 (strain 1131) complements H5*ts*D17, E20, A1, F2, B3, and C4 (all with defects in hexon production), as well as 2 other late mutants H18 and L24. No evidence for recombination was found, although in some crosses phenotypic mixing took place. Ad 12 does not complement, or complements poorly, the late mutants H5*ts*J5, G9, K22 (all fiber-deficient), and I19 and M31. Significantly, Ad 12 does not complement the Ad 5 early mutants H5*ts*N36 or Q125. Complementation implies that an Ad 12 protein can replace the defective Ad 5 mutant protein. Ad 12, whose genome is less than 15% homologous with Ad 5, apparently produces several late proteins (but no known early proteins), that are functionally similar to Ad 5 proteins.

C. Ad 12 *ts* Mutants

A number of groups have isolated Ad 12 *ts* mutants [147, 164, 250, 251]. Ledinko [147] isolated 10 *ts* mutants in 6 complementation groups A to F. H12*ts*B401 and H12*ts*E405 are early mutants of 2 groups. These are viral DNA-negative, and are defective in production of late proteins as shown by fluorescent antibody tests and SDS polyacrylamide gel electrophoresis. H12*ts*B401 transforms hamster cells at 2-8 times greater frequency than wild-type at NPT, while H12*ts*E405 transforms like wild-type. Both mutants shut off host DNA synthesis, and are not defective in thymidine uptake or induction of thymidine kinase activity.

Shiroki et al. [250] isolated 88 Ad 12 *ts* mutants, and classified 34 into 13 complementation groups, A to M. None were defective in viral DNA replication or synthesis of T-antigen at NPT. Serological tests and SDS polyacrylamide gel electrophoresis suggested that mutants in groups A to G were defective in the production of 2 or more capsid components (hexon, fiber, penton base), mutants in groups H, I, and J were defective in production of 1 capsid component, and mutants in groups K, L, and M produced all capsid components. However, serological and polyacrylamide gel analyses of *ts* proteins must be interpreted with caution, and may not represent the true nature of the *ts* defect.

Shiroki and Shimojo [251] have subsequently isolated 10 Ad 12 *ts* mutants defective in viral DNA replication at NPT, and have arranged 7 of these into 3 complementation groups A, B, and C (not to be confused with complementation groups A to M of late mutants described in [250]). A *ts* mutant of Ad 31, H31*ts*A13 (formerly *ts*13 [269]) reacts in complementation tests as an Ad 12 group A mutant. All 3 mutants (H12*ts*A275, H12*ts*B221, and H12*ts*C295) appear defective in initiation of viral DNA synthesis at NPT as revealed by [^{3}H] thymidine pulse/shift-up/chase experiments, and shift-up/density label experiments. All mutants, especially H12*ts*A275, are defective in formation of a viral DNA replication complex isolated by the "M-band" technique (see Section VII). All mutants induced T-antigen in HEK cells, and cellular DNA synthesis in resting hamster embryo cells (which are nonpermissive for Ad 12) at NPT. Thus 3 Ad 12 genes, and 1 or possibly 2 Ad 5 genes, have so far been implicated in the initiation of viral DNA synthesis.

As described above, the H5*ts*125 lesion is in the gene encoding a 72,000 single-stranded DNA binding protein. Rosenwirth et al. [226] have recently isolated a 60,000 protein from Ad 12-infected cells that binds only to single-stranded DNA and which presumably has the same function as the Ad 5 72,000 protein in viral DNA replication. An Ad 12 48,000 single-stranded DNA-binding protein was also isolated, but as with the Ad 5 48,000 DNA-binding protein, this is apparently a proteolytic degradation product of the larger DNA-binding protein. The early DNA-negative Ad 12 mutants isolated by Shiroki and Shimojo [251] were tested for the production of these proteins at NPT; H12*ts*A275 synthesized reduced levels of both 60,000 and 48,000 proteins [226]. H12*ts*B221 and H12*ts*C295 produced normal amounts of both proteins at NPT. Since H31*ts*A13 (see below) does not complement H12*ts*A275, it may also be defective in the DNA-binding protein

Shiroki et al. [372] have recently reported that, at NPT, group B and C mutants, but not group A mutants, are able to grow in H5 cells, a monkey cell line transformed by Ad 7-SV40 hybrid virus. Ad 7 complemented growth of all mutants at NPT. Thus H5 cells may express an Ad 7 coded protein that can replace the Ad 12 group B and C defective protein [i.e., H5 cells may be analogous to the Ad 5 transformed HEK cells that support the growth of H5*hr*1 (see above)]. Since H5 cells presumably synthesize Ad 7- coded-protein(s) responsible for maintenance of cell transformation (and possibly other nontransformation proteins), either group B or C Ad 12 could be defective in transforming gene(s).

D. Ad 31 *ts* Mutants

Suzuki et al. [270] isolated 17 *ts* mutants of Ad 31 (closely related to Ad 12), and classified 12 of these into 8 complementation groups, I to VIII. A mutant in group I, *ts*13 [269] now named H31*ts*A13 [251], is defective at NPT in synthesis of viral DNA, capsid proteins, and inclusion bodies, and does not produce typical CPE. However, it induces T-antigen, DNA polymerase, and thymidine kinase in HEK cells, and cellular DNA synthesis in nongrowing hamster cells [269]. At NPT this mutant is unable to form the viral DNA replication complex isolated by the M-band technique [271], suggesting that it is defective in a protein necessary to form this complex. The lesion in

H31*ts*A13 seems very similar to that of the Ad 12 group A DNA-negative mutants [251] discussed above, which are apparently defective in the gene for DNA binding protein. H31*ts*A13 supports the growth of Ad-associated virus at NPT [123, 171]. Regarding the other Ad 31 complementation groups, at NPT mutants in groups II and III have reduced rates of viral DNA and capsid protein synthesis, those in groups IV and V are defective in fiber and hexon, but not penton base, a mutant in group VI is defective only in fiber, and mutants in group VII and VIII are not defective in any of the above.

E. CELO *ts* Mutants

Ishibashi [118] isolated *ts* mutants of CELO avian Ad that exhibited peculiarities in transport of capsid antigens. Forty-nine *ts* mutants were classified into 5 groups (not complementation groups) with respect to cellular distribution of capsid antigens, as measured by immunofluorescence [119]. At NPT the mutants accumulated capsid antigens as follows: (1) in nucleus like wild-type (group I); (2) in both nucleus and cytoplasm (group II); and (3) only in cytoplasm (group III). Group IV mutants produced little or no antigens. A group V mutant (*ts*22) produced no viral antigens or intranuclear inclusions, and synthesized little viral DNA, and therefore may be a DNA-negative early mutant. Group III mutants, which accumulate capsid antigens in the cytoplasm, are reminiscent of the Ad 5 *ts* mutants in serological group V [323], and Ginsberg's complementation group V [75, 76].

F. Ad 2 *ts* Mutants

Ad 2 *ts* mutants have recently been isolated [10]. Of 36 mutants, 15 were classified into 13 complementation groups, and based on recombination frequencies aligned into a genetic sequence: *ts*9-53-7-1-11-4-6-74-5-3-2-8 [311]. All members of the complementation groups are late mutants since at NPT they are normal in adsorption, viral inclusion formation, and viral DNA synthesis [311]. Three classes were observed by immunodiffusion tests: 1 group (H2*ts*3) negative for hexon, penton, and fiber; 3 groups (H2*ts*11, 4, and 7) negative for penton only; and 8 groups positive for all capsid antigens. Three groups (H2*ts*3, 4, and 5) are defective in assembly [311]. H2*ts*1 is defective in cleavage of P-VI, P-VII, and P-VIII, precursors to the VI, VII, and VIII virion proteins [310].

G. Host-Range Mutants of Ad_2-SV40 Hybrid Viruses

$Ad_2{}^+ND_1$, 1 of several Ad_2-SV40 hybrids [155], (Section IX) grows well on both human and monkey cells, whereas Ad 2 grows very poorly on monkey cells. $Ad_2{}^+ND_1$ contains an insertion of 17% of the SV40 genome at the location of the cleavage site between Eco RI-D and E fragments (see Section X) on the Ad 2 genome [31, 106] and a deletion of 5.5% of the Ad 2 genome at the site of the SV40 insertion [175]. The ability of Ad_2-SV40 hybrids to grow well on monkey cells is presumably due to the SV40 insertion. Grodzicker et al. [97] isolated several host-range mutants of $Ad_2{}^+ND_1$, based on their ability to grow on human, but not monkey cells. One $Ad_2{}^+ND_1$ *hr* mutant, H39, has been studied in detail [97]. It is not *ts*, and the SV40 segment has not been deleted. H39 interacts with monkey cells like Ad 2; late proteins are synthesized in reduced amounts, and growth is enhanced by SV40. Unlike parental $Ad_2{}^+ND_1$, a 30,000 protein specific for the hybrid is not synthesized. Also, induction of the SV40-specific U-antigen differs. Thus H39 behaves as if it has lost the function supplied by the SV40 segment in $Ad_2{}^+ND_1$. The nature of this function is unknown. $Ad_2{}^+ND_1$ *ts* mutants have also been isolated, and although their biochemistry has not been studied, they have been used successfully in mapping the physical location on the genome of Ad 5 *ts* mutants [98, 238]. The $Ad_2{}^+ND_1$ *ts* mutants presumably have lesions in the Ad portion of their genome, since they fail to grow in human cells at NPT. The location of the Ad 5 and $Ad_2{}^+ND_1$ *ts* mutations are shown in Fig. 12.9.

Clearly, although many Ad mutants are available, and as many as 19 Ad 5 complementation groups have been discovered, Ad genetics is in its infant stages. Undoubtedly more complementation groups will be found, especially with defects in early genes. A great deal of biochemical characterization of these mutants remains to be done. Other classes of mutants, like the *hr* mutants already available, should prove extremely valuable. Unfortunately, Ad deletion mutants have not yet been described, as these should assist in relating specific proteins to specific phenotypes. Most interesting, an initial report has appeared which deals with the replication of Ad 2 in *ts* mutants of hamster cells [184]. Such studies should enhance our understanding of the role of host cell functions in Ad replication.

VI. Adenovirus Replicative Cycle (Productive Infection)

A. Early Events

To ensure that all cells are infected simultaneously, which is essential for biochemical analyses, cells are generally infected with 10-200 plaque forming units (PFU) of Ad per cell. Not all particles are infectious, and the ratio of physical to infectious particles ranges from 10-2000, depending upon the serotype [93]. Electron microscopy and biochemical studies of cells infected with radioactively tagged virions have indicated the following series of events, although the detailed mechanisms and intermediates are not established [23, 209]. Virus particles are adsorbed to specific surface receptors and enter the cell either by pinocytosis or direct penetration. The virion is rapidly uncoated, with the loss of penton capsomeres [266], and transported to nuclear pores, perhaps via microtubules [32]. The partially uncoated viral capsid is stripped during transport into the nucleus. The virus core enters the nucleus, possibly by an ATP-dependent process [25], with the capsid proteins remaining behind. There the removal of core proteins from the DNA is accomplished [160]. At this stage, transcription of the free DNA commences, presumably by host RNA polymerase II, since viral RNA synthesis is inhibited by α-amanitin [216, 305].

B. The Molecular Events of Productive Infection

Viral mRNA was first demonstrated in Ad-infected cells in 1966 [280], leading to much subsequent work on the transcription of the Ad genome during productive infection. The most widely studied experimental system has proven to be infection of exponentially growing human KB cells by Ad 2 and most of our remarks will concern this system [85]. The time course of synthesis of viral RNA, the early 70,000-75,000 single-stranded DNA binding protein, viral DNA, virion proteins, and infectious virus during productive infection is illustrated in Fig. 12.10. Ad 2 infection blocks division of exponentially growing KB cells [85]. The infection proceeds in 2 temporal stages of gene expression, early (before viral DNA synthesis commences) and late [83, 281]. Only a limited fraction of viral genome is expressed early [60]. Current work suggests that there are 6-11 early genes. At early stages (up to 6 hr), 2% of pulse labeled polyribosomal RNA [280] and up to 18% of poly(A)-terminated mRNA [159] is viral-specific. Transcription of the viral genome and generation of viral mRNA is accomplished by host enzymes since cycloheximide does not block early viral mRNA production [195], and since virions contain no

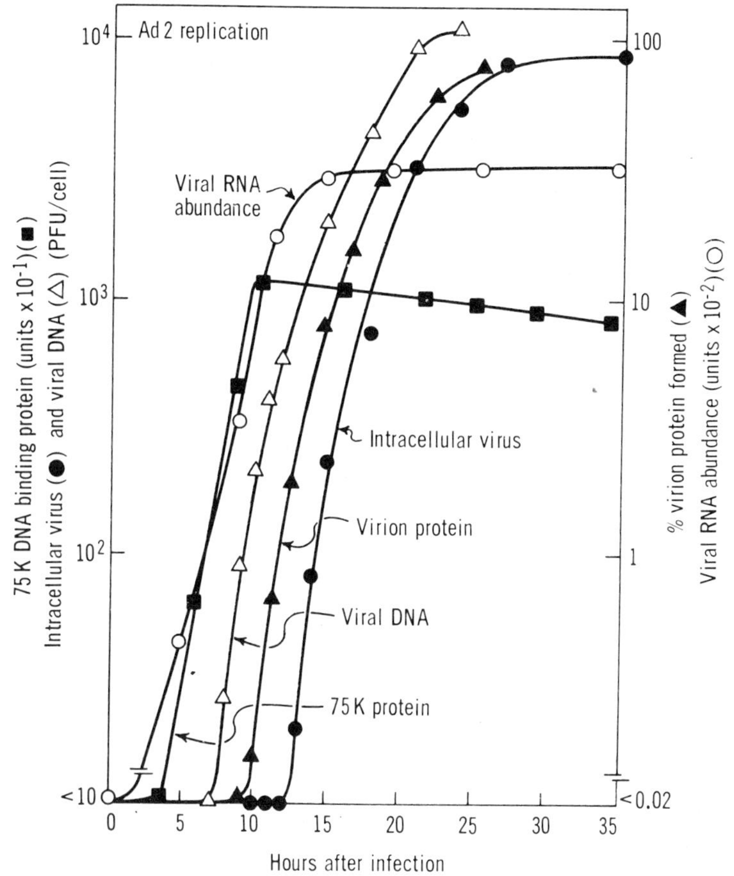
Ad 2 replication
10⁴
10³
10²
<10
100
10
1
<0.02
Viral RNA abundance
Intracellular virus
Virion protein
Viral DNA
75 K protein
0
5
10
15
20
25
30
35
Hours after infection
75 K DNA binding protein (units x 10⁻¹)(■)
Intracellular virus (●) and viral DNA (△) (PFU/cell)
% virion protein formed (▲)
Viral RNA abundance (units x 10⁻²)(○)

Fig. 12.10 Ad 2 replication cycle. Time course of synthesis of viral RNA, early viral 75,000 protein, viral DNA, virion protein, and intracellular virus. Data for viral DNA and intracellular virus from Green [81], virion protein from Polasa and Green [368], viral RNA abundance from Wold et al. [326], and 75,000 protein from Gilead et al. [72].

detectable RNA polymerase. Cellular DNA, RNA, and protein synthesis proceeds normally early after infection [81]. The most abundant early viral-coded protein, the 70,000-75,000 single-stranded DNA binding protein [243, 294], is detected as early as 4 hr postinfection and reaches a plateau at about 9 hr (Fig. 12.10) [72].

There is a dramatic change in macromolecular synthesis at 6-7 hr postinfection, when the cells switch from early to late stages of infection. This switch is marked by the onset of viral DNA replication. The synthesis of host cell protein [74], DNA [74, 214], and cytoplasmic rRNA [219] are dramatically inhibited late after infection, whereas tRNA synthesis is unaffected [219]. The mechanism of inhibition is not established. Philipson et al. [210] have determined that rRNA is actually transcribed in abundant quantities late, but mature 18S and 28S rRNA is not found in the cytoplasm. Since several early Ad 5 and Ad 12 DNA-negative *ts* mutants block host DNA synthesis at NPT (Section V), this block may be the result of an early viral gene product, or some other consequence of cell-virus interaction.

At late stages of infection the cell becomes directed to the synthesis of mainly viral macromolecules. The vast majority of mRNA transcribed late is viral, accounting for 40% of pulse labeled total RNA [280], and up to 85% of poly(A)-terminated mRNA [159]. DNA synthesis at late stages is exclusively viral. Virion protein synthesis begins about 10 hr postinfection, reaching maximum levels at 24 hr (Fig. 12.10). An abrupt arrest of the early 75,000 protein occurs with the onset of late protein synthesis [72]. Most viral proteins synthesized are structural proteins, although regulatory proteins probably are produced also. Structural proteins are produced in excess (especially hexon); only 5-6% of these are incorporated into mature virions [81]. Ad proteins are synthesized in the cell cytoplasm, as shown by the detection of viral mRNA in polyribosomes both early and late after infection [280] and of immunologically active viral proteins associated with polyribosomes [299]. Viral proteins are rapidly transported to the nucleus where they assemble into virions [300]. Infectious virus particles are first detected at 13 hr postinfection. The final virus yield is 10,000 PFU per cell or about 200,000 physical particles per cell [81].

Ad assembly events are unclear because "true" assembly intermediates are difficult to identify (only 5-6% of capsid proteins enter virions, and intermediates may break apart upon extraction). Initial studies suggested that empty capsids were first formed that became filled with viral DNA cores [120]. Recently Edvardsson et al. [346] suggested that a fragile nuclear intermediate (550-670S) is first formed that contains capsid, cores, P-VI and P-VII, but not VI and VII. Next nuclear "young virions" are formed that contain both precursor and mature VI and VII. Finally, mature virions are formed that contain mature VI and VII. Other studies have shown that viral-coded proteins are involved in virus assembly (Section V).

Doerfler and colleagues have reported that Ad DNA becomes integrated into host cell DNA during productive infection [41]. It is doubtful that these integrated sequences participate in Ad multiplication.

C. Viral Gene Transcription

Several early studies have shown that only a limited portion of the Ad 2 genome is expressed as stable mRNA early after infection. Using methods able to detect only abundant viral RNA sequences, it was established that 8-20% of the genome is expressed early [60] and 80-100% late after infection [61]. Subsequent studies (reviewed in ref. 347) employed strand separation, dissection of the viral genome by restriction endonucleases, and electron microscopy. Using intact *r* and *l* strands prepared by the ribocopolymer affinity method (see Sections VII and X) [142], it was found that viral RNA was transcribed from both strands early and late during productive infection and in transformed cells [88, 143]. Pettersson and Philipson [204] have recently obtained evidence that most mRNA transcribed at 18 hr originates from *n*-strand. Early genes lie in 2 noncontiguous blocks on each DNA strand; most but not all late genes are on *r* strand (Fig. 12.11). By hybridization of labeled viral DNA strands with unlabeled RNA, it was estimated that early mRNA arose from 9-15% and 14-19% and late mRNA from 20-25% and 60-75% of *l*- and *r*-strand, respectively [245, 283, 367]. Early and late mRNA refer to virus-specific RNA present in the cytoplasm early and late, respectively, after infection. Electron microscopy studies, using the "R-loop technique," indicate that early mRNA is derived from 10.8% of *l* strand (map units 62.4-67.9, 91.5-96.8) and 17.5% of *r* strand (units 1.3-11.1, 78.6-86.2) (see page 893). By computer nonlinear regression analysis of the kinetics of hybridization of early polysomal RNA to separated [^{32}P] DNA strands, Wold et al. [326] concluded that early mRNA arose from 10-12% and 14-20% of *l*- and *r*-strand, as abundant and scarce classes complementary to about 6% and 5% of *l*-strand, and 7% and 10% of *r*-strand. Flint and Sharp [349] quantitated viral mRNAs from the 4 gene blocks, and concluded that the mRNAs were present at 300 (mRNA for the DNA binding protein) to 1000 copies per cell.

Late cytoplasmic RNA hybridizes to nearly 100% of the viral genome, assuming mRNA originates from 1 strand of each gene. But all these sequences may not be transcribed late. For example, late cytoplasmic RNA may contain early mRNA sequences that were transcribed early and persist in the cytoplasm at late times (i.e., class I RNA; see below), or "non-mRNA" contaminants from the nucleus. To estimate how much of the genome is transcribed late, Wold et al. [327] carried out an abundance analysis of late RNA hybridization kinetics, assuming that mRNA synthesized late would be in much higher concentration than early class I RNA (see Fig. 12.10) and hence would anneal much more rapidly. From this, they concluded that mRNA *synthesized late* originates from 7% and 45-55% of *l*- and *r*-strand, respectively. Therefore, since early mRNA is derived from about 28% of the genome and late mRNA from 55-60%, then roughly 15% of the genome may not code for structural genes.

Viral mRNA molecules synthesized early and late after infection have been investigated by size fractionation, polyacrylamide gel electrophoresis, and molecular hybridization. Three broad size classes of early Ad 2 mRNA have been identified with sedimentation values of 11-16S, 20-23S, and larger than 24S [159, 195]. Büttner et al. [21] have preparatively isolated early viral mRNA by hybridization to separated DNA strands and shown that the major 20-23S size class is composed of at least 2 viral mRNA species, one a transcript of the Ad 2 *l*-strand and the second of the Ad 2 *r*-strand. Late RNA sequences have been resolved into as many as 5-6 size classes, ranging from 10-29S [29, 88, 159, 194, 210, 276]. The identification and mapping of additional size classes of early and late viral·mRNA using restriction DNA fragments and electron microscopy are described in Section X.

D. Virus-Associated RNA

At late stages of infection Ad 2-infected cells synthesize large quantities of an unusual RNA species, termed virus-associated RNA (VA-RNA) [221]. Very recently, it has been established that there is 1 major and 1 (5.2S) [205, 209, 373] or 3 [369, 371] minor VA-RNA species, all of which map at approximately 0.3 on the Ad 2 genome. The major VA-RNA sediments at 5.5S, is 156 nucleotides long, and has been sequenced [187, 188, 189]. VA-RNAs are synthesized in vitro by isolated nuclei and are transcribed by RNA polymerase III [216, 313].

The functions of the VA-RNAs are unknown. The major VA-RNA is made early but is more abundant late. It is found in the cytoplasm, mainly unattached to polyribosomes. Its nucleotide sequence, which shows internal duplications and implies a high degree of secondary structure, rules out a mRNA role. It has been pointed out that the VA-RNA genes (and presumably the promoter site for RNA polymerase III) lie at or near the 3′ end of a long stretch of late structural genes (see mRNA maps in Section X), which are transcribed by RNA polymerase II, and thus VA-RNA could play a role in regulating transcription [169, 205].

E. Biogenesis of Adenovirus mRNA

It is well established that mRNA genesis and the regulation of gene expression in prokaryotic cells occurs mainly at the transcription level. With eukaryotic cells, however, a considerable aspect of mRNA production and perhaps of gene regulation, may be posttranscriptional [34, 151], although there is some controversy on this point [35]. The majority of eukaryotic mRNA molecules contain poly(A) tracts of about 200 nucleotides located at the 3′-OH end (see review by Darnell et al. [34]). A large body of evidence suggests that the primary unit of transcription is a high molecular weight precursor, of from 4,000-50,000 nucleotides, termed heteronuclear RNA (HnRNA). HnRNA also contains poly (A) at the 3′ end.

It has recently been demonstrated that mRNA from animal cells [199], and viruses that replicate in eukaryotic cells [64] is methylated. With viruses that replicate without a nuclear phase the methylation is exclusively at the 5′ terminus in an oligonucleotide of the structure $m^7G(5')ppp(5')N^m pNp\cdots$ (N^m represents a ribose 2′-O methylation) [227]. This structure has been termed a "cap", and it may be involved in binding of mRNA to polyribosomes [16]. Animal cell mRNA contains the above cap (cap 1), plus an additional cap (cap 2) of the structure $m^7G(5')ppp(5')N^m pN^m pNp\cdots$; but unlike cytoplasmic viruses, animal cell mRNA is also methylated at internal positions, mainly (or exclusively) as N^6-methyl adenosine (N^6mA) [1, 37, 65, 312]. HnRNA is also methylated, as internal N^6mA, and 5′ termini with cap 1 structures [200].

It is considered that following transcription, HnRNA is cleaved, polyadenylated, and methylated (order unknown), mRNA sequences selectively transported to the cytoplasm, and the remaining RNA portions degraded in the nucleus. However, it has not been possible to prove this model for eukaryotic mRNA production.

Ads are excellent probes of eukaryotic mRNA genesis, because they replicate in the nucleus utilizing mainly host cell enzymes, and because specific viral mRNA sequences can be readily assayed in an RNA population. As with host cell mRNA, both nuclear and cytoplasmic viral RNA contain 3′-poly(A) [211]. Ad mRNA is also methylated [52, 103, 312, 328]. Late after infection, both nuclear and cytoplasmic viral RNA contain internal base methylations, and cap 1 and cap 2 type 5′ termini [328]. Recently, Gelinas and Roberts [350], confirming and extending the work of Sommer et al. [374], found that Ad 2 late mRNA contains one predominant 5′capped T_1 oligonucleotide, $m^7G(5')ppp(5')A^mC^{(m)}U(C_4,U_3)G$. In contrast early Ad 2 mRNA synthesized in the presence of cycloheximide (to enhance the yield of early viral mRNA) contains internal N^6mA, plus cap 1 and cap 2 5′ termini with m^7G and all 4 common 2′-O-methylribonucleotides, and possibly N^6,2′-O-dimethyladenosine [103]. Thus studies on Ad mRNA transcription and processing should tell us much about both viral and host cell mRNA synthesis.

Results of detailed studies during the past 4 years have shown the following.

1. Large viral RNA molecules are present in the nucleus, but only smaller mRNA sequences are detected in the cytoplasm early [303], and late [88, 174, 194, 303].
2. Nuclei contain viral RNA sequences that are not present in the cytoplasm both early and late after infection [30, 204, 245, 326, 327].
3. Much or all of both DNA strands are transcribed (i.e., to form complementary RNAs) early after infection, but only mRNA sequences are exported to the cytoplasm, and anti-mRNA remains behind in the nucleus [245, 326, 382].

Complementary RNAs are also found late after infection [163, 204, 210, 245, 327]. However, it was not known whether the *l*-strand transcripts present late are transcribed late or represent persisting early transcripts [204]. These observations are illustrated in Fig. 12.11.

F. The Switch from Early to Late Viral Gene Expression

The mechanisms regulating the switch from early to late viral gene expression are not known. It is likely that an early viral protein, either directly or indirectly, is involved because cycloheximide (which inhibits protein synthesis) prevents the expression of late genes. Initiation of viral DNA replication is also necessary, because cytosine arabinoside (which inhibits DNA replication) prevents the transition, and all Ad DNA-negative mutants do not synthesize late RNA [340, 342] (Section V). Associated with the transition is not only the "turn on" of late genes, but also the "shut off" of some early genes, for there is a class of early mRNA [60], termed class I RNA [28, 163], that may not be synthesized late [26]. The shut-off of class I genes does not occur with DNA-negative mutants at NPT [342], and thus may be a late function.

Although the productive infection is conveniently divided into early and late stages of infection (before and after initiation of viral DNA replication), the possibility that intermediate "stages" of gene regulation exist. For example, the maximum rate of synthesis of the 73,000 DNA building protein (coded by an early gene in Eco RI-B fragment) occurs after initiation of DNA synthesis at 6-7 hr. Synthesis (accumulation) of this protein is then curtailed at 11-12 [72, 375]. Consistent with this, the mRNA encoding this protein increases 10-fold in concentration after viral DNA replication has begun [349]. Thus, regulation of DNA binding protein production may be effected at a "middle" phase (8-12 hr) in the productive infection. Similarly, the concentration of mRNAs encoding viral structural proteins increases 10-fold between 18 and 32 hr, perhaps suggesting a "very late" stage of infection [349].

Regulation of genes could occur at the transcriptional level (as with bacteriophage), e.g., by modifications of RNA polymerase II so that it recognizes late but not some early gene promotors. Indeed Pettersson and Philipson [204] have presented evidence that *r*-strand is transcribed much more efficiently late, and that *l*-strand may not be transcribed at all. A second possibility is that late gene expression results from a change in RNA processing. Supporting this model is the fact that at least 70% of the *r*-strand is transcribed early, which includes a substantial proportion of late mRNA sequences that do not mature into mRNA in the cytoplasm at early times. Thus the switch to late gene expression could be a consequence of an alteration in RNA processing so that late gene transcripts are now "recognized" as mRNA, and consequently are exported to the cytoplasm. The current data, therefore, would suggest that regulation may occur at both transcriptional and post-transcriptional levels, and that the regulation of Ad gene expression may be very complex.

G. Biosynthesis of Adenovirus Proteins

1. *Early Viral Proteins*

Initial studies on early proteins employed serological techniques for their detection. Sera from hamsters bearing tumors induced by Ads will react not only with proteins of homologous tumor cells, but also with those of early productively infected cells [113, 242]. These studies detected a class of early proteins called tumor (T) antigens that are synthesized in Ad tumor and transformed cells. Early attempts to purify Ad 12 T-antigens resulted in isolation of from one [71] to several proteins [285]. By definition, T-antigens are viral-induced proteins that are expressed in tumors and are immunogenic in the tumor-bearing animal, and therefore could include the entire spectrum of early proteins depending upon the integrated and expressed viral genes in a particular tumor or transformed cell line. "Transformation proteins" would be those T-antigens whose functions are essential for initiating or maintaining the transformed cell. Early proteins are important because in addition to cell transformation, they regulate viral DNA replication and possibly transcription.

Ad 2 (or Ad 5)-induced early proteins have been studied by labeling infected cells with ^{35}S-methionine followed by SDS polyacrylamide gel electrophoresis and autoradiography [306]. To facilitate identification of

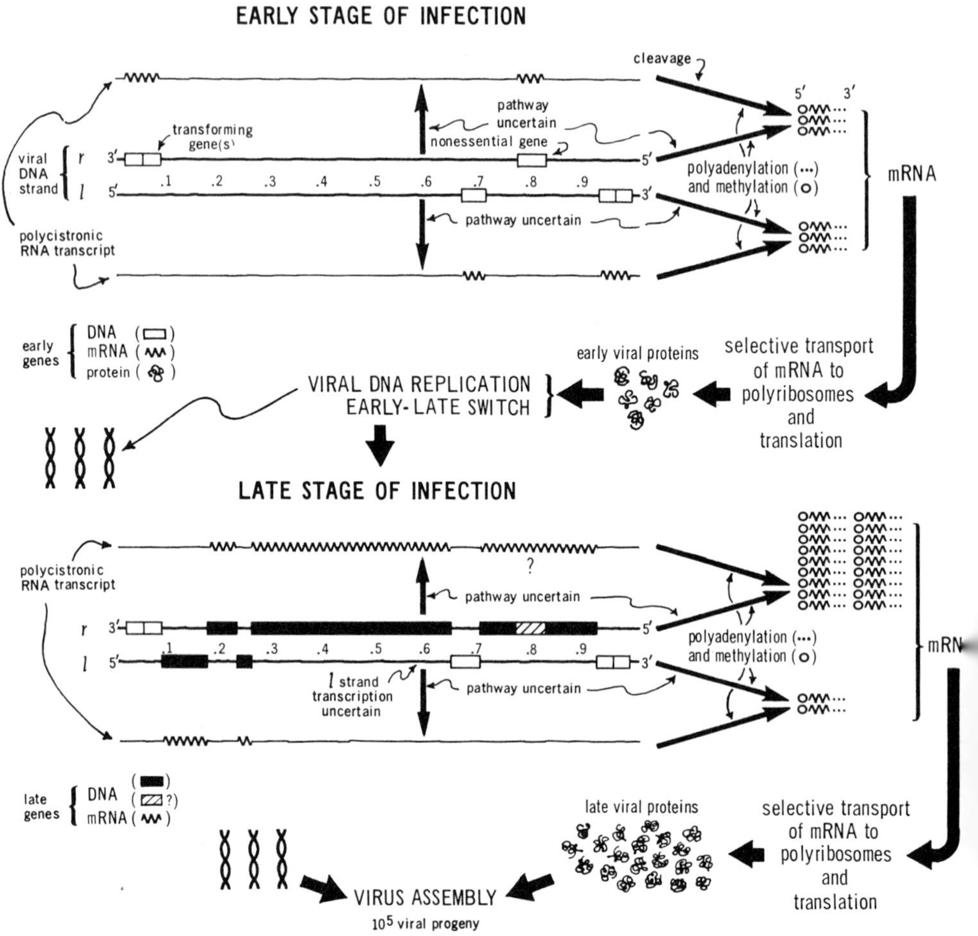

EARLY STAGE OF INFECTION
cleavage
pathway uncertain
transforming gene(s)
nonessential gene
viral DNA strand
polyadenylation (...) and methylation (o)
mRNA
pathway uncertain
polycistronic RNA transcript
early genes
DNA
mRNA
protein
early viral proteins
selective transport of mRNA to polyribosomes and translation
VIRAL DNA REPLICATION EARLY-LATE SWITCH
LATE STAGE OF INFECTION
polycistronic RNA transcript
pathway uncertain
l strand transcription uncertain
late genes
late viral proteins
VIRUS ASSEMBLY
10⁵ viral progeny

Fig. 12.11 Schematic representation of current concepts regarding early and late stages of gene expression during productive infection of human KB cells by Ad 2. Briefly, the following steps occur. (1) The virus penetrates the cell and the DNA is eventually uncoated in the nucleus. (2) Early genes (light blocks) are transcribed into mRNA (sawtooth lines), as discussed below, which is polyadenylated, methylated, and exported to the cytoplasmic polyribosomes where it is translated into early viral proteins. (3) The early proteins are transported back into the nucleus, where they initiate viral DNA replication, and presumably turn on late genes and turn off early class I genes (light blocks shown in late stages of infection), and probably carry out other unknown functions that may include blocking host cell DNA synthesis. (4) After viral DNA replication commences, late genes (black blocks) and class II early genes (hatched block) are expressed as late viral proteins, many of which are virion structural proteins. (5) Virion proteins are transported back into the nucleus, where progeny virions are assembled.

early proteins within the large background of host polypeptides, cells have been coinfected with poliovirus in the presence of guanidine [7] (which minimizes host protein synthesis), pretreated with cycloheximide [102] (cycloheximide increases the synthesis of Ad RNA [195, 326], so that after release from cycloheximide inhibition Ad early proteins are enhanced relative to host), or labeling cells in hypertonic medium [345] (for unknown reasons, synthesis of viral proteins is enhanced relative to host). Work done in several laboratories have resulted in the identification of from 6 [102, 343] to as many as 11 (Jeng, Wold, and Green, unpublished data) or 13 [371] Ad 2-induced polypeptides. In general, the polypeptides that are easily identified have molecular weights of 70,000-75,000, 19,000-21,000, 19,000, 15,000, 11,500, and 11,000. Polypeptides of 40,000-50,000 are often observed, but at least some of these are subspecies of the 75,000; it is not known whether these represent functional proteins. Polypeptides of 53,000 (58,000 with Ad 5) and 15,000 daltons can be immunoprecipitated by sera against Ad 2-transformed cells, i.e., these proteins are synthesized by the transformed cells; these are excellent candidate Ad-coded transformation proteins (Section VIII). Polypeptides of 17,000-18,000, 12,000-14,500, and 8,000-9,000 are often seen as well ([371]; Jeng et al., unpublished data).

Ad 2-induced early polypeptides have also been identified by cell-free translation of mRNA extracted from early infected cells. Saborio and Öberg [371] identified 15 polypeptides (72, 67, 60, 45, 42, 40, 35, 19, 18.5, 17.5,

---

Fig. 12.11 (continued)

Early and late viral mRNA is transcribed by host cell RNA polymerase II. Transcription takes place from the 3′-5′ direction on each DNA strand. The RNA sequences complementary to structural genes (i.e., mRNA) are represented by sawtooth lines. The mechanisms of early and late mRNA genesis are not known, and include the following possibilities. (1) All or most of both DNA strands is transcribed into large polycistronic RNA molecules that are processed (cleaved, polyadenylated, methylated, and physically transported to polyribosomes) into mRNA. In this case, mRNA arises in an obligatory manner from large precursors. Non-mRNA sequences are retained in the nucleus, and presumably are degraded. (2) mRNA does not arise from large precursors, but rather as a unicistronic primary transcript of the structural genes containing no or little non-mRNA sequences, that is then processed into polyribosomal mRNA. (3) As discussed in Recent Developments, early and late mRNAs may be processed from polycistronic primary transcripts representing 4 early eukaryotic "operons" and 3 late (2 on *r* strand, 1 on *l* strand) "operons."

15.5, 15, 14.5, 12.5, and 10,500 daltons), by in vitro translation followed by immunoprecipitation with antisera prepared against Ad 2 early infected cells. Lewis et al. [363] identified 6 polypeptides (72, 44-50, 19, 15.5, 15, and 11,000 daltons) by translation of early mRNA purified by hybridization to Ad 2 DNA and DNA restriction fragments, a method that proves the polypeptides are viral coded, and also maps the location of their genes (Section X and Fig. 9A). Excepting the 72,000, the equivalence between the proteins identified in vivo and in vitro has not been proved chemically (e.g., peptide maps).

Except for the 75,000 and 40,000-50,000 polypeptides, rigorous peptide mapping experiments have not been done to determine whether the small polypeptides are subspecies of the larger ones. Except for the 75,000 protein, there is no formal proof that any of these polypeptides are viral coded (and not cell coded and viral induced). Nevertheless, much is being learned about these important proteins. The 75,000 is the well-characterized Ad-coded single-stranded DNA binding phosphoprotein, that is believed to function in Ad DNA replication. The 19,000-21,000 is a glycoprotein ([357]; Jeng, Wold, and Green unpublished data), and has been reported to be a component of plasma membranes [343]. The 11,000 is localized exclusively in the nucleus ([371]; Jeng et al., unpublished data), perhaps as a component of the nuclear matrix [344]. The 75,000, 19,000-21,000, 15,000 and 11,000 have been identified in a soluble Ad 2 DNA replication complex (Rho, Jeng, Wold, and Green, BBRC, in press), suggesting a possible role for these proteins in Ad DNA replication. The 53,000, 19,000-21,000 (glycoprotein), 15,000, 11,500, 11,000 are probably viral coded, because they can be precipitated from early infected human cells by antisera against various Ad 2 transformed cells (F17, T2C4, 8617, F4) (53,000, and 15,000, are precipitated by all sera).

It is interesting that at least 1 early gene product appears to be nonessential for the replication of Ad 2. Two nondefective Ad 2-SV40 hybrid viruses contain a deletion of about 5% of Ad 2 DNA that codes for early cytoplasmic RNA, yet they grow in KB and HeLa cells [54].

2. *Late Adenovirus Proteins*

Studies on late Ad proteins are simplified because host-cell protein synthesis is curtailed late and viral proteins are synthesized in large quantities. As many as 22 virus-specific proteins have been observed in Ad 2-infected cells at late stages of infection (see Fig. 12.7 and Table 12.5) [5]. Thirteen of these have electrophoretic mobilities identical to proteins from labeled virus particles [157] while 3 are cleavage products. Russell and Blair [230] have recently reported that the 100,000 and 66,000 (IIIa) may be phosphoproteins. Fiber (IV) is a glycoprotein [357]. Cell-free translation of late Ad 2 mRNA is discussed in Section X.

Table 12.5 Adenovirus 2-Induced Proteins[a]

Band designation	Molecular weight	Relationship to virion
II	120,000	Hexon
100K	100,000	Nonvirion late protein
III	85,000	Penton base
70-75K	73,000[b]	(Single-stranded DNA binding phosphoprotein)
IIIa	66,000	Virion component
IV	62,000	Fiber (glycoprotein)
IVa_1	60,000	Virion component
IVa_2	56,000	Virion component
53K	53,000[c]	Early protein[c]
50K	50,000	Nonvirion late protein
V	48,500	Core
P-VI	27,000	Precursor to VI[d]
P-VIII	26,000	Precursor to VIII[d]
VI	24,000	Hexon-associated
P-VII	20,000	Precursor to major core (VII)[d]
-	19-21,000	E_2-glycosylated[e,g] (early protein)
VII	18,500	Major core (AAP)
18K	18,000	Early protein[f,g]
15K	15,000	Early protein[f,g]
14K	14,500	Early protein [g]
13.5K	13,500	Early protein[g]
VIII	13,000	Hexon-associated (appears during chase)
IX	12,000	Hexon-associated
11.5K	11,500	Early protein[f,g]
10-11K	11,000	(E_3) Early protein[f,g]
X	6,500	Virion component (appears during chase)
XI	6,000	Virion component (appears during chase)
XII	5,000	Virion component (appears during chase)
8 to 9K		Additional early proteins

[a]Adapted from [157]. Apparent molecular weights determined by SDS polyacrylamide gel electrophoresis.
[b]Molecular weight of highly purified native Ad 2 DNA binding protein [263].
[c]Candidate transformation protein [351], probably equivalent to an Ad 5 58K [360].
[d]Precursor-product relationships between P-VI and VI, P-VII and VII, and (with less certainty) P-VIII and VIII, have been established by pulse-chase experiments [5, 310], peptide mapping [5, 346], cell-free translation [6, 158, 366], immunological procedures [366], and genetic studies [310].
[e]Ishibashi and Maizel [357].
[f]Chin and Maizel [343]; [102].
[g]Jeng, Wold, and Green, unpublished data.

VII. Adenovirus DNA Replication

A. A Model for DNA Replication in Mammalian Cells

The synthesis of DNA in prokaryotic and eukaryotic cells is complex, involving numerous enzymes and other proteins that may function in multi-enzyme systems. A molecular description of DNA replication in *Escherichia coli* is only beginning to be realized, after fifteen years of intensive studies using small bacteriophage DNA molecules as probes [239]. Full details, however, will no doubt require many years of additional study. Several of the approximately fifteen proteins that are involved in DNA replication in *E. coli* (excluding DNA precursor enzymes), as indicated by biochemical and/or genetic studies, have not yet been identified. Moreover, functions have been established for only some of these proteins, and their complex interactions have only recently begun to be appreciated.

Much less is known about DNA replication in mammalian cells. The complexity of the mammalian genome limits the use of cell DNA synthesis to understand the mechanism of DNA replication. On the other hand, cells replicating Ad genomes are particularly useful models to study DNA replication and its regulation, for the following reasons. (1) Cell DNA synthesis is blocked late after infection with human Ads [74, 214] thus permitting the unambiguous analysis of viral DNA synthesis. (2) Isolated nuclei [268] and subnuclear complexes [249, 331-333] have been prepared from Ad 2, 5, and Ad 12-infected cells that synthesize almost exclusively viral DNA sequences in vitro. (3) DNA-negative Ad 5 and Ad 12 *ts* mutants (Section V) are available which are very useful in understanding DNA replication. (4) With the exception of viral-coded proteins involved in initiation, Ad DNA (papovavirus DNA as well) is replicated mainly by cellular enzymes and mechanisms. Thus, the study of Ad DNA replication illuminates cellular mechanisms of DNA replication.

B. Viral-Coded Proteins Involved in Adenovirus DNA Replication

In contrast to the replication of mammalian cell DNA which requires continuous protein synthesis, the replication of Ad DNA occurs in the presence of cycloheximide, an inhibitor of protein synthesis [112, 333]. Viral proteins synthesized early after infection, in the absence of viral DNA replication, are required for viral DNA synthesis late after infection. At least 3 viral-coded proteins are required for replication of group C Ad DNA, since 3 DNA-negative complementation groups (H5*ts*125, H5*ts*36, H5*hr*1) have been identified for Ad 5 mutants [75, 323, J. Williams, personal communication]. H5*ts*125 seems unable to initiate viral DNA synthesis at NPT [297]. Three DNA-negative complementation groups, apparently also defective in the initiation of DNA synthesis, were reported for Ad 12 *ts* mutants [251].

Polypeptides of 70-75,000 daltons, that bind to single-stranded DNA, have been isolated from cells infected with Ad 5 and Ad 2 [243, 294, 295]. The DNA binding protein is an Ad-coded early protein (Sections V and X). The Ad 2 DNA binding protein has been purified to homogeneity and several of its properties studied [263]. Sedimentation velocity and gel filtration studies indicate that it has a molecular weight of 73,000 (73K) and that it is a fibrous rather than a globula protein. The pure protein binds cooperatively to single-stranded but not to native Ad 2 DNA, in a manner molecularly analogous to that of the T4 gene 32 DNA binding protein that is required for T4 DNA replication and genetic recombination [2]. One molecule of 73K binds to about 25 nucleotides of single-stranded Ad 2 DNA. The 73K is a phosphoprotein [124, 230, 362], and contains phosphoserine but not phosphothreonine [124]. Hydrolysis of the phosphate groups using alkaline phosphatase apparently does not grossly affect the protein's affinity for single-stranded DNA [362].

Temperature-shift experiments with H5*ts*125 suggest that this protein functions in initiation of DNA replication [294, 295]. Consistent with this, 73K is found in Ad 2 nuclear membrane [243, 331, 333] and soluble ([332]; Rho, et al., BBRC, in press) DNA replication complexes. Immunofluorescence studies have shown that 73K is localized in the nucleus during the active period of Ad 2 DNA replication [375]. The 73K is produced in large amounts, suggesting a stoichiometric rather than catalytic role. At NPT, H5*ts*125 transforms cells more efficiently than wild type Ad 5; thus, the DNA binding protein may play an indirect role in cell transformation [75]. Although some transformed cells produce the protein, most do not, indicating that it is not a transformation-maintenance protein [361, 375].

C. Replicative Forms and Viral DNA Intermediates

Intermediates in Ad DNA replication have been identified by labeling intact cells, isolated nuclei, and subnuclear complexes. In vitro preparations of nuclei and subnuclear fractions, although not shown to initiate DNA synthesis in vitro, provide the advantage of rapid equilibration into unlabeled pools of added label and inhibitors, and facilitate the identification of enzymatic and other components involved in viral DNA synthesis. One disadvantage is that the DNA synthesis machinery may be altered in vitro.

Ad DNA replication is semiconservative [12, 296, 331, 332], and may be discontinuous, proceding via 9-11S "Okazaki" fragments [12, 331], that are derived from both strands of most of the Ad genome [301, 325]. Replicating Ad DNA consists of linear genome-length molecules with single-stranded regions; no circular or linear concatomers have been detected [12, 110, 197, 201, 223, 266, 292]. The single-stranded DNA is derived from both strands of most of the viral genome [144, 286, 348, 377]. Temperature-shift studies with H5*ts*125 suggest that the single-stranded DNA is generated as the Ad genome is replicated [348]. Electron microscope studies of replicating viral DNA have detected both branched and linear forms with single-stranded regions [267, 268].

Within the past 2 years, the termini and origins of Ad DNA replication have been localized by several groups, using basically the method of Danna and Nathans [33]. In this experiment, viral DNA is pulse-labeled for a period (e.g., 10 min), less than required to synthesize full length strands. Mature native DNA molecules are then isolated, and the distribution of radioactivity in restriction fragments determined. With Ad DNA, much more label was found in terminal than in middle fragments, indicating that termini are located at both ends of the genome [240, 286, 356, 376, 379]. Analyses of the strand specificity of the radioactivity in the terminal fragments (either by hybridization to separated fragment strands, or by direct separation of the labeled strands), showed that the label was in the *l* strand of right hand terminal fragments, and the *r* strand of left hand fragments [356, 376, 379]. These results indicate that initiation of synthesis of *r* and *l* strands occurs at the right and left ends respectively of the parental strands. Consistent with this, Flint et al. [348] have found that *r* strand is at a higher concentration than *l* strand in genome regions to the right of map position 28-41, whereas *l* strand predominates to the left of position 28-41. The above studies are in accord with the basic idea of Horwitz [111], that Ad replication proceeds in a 5′ to 3′ direction on each strand, with initiation and termination of replication occurring at or very near the ends of both strands. The problem of completing the 5′ ends is discussed below.

D. Enzymes Involved in Adenovirus DNA Replication

Because multiple DNA polymerase activities are present in mammalian host cells and isolated nuclei, it is difficult to identify the specific enzymes involved in Ad DNA replication. Three major DNA polymerase enzymes, not including mitochondrial DNA polymerase, have been identified in mammalian cells: DNA polymerase α, β, and γ [314]. The major DNA polymerase α (molecule weight 1.2-2.2 × 10^5) as well as β (molecular weight 4-5 × 10^4) have been purified to homogeneity. The minor DNA polymerase γ appears to represent several species with a molecular weight of about 1.1 × 10^5. The functions of these polymerases in DNA replication has not been established (see review by Edenberg and Huberman [44]). (see review by Edenberg and Huberman [44]).

The analysis of subnuclear replication complexes isolated from Ad-infected cells may help clarify the role of various enzymes in DNA replication. The major DNA polymerase in a nuclear membrane complex capable of synthesizing Ad 2 DNA Okazaki fragments in vitro has been purified about 900-fold from Ad 2-infected KB cells [121]. The enzyme was characterized as belonging to the class of mammalian DNA polymerases (DNA polymerase γ) that can utilize poly(A)·oligo(dT) as template primer [121]. This suggests that γ DNA polymerase may play a role in Ad DNA synthesis. A detailed comparison of the complex from uninfected and Ad 2-infected cells revealed no differences in the DNA polymerase γ species with regard to chromatographic properties and template specificity. A minor DNA polymerase species has also been identified [122].

A second subnuclear complex solubilized from nuclei of Ad 2-infected cells can complete the synthesis of viral DNA molecules in vitro [332]. The soluble complex appears to contain at least 5 components, DNA polymerase, RNA polymerase, DNA ligase, the 73K single-stranded DNA binding protein and RNase [338]. The role of these various enzymes in the synthesis of Ad DNA is not yet established.

E. Current Model for Adenovirus DNA Replication

Recent studies have begun to clarify the mechanism of Ad DNA replication. The Ad genome is replicated in the cell nucleus, possibly in association with a specific replication complex, as suggested by both biochemical and genetic studies. Several types of subnuclear preparations that can synthesize Ad DNA sequences in vitro have been described, although none have yet been shown to both initiate and complete the synthesis of mature viral DNA. Some of these preparations contain nuclear membrane material that probably becomes associated during isolation, since recent studies have suggested that Ad DNA is synthesized in the interior of the nucleus and not on the nuclear membrane [252, 254, 302].

Evidence that both daughter strands of DNA are synthesized discontinuously as relatively large (9-11S) Okazaki fragments has been presented. Evidence that the origin of DNA replication is at the right- and left-hand ends is both strong and is attractive, since the inverted terminal repetition and covalently bound protein at each terminus could serve as recognition sites for a DNA polymerase or for specific initiation proteins.

A perplexing problem is a mechanism for completion of the 5′ end of the daughter strands of linear Ad DNA. Since Ad DNA molecules contain neither duplex terminal repetitions nor cohesive ends, they cannot form the type of circular or concatemeric structures that are intermediates in the replication of linear bacteriophage DNA molecules [308]. Three possible ways to overcome this dilemma have been suggested. Robinson et al. [225] suggested that replicating viral DNA is a circular DNA-protein complex. Cavalier-Smith [24] proposed a general model for completion of linear DNA molecules which requires palindromic sequences at DNA termini. According to the Cavalier-Smith model, after removal of the RNA primer at the 5′ termini of daughter DNA molecules, the self-complementary 3′ terminus (containing the palindrome) of the parental strand could form a hairpin loop, the 3′-OH then serving as a primer to fill the gap left by removal of the RNA primer. After sealing by polynucleotide ligase, the action of a specific endonuclease and DNA polymerase could then complete the maturation of the duplex Ad DNA. Electron microscope studies [372] have detected internal secondary structures within single-stranded inverted terminal repetitions. Conceivably, this structure functions in completing the synthesis of Ad 2 DNA. However, recent findings argue against this notion. See Recent Developments for a third proposed mechanism.

VIII. Cell Transformation

The major goals of tumor virology are to identify viral genes that initiate and maintain cell transformation and to isolate and understand the functions of those viral proteins that are coded by transforming genes. These are realistic possibilities in the case of the oncogenic Ads because the technology has recently become available to identify viral transforming genes and the mRNA and protein products they specify, and to purify the protein in large quantities for their biochemical characterization. Furthermore since the human Ads form 3 oncogenic groups (see Section II), they provide unique opportunities to make functional comparisons of viral-transforming genes and proteins.

Human Ads transform hamster, rat, and rabbit cells, giving rise to typical epithelial cells that grow to high cell density and form colonies in "soft agar" [83]. The efficiency of transformation is low, and only a small fraction of cells are transformed (i.e., transformation is not part of the Ad life cycle). It seems likely that the major factor dictating the efficiency of transformation is the species of host cell, and that most if not all human Ad serotypes can transform cells, providing that viral DNA segments containing transforming genes are integrated and expressed. As a general rule of thumb, viruses transform cells that are nonpermissive, or semipermissive for virus replication. For example, Ad 12 does not replicate in hamster cells and readily induces tumors in newborn hamsters and transforms hamster cells in vitro, while Ad 2 and Ad 5 replicate in hamster cells, do not induce tumors in newborn hamsters, and transform hamster cells in vitro with low frequency (see Section IX.D) [152, 153, 319]. The Ad 2 and Ad 5-transformed hamster cells do induce tumors when injected into newborn hamsters. As another example, Ad 2 and Ad 5 replicate very poorly in rat cells (i.e., rat cells are semipermissive for Ad 2 and Ad 5 replication), transform cells with low efficiency in vitro, and do not induce tumors in rats [173]. Ad 2-transformed rat cells vary in their ability to form tumors when injected into syngeneic or immunosuppressed rats, or nude mice. Finally, Graham (personal communication) has isolated a line of human embryo kidney cells transformed by transfection with sheared Ad 5 DNA. The implication of viral transformation of cells permissive for replication is important because this raises the possibility that, for example, human Ads may occasionally undergo abortive infections in "natural" human infection, and thus could be involved in human cancer.

Progress during the past 10 years in understanding the molecular events of cell transformation by oncogenic DNA viruses has been impressive. The main event of virus-induced cell transformation appears to be a change in the regulation of cell growth brought about by the functioning of protein products from integrated viral genes. This was by no means clear 10 years ago when no infectious virus or infectious viral DNA could be detected in tumors induced by Ads or papovaviruses [82]. The possibility that cell transformation by these oncogenic DNA viruses was indirect, a "hit and run" mechanism not involving viral genes, could not be ruled out. Huebner et al. [116] had detected T-antigens in Ad- and papovavirus-transformed cells that were specific for each virus type, consistent with their identity as viral-coded proteins, but not excluding the possibility that they were derepressed cell proteins. The presence of viral information was demonstrated more directly by Fujinaga and Green [56], who found viral mRNA in the polyribosomes of Ad 12-transformed cells, and by Benjamin [13] who detected virus-specific RNA in polyomavirus-transformed cells. Proof that the transformed cell phenotype resulted from the expression of persisting viral genes was suggested by the isolation of *ts* mutants (early region) of SV40 that seem defective in maintaining transformation [18, 135, 190, 278, 279]. Analogous Ad mutants have not yet been isolated. However, as described below, biochemical studies with Ad-transformed cells strongly support the view that Ad gene products are responsible for initiating and maintaining the transformed state.

A. Viral DNA Sequences in Adenovirus-Transformed Cells

Cells transformed by Ad 2, 7, and 12, members of oncogenic groups C, B, and A, respectively, contain multiple copies of viral DNA sequences, as determined by hybridization of transformed cell DNA on filters to radioactive viral complementary RNA (cRNA) [88]. Ad 2, 7, and 12 cRNA synthesized in vitro by the *E. coli* RNA polymerase contained 80% early (and transformed) and 20% late viral RNA sequences [87]. Although RNA sequences derived from the entire Ad genome are transcribed in vitro [207], segments containing early genes are copied more frequently [87, 265].

By hybridization of viral cRNA with Ad 2-transformed rat cell DNA (8617 cell line), estimates of 8-12 viral DNA equivalents per diploid quantity of cell DNA (3.9×10^{12} daltons) were reported [88]. On the other hand, reassociation kinetics of labeled Ad 2 DNA in the presence of 8617 DNA indicated 1 copy of Ad 2 DNA per cell [206]. However, as later shown by Sambrook et al. [236], 8617 cells contain only 46% of the viral genome, so that reassociation kinetics using the whole Ad 2 genome leads to an underestimation of copy number. Reassociation measurements with Ad 2 restriction fragments revealed that different segments of the viral genome are present in different amounts, and that 13 copies of a small segment at the left block end (early transforming gene block) are present in 8617 cells [236], in surprisingly good agreement with the results by cRNA hybridization [88]. Ten Ad 2-transformed rat cell lines and 2 Ad 5-transformed hamster cell lines were found to contain 4-16 copies of different viral DNA fragments but no complete viral genomes; the left 14% of the viral genome was common to all group C transformed cell lines examined [66, 236, 246]. Most exciting, the endonuclease Hind III-G fragment (molecular weight 1.6×10^6) from the left end of Ad 2 and Ad 5 genomes could transform cells by transfection; other restriction fragments were inactive [78]. By digestion of Ad 5 DNA with exonuclease III and single-strand specific nuclease, Graham et al. [78] also determined that about 1% of the genome could be removed at the ends without loss of transforming activity. These studies localized the transforming genes at Ad map positions around 0.01 to 0.075.

A new method for analysis of the reassociation kinetics of viral DNA was used to estimate copy number and the fraction of the viral genome present in a tumor cell line induced by Ad 7, a group B weakly oncogenic Ad. Several hundred copies of approximately 20% of the viral genome were detected [63].

The pattern of viral genes in nonpermissive hamster cells transformed by Ad 12 (the HE C19 cell line), a member of highly oncogenic group A, is uniquely different from that described for cells transformed by group C and B human Ads. HE C19 contained all or nearly all (93-100%) of the viral genome [94].

Moreover, all portions of the viral genome, as defined by endonuclease Eco RI restriction fragments, are present in nearly equimolar quantities, 8-10.5 copies per cell. An Ad 12 hamster tumor cell line (HT2) was also shown to contain 18 copies of most of the viral genome [94]. The presence of multiple copies of most of the Ad 12 genome was demonstrated also in Ad 12-induced hamster tumors [96]. Since in vitro Ad 12-transformed cells closely resemble cells from Ad 12-induced tumors in the manner in which the Ad 12 genome is integrated, Ad 12 in vitro transformed cells provide realistic models for analyzing viral tumorigenesis. Cell transformation and tumor induction by highly oncogenic Ad 12 may involve a different mechanism of integration than that for group B and C Ads (which integrate only portions of their genomes), possibly explaining the increased oncogenicity of Ad 12.

The integration of Ad DNA into the transformed cell genome is suggested by several studies. Ad 7 and Ad 12 sequences are present in similar quantities in DNA isolated from chromosomes and from nuclei of transformed cells [83, 88]. By in situ hybridization, viral DNA was associated with chromosomes of Ad 2-, 7-, and 12-transformed cells [43, 161]. Most significant, viral RNA molecules containing covalently attached cellular RNA sequences are present in Ad 2- [289, 304] and Ad 7-transformed [289] cells; these molecules most likely arise from the uninterrupted transcription of integrated viral and contiguous cell DNA sequences. Recently, direct evidence for integration was provided by the demonstration that cell DNA networks prepared from Ad 12 [94] and avian CELO-transformed [11] hamster cells contain most and probably all of the viral DNA sequences in the cell (8 copies of Ad 12 and 2.5 copies of CELO DNA). Cell DNA networks are formed by reannealing high molecular weight cell DNA which contains interspersed reiterated DNA sequences; brief centrifugation separates cell DNA with integrated viral DNA from free viral DNA molecules [298].

B. Transcription of Viral RNA in Adenovirus-Transformed Cells

Fujinaga and Green [56] showed that 2-5% of pulse-labeled (20 min) mRNA in polyribosomes of Ad 12-transformed hamster embryo cells was virus-specific. It was suggested that viral RNA sequences were preferentially transcribed and/ or processed to mRNA. From 1966-1969, viral mRNA was demonstrated in virus-free tumors and transformed cells induced by 10 human Ads, 3 members of groups A, 4 of group B, and 3 of group C [56-59, 62, 91]. Of RNA labeled during a 3 hr exposure to [^{3}H]uridine (cell mRNA, rRNA, and tRNA are also labeled), 0.1-0.3% hybridized to homologous viral DNA, levels 5-20 times higher than found with most SV40- and polyoma-transformed cells [83].

Viral RNA from group A-transformed cells hybridized only with the 3 group A viral DNAs and not with groups B and C DNAs. Likewise, RNA from group B- or group C-transformed cells hybridized only with viral DNAs of their respective groups. These relationships, confirmed with highly purified viral RNA preparations [59], showed that virus-specific RNA sequences expressed in transformed cells are group-specific and that different viral coded information is involved in transformation by the 3 oncogenic Ad groups. These early findings are consistent with recent DNA homology measurements showing that Ad 2 and Ad 12 restriction fragments containing the transforming genes hybridize exclusively with Ad DNAs of the same group (see Table 12.2).

The base composition of highly purified ^{32}P-labeled virus-specific RNA isolated from Ad tumor and transformed cells by DNA-RNA hybrid formation and elution was determined [59, 60]. Although Ad 12- and Ad 18-specific RNA have G + C contents of 47-48%, values similar to those of viral DNA, the G + C contents of Ad 7 and Ad 16 (group B) RNAs are 3-4% lower than the corresponding DNAs, while the G + C content of Ad 2, 5, and 6 (group C) RNAs are 8% lower than their viral DNAs. These data indicate that viral DNA regions with an average G + C content of 47-50% are integrated or transcribed selectively in Ad-transformed cells. Thus 2 lines of evidence suggested that only a portion of the viral genome functions in Ad-transformed cells: virus-specific RNA from cells transformed by group A, B, and C Ads do not share base sequences, yet the different transforming viruses share 5-25% of their DNA sequences; and the G + C content of viral RNAs isolated from group B- and C-transformed cells are lower than those of viral DNAs.

The fraction of the Ad 2 and 7 genome transcribed early and late during productive infection and in transformed cells and the relationship between the 3 viral RNA species were determined by hybridization competition measurements with the following results [60, 88]: (1) viral mRNA sequences expressed in Ad 2- (8617 cell line) and Ad 7- (5728 cell line) transformed cells are transcribed early during productive infection, i.e., they derive from early genes, and they are present late after infection; (2) 4-10% of the Ad 2 genome, i.e., 1-5 genes, and about 20% of the Ad 7 genome, are transcribed in the transformed cell lines; (3) early Ad 2 genes expressed in 8617 cells are not transcribed efficiently late during productive infection; and (4) early Ad 7 genes expressed in Ad 7-transformed cells are transcribed late during productive infection. The most significant finding is that the transcription of only a small fraction of the Ad genome occurs in the transformed cell.

The sequences of Ad 2 DNA expressed as cytoplasmic RNA, presumably viral mRNA, in 5 Ad 2-transformed cell lines were analyzed by liquid phase hybridization between ^{32}P-labeled strands of restriction fragments and unlabeled RNA in excess [53, 245]. All lines contained viral mRNA corresponding to 7% of the viral genome and complementary to the *r*-strand of the far left end of the Ad 2 genome (Figs. 12.12 and 12.9), i.e., the Hind III-G fragment and the Hpa I-E and to some extent the Hpa I-C fragment. Two cell lines (F17 and F18) contained only these left-end RNA sequences; 2 others (8617 and REM) contained in addition mRNA complementary to about 7% of the right end of Ad 2 DNA; a fifth line (T2C4) contained these and additional viral RNA sequences. The viral RNA sequences found in all Ad 2-transformed cell lines correspond to early gene sequences [53]. These findings confirm the hybridization-competition studies of Fujinaga and Green [60] with 8617 cell RNA, which showed that only early RNA sequences derived from 4-10% of the Ad 2 genome are expressed in the 8617 cell line.

Ad 2-specific RNA in the nucleus of 8617 cells was found to be larger than that in the cytoplasm [195, 304], leading to the suggestion that nuclear viral RNA sequences are covalently linked to host cell RNA [195]. The existence of covalently attached host-cell RNA was proven by isolation of viral RNA from transformed cells using multiple hybridizations and elutions under mild conditions; RNA selected in this manner from Ad 2- [289, 304] and Ad 7- [289] transformed cells hybridized to both viral and cell DNA with high efficiency, suggesting that viral DNA is integrated adjacent to highly reiterated cell DNA sequences. Ad 2 and Ad 7 RNA isolated from productively infected cells did not hybridize significantly to cell DNA [289]. Most of the cell sequences in 8617 cells are removed during processing since viral RNA from polyribosomes contains little if any cell-specific sequences [304]. Virus-specific nuclear and cytoplasmic RNA both contain poly(A) sequences, presumably at the 3′ terminus, and mostly the same viral RNA sequences [248] consistent with a precursor-product relationship [304]. However, it has been difficult to prove such a relationship since most pulse labeled nuclear virus-specific RNA labeled is 20S, the size of the major virus-specific mRNA species [69].

Three size classes of virus-specific RNA were found in the cytoplasm of 8617 cells by zonal centrifugation in sucrose density gradients, a major 20S and minor 26 and 16S RNA species [304]. The combined molecular weight of these species, 2.8×10^6, is much higher than the viral RNA sequence content of 8617 cells [53, 60]. Recent studies preparatively isolated 22S (MW 1.1×10^6 daltons) and 14S (MW 5.5×10^5) (gel electrophoresis S values) viral mRNA from 8617 cells by hybridization to Hind III-G fragment or Hpa I-E fragment [26]. Similar results were obtained by Bachenheimer and Darnell [339] who identified analogous 20S and 15S RNA species (sucrose gradient S values) from 8617 cells that were complementary to the left end of the Ad 2 genome. Several considerations argue for a precursor-product relationship between the 22S and 14S RNA species [26]. (1) Hpa-E has a single strand equivalent molecular weight of 4.5×10^5, which is insufficient to code both 22S and 14S RNA. (2) The size of the 22S species alone is sufficient to account for the hybridization of 8617 mRNA to the left end of

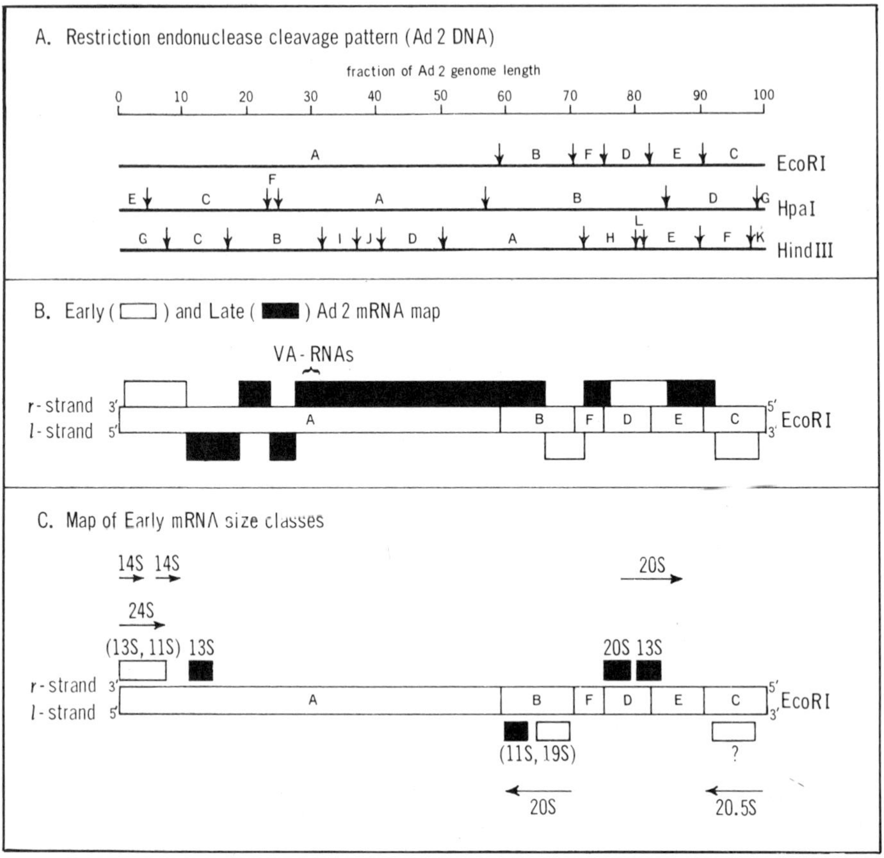

A. Restriction endonuclease cleavage pattern (Ad 2 DNA)
fraction of Ad 2 genome length
0 10 20 30 40 50 60 70 80 90 100
A B F D E C EcoRI
E C F A B D G HpaI
G C B I J D A H L E F K HindIII
B. Early ( ) and Late ( ) Ad 2 mRNA map
VA-RNAs
r-strand 3′
l-strand 5′
A B F D E C
5′ 3′ EcoRI
C. Map of Early mRNA size classes
14S 14S
24S
(13S, 11S) 13S
20S
20S 13S
r-strand 3′
l-strand 5′
A B F D E C
5′ 3′ EcoRI
(11S, 19S)
?
20S
20.5S

Fig. 12.12 Mapping of early and late Ad 2 mRNA. (A) Eco RI, Hpa I, and Hind III restriction endonuclease cleavage patterns of Ad 2 DNA. Cleavage sites for Eco RI and Hpa I were taken from Mulder et al. [177], and those for Hind III from R. J. Roberts, J. Sambrook, and P. Sharp (personal communication). (B) DNA coding early (light bars) and late (black bars) mRNA. This is the least complicated map possible, obtained by hybridizing early and late poly(A)-terminated cytoplasmic RNA to separated strands of Eco RI, Hpa I, and Hind III restriction fragments. There may still be uncertainty in the regions spanned by Hind III-B and C fragments. This map was taken from Pettersson et al. [367]. Similar studies have been carried out by Sharp et al. [245]. The location of VA-RNA (major and minor species) was determined by Mathews [169] and Pettersson and Philipson [205]. (C) Ad 2 DNA regions encoding size classes of early viral mRNA. Light bars represent class I RNA species, and black bars represent class II [27, 30]. The strand specificity of the RNA species was deduced from the map shown in (B). The arrows represent the early mRNA species observed by Buttner et al. [341]. The strand specificity was determined by hybridization of RNA species to *r* and *l* strands.

the viral genome [53]. (3) The ratio of 22S to 14S RNA varied in different experiments. Thus it is possible that 14S RNA arises by cytoplasmic processing of the 22S primary transcript, analogous to the late 19S and 16S species of SV40 [3].

The accumulation of Ad 2-specific RNA was shown to vary with the cell cycle of 8617 cells, being higher in early S phase than in late S or G2 [108]. The accumulation of RNA complementary to the Ad 2 Hind III-G and Eco RI-C fragments, the left and right ends of the viral genome, paralled that of total viral RNA throughout the cell cycle (Chinnadurai and Green, unpublished data). Therefore, there is no differential expression of different portions of the viral genome during the cell cycle.

Ad 12 abortively infects nonpermissive hamster cells leading to the establishment of transformation in a small fraction of cells. Abortively infected hamster cells synthesized early mRNA sequences [167, 262] and T-antigen but no viral DNA or virion proteins (reviewed in [262]). The same size classes of viral mRNA were detected in Ad 12 abortively infected and Ad 12-transformed cells with molecular weights 0.9×10^6 (major species), 0.66×10^6 (minor species); it is not established that these are discrete mRNA species.

Recent studies suggest that Ad 12-transformed hamster embryo cells may provide an exciting system to define the role of posttranscriptional mechanisms in gene regulation and RNA processing in eukaryotic cells [95]. Ad 12 DNA labeled to very high specific radioactivity in vitro (2-4 $\times$ 10^8 cpm/μg was used as probe in saturation-hybridization analysis of cell RNA. A minimum of 80% of the single-stranded viral genome equivalent was expressed as nuclear RNA, but only 20-22% as cytoplasmic RNA. This is the first demonstration that viral RNA sequences are present in the nucleus of the transformed cell that are absent from the cytoplasm. These findings suggest that strong posttranscriptional controls must operate to transport selectively viral mRNA sequences to the cytoplasm and to restrict non-mRNA sequences to the nucleus. This is analogous to mRNA biogenesis in uninfected eukaryotic cells in which most nuclear RNA sequences do not reach the cytoplasm. Of significance, this finding provides evidence for posttranscriptional regulation of gene expression as distinguished from RNA processing since nearly all the Ad 12 genome specifies mRNA, i.e., late after infection, mRNA sequences complementary to all or nearly all of the asymmetric genome, are present in the cytoplasm (Green and Green, unpublished data). By contrast, most of the eukaryotic genome does not code for mRNA. Since Ad 12-transformed hamster embryo cells contain multiple copies of all or most of the viral genome, and different segments of the genome are present in equimolar amounts, the Ad 12 system may provide the opportunity to study regulation of transcription from integrated multigenic sequences.

C. Adenovirus-Induced Proteins in Transformed Cells

Ad-transformed cells synthesize T-antigens, which include any early viral protein that is expressed in the tumor cell and is immunogenic (and not necessarily involved in transformation); and a transplantation antigen (TSTA) present on the cell surface and responsible for tumor rejection [255]. Viral transformation proteins are coded by the left end of group C viral DNA and probably have T-antigen specificity. Although the 70,000-75,000 single-stranded DNA-binding protein is an early viral coded product [295], and is a T-antigen [70], it is not a transformation protein since it maps in the middle of the Ad 2 genome (Fig. 12.9). To identify transforming proteins, Gilead et al. [351] prepared antisera in rats against the Ad 2-transformed rat cell line, F17, which expresses only 7% of the Ad genome in the region of transforming genes [53]. This antisera precipitated 53,000 and 15,000 dalton polypeptides from Ad 2 early infected KB cells [351]. Rat antisera antisera other Ad 2-transformed rat cell lines (T2C4, F4, 8617) also precipitated the 53,000 and the 15,000 (Wold and Green, unpublished data). In similar types of studies, using a hamster cell line, 14b (which expresses mainly transforming genes as mRNA), Levinson and Levine [360] identified a 58,000 polypeptide. Most interesting, Lewis et al. [363] observed 44,000-50,000 and 15,000 polypeptides, after translation of mRNA selected by hybridization of early infected or transformed cell (F17, 8617, 14b) RNA to transforming fragments (e.g., HindIII-G, see Section X). These exciting findings suggest that the 53,000 and/or the 15,000 polypeptides may be viral early-coded transformation proteins that maintain cell transformation. Current efforts are focused on the functional characterization of these proteins, which should tell us much about Ad cell transformation.

IX. Adenovirus-Simian Virus 40 (SV40) Hybrids

A. Simian Virus 40 Enhancement of Adenovirus Replication in Monkey Cells

Ads do not generally grow in monkey cells, but coinfection with SV40 permits their growth [217], presumably because of some function supplied by SV40. This phenomenon is termed "enhancement." The nature of the block to Ad replication in monkey cells is not understood. Early Ad genes are expressed in unenhanced cells (see [9]), but the synthesis of some late Ad proteins, especially capsid proteins, is greatly reduced. The block in synthesis of late proteins is posttranscriptional [9]. It has been suggested that late Ad mRNA does not associate properly with ribosomes [9, 104], and that the SV40 function enhances by altering ribosomes to enable them to translate late Ad mRNA [180]. In contrast, Eron et al. [48] found comparable amounts of Ad 2 mRNA on polyribosomes from both enhanced and unenhanced cells, and that both enhanced and unenhanced mRNA is translated with equal efficiency in a cell-free translation reaction into a number of proteins, including fiber, penton base, and hexon. Different results were obtained by Klessig and Anderson [136] who found a marked deficiency in the amount of fiber and 11,500 protein, and a reduction in several other viral proteins, upon translation of unenhanced mRNA compared to enhanced mRNA. They also found a reduction in concentration of mRNA complementary to the sections of the Ad 2 genome known to encode some of these proteins (Fig. 12.9), and suggested that the block to Ad 2 replication in monkey cells occurs at the level of mRNA processing. Clearly, much more work is necessary to clarify the Ad abortive infection in monkey cells.

Little is known about the SV40 helper function that enhances Ad replication. Experiments with SV40 *ts* mutants indicate that an early gene (i.e., before SV40 DNA replication commences) function is either directly or indirectly involved [125, 126, 134]. It is relevant that the only known early SV40 gene product is a 70,000-100,000 protein that binds to double-stranded DNA, and is necessary for initial viral DNA replication and maintenance of cell transformation (see below). Treatment of monkey cells with iododeoxyuridine also enhances (in the absence of SV40) Ad replication [126, 259] suggesting that an "inducible" host cell factor is involved. Perhaps the SV40 early gene product affects the induction of this factor.

B. Ad-SV40 Hybrids: Isolation and Properties

Ad-SV40 hybrids are recombinant viruses with covalently linked Ad and SV40 DNA sequences enclosed within an Ad capsid [209, 218]. All hybrids contain deletions in the Ad genome, and insertions of all or part of the SV40 genome (see Fig. 12.13 and Table 12.6). These viruses were originally isolated when Ads were adapted to grow on monkey cells (Ad 1-5, 7) for the purpose of Ad vaccine production, or when Ad 12 and SV40 were propagated together on monkey cells. Hybrids fall into 2 categories: (1) defective and unable to replicate without a helper wild-type virus that is always present in the virus stocks, and (2) nondefective and able to replicate normally without helper virus. Both defective and nondefective hybrids are being actively studied, and have provided the following types of information.

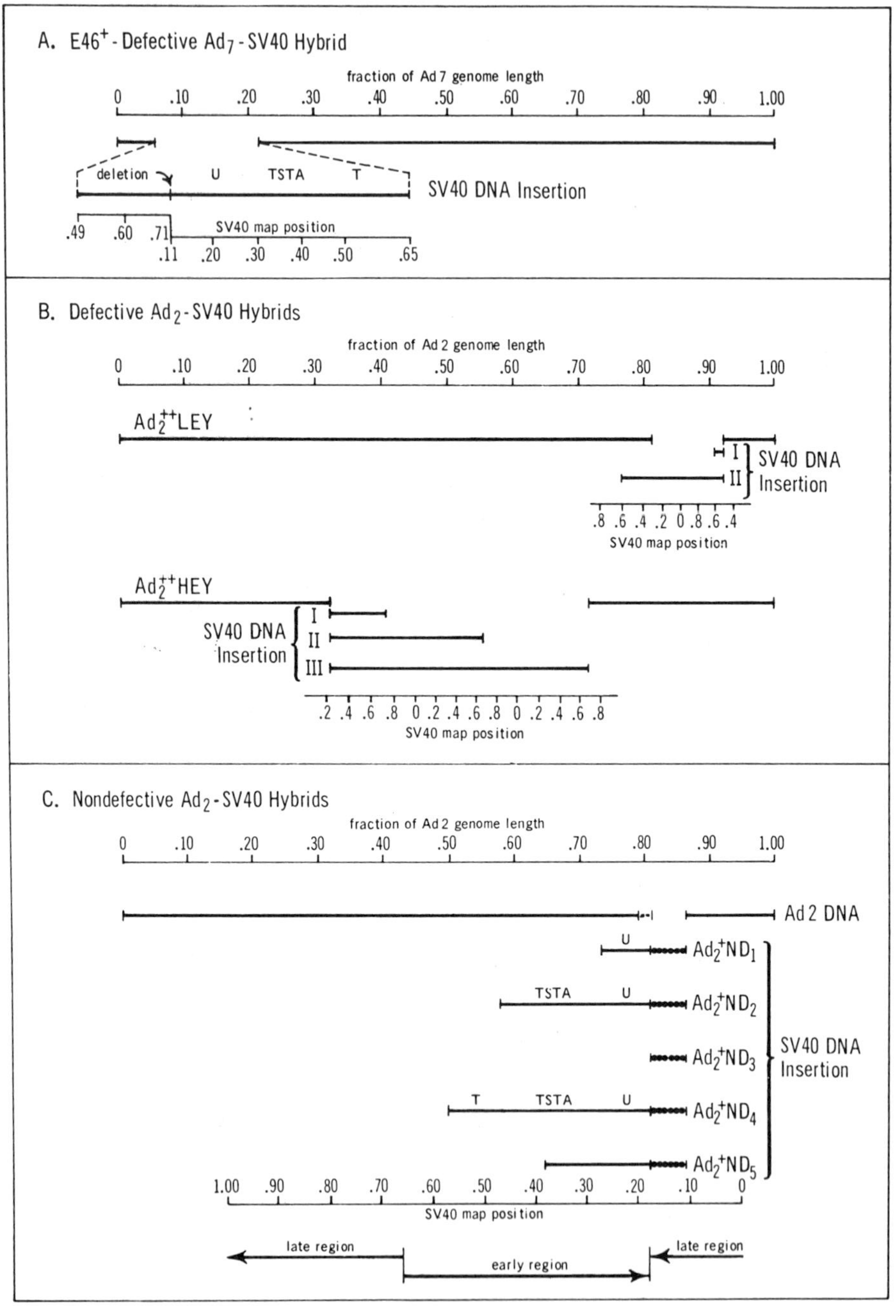
A. E46+ - Defective Ad7 - SV40 Hybrid
fraction of Ad 7 genome length
0 .10 .20 .30 .40 .50 .60 .70 .80 .90 1.00
deletion
U
TSTA
T
SV40 DNA Insertion
.49 .60 .71
SV40 map position
.11 .20 .30 .40 .50 .65
B. Defective Ad2 - SV40 Hybrids
fraction of Ad 2 genome length
0 .10 .20 .30 .40 .50 .60 .70 .80 .90 1.00
Ad2++LEY
I
II
SV40 DNA Insertion
.8 .6 .4 .2 0 .8 .6 .4
SV40 map position
Ad2++HEY
SV40 DNA Insertion
I
II
III
.2 .4 .6 .8 0 .2 .4 .6 .8 0 .2 .4 .6 .8
SV40 map position
C. Nondefective Ad2 - SV40 Hybrids
fraction of Ad 2 genome length
0 .10 .20 .30 .40 .50 .60 .70 .80 .90 1.00
Ad 2 DNA
U
Ad2+ND1
TSTA
U
Ad2+ND2
Ad2+ND3
T
TSTA
U
Ad2+ND4
Ad2+ND5
SV40 DNA Insertion
1.00 .90 .80 .70 .60 .50 .40 .30 .20 .10 0
SV40 map position
late region
early region
late region

1. Identification of regions in SV40 DNA involved in induction of SV40 U-, TSTA-(SV40-specific transplantation antigen), and T-antigens.
2. Sequences at SV40 map positions around 0.18 specify the 3′ end of early mRNA.
3. Identification of possible early SV40-induced polypeptides.
4. The functions of part of the Ad 2 early gene block that spans the Eco RI-D and E fragments (Fig. 12.12) are not essential for Ad 2 replication.
5. Regions between Ad 2 map positions 0.326-0.719 and 0.825-0.934 are essential for Ad 2 replication.

Fig. 12.13 Ad-SV40 genome structure (see Table 12.6). (A) Structure of $E46^+$, a defective Ad_7-SV40 hybrid. Stocks of $E46^+$ also contain helper wild-type Ad 7. Ad 7 DNA from 0.05-0.21 is deleted. The SV40 DNA insertion ranges from map position 0.49 on the left to 0.65 on the right. Sequences at position 0.71 are linked to those at 0.11, with the sequences 0.71-0.11 deleted. SV40-specific antigens T, TSTA, and U are induced by $E46^+$, and the SV40 DNA sequences apparently involved in their expression are shown. Probably only early SV40 DNA sequences are expressed as mRNA, and the transcription of these is right to left in the diagram. Adapted from Kelly [127] and Lebowitz and Khoury [146]. (B) Structure of Ad_2^{++}LEY- and Ad_2^{++}HEY-defective hybrids. Both LEY and HEY contain insertions of the complete SV40 genome. LEY is a mixture of wild-type Ad 2, and 2 different hybrids LEY-I and LEY-II. HEY is a mixture of wild-type Ad 2, and 3 different hybrids, HEY-I, HEY-II, and HEY-III. The SV40 minus (early) strand is integrated into the Ad 2 *r*-strand with LEY, and into the Ad 2 *h*-strand with HEY. Adapted from Kelly et al. [129, 130]. (C) Structure of 5 nondefective Ad_2-SV40 hybrids. Depending upon the hybrid, 4.5-7.1% of the Ad 2 genome is deleted, to the left of Ad 2 map position 0.86. All hybrids have the SV40 minus (early) strand integrated into Ad 2 *r*-strand, terminating at the right of the SV40 insertion at SV40 map position 0.11. Transcription of the SV40 minus strand proceeds from left to right in the diagram. Only early SV40 regions are expressed in both early and late stages of productive infection, and the 3′ end of the SV40 RNA transcripts maps at SV40 map position around 0.18. That is, SV40 regions 0.18-0.11 are not represented as "cytoplasmic mRNA". With $Ad_2{}^+ND_1$ and $Ad_2{}^+ND_3$, Ad 2 *r*-strand DNA sequences at least 1000 bases to the left of the SV40 insertion are represented as abundant cytoplasmic RNA sequences early and late after infection. Adapted from Walter and Martin [307], and from the results of a number of studies (see Table 12.6).

Table 12.6 Properties of Adenovirus-SV40 Hybrids

	Growth on monkey cells	SV40 insertion (map position)	Adenovirus deletion (map position)	SV40 Antigens induced	Hybrid specific polypeptides
Defective					
$E46^+$ (Ad_7-SV40)	+	See text[a]	16% (0.05-0.21)[b]	U, TSTA, T	ND[c]
Ad_2^{++}HEY-I[d]	+	0.45 genomes[d]	39.3% (0.326-0.719)[d]	Yields infectious SV40[e]	ND
HEY-II	+	1.43 genomes			
HEY-III	+	2.39 genomes			
Ad_2^{++}LEY-I	+	0.03 genomes	10.9%[d] (0.825-0.934)	U, TSTA, T[e] No capsid antigens	ND
LEY-II	+	1.05 genomes			
Nondefective[h]					
$Ad_2^+ND_1$	+	20%[f] (0.110-0.28)	5.4%[d] (0.806-0.860)	U	28K[g]
$Ad_2^+ND_2$	+	36.5%[f] (0.110-0.475)	6.1%[d] (0.799-0.860)	U, TSTA	42', 56K[g]

$Ad_2^+ND_3$	–	6.4%[f] (0.110-0.174)	5.3%[d] (0.809-0.860)	None	None[g]
$Ad_2^+ND_4$	+	46.4%[f] (0.110-0.573)	4.5%[d] (0.807-0.860)	U, TSTA, T	56K[g]
$Ad_2^+ND_5$	–	27.3%[f] (0.110-0.383)	7.1%[d] (0.789-0.860)	None	42K[g]

[a]Kelly, [127].
[b]Kelly and Lewis, [128].
[c]Not done.
[d]Kelly et al., [129, 130].
[e]Siegel et al., [253].
[f]Lebowitz et al., [145].
[g]Walter and Martin, [307].
[h]All defective and nondefective hybrids were isolated from Ad_2^{++} (pool 2).

In addition, these hybrids are being studied as model systems to understand the transcription of the SV40 genome in transformed cells, since both hybrids and transformed cells have linear DNA molecules with integrated early regions of SV40 DNA.

C. Defective Ad-SV40 Hybrids

1. $E46^+$

The first hybrid discovered was $E46^+$, which is also referred to as PARA (particles aiding replication of Ad) [228]. Stocks of $E46^+$ contain Ad_7-SV40 hybrid virus and wild-type Ad 7 helper virus. The hybrid DNA contains an SV40 insertion, beginning at 0.05 fractional lengths from 1 end of the Ad 7 genome, and a deletion of about 16% of Ad 7 DNA (Fig. 12.13, Table 12.6). The inserted SV40 DNA segment is a recombinant of 2 SV40 genomes, as follows: 0.49 – 0.71-0.11 – 0.65. Therefore, the left end of the inserted SV40 segment lies at SV40 map position 0.49 and the right end at 0.65, and the segments between 0.71 and 0.11 are deleted. As a consequence, DNA segments extending from map position 0.49 to 0.65 are repeated [127]. Both the SV40 and Ad 7 genomes in the hybrid are defective, and therefore replication of $E46^+$ requires Ad 7 wild-type helper to supply necessary gene functions for replication in human and monkey cells. $E46^+$ plaques with 2 hit kinetics on monkey cells, reflecting replication of wild-type Ad 7, whose growth in monkey cells is allowed by the SV40 functions in the hybrid, and replication of the hybrid, whose growth depends upon wild-type Ad 7 functions. On human cells the virus plaques with 1 hit kinetics, and the hybrid is unable to replicate in cells that are not coinfected with wild-type Ad 7.

SV40 U- and T-antigens, but not capsid antigens, are synthesized in both monkey and human cells. During productive infection by $E46^+$ most stable cytoplasmic SV40 RNA is derived from the early regions of the SV40 minus (early) strand [146]. Early SV40 mRNA is transcribed from the minus strand in a counterclockwise direction from SV40 map position 0.66-0.17 [131]. $E46^+$ cytoplasmic RNA complementary to late regions on the minus strand (i.e., anti-late RNA) is present in low amounts during productive infection, but these probably do not function as mRNA. Transcripts from the plus (late) DNA strand are not present. These results, along with the production of exclusively early SV40 antigens, indicate that only early SV40 genes are expressed in $E46^+$.

The original $E46^+$ was reported to be highly oncogenic in newborn hamsters [114] and the tumor-bearing animals produced antibodies to SV40 T-antigen, and usually Ad 7 T-antigen. In vitro transformation of human and hamster cells has also been reported [15].

2. *Defective Ad_2-SV40 Hybrids*

Defective Ad_2-SV40 hybrids have been isolated from stocks of Ad 2 adapted to grow in monkey cells. This "pool", Ad_2^{++} contains a mixture of wild-type Ad 2, complete SV40 virions, and a heterogeneous population of Ad_2-SV40 hybrids. Two genetically stable hybrid populations that have been isolated from Ad_2^{++} are Ad_2^{++}HEY and Ad_2^{++}LEY, which are mixtures of wild-type Ad 2 and several hybrid viruses. The hybrids are defective in growth on human cells, and replication requires wild-type Ad 2. Both HEY and LEY yield infectious SV40 when grown on monkey cells, indicating that they contain insertions of complete nondefective SV40 genomes. However, the SV40 yield is 10^3-10^4 times greater from HEY than from LEY. HEY synthesizes both early and late SV40-specific RNA, as well as viral DNA and the late V-antigen [253]. LEY synthesizes SV40 RNA, induces U- and T-antigens, but does not produce experimentally detectable viral DNA, late mRNA, and V-antigen. These results are consistent with the high yield of infectious SV40 by HEY and the very low yield by LEY, since LEY does not appear to carry out late SV40 functions in quantity. However, since LEY does produce infectious SV40, expression of late genes must occur.

HEY stocks contain 3 hybrids, HEY-I, HEY-II, and HEY-III, with 0.45, 1.43, and 2.39 SV40 genomes, respectively [129, 130] (Fig. 12.13, Table 12.6). In all 3 hybrids 39.3% of Ad 2 DNA is deleted at positions 0.326-0.719 on the Ad 2 genome. Thus a very large block of late Ad 2 genes are deleted, including the hexon structural gene, plus the early gene encoding the Ad 2 single-stranded DNA-binding protein (see Figs. 12.9 and 12.12). HEY-II hybrids are present in Ad_2^{++}HEY stocks at about 5-10% frequency of wild-type Ad 2. HEY-III hybrids are present at about one-half the frequency of HEY-II, and HEY-I are very rare. LEY contains 2 hybrids, LEY-I and LEY-II, with 0.03 and 1.05 SV40 genomes. Both hybrids have 10.9% of the Ad 2 genome deleted, at positions 0.825-0.934 on the Ad 2 genome, so the structural gene for Ad 2 fiber polypeptide is probably deleted from LEY (Fig. 12.9). LEY-II hybrids are present at about 10% the frequency of Ad 2 wild-type, while LEY-I are very rare.

In hybrids with more than 1 SV40 genome the SV40 sequences are organized in tandem repetition, and it was suggested [129, 130] that HEY-II and HEY-III yield more infectous SV40 in monkey cells than LEY because their genomes contain longer SV40 tandem repetitions. That is, with HEY the probability of excision of SV40 from the hybrid genome is greater than with LEY. In SV40-transformed cells the plus strand does not appear to be transcribed, yet infectious virus can be rescued by such techniques as cell fusion [129, 130]. This suggests that excision of the integrated SV40 genome is necessary for late SV40 gene transcription, and therefore rescue of infectious virus.

The polarity of the SV40 genome is reversed in the 2 hybrids: in HEY the SV40 minus (early) strand is on Ad 2 *l*-strand, and in LEY the SV40 minus strand is on Ad 2 *r*-strand [129, 130]. This polarity difference may also explain why in monkey cells HEY yields more infectious SV40 than LEY. Whereas both *l*- and *r*-strands are transcribed with about equal frequency in early stages of Ad 2 infection (see Section VI), there is evidence that the Ad 2 *l*-strand is not transcribed late, or is transcribed late with much lower frequency than *r*-strand [204]. Because of this, the late SV40 genes on the SV40 plus (late) strand, which is integrated into the Ad 2 *l*-strand, may not be expressed at late stages in LEY replication. This would also explain why LEY synthesizes early but not late SV40 antigens, and why the stable SV40-specific RNA synthesized by LEY is complementary to the SV40 minus strand [129, 130]. On the other hand, early SV40 genes in HEY replication would be expressed early, and late genes would be expressed late.

D. Nondefective Ad_2-SV40 Hybrids

In addition to the defective hybrids described above, a series of nondefective Ad_2-SV40 hybrids have been isolated from stocks of Ad_2^{++} [154, 155]. All have deletions of about 5% of Ad 2 DNA at positions to the left of 0.86 on the Ad 2 map (Fig. 12.12), and contain insertions of from 6-46% of the SV40 map and extend into the early region, i.e., towards map position 0.66 [145, 176]. Since the early region of the SV40 genome ranges from position 0.66 to position 0.17 in direction of transcription of the early strand [131], hybrids contain some late regions, plus overlapping sequences in early regions [145, 150]. $Ad_2^+ND_3$ does not appear to contain any early DNA sequences. The SV40 minus strand is on the Ad 2 *r*-strand [196].

SV40-specific antigens U, TSTA, and T are induced in early phases of SV40 productive infection. Little is known about the chemical and biological properties of U and TSTA. T-antigen is believed to bind to double-stranded DNA [22] at the origin of SV40 DNA replication [220]. Genetic studies suggest that SV40 T-antigen is coded by the viral genome [4, 279] and that it is a 70,000-100,000 polypeptide [36, 279] which is involved in the initiation of SV40 DNA replication [278] and maintenance of transformation (see Section V.B.1). The various nondefective hybrids express U-, TSTA-, and T-antigens according to the amount of SV40 DNA inserted (Table 12.6). This has allowed the mapping of the SV40 genome sections responsible for induction of these antigens [128] (these data do not establish whether the SV40 genome encodes or induces these antigens).

$Ad_2^+ND_1$, $Ad_2^+ND_2$, and $Ad_2^+ND_4$ grow on both human and monkey cells, while $Ad_2^+ND_3$ and $Ad_2^+ND_5$ grow only on human cells and not on monkey cells. The ability of these hybrids to grow on monkey cells presumably results from expression of inserted SV40 DNA, since SV40 enhances Ad growth on monkey cells. This conclusion is supported by genetic evidence, since H39, a *hr* mutant of $Ad_2{}^+ND_1$ that grows in human cells but has lost the ability to grow in monkey cells, produces U-antigen differently from parental $Ad_2{}^+ND_1$ and like Ad 2 is defective in synthesis of late proteins in monkey cells [97]. The hybrids synthesize polypeptides not produced by wild-type Ad 2 [307] (Table 12.6), as follows: 28,000 by $Ad_2{}^+ND_1$ [97, 162]; 42,000 and 56,000 by $Ad_2{}^+ND_2$; 56,000 by $Ad_2{}^+ND_4$; and 42,000 by $Ad_2{}^+ND_5$. It is possible that these polypeptides are responsible for the induction of SV40-specific antigens and the growth of the hybrids on monkey cells. However, these could also be "nonsense" polypeptides resulting from, for example, translation of SV40-Ad 2 hybrid RNA molecules that are apparently synthesized by these hybrids [54, 191]. As expected, $Ad_2{}^+ND_3$, which contains no early SV40 DNA, does not grow on monkey cells, induce SV40 antigens, or synthesize a polypeptide distinct from wild-type Ad 2. For unknown reasons, although $Ad_2{}^+ND_5$ contains 28% of the SV40 genome and produces a unique 42,000 polypeptide, it does not grow on monkey cells or induce SV40-specific antigens.

These nondefective hybrids are providing some unusual insights into the molecular biology of Ad 2, SV40, and both human and monkey cells. The Ad 2 early gene block that spans the Eco RI-D and E fragments (Fig. 12.12) is partially deleted from the hybrids; thus, an Ad 2 gene(s) may not be essential for Ad 2 replication [54]. Instead, $Ad_2{}^+ND_1$ and $Ad_2{}^+ND_3$ transcribe a "novel" RNA species, whose 5′ end lies 1000 bases to the left of the SV40 insertion, and whose 3′ end terminates in the inserted SV40 segment at SV40 map position around 0.18. The 3′ end of early SV40 mRNA synthesized during productive infection is believed to map at SV40 position 0.17-0.18 [38, 131]. Therefore $Ad_2{}^+ND_1$ expresses only early SV40 sequences as mRNA [132] and the novel mRNA is terminated at the same site as early SV40 mRNA [54]. The base sequences at SV40 map position around 18 could act as a termination signal for transcription or as a recognition site for mRNA processing enzymes. Although hybrid Ad-SV40 RNA molecules are synthesized in $Ad_2{}^+ND_4$ infection [191], there is no proof that these function as mRNA. However, the finding that transcription of the novel gene containing SV40 DNA begins at least 1000 bases to the left of the SV40 insertion supports the conclusion of Oxman et al. [191], based on the different sensitivity of SV40 T-antigen induction by $Ad_2{}^+ND_4$ and SV40 to interferon, that transcription of SV40 DNA in $Ad_2{}^+ND_4$ is controlled by the Ad 2 section of the genome.

The novel Ad_2-SV40 hybrid gene is transcribed at both early and late stages of infection, and the cytoplasmic RNA from this gene is much more abundant late than early. This result is consistent with other studies which showed that the putative SV40-specific polypeptides are synthesized by these hybrids mainly, or exclusively, at late times [307], and that SV40 U-antigen is expressed early after $Ad_2{}^+ND_1$ infection, but the amount is greatly increased late [156]. Late after infection anti-late RNA from the minus strand of the late region in the SV40 insertion is found in both nucleus and cytoplasm, but the cytoplasmic anti-late RNA is in much lower concentration than the SV40 mRNA sequences complementary to DNA to the left of 0.18 (i.e., the putative mRNA sequences), and probably does not function as mRNA. SV40 RNA from the plus strand is also present at very low concentrations late, but only in the nucleus. The transcription of non-mRNA sequences by these hybrids is not surprising, since transcripts from most or all of both *l* and *r* DNA strands are present in nuclei late after Ad 2 infection (Section VI).

Since this novel hybrid gene is transcribed during both early and late stages of hybrid infection, it is a class II early gene (Section VI). Curiously, the wild-type Ad 2 "gene" that has been deleted is also a class II gene [30]. The Ad 2 sequences represented in this novel gene are not transcribed in early stages of Ad 2-productive infection, but are transcribed late (Section VI). The elucidation of why late Ad 2 DNA sequences are transcribed early by these hybrids should prove interesting with regard to mechanisms that regulate the transition from early to late stages of Ad 2 gene expression, and possible mRNA processing.

The nondefective Ad 2 hybrids and $E46^+$ [54, 146] appear to express only early regions of the SV40 genome. This is apparently also the case with SV40-transformed cells [132, 192, 237]. Thus the Ad-SV40 hybrids would appear to be good models to study the expression of SV40 genetic material integrated into host DNA molecules. It is of great interest that with all nondefective Ad_2-SV40 hybrids and with the defective Ad_7-SV40 hybrid $E46^+$ [146], the SV40 DNA is inserted at map position 0.11, suggesting that this is a preferred site of integration. It may be of significance that this is the approximate location of the terminal point of wild-type SV40 DNA replication [33]. The defective Ad_2-SV40 hybrids HEY and LEY have different patterns of SV40 insertion [129, 130], but these contain complete SV40 genomes, and therefore may not be strictly comparable to hybrids with only fractions of SV40 genomes. It is also of interest that all the Ad 2 nondefective hybrids have deletions to the left of map position 0.86. However, this may not be of significance regarding Ad 2 recombination with foreign DNA, since: (1) the hybrids originated from the same Ad 2 pool ($Ad_2{}^{++}$); (2) many Ad 2-transformed rat cells contain only the left end of the Ad 2 genome integrated (Section VIII); and (3) the defective hybrids $Ad_2{}^{++}$HEY and $Ad_2{}^{++}$LEY have deletions elsewhere.

The nondefective hybrids, as well as wild-type Ad 2, have been reported to transform hamster kidney cells in vitro using very high multiplicity of infection, and the transformed cell lines produce tumors when injected into newborn hamsters [152, 153]. This is interesting because Ad 2 (oncogenic group C) and the nondefective Ad_2-SV40 hybrids do not induce tumors in newborn hamsters and are able to replicate in hamster cells [152]; apparently a rare abortive infection may lead to transformation. Ad 5 (group C) also transforms hamster cells [319]. Transformation by the hybrids is the result of the Ad 2, rather than the SV40 sections of the genome.

Mutants of $Ad_2{}^+ND_1$, both *hr* and *ts*, have been isolated [97, 98]. The *hr* mutants, which are able to grow in human but not monkey cells, probably have defects in the SV40 section of the hybrid genome. The *ts* mutants, which grow both on human and monkey cells at permissive temperatures, and do not grow on human cells at nonpermissive temperatures, probably have defects in the Ad section of the genome.

X. Mapping the Adenovirus Genome

Recent technological breakthroughs have permitted the detailed structural analysis of the Ad genome, the identification of regions encoding mRNA, and the localization of specific genes in the DNA molecule. The first advance was the preparative separation of the viral DNA strands [142, 196, 284]. The principle of this method is that 1 viral DNA strand (*l*) (heavy) has greater affinity for a poly(U,G) ribocopolymer than the other strand (*r*) (light); thus strands can be separated and purified by CsCl equilibrium density centrifugation. DNA strands separated in this manner have been used to assess the fraction of each strand transcribed into mRNA [210, 326, 327], as well as to study other aspects of transcription and DNA replication.

The second breakthrough was the discovery and immediate application of restriction endonucleases to the analysis of viral DNA genomes (reviewed in [181]). These remarkable enzymes cleave DNA molecules at specific sequences to generate unique DNA fragments that can be separated on the basis of size by electrophoresis in agarose gels; the fragments can be eluted from gels for use in subsequent experiments. The most commonly used enzyme for Ad studies is endonuclease Eco RI which cleaves Ad 2 DNA into 6 and Ad 5 DNA into 3 fragments [178, 203]. The cleavage maps of Ad 2 DNA by Eco RI and 2 other important enzymes, Hpa I and Hind III, are shown in Fig. 12.12). The linear order of the fragments has been deduced by partial denaturation mapping, cleavage of partial digest products, cleavage of specific fragments with other endonucleases, and cleavage of end-labeled DNA molecules (see [177] for the establishment of cleavage maps for Ad 2 and Ad 5). More than 40 different restriction endonucleases from many types of organisms have been purified so far. Obviously, by cleaving the viral DNA molecule with appropriate combinations of restriction endonucleases, the genome can be dissected into very small defined segments. Separation of the segment DNA strands provides further dissection. Strands of denatured Ad 2 restriction fragments have been separated, either by electrophoresis in agarose gels [53, 245], or by annealing to an excess of purified intact strand (prepared by the ribocopolymer affinity method), and purifying the unannealed fragment strand by hydroxyapatite chromatography [283]. The purified fragment strands were identified as *l* or *r* by hybridization to purified intact *l*- and *r*-strands.

Restriction fragments have been extremely useful in mapping the viral genome. A number of different approaches have been used, and below we will discuss briefly the principle of each. The resulting maps are illustrated in Figs. 12.9 and 12.12 along with Williams' Ad 5 genetic map (the Ad 5 *ts* phenotypes are summarized in Tables 12.3 and 12.4).

A. Mapping DNA Regions Encoding mRNA

The location and size of DNA regions encoding mRNA (i.e., genes) can be established by hybridizing (to completion) separated radioactive DNA strands of restriction fragments to vast excesses of unlabeled RNA extracted from the cytoplasm or polyribosomes of cells early (early genes) or late (late genes) after infection [53, 209, 210, 245, 283]. The larger the number of different restriction endonucleases used, the more precisely the location and size of each gene can be ascertained. The early and late Ad 2 mRNA map is shown in Fig. 12.12(B), and is taken from Philipson et al. [209] and Pettersson et al. [367], who used 18 different restriction fragments to construct these maps. Although the map indicates that structural genes are distributed over 100% of the asymmetric genome, other studies [327] suggest that about 15% of the genome does not encode structural genes. It should also be noted that this is an mRNA map, and not a transcription map, because regions of the genome are transcribed into sequences that do not function as mRNA (Section VI).

The map (Fig. 12.12, part B) also shows the position of the 2 known VA-RNAs (Section VI) as determined by Pettersson and Philipson [205], and Mathews [169]. Pettersson and Philipson hybridized ^{125}I-labeled VA-RNA to fragments generated by 3 restriction endonucleases. Mathews used a novel technique developed by Southern [258]: restriction fragments were separated electrophoretically, but instead of eluting them from the agarose gels, the DNA fragments were transferred directly onto nitrocellulose filters in a manner which preserves their electrophoretic pattern. [^{32}P]VA-RNA was annealed to the permanently bound fragments, and hybridization monitored by autoradiography. Thus, the VA-RNA genes were localized with relatively little effort.

B. Mapping Size-Fractionated mRNA

The annealing of unlabeled RNA to ^{32}P-DNA strands does not establish whether early mRNA sequences are also transcribed late. Fujinaga and Green [60] reported that a subclass of early mRNA corresponding to that expressed in the Ad 2-transformed rat cell line 8617, is not detected by hybridization-competition in pulse labeled late RNA. Lucas and Ginsberg [163] have also described a class of early mRNA (termed class I RNA), that is present in greatly reduced quantities late, and presumably is not actively transcribed late. Class II early mRNA is transcribed both early and late. The above method also does not establish the number of genes within each gene block. These questions can be answered by labeling viral mRNA molecules, fractionating them into size classes by polyacrylamide gel electrophoresis, and then determining the DNA fragment from which they originated. The fragment specificity can be determined by (1) size-fractionation by gel electrophoresis followed by hybridization to various restriction fragments, or (2) hybridization to restriction enzyme fragments (under conditions that do not degrade the RNA, such as 50% formamide at 37°C), followed by elution from the fragment and size determination by gel electrophoresis.

Method 1 (above) has been used to map early Ad 2 mRNA [29, 277]. To identify class I and class II sequences, each mRNA size class was further analyzed by a 2-step RNA hybridization-competition procedure (early [^{3}H]-mRNA versus unlabeled late RNA, and late [^{3}H]mRNA versus unlabeled early RNA) [27, 28, 30]. The current status of these studies is summarized in Fig. 12.12(C), which shows the distribution of class I (open rectangles) and class II (filled rectangles) mRNA size classes within each early gene block. The existence of the 11S RNA from the Eco R1-A fragment is uncertain, and they have not yet resolved the mRNA size classes from the Eco R1-C fragment. The strand specificities of the mRNA classes were deduced from the mRNA maps of Sharp et al. [245] and Philipson et al. [210]. These data suggest that there are 7-8 early genes [27], assuming 1 mRNA species is derived from Eco R1-C. However, since mRNA from Eco R1-C can be translated in vitro into 2 distinct polypeptides (see below), there may be 9 or more early Ad 2 genes. Late mRNA has also been mapped in this manner [277] but the studies are in preliminary stages.

Buttner et al. [341] have used method 2 (above) to map early Ad 2 mRNA, and their results are shown in Fig. 12.12(C) (arrows). Cross-hybridization of the fragment-specific mRNA size classes with adjacent DNA fragments has established that each mRNA species is unique. The relationship between the 2 Eco R1-A species is unclear, and it is possible that they share part of the same sequences. The strand specificity of each [^{3}H]-mRNA species was determined by hybridization to *l* and *r* DNA strands. Chinnadurai et al. [26] have also obtained evidence for 2 RNAs derived from *r*-strand of the left end of the genome during the early stages of lytic infection and in the Ad 2-transformed rat cell line 8617.

In principle, by using suitable combinations of restriction enzymes, both these methods are capable of determining the number of distinct mRNA species. The second method has the added advantage that mRNA can be purified in quantity, and thus further analyzed by RNA sequencing procedures. The mRNA can also be translated in vitro, thereby establishing a direct correspondence between a specific DNA sequence, mRNA, and protein. Both these methods have 2 drawbacks. First, it is rather difficult to isolate mRNA reproducibly without *any* degradation. Second, some mRNAs may be derived from precursor molecules on polyribosomes. For example, with the SV40 system, the 16S mRNA synthesized at late stages of infection is processed in the cytoplasm from a 19S precursor [3]. Thus it is difficult to distinguish between discrete mRNA molecules, precursor molecules, and partially degraded molecules.

C. Mapping Viral-Coded Proteins by Cell-Free Translation of Restriction Fragment-Specific mRNA

The methods in Sections X.A and X.B provide estimates of the number and location of viral structural genes. However, except for size, they reveal nothing about the nature of the gene products. An approach to this is the cell-free translation of mRNA, and the correlation of the translation products with specific mRNAs, and with viral-induced proteins observed by in vivo labeling experiments (Section VI). For example, late Ad 2 mRNA has been size-fractionated by rate-zonal centrifugation in sucrose gradients, each fraction translated in vitro, and the translation products visualized by SDS polyacrylamide gel electrophoresis and autoradiography [6, 47, 157, 316]. Thus specific viral polypeptides were correlated with mRNA size.

Lewis et al. [158] have devised an elegant procedure to identify the DNA segment encoding specific viral polypeptides, based on the observation that Ad 2 mRNA purified by hybridization to and elution from Ad 2 DNA is translatable in vitro [47]. Ad 2 mRNA specific to various DNA restriction fragments were hybridization-purified, translated in vitro, and the ^{35}S-labeled polypeptides visualized after gel electrophoresis and autoradiography. Figure 12.9 (A and B) shows the results obtained by translation of early and late mRNA [158, 363]. (Also see [371] for recent in vitro proteins.) The location and strand specificity of the translated polypeptides was deduced from the linear order of the restriction fragments used, and from the mRNA maps (Fig. 12.12, parts A and B). They have identified 6 early proteins, ranging in size from 11,000-72,000 daltons. By translation of mRNA from 3 transformed cell lines they have identified at least 2 proteins. The 44,000-50,000 protein (which appears as a doublet or a broad band) is specific to the Hpa I-E fragment (see Fig. 12.12, part A) but not the Hpa I-C fragment. The 15,000 protein is specific to the Hpa I-C fragment. Both proteins are specific to Hind III-G fragment. The 15,000 protein was synthesized in much larger amounts than the 44,000-50,000 protein. Cell-free translation of early mRNA selected by hybridization to the Hpa I-E, Hpa I-C, and Hind III-G fragments gave results identical to transformed cell mRNA. We point out that although translation of Hpa I-E and not Hpa I-C fragment mRNA gave rise to the 44,000-50,000 polypeptides, the Hpa I-E fragment may be too small to encode a 44,000-50,000 polypeptide.

These important data suggest that either the 15,000 or 44,000-50,000 proteins (or both) are responsible for both initiation and maintenance of cell transformation. Evidence favoring this conclusion is as follows.

1. mRNA synthesized by Ad 2-transformed cells is a subset of early mRNA [88].
2. Hind III-G fragment harbors information for both initiation and maintenance of cell transformation, since transfection by G fragment alone transforms cells [78].
3. All Ad 2-transformed cell lines tested contain at least 14% of the left end of the Ad 2 genome [236].
4. All Ad 2-transformed cell lines tested synthesize mRNA complementary to about 6-7% of the *r*-strand of the very left end of the genome (i.e., Hind III-G fragment) [53].

Translation of Eco R1-C fragment-specific mRNA from the transformed cell line 8617 gave a 19,000 polypeptide. Early mRNA from this fragment also yields a 19,000 polypeptide (plus an 11,000). This result is in accord with the findings that 8617 both contains and transcribes Eco R1-C fragment information [53, 236].

The results of translation of hybridization-purified fragment-specific late Ad 2 mRNA are shown in Fig. 12.9(B). The order of the major late polypeptides is: IX, $1Va_2$, IIIa, (III, V), p-VII, II, 100,000, IV. The order of the polypeptides in parentheses is not known.

The procedure is important in that it not only maps proteins synthesized by Ad 2-infected cells, but it also establishes that the proteins are encoded by the virus, and not by the cell and induced by the virus. The drawback to this method is that the proteins observed may not be real (for example, because of premature termination of translation). However, there is excellent agreement between sizes of both early and late viral proteins observed after in vivo labeling (see Section VI for characterization of viral proteins labeled in vivo) and in vitro translation. Therefore, this procedure is very promising indeed.

D. Physical Mapping of Adenovirus *ts* Mutants

Ad 5 *ts* mutants have been discussed in Section V, and the known phenotypes summarized in Tables 12.3 and 12.4. A genetic map of these mutants has been constructed (Fig. 12.9, part D), based upon recombination frequencies [323]. The genetic map, however, does not locate the *ts* mutations on the physical map. Grodzicker et al. [98] have recently developed a method, based on restriction enzyme cleavage patterns, to align the genetic and physical maps. The features of the method [321] are illustrated in Fig. 12.14 and outlined below.

1. Recombinants are isolated from crosses of *ts* mutants of different Ad serotypes. Of course, the serotypes must be sufficiently homologous so that recombination can occur, and Grodzicker et al. [98] used *ts* mutants of Ad 5 and $Ad_2{}^+ND_1$. The isolation of recombinants with ts^+ phenotype implies that the recombinant contains $Ad_2{}^+ND_1$ DNA sequences at the location of the Ad 5 *ts* lesion, and Ad 5 sequences at the $Ad_2{}^+ND_1$ *ts* lesion.
2. The Ad serotypes must differ in their restriction endonuclease cleavage patterns. This is satisfied with Ad 5 and $Ad_2{}^+ND_1$ DNAs, which are cleaved into 3 and 5 fragments, respectively, by Eco R1. They are also cleaved differently by other restriction endonucleases.
3. Since the serotypes differ in cleavage pattern, digestion of recombinants with a battery of restriction endonucleases, and analysis of the fragment patterns generated, can establish which regions of the genome are composed of Ad 5 and $Ad_2{}^+ND_1$ DNA, and thereby pinpoint the cross-over point. Determination of the cross-over point of a number of recombinants from 1 set of mutants (each recombinant will have a different cross-over point) will establish the general location of both the Ad 5 and $Ad_2{}^+ND_1$ *ts* lesions.

The results of such experiments, taken from Sambrook et al. [238] and Grodzicker et al. [98] are shown in Fig. 12.9(C).

The fact that recombinants contain DNA from both serotypes makes this approach very powerful, because Ad 2 protein are synthesized from $Ad_2{}^+ND_1$

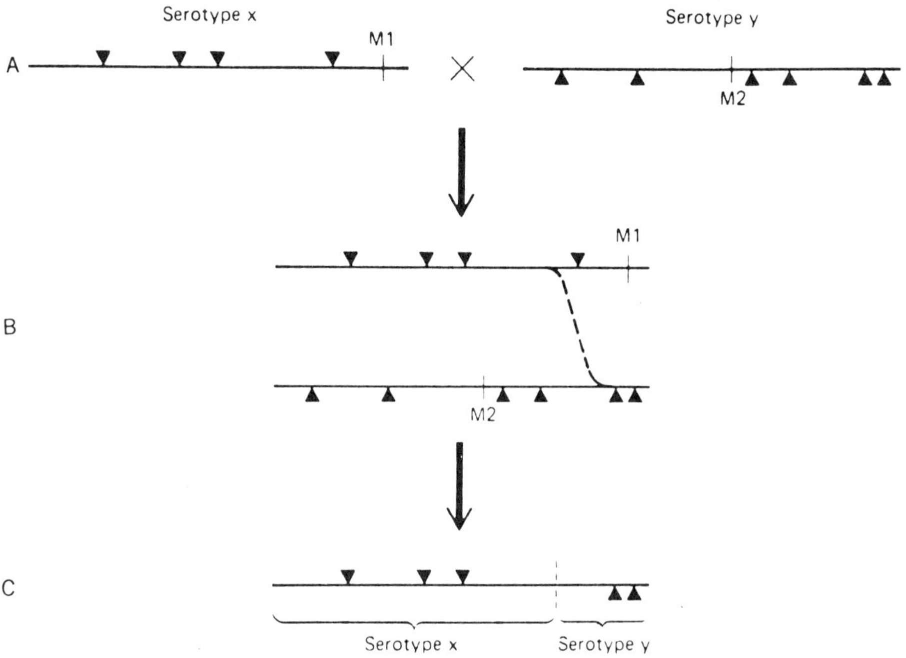
Serotype x
M1
A
Serotype y
M2
M1
B
M2
C
Serotype x
Serotype y

Fig. 12.14 Outline of a method for correlation of the genetic and physical maps of Ads. (A) x and y are 2 serotypes of Ad whose DNAs differ in their cleavage patterns with restriction endonucleases. A restriction endonuclease cleaves the DNA of serotype x at 4 specific sites (▼), generating 5 unique fragments. The same enzyme cleaves the DNA of serotype y at 6 sites (▲), generating 7 unique fragments. A *ts* mutant of serotype x bears a mutation at site M1 and a *ts* mutant of serotype y carries a mutation at site M2. (B) A crossover between the sites of the two mutations, M1 and M2, leads to the formation of a wild-type recombinant. (c) The recombinant contains DNA sequences from each of the parental serotypes. It contains serotype x sequences in the region of the serotype y mutation and serotype y sequences in the region of the serotype x mutation. The recombinant illustrated contains the 3 left-most cleavage sites of serotype x and the 2 right-most cleavage sites of serotype y. Restriction enzyme analysis of the recombinant reveals 6 unique fragments: 3 from serotype x, 2 from serotype y, and a new fragment derived from both x and y. Taken from Grodzicker et al. [98].

sections of the genome, and Ad 5 proteins from Ad 5 sections. Exploiting this fact, Mautner et al. [170] have localized the structural genes for Ad 5 hexon (H5*ts*F2) and fiber (H5*ts*K22). That is, all ts^+ recombinants of H5*ts*F2 (hexon-deficient) and H5*ts*K22 (fiber-deficient) with $Ad_2{}^+ND_1$ express Ad 2 hexon and fiber type-specific antigens, respectively. Dorsett and Ginsberg [40] have recently determined that Ad 5 fiber consists of 3 polypeptide chains, each of 61,000 daltons. Tryptic peptide mapping and electrophoretic separation of the 3 polypeptides have suggested that 2 of the polypeptides are identical and the third is unique. If this is so, then H5*ts*K22 presumably is defective in the structural gene for only 1 of these polypeptides, the 1 that confers antigenic type specificity on the fiber capsomere.

Since ts^+ recombinants of H5*ts*A1 (also hexon-deficient) expressed Ad 5 type-specific hexon antigen, it may be concluded that the hexon-minus phenotype of H5*ts*A1 is not the result of a lesion in hexon structural gene. It is clear that once the genetic composition of a series of Ad 5-$Ad_2{}^+ND_1$ recombinants has been established, these recombinants can be used to locate the gene for *any* serotype-specific protein (e.g., Ad 2 and Ad 5 proteins distinguishable by SDS polyacrylamide gel electrophoresis), as was done through immunological tests by Mautner et al. for hexon and fiber.

Although Ad genomic mapping is in its infant stages, these studies show great promise. The mRNA maps, the mRNA translation maps, and the ts^+ recombinant cleavage maps are entirely consistent with each other, and with the genetic map of Ad 5 *ts* mutants. Clearly, restriction enzyme technology, the basis for these methods (except the genetic map), holds the key to future progress. Methods used to map mutants of other DNA viruses, such as "marker-rescue" [105, 117, 141] and S1-nuclease digestion of heteroduplex DNA molecules [247] have not yet been applied to Ads. In addition, Ad deletion mutants, which would allow for the immediate identification of the altered protein product, have not been described as yet. Finally, it has recently become possible to localize mRNA on the viral genome by hybridization followed by direct visualization of the RNA-DNA hybrid by electron microscopy (Recent Developments).

Other novel mapping methods may be developed. Westphal and Crouch [315] have described the nonrandom cleavage of Ad 2 mRNA by *E. coli* RNase III (specific for double-stranded RNA), and have pointed out the potential use of this enzyme to dissect mRNA for analysis of specific intramolecular regions, e.g., ribosome binding site, which can then be positioned on the viral genome by hybridization. The value of this type of approach remains to be established.

Acknowledgments

We wish to thank Gloria Llanas for her expert help in preparing and executing the final manuscript and Patricia Llanas for her patience. We also want to thank those authors who have contributed their results prior to publication. The work of the authors reported in this chapter was supported by Contract NO 1 CP 43359 from the Virus Cancer Program of the National Cancer Institute and from Public Health Service Grant AI-01725 from the National Institute of Allergy and Infectious Diseases. WSMW is a Research Fellow of the Canadian Medical Research Council. MG is a NIH Research Career Awardee.

References

1. J. M. Adams and S. Corey, Nature, 255, 28-33 (1975).
2. B. M. Alberts and L. Frey, Nature, 227, 1313-1318 (1970).
3. Y. Aloni, M. Shani, and Y. Reuveni, Proc. Natl. Acad. Sci. USA, 72, 2587-2591 (1975).
4. J. C. Alwine, S. I. Reed, J. Ferguson, and G. R. Stark, Cell, 6, 529-533 (1975).
5. C. W. Anderson, P. R. Baum, and R. F. Gesteland, J. Virol., 12, 241-252 (1973).
6. C. W. Anderson, J. B. Lewis, J. F. Atkins, and R. F. Gesteland, Proc. Natl. Acad. Sci., USA, 71, 2756-2760 (1974).
7. R. Bablanian and W. C. Russell, J. Gen. Virol., 24, 261-279 (1974).
8. S. Bachenheimer and J. E. Darnell, Proc. Natl. Acad. Sci. USA, 72, 4445-4449 (1975).
9. S. G. Baum and R. I. Fox, Cold Spring Harbor Symp. Quant. Biol., 39, 567-573 (1974).
10. M. Bégin and J. Weber, J. Virol., 15, 1-17 (1975).
11. A. J. D. Bellett, Virology, 65, 427-435 (1975).
12. A. J. D. Bellett and H. B. Younghusband, J. Mol. Biol., 72, 691-709 (1972).
13. T. L. Benjamin, J. Mol. Biol., 16, 359-373 (1966).
14. K. I. Berns and T. J. K. Kelly, Jr., J. Mol. Biol., 82, 267-271 (1974).
15. P. H. Black and G. J. Todaro, Proc. Natl. Acad. Sci., USA, 54, 374-381 (1965).
16. G. W. Both, A. K. Banerjee, and A. J. Shatkin, Proc. Natl. Acad. Sci. USA, 72, 1189-1193 (1975).
17. D. T. Brown, M. Westphal, B. T. Burlingham, U. Winterhoff, and W. Doerfler, J. Virol., 16, 366-387 (1975).
18. J. S. Brugge and J. S. Butel, J. Virol., 15, 619-635 (1975).
19. J. P. Burnett and J. A. Harrington, Proc. Natl. Acad. Sci. USA, 60, 1023-1029 (1968).
20. J. P. Burnett and J. A. Harrington, Nature (London), 220, 1245-1246 (1968).
21. W. Büttner, Z. Veres-Molnár, and M. Green, Proc. Natl. Acad. Sci. USA, 71, 2951-2955 (1974).
22. R. B. Carroll, L. Hager, and R. Dulbecco, Proc. Natl. Acad. Sci. USA, 71, 3754-3757 (1974).
23. S. Casjens and J. King, Ann. Rev. Biochem., 44, 555-611 (1975).
24. T. Cavalier-Smith, Nature, 250, 467-470 (1974).
25. Y. Chardonnet and S. Dales, Virology, 48, 342-359 (1972).
26. G. Chinnadurai, H. M. Rho, R. B. Horton, and M. Green, J. Virol., 20, 255-263 (1976).
27. E. A. Craig, M. McGrogan, C. Mulder, and H. J. Raskas, J. Virol., 16, 905-912 (1975).
28. E. A. Craig and H. Raskas, J. Virol., 14, 751-757 (1974).
29. E. A. Craig, J. Tal, T. Nishimoto, S. Zimmer, M. McGrogan, and H. J. Raskas, Cold Spring Harbor Symp. Quant. Biol., 39, 483-493 (1974).
30. E. A. Craig, S. Zimmer, and H. J. Raskas, J. Virol., 15, 1202-1213 (1975).
31. C. S. Crumpacker, P. H. Henry, T. Kakefuda, W. P. Rowe, M. J. Levin, and A. M. Lewis, Jr., J. Virol., 7, 352-358 (1971).

32. S. Dales and Y. Chardonnet, Virology, 56, 465-483 (1973).
33. K. J. Danna and D. Nathans, Proc. Natl. Acad. Sci. USA, 69, 3097-3100 (1972).
34. J. E. Darnell, W. Jelinek, and J. Molloy, Science, 181, 1215-1221 (1973).
35. H. E. Davidson and R. J. Britton, Q. Rev. Biol., 48, 565-613 (1973).
36. B. C. DelVillano and V. Defendi, Virology, 51, 34-46 (1973).
37. R. C. Desrosiers, K. H. Friderici, and F. Rottman, Biochemistry, 14, 4367-4374 (1975).
38. R. Dhar, S. Zain, S. M. Weissman, J. Pan, and K. Subramanian, Proc. Natl. Acad. Sci. USA, 71, 371-375 (1974).
39. W. Doerfler and A. K. Kleinschmidt, J. Mol. Biol., 50, 579-593 (1970).
40. P. H. Dorsett and H. S. Ginsberg, J. Virol., 15, 208-216 (1975).
41. W. Doerfler, H. Burger, J. Ortin, E. Fanning, D. T. Brown, M. Westphal, U. Winterhoff, B. Weiser, and J. Schick, Cold Spring Harbor Symp. Quant. Biol., 39, 505-521 (1974).
42. R. Dulbecco and M. Vogt, Cold Spring Harbor Symp. Quant. Biol., 18, 273-279 (1953).
43. A. R. Dunn, P. H. Gallimore, K. W. Jones, and J. K. McDougall, Int. J. Cancer, 11, 628-636 (1973).
44. H. J. Edenberg and J. A. Huberman, Ann. Rev. Genet., 9, 245-284 (1975).
45. M. J. Ensinger, Dissertation, Faculty Graduate School of Arts and Sciences, University of Pennsylvania, 1973.
46. M. J. Ensinger and H. S. Ginsberg, J. Virol., 10, 328-339 (1972).
47. L. Eron and H. Westphal, Proc. Natl. Acad. Sci. USA, 3385-3389 (1974).
48. L. Eron, H. Westphal, and G. Khoury, J. Virol., 15, 1256-1261 (1975).
49. E. Everitt, L. Lutter, and L. Philipson, Virology, 67, 197-208 (1975).
50. E. Everitt, B. Sundquist, U. Pettersson, and L. Philipson, Virology, 52, 130-147 (1973).
51. H. Ezoe and S. Mak, J. Virol., 14, 733-739 (1974).
52. S. M. Fernandez and H. J. Raskas, Biochem. Biophys. Res. Commun., 66, 67-74 (1975).
53. S. J. Flint, P. H. Gallimore, and P. A. Sharp, J. Mol. Biol., 96, 47-68 (1975).
54. S. J. Flint, Y. Wewerka-Lutz, A. S. Levine, J. Sambrook, and P. A. Sharp, J. Virol., 16, 662-673 (1975).
55. A. E. Freeman, P. H. Black, E. A. Vanderpool, P. H. Henry, J. B. Austin, and R. J. Huebner, Proc. Natl. Acad. Sci. USA, 58, 1205-1212 (1967).
56. K. Fujinaga and M. Green, Proc. Natl. Acad. Sci. USA, 55, 1567-1574 (1966).
57. K. Fujinaga and M. Green, Proc. Natl. Acad. Sci. USA, 57, 806-812 (1967).
58. K. Fujinaga and M. Green, J. Virol., 1, 576-582 (1967).
59. K. Fujinaga and M. Green, J. Mol. Biol., 31, 63-73 (1968).
60. K. Fujinaga and M. Green, Proc. Natl. Acad. Sci. USA, 65, 375-382 (1970).
61. K. Fujinaga, S. Mak, and M. Green, Proc. Natl. Acad. Sci. USA, 60, 959-966 (1968).
62. K. Fujinaga, M. Pina, and M. Green, Proc. Natl. Acad. Sci. USA, 64, 255-262 (1969).
63. K. Fujinaga, K. Sekikawa, H. Yamazaki, and M. Green, Cold Spring Harbor Symp. Quant. Biol., 39, 633-636 (1974).
64. Y. Furuichi, Nucl. Acid Res., 1, 809-818 (1974).
65. Y. Furuichi, M. Morgan, A. J. Shatkin, W. Jelinek, M. Solditt-Georgieff, and J. E. Darnell, Proc. Natl. Acad. Sci. USA, 72, 1904-1908 (1975).
66. P. H. Gallimore, P. A. Sharp, and J. Sambrook, J. Mol. Biol., 89, 49-72 (1974).

67. C. F. Garon, K. W. Berry, and J. A. Rose, Proc. Natl. Acad. Sci. USA, 69, 2391-2395 (1972).
68. C. F. Garon, K. W. Berry, and J. A. Rose, Proc. Natl. Acad. Sci. USA, 72, 3039-3043 (1975).
69. M. Georgieff, S. Bachenheimer, and J. E. Darnell, Cold Spring Harbor Symp. Quant. Biol., 39, 475-482 (1974).
70. Z. Gilead, M. Q. Arens, S. Bhaduri, G. Shanmugam, and M. Green, Nature, 254, 533-536 (1975).
71. Z. Gilead and H. S. Ginsberg, J. Virol., 2, 15-20 (1968).
72. Z. Gilead, K. Sugawara, G. Shanmugam, and M. Green, J. Virol., 18, 454-460 (1976).
73. H. S. Ginsberg, in The Biochemistry of Viruses (H. B. Levy, ed.), Dekker, New York, 1969, pp. 329-359.
74. H. S. Ginsberg, L. J. Bello, and A. J. Levine, in The Molecular Biology of Viruses (J. S. Coulter and W. Paranchych, eds.). Academic, New York. 1967, pp. 547-572.
75. H. S. Ginsberg, M. J. Ensinger, R. S. Kauffman, A. J. Mayer, and U. Lundholm, Cold Spring Harbor Symp. Quant. Biol., 39, 419-426 (1974).
76. R. S. Kauffman and H. S. Ginsberg, J. Virol., 19, 643-658 (1976).
77. H. S. Ginsberg, J. F. Williams, W. Doerfler, and H. Shimojo, J. Virol., 12, 663-664 (1973).
78. F. L. Graham, J. Abrahams, C. Mulder, H. L. Heijneker, S. O. Warnaar, F. A. J. deVries, and A. J. van der Eb, Cold Spring Harbor Symp. Quant. Biol., 39, 637-650 (1974).
79. F. L. Graham and A. J. van der Eb, Virology, 52, 456-467 (1973).
80. F. L. Graham, and A. J. van der Eb, Virology, 54, 536-539 (1973).
81. M. Green, Cold Spring Harbor Symp. Quant. Biol., 27, 219-235 (1962).
82. M. Green, Ann. Rev. Microbiol., 20, 189-222 (1966).
83. M. Green, Ann. Rev. Biochem., 39, 701-756 (1970).
84. M. Green in Oncology, Vol. V (Clark, Cumley, McCay, and Copeland, eds.), Year Book Medical, Chicago. 1971, pp. 156-165.
85. M. Green and G. E. Daesch, Virology, 13, 169-176 (1961).
86. M. Green and G. F. Gerard, Prog. Nucl. Acid Res. Mol. Biol., 14, 187-334 (1974).
87. M. Green and M. Hodap, J. Mol. Biol., 64, 305-309 (1972).
88. M. Green, J. T. Parsons, M. Pina, K. Fujinaga, H. Caffier, and I. Landgraf-Leurs, Cold Spring Harbor Symp. Quant. Biol., 35, 803-818 (1970).
89. M. Green and M. Pina, Proc. Natl. Acad. Sci. USA, 50, 44-46 (1963).
90. M. Green and M. Pina, Proc. Natl. Acad. Sci. USA, 1251-1259 (1964).
91. M. Green, M. Pina, K. Fujinaga, S. Mak, and D. Thomas, in Perspectives in Virology, Vol. VI (M. Pollard, ed.), Academic, New York. 1968, pp. 15-38.
92. M. Green, M. Pina, R. Kimes, P. C. Wensink, L. A. MacHattie, and C. A. Thomas, Proc. Natl. Acad. Sci. USA, 57, 1302-1309 (1967).
93. M. Green, M. Pina, and R. L. Kimes, Virology, 31, 562-566 (1967).
94. M. R. Green, J. K. Mackey, G. Chinnadurai, and M. Green, Cell, 7, 419-428 (1976).
95. M. R. Green, M. Green, and J. K. Mackey, Nature (London), 261, 340-342 (1976).
96. M. R. Green, J. K. Mackey, and M. Green, J. Virol., 22, 238-242 (1976).

97. T. Grodzicker, C. Anderson, P. A. Sharp, and J. Sambrook, J. Virol., 13, 1237-1244 (1974).
98. T. Grodzicker, J. Williams, P. Sharp, and J. Sambrook, Cold Spring Harbor Symp. Quant. Biol., 39, 439-446 (1974).
99. M. Gruetter and R. F. Franklin, J. Mol. Biol., 89, 163-178 (1974).
100. H. Handa, K. Shiroki, and H. Shimojo, J. Gen. Virol., 29, 239-242 (1975).
101. T. Harrison, F. Graham, and J. Williams, Virology, 77, 319-329 (1977).
102. M. Harter, G. Shanmugam, W. S. M. Wold, and M. Green, J. Virol., 19, 232-242 (1976).
103. S. Hashimoto and M. Green, J. Virol., 20, 425-435 (1976).
104. K. Hashimoto, K. Nakajima, K.-I. Oda, and H. Shimojo, J. Mol. Biol., 81, 207-223 (1973).
105. M. N. Hayashi and M. Hayashi, J. Virol., 14, 1142-1151 (1974).
106. P. H. Henry, L. E. Schnipper, R. J. Samaha, C. S. Crumpacker, A. M. Lewis, Jr., and A. S. Levine, J. Virol., 11, 665-671 (1973).
107. M. R. Hilleman and J. H. Werner, Proc. Soc. Exp. Biol. Med., 85, 183-188 (1954).
108. P. R. Hoffman and J. E. Darnell, Jr., J. Virol., 15, 806-811 (1975).
109. R. W. Horne, S. Brenner, A. P. Waterson, and P. Wildy, J. Mol. Biol., 1, 84-86 (1959).
110. M. S. Horwitz, J. Virol., 8, 675-683 (1971).
111. M. S. Horwitz, J. Virol., 13, 1046-1054 (1974).
112. M. S. Horwitz, C. Brayton, S. G. Baum, J. Virol., 11, 544-551 (1973).
113. R. J. Huebner, in Perspectives in Virology, Vol. V (M. Pollard, ed.), Academic, New York. 1967, pp. 147-166.
114. R. J. Huebner, H. G. Pereira, A. C. Allison, A. C. Hollinshead, and H. C. Turner, Proc. Natl. Acad. Sci. USA, 51, 532-539 (1964).
115. R. J. Huebner, W. P. Rowe, and W. T. Lane, Proc. Natl. Acad. Sci. USA, 48, 2051-2058 (1962).
116. R. J. Huebner, W. P. Rowe, H. C. Turner, and W. T. Lane, Proc. Natl. Acad. Sci. USA, 50, 379-389 (1963).
117. C. A. Hutchinson, III, and M. H. Edgel, J. Virol., 8, 181-189 (1971).
118. M. Ishibashi, Proc. Natl. Acad. Sci. USA, 65, 304-309 (1970).
119. M. Ishibashi, Virology, 45, 42-52 (1971).
120. M. Ishibashi and J. V. Maizel, Jr., Virology, 57, 409-424 (1974).
121. K. Ito, M. Arens, and M. Green, J. Virol., 15, 1507-1510 (1975).
122. K. Ito, M. Arens, and M. Green, Biochim. Biophys. Acta, 447, 340-452 (1976).
123. M. Ito and E. Suzuki, J. Gen. Virol., 9, 243-245 (1970).
124. Y. H. Jeng, W. S. M. Wold, K. Sugawara, Z. Gilead, and M. Green, J. Virol., 22, 402-411 (1977).
125. M. Jerkofsky and F. Rapp, Virology, 51, 466-473 (1973).
126. M. Jerkofsky and F. Rapp, J. Virol., 15, 253-258 (1975).
127. T. J. Kelly, Jr., J. Virol., 15, 1267-1272 (1975).
128. T. J. Kelly, Jr., and A. M. Lewis, Jr., J. Virol., 12, 643-652 (1973).
129. T. J. Kelly, Jr., A. M. Lewis, Jr., A. S. Levine, and S. Siegel, J. Mol. Biol., 89, 113-126 (1974).
130. T. J. Kelly, Jr., A. M. Lewis, Jr., A. S. Levine, and S. Siegel, Cold Spring Harbor Symp. Quant. Biol., 39, 409-417 (1974).
131. G. Khoury, P. Howley, D. Nathans, and M. Martin, J. Virol., 15, 433-437 (1975).

132. G. Khoury, A. M. Lewis, Jr., M. N. Oxman, and A. S. Levine, Nature New Biol., 246, 204-205 (1973).
133. R. Kimes and M. Green, J. Mol. Biol., 50, 203-206 (1970).
134. G. Kimura, Nature, 248, 590-592 (1974).
135. G. Kimura and A. Itagaki, Proc. Natl. Acad. Sci. USA, 72, 673-677 (1975).
136. D. F. Klessig and C. W. Anderson, J. Virol., 16, 1650-1668 (1975).
137. F. V. Koczot, B. J. Carter, C. F. Garon, and J. A. Rose, Proc. Natl. Acad. Sci. USA, 70, 215-219 (1973).
138. S. Lacy, Sr. and M. Green, Proc. Natl. Acad. Sci. USA, 52, 1053-1059 (1964).
139. S. Lacy, Sr., and M. Green, Science, 150, 1296-1298 (1965).
140. S. Lacy, Sr. and M. Green, J. Gen. Virol., 1, 413-418 (1967).
141. C.-J. Lai and D. Nathans, J. Mol. Biol., 89, 179-193 (1974).
142. M. Landgraf-Leurs and M. Green, J. Mol. Biol., 60, 185-202 (1971).
143. M. Landgraf-Leurs and M. Green, Biochim. Biophys. Acta, 312, 667-673 (1973).
144. G. Lavelle, C. Patch, G. Khoury, and J. Rose, J. Virol., 16, 775-782 (1975).
145. P. Lebowitz, T. J. Kelly, Jr., D. Nathans, T. N. H. Lee, and A. M. Lewis, Jr., Proc. Natl. Acad. Sci. USA, 71, 441-445 (1974).
146. P. Lebowitz and G. Khoury, J. Virol., 15, 1214-1221 (1975).
147. N. Ledinko, J. Virol., 14, 457-468 (1974).
148. J. Leibowitz and M. S. Horwitz, Virology, 66, 10-24 (1975).
149. A. J. Levine, P. C. van der Vliet, B. Rosenwirth, J. Rabek, G. Frenkel and M. Ensinger, Cold Spring Harbor Symp. Quant. Biol., 39, 559-566 (1974).
150. A. S. Levine, M. J. Levin, M. N. Oxman, and A. M. Lewis, Jr., J. Virol., 11, 672-681 (1973).
151. B. Lewin, Cell, 4, 11-20 (1975).
152. A. M. Lewis, Jr., J. H. Breeden, Y. L. Wewerka, L. E. Schnipper, and A. S. Levine, Cold Spring Harbor Symp. Quant. Biol., 39, 651-656 (1974).
153. A. M. Lewis, Jr., A. S. Rabson, and A. S. Levine, J. Virol., 13, 1291-1301 (1974).
154. A. M. Lewis, Jr., M. J. Levin, W. H. Wiese, C. S. Crumpacker, and P. H. Henry, Microbiology, 63, 1128-1135 (1969).
155. A. M. Lewis, Jr., A. S. Levine, C. S. Crumpacker, M. J. Levin, R. J. Samaha, and P. H. Henry, J. Virol., 11, 655-664 (1973).
156. A. M. Lewis, Jr., and W. P. Rowe, J. Virol., 7, 189-197 (1971).
157. J. B. Lewis, C. W. Anderson, J. F. Atkins, and R. F. Gesteland, Cold Spring Harbor Symp. Quant. Biol., 39, 581-590 (1974).
158. J. B. Lewis, J. F. Atkins, C. W. Anderson, P. R. Baum, and R. F. Gesteland, Proc. Natl. Acad. Sci. USA, 72, 1344-1348 (1975).
159. U. Lindberg, T. Persson, and L. Philipson, J. Virol., 10, 909-919 (1972).
160. K. Lonberg-Holm and L. Philipson, J. Virol., 4, 323-338 (1969).
161. M. C. Loni and M. Green, J. Virol., 12, 1288-1292 (1973).
162. R. Lopez-Revilla and G. Walter, Nature New Biol., 244, 165-167 (1973).
163. J. J. Lucas and H. S. Ginsberg, J. Virol., 8, 203-213 (1971).
164. U. Lundholm and W. Doerfler, Virology, 45, 827-829 (1971).
165. J. V. Maizel, Jr., D. O. White, and M. D. Scharff, Virology, 36, 115-125 (1968).
166. J. V. Maizel, Jr., D. O. White, and M. D. Scharff, Virology, 36, 126-136 (1968).
167. S. Mak, Virology, 66, 474-480 (1975).

168. R. G. Martin and J. Y. Chow, J. Virol., 15, 599-612 (1975).
169. M. B. Mathews, Cell, 6, 223-229 (1975).
170. V. Mautner, J. Williams, J. Sambrook, P. A. Sharp, and T. Grodzicker, Cell, 5, 93-99 (1975).
171. H. D. Mayor and J. Ratner, Biochim. Biophys. Acta, 299, 189-195 (1973).
172. R. M. McAllister, M. O. Nicholson, G. Reed, J. Kern, R. V. Gilden, and R. J. Huebner, J. Gen. Virol., 4, 29-36 (1969).
173. J. K. McDougall, A. R. Dunn, and P. H. Gallimore, Cold Spring Harbor Symp. Quant. Biol., 39, 591-600 (1974).
174. P. M. McGuire, C. Swart, and L. D. Hodge, Proc. Natl. Acad. Sci. USA, 69, 1578-1582 (1972).
175. J. F. Morrow and P. Berg, Proc. Natl. Acad. Sci. USA, 69, 3365-3369 (1972).
176. J. F. Morrow, P. Berg, T. J. Kelly, Jr., and A. M. Lewis, J. Virol., 12, 653-658 (1973).
177. C. Mulder, J. R. Arrand, H. Delius, W. Keller, U. Pettersson, R. J. Roberts, and P. A. Sharp, Cold Spring Harbor Symp. Quant. Biol., 39, 397-400 (1974).
178. C. Mulder, P. A. Sharp, H. Delius, and U. Pettersson, J. Virol., 14, 68-77 (1974).
179. R. E. Murray and M. Green, J. Mol. Biol., 74, 735-738 (1973).
180. K. Nakajima and K. Oda, Virology, 67, 85-93 (1975).
181. D. Nathans and H. O. Smith, Ann. Rev. Biochem., 44, 273-293 (1975).
182. M. V. Nermut, Virology, 65, 480-495 (1975).
183. M. O. Nicholson and R. M. McAllister, Virology, 48, 14-21 (1972).
184. T. Nishimoto, H. J. Raskas, and C. Basilico, Proc. Natl. Acad. Sci. USA, 72, 328-332 (1975).
185. E. Norrby, Virology, 28, 236-248 (1966).
186. E. Norrby, J. Virol., 5, 221-236 (1969).
187. K. Ohe, Virology, 47, 726-733 (1972).
188. K. Ohe and S. Weissman, Science, 167, 879-881 (1970).
189. K. Ohe and S. Weissman, J. Biol. Chem., 246, 6991-7009 (1971).
190. M. Osborn and K. Weber, J. Virol., 15, 636-644 (1975).
191. M. N. Oxman, M. J. Levin, and A. M. Lewis, Jr., J. Virol., 13, 322-330 (1974).
192. B. Ozanne, P. A. Sharp, and J. Sambrook, J. Virol., 12, 90-98 (1973).
193. R. Padmanabhan, R. Padmanabhan, and M. Green, Biochem. Biophys. Res. Commun., 69, 860-867 (1976).
194. J. T. Parsons, J. Gardner, and M. Green, Proc. Natl. Acad. Sci. USA, 68, 557-560 (1971).
195. J. T. Parsons and M. Green, Virology, 45, 154-164 (1971).
196. C. T. Patch, A. M. Lewis, Jr., and A. S. Levine, Proc. Natl. Acad. Sci. USA, 69, 3375-3379 (1972).
197. G. D. Pearson and P. C. Hanawalt, J. Mol. Biol., 62, 65-80 (1971).
198. H. G. Pereira and N. G. Wrigley, J. Mol. Biol., 85, 617-631 (1974).
199. R. P. Perry and D. E. Kelley, Cell, 1, 37-42 (1975).
200. R. P. Perry, D. E. Kelley, K. H. Friderici, and F. M. Rottman, Cell, 6, 13-19 (1975).
201. U. Pettersson, J. Mol. Biol., 81, 521-527 (1973).
202. U. Pettersson and S. Hoglund, Virology, 39, 90-106 (1969).
203. U. Pettersson, C. Mulder, H. Delius, and P. A. Sharp, Proc. Natl. Acad. Sci. USA, 70, 200-204 (1973).

204. U. Pettersson and L. Philipson, Proc. Natl. Acad. Sci. USA, 71, 4887-4891 (1974).
205. U. Pettersson and L. Philipson, Cell, 6, 1-4 (1975).
206. U. Pettersson and J. Sambrook, J. Mol. Biol., 73, 125-130 (1973).
207. U. Pettersson, J. Sambrook, H. Delius, and C. Tibbetts, Virology, 59, 153-167 (1974).
208. L. Philipson and U. Lindberg, in Comprehensive Virology, Vol. 3 (H. Fraenkel-Conrat and R. Wagner, eds.), Plenum, New York. 1974, pp. 143-207.
209. L. Philipson, U. Pettersson, and U. Lindberg, Molecular Biology of Adenoviruses, Springer-Verlag, New York. 1975, pp. 1-115.
210. L. Philipson, U. Pettersson, U. Lindberg, C. Tibbetts, B. Vennström, and T. Persson, Cold Spring Harbor Symp. Quant. Biol., 39, 447-456 (1974).
211. L. Philipson, R. Wall, G. Glickman, and J. E. Darnell, Proc. Natl. Acad. Sci. USA, 68, 2806-2809 (1971).
212. M. Pina and M. Green, Proc. Natl. Acad. Sci. USA, 54, 547-551 (1965).
213. M. Pina and M. Green, Virology, 36, 321-323 (1968).
214. M. Pina and M. Green, Virology, 38, 573-586 (1969).
215. L. Prage, U. Pettersson, S. Höglund, K. Lonberg-Holm, and L. Philipson, Virology, 42, 341-358 (1970).
216. R. Price and S. Penman, J. Virol., 9, 621-626 (1972).
217. A. S. Rabson, G. T. O'Conor, I. K. Berezesky, and F. J. Paul, Proc. Soc. Exp. Biol., 116, 187-190 (1964).
218. F. Rapp, Prog. Exp. Tumor Res., 18, 104-137 (1973).
219. H. J. Raskas, D. C. Thomas, and M. Green, Virology, 40, 893-902 (1970).
220. S. I. Reed, J. Ferguson, R. W. Davis, and G. R. Stark, Proc. Natl. Acad. Sci. USA, 72, 1605-1609 (1975).
221. P. R. Reich, S. G. Baum, J. A. Rose, W. P. Rowe, and S. M. Weisman, Proc. Natl. Acad. Sci. USA, 55, 336-341 (1966).
222. R. J. Roberts, J. R. Arrand, and W. Keller, Proc. Natl. Acad. Sci. USA, 71, 3829-3833 (1974).
223. J. Robin, D. Bourgaux-Ramoisy, and P. Bourgaux, J. Gen. Virol., 20, 233-237 (1973).
224. A. J. Robinson and A. J. D. Bellett, Cold Spring Harbor Symp. Quant. Biol., 39, 523-531 (1974).
225. A. J. Robinson, H. B. Younghusband, and A. J. D. Bellett, Virology, 56, 54-69 (1973).
226. B. Rosenwirth, K. Shiroki, A. J. Levine, and H. Shimojo, Virology, 67, 14-23 (1975).
227. F. Rottman, A. J. Shatkin, and R. P. Perry, Cell, 3, 197-199 (1974).
228. W. P. Rowe and S. G. Baum, Proc. Natl. Acad. Sci. USA, 52, 1340-1347 (1964).
229. W. P. Rowe, R. J. Huebner, L. K. Gillmore, H. Parrott, and T. G. Ward, Proc. Soc. Exp. Biol. Med., 84, 570-573 (1953).
230. W. C. Russell and G. E. Blair, J. Gen. Virol., 34, 19-35 (1977).
231. W. C. Russell, K. Hayashi, P. J. Sanderson, and H. G. Pereira, J. Gen. Virol., 1, 495-507 (1967).
232. W. C. Russell, C. Newman, and J. F. Williams, J. Gen. Virol., 17, 265-279 (1972).

233. W. C. Russell and J. J. Skehel, J. Gen. Virol., 15, 45-57 (1972).
234. W. C. Russell, J. J. Skehel, and J. F. Williams, J. Gen. Virol., 24, 247-259 (1974).
235. S. Salzberg, M. S. Robin, and M. Green, Virology, 53, 186-195 (1973).
236. J. Sambrook, M. Botchan, P. Gallimore, B. Ozanne, U. Pettersson, J. Williams, and P. A. Sharp, Cold Spring Harbor Symp. Quant. Biol., 39, 615-632 (1974).
237. J. Sambrook, B. Sugden, W. Keller, and P. A. Sharp, Proc. Natl. Acad. Sci. USA, 70, 3711-3715 (1973).
238. J. Sambrook, J. Williams, P. A. Sharp, and T. Grodzicker, J. Mol. Biol., 97, 369-390 (1975).
239. R. Schekman, A. Weiner, and A. Kornberg, Science, 186, 987-993 (1974).
240. R. Schilling, B. Weingartner, and E. L. Winnacker, J. Virol., 16, 767-774 (1975).
241. A. L. Schinkaryl, and W. K. Joklik, Virology, 56, 532-548 (1973).
242. R. W. Schlesinger, in Adv. Virus Res., 69, 1-61 (1969).
243. G. Shanmugam, S. Bhaduri, M. Arens, and M. Green, Biochemistry, 14, 332-337 (1975).
244. P. A. Sharp, B. Sugden, and J. Sambrook, Biochemistry, 12, 3055 (1973).
245. P. A. Sharp, P. H. Gallimore, and S. J. Flint, Cold Spring Harbor Symp. Quant. Biol., 39, 457-474 (1974).
246. P. A. Sharp, U. Pettersson, and J. Sambrook, J. Mol. Biol., 86, 709-726 (1974).
247. T. E. Shenk, C. Rhodes, P. W. J. Rigby, and P. Berg, Proc. Natl. Acad. Sci. USA, 989-993 (1975).
248. K. Shimada, K. Fujinaga, S. Hama, K. Sekikawa, and Y. Ito, J. Virol., 10, 648-652 (1972).
249. H. Shimojo, K. Shiroki, and K. Yamaguchi, Cold Spring Harbor Symp. Quant. Biol., 39, 533-538 (1974).
250. K. Shiroki, J. Irisawa, and H. Shimojo, Virology, 49, 1-11 (1972).
251. K. Shiroki and H. Shimojo, Virology, 61, 474-485 (1974).
252. K. Shiroki, H. Shimojo, and K. Yamaguchi, Virology, 60, 192-199 (1974).
253. S. E. Siegel, C. T. Patch, A. M. Lewis, Jr., and A. S. Levine, J. Virol., 16, 43-52 (1975).
254. T. Simmons, P. Heywood, and L. D. Hodge, J. Mol. Biol., 89, 423-433 (1974).
255. H. O. Sjogren, J. Minowada, and J. Ankerst, J. Exp. Med., 125, 689-701 (1967).
256. K. O. Smith, W. D. Gehle, and M. D. Trousdale, J. Bacteriol., 90, 254-261 (1965).
257. R. Sohier, Y. Chardonnet, and M. Prunieras, Prog. Med. Virol., 7, 253-325 (1965).
258. E. M. Southern, J. Mol. Biol., 98, 503-517 (1975).
259. S. P. Staal and W. P. Rowe, Virology, 64, 513-519 (1975).
260. J. T. Stasny, A. R. Neurath, and B. A. Rubin, J. Virol., 2, 1429-1442 (1968).
261. P. H. Steenbergh, J. S. Sussenbach, R. J. Roberts, and H. S. Jansz, J. Virol., 15, 268-272 (1975).
262. W. A. Strohl, Prog. Exp. Tumor Res., 18, 199-239 (1973).
263. K. Sugawara, Z. Gilead, and M. Green, J. Virol., 21, 338-346 (1977).

264. B. Sundquist, U. Pettersson, L. Thelander, and L. Philipson, Virology, 51, 252-256 (1973).
265. S. Surzycki, J. A. Surzycki, W. DeLorbe, and G. N. Gussin, Cold Spring Harbor Symp. Quant. Biol., 39, 501-504 (1974).
266. J. S. Sussenbach, D. J. Ellens, and H. S. Jansz, J. Virol., 12, 1131-1138 (1973).
267. J. S. Sussenbach, D. J. Ellens, P. Ch. van der Vliet, M. G. Kuijk, P. H. Steenbergh, J. M. Vlak, H. Ruzijn, and H. S. Jansz, Cold Spring Harbor Symp. Quant. Biol., 39, 539-545 (1974).
268. J. S. Sussenbach, P. C. van der Vliet, D. J. Ellens, and H. S. Jansz, Nature New Biol., 239, 47-49 (1972).
269. E. Suzuki and H. Shimojo, Virology, 43, 488-494 (1971).
270. E. Suzuki, H. Shimojo, and Y. Moritsugu, Virology, 49, 426-438 (1972).
271. E. Suzuki and H. Shimojo, J. Virol., 13, 538-540 (1974).
272. M.Takahashi, Virology, 49, 815-817 (1974).
273. N. Takemori,Virology, 47, 157-167 (1972).
274. N. Takemori, J. L. Riggs, and C. Aldrich, Virology, 36, 575-586 (1968).
275. N. Takemori, J. L. Riggs, and C. D. Aldrich, Virology, 38, 8-15 (1969).
276. J. Tal, E. A. Craig, and H. J. Raskas, J. Virol., 15, 137-144 (1975).
277. J. Tal, E. A. Craig, S. Zimmer, and H. J. Raskas, Proc. Natl. Acad. Sci. USA, 10, 4057-4061 (1974).
278. P. Tegtmeyer, J. Virol., 10, 591-598 (1972).
279. P. Tegtmeyer, J. Virol., 15, 613-618 (1975).
280. D. Thomas and M. Green, Proc. Natl. Acad. Sci. USA, 56, 243-246 (1966).
281. D. C. Thomas and M. Green, Virology, 39, 205-210 (1969).
282. C. A. Thomas, Jr., and L. A. MacHattie, Ann. Rev. Biochem., 36, 485-518 (1967).
283. C. Tibbetts and U. Pettersson, J. Mol. Biol., 88, 767-784 (1974).
284. C. Tibbetts, U. Pettersson, K. Johansson, and L. Philipson, J. Virol., 13, 370-377 (1974).
285. C. Tockstein, H. Polasa, M. Pina, and M. Green, Virology, 36, 377-386 (1968).
286. A. Tolun and U. Pettersson, J. Virol., 16, 759-766 (1975).
287. J. Tooze, The Molecular Biology of Tumour Viruses, Cold Spring Harbor Laboratory, New York. 1973.
288. J. J. Trentin, Y. Yabe, and G. Taylor, Science, 137, 835-841 (1962).
289. D. Tsuei, K. Fujinaga, and M. Green, Proc. Natl. Acad. Sci. USA, 69, 427-430 (1972).
290. S. Ustacelebi and J. F. Williams, Nature, 235, 52-53 (1972).
291. R. C. Valentine and H. G. Pereira, J. Mol. Biol., 13, 13-20 (1965).
292. A. J. van der Eb, Virology, 51, 11-23 (1973).
293. A. J. van der Eb, L. W. Kesteren, and E. F. J. Van Bruggen, Biochim. Biophys. Acta, 182, 530 (1969).
294. P. C. van der Vliet and A. J. Levine, Nature New Biol., 246, 170-174 (1973).
295. P. C. van der Vliet, A. J. Levine, M. J. Ensinger, and H. S. Ginsberg, J. Virol., 15, 348-354 (1975).
296. P. C. van der Vliet and J. S. Sussenbach, Eur. J. Biochem., 30, 548-592 (1972).
297. P. C. van der Vliet and J. S. Sussenbach, Virology, 67, 415-426 (1975).
298. H. E. Varmus, P. K. Vogt, and J. M. Bishop, Proc. Natl. Acad. Sci. USA, 70, 3067-3071 (1973).
299. L. Velicer and H. Ginsberg, Proc. Natl. Acad. Sci. USA, 61, 1263-1271 (1968).

300. L. Velicer and H. Ginsberg, J. Virol., 5, 338-352 (1970).
301. J. M. Vlak, Th. H. Rozijn, and J. S. Sussenbach, Virology, 63, 168-175 (1975).
302. J. M. Vlak, Th. H. Rozijn, and F. Spies, Virology, 65, 535-545 (1975).
303. R. L. Wall, L. Philipson, and J. E. Darnell, Virology, 50, 27-34 (1972).
304. R. Wall, J. Weber, Z. Gage, and J. E. Darnell, J. Virol., 11, 953-960 (1973).
305. R. D. Wallace and J. Kates, J. Virol., 9, 627-635 (1972).
306. G. Walter and J. V. Maizel, Virology, 57, 402-408 (1974).
307. G. Walter and H. Martin, J. Virol., 16, 1236-1247 (1975).
308. J. D. Watson, Nature New Biol., 239, 197-201 (1972).
309. J. D. Watson and F. H. C. Crick, Cold Spring Harbor Symp. Quant. Biol., 18, 123-131 (1953).
310. J. Weber, J. Virol., 17, 462-471 (1976).
311. J. Weber, M. Bégin, and G. Khittoo, J. Virol., 15, 1049-1056 (1975).
312. C.-M. Wei, A. Gershowitz, and B. Moss, Nature, 257, 251-253 (1975).
313. R. Weinmann, H. J. Raskas, R. G. Roeder, Proc. Natl. Acad. Sci. USA, 71, 3426-3430 (1974).
314. A. Weissbach, Cell, 3, 101-108 (1975).
315. H. Westphal and R. J. Crouch, Proc. Natl. Acad. Sci. USA, 72, 3077-3081 (1975).
316. H. Westphal, L. Eron, F. J. Ferdinand, R. Callahan, and S. P. Lai, Cold Spring Harbor Symp. Quant. Biol., 39, 575-579 (1974).
317. W. C. Wilcox, H. S. Ginsberg, and T. F. Anderson, J. Exp. Med., 18, 307-314 (1963).
318. N. M. Wilkie, S. Ustacelebi, and J. F. Williams, Virology, 51, 499-503 (1973).
319. J. F. Williams, Nature, 243, 162-163 (1973).
320. J. F. Williams, M. Gharpure, S. Ustacelebi, and S. McDonald, J. Gen. Virol., 11, 95-102 (1971).
321. J. Williams, T. Grodzicker, P. Sharp, and J. Sambrook, Cell, 4, 113-119 (1975).
322. J. E. Williams and S. Ustacelebi, J. Gen. Virol., 13, 345-348 (1971).
323. J. F. Williams, C. S. Young, and P. E. Austin, Cold Spring Harbor Symp. Quant. Biol., 39, 427-437 (1974).
324. E. J. Wills, W. C. Russell, and J. F. Williams, J. Gen. Virol., 20, 407-412 (1973).
325. E. L. Winnacker, J. Virol., 15, 744-758 (1975).
326. W. S. M. Wold, M. Green, K. H. Brackmann, C. Devine, and M. A. Cartas, J. Virol., 23, 616-625 (1977).
327. W. S. M. Wold, M. Green, K. H. Brackmann, M. Cartas, and C. Devine, J. Virol., 20, 465-477 (1976).
328. W. S. M. Wold, M. Green, and T. Munns, Biochem. Biophys. Res. Commun., 68, 643-649 (1976).
329. J. Wolfson and D. Dressler, Proc. Natl. Acad. Sci. USA, 69, 3054-3057 (1972).
330. H. Yamamoto, H. Shimojo, and C. Hamada, Virology, 50, 743-752 (1972).
331. T. Yamashita, M. Arens, and M. Green, J. Biol. Chem., 250, 3272-3279 (1975).
332. T. Yamashita, M. Arens, and M. Green, J. Biol. Chem., in press (1977).
333. T. Yamashita and M. Green, J. Virol., 3, 412-420 (1974).
334. C. S. H. Young and J. F. Williams, J. Virol., 15, 1168-1175 (1975).
335. H. B. Younghusband and A. J. D. Bellett, J. Virol., 8, 265-274 (1971).
336. H. B. Younghusband and A. J. D. Bellett, J. Virol., 10, 855-857 (1972).
337. Adenovirus strand nomenclature, a proposal, J. Virol., 22, 830-831 (1977).

338. M. Arens, T. Yamashita, R. Padmanabhan, T. Tsuruo, and M. Green, J. Biol. Chem., in press (1977).
339. S. Bachenheimer and J. E. Darnell, J. Virol., 19, 286-289 (1976).
340. S. M. Berget, S. J. Flint, J. F. Williams, and P. A. Sharp, J. Virol., 19, 879-889 (1976).
341. W. Büttner, Z. Veres-Molnár, and M. Green, J. Mol. Biol., 107, 93-114 (1976).
342. T. H. Carter and H. S. Ginsberg, J. Virol., 18, 156-166 (1976).
343. W. W. Chin and J. V. Maizel, Jr., Virology, 71, 518-530 (1976).
344. W. W. Chin and J. V. Maizel, Jr., Virology, 76, 79-89 (1977).
345. G. Chinnadurai, Y.-H. Jeng, Z. Gilead, and M. Green, Biochem. Biophys. Res. Commun., 74, 1199-1205 (1977).
346. B. Edvardson, E. Everitt, H. Jörnvall, L. Prage, and L. Philipson, J. Virol., 19, 533-547 (1976).
347. J. Flint, Cell, 10, 153-166 (1977).
348. S. J. Flint, S. M. Berget, and P. A. Sharp, Cell, 9, 559-571 (1976).
349. S. J. Flint and P. A. Sharp, J. Mol. Biol., 106, 749-771 (1976).
350. R. E. Gelinas and R. J. Roberts, Cell, 11, 533-544 (1977).
351. Z. Gilead, Y.-H. Jeng, W. S. M. Wold, K. Sugawara, H. M. Rho, M. L. Harter, and M. Green, Nature, 264, 263-266 (1976).
352. H. S. Ginsberg and C. S. H. Young, Adv. Cancer Res., 23, 91-130 (1976).
353. M. Green, Proc. Natl. Acad. Sci. USA, 69, 1036-1041 (1972).
354. M. Green and J. K. Mackey, Cold Spring Harbor Conf. on Cell Prolif., Vol. 4, in press (1977).
355. M. Green and W. S. M. Wold, Seminars in Oncology, 3, 65-79 (1976).
356. M. S. Horwitz, J. Virol., 18, 307-315 (1976).
357. M. Ishibashi and J. V. Maizel, Jr., Virology, 345-361 (1974).
358. G. G. Jackson and R. L. Muldoon, The University of Chicago Press, Chicago and London, 1975, pp. 165-197.
359. W. Keegstra, P. S. van Wielink, and J. S. Sussenbach, Virology, 76, 444-447 (1977).
360. A. Levinson and A. J. Levine, Virology, 76, 1-11 (1977).
361. A. Levinson, A. J. Levine, S. Anderson, M. Osborn, B. Rosenwirth, and K. Weber, Cell, 7, 575-584 (1976).
362. A. D. Levinson, E. H. Postel, and A. J. Levine, Virology, 79, 144-159 (1977).
363. J. B. Lewis, J. F. Atkins, P. R. Baum, R. Solem, R. F. Gesteland, and C. W. Anderson, Cell, 7, 141-151 (1976).
364. J. K. Mackey, K. H. Brackmann, M. R. Green, and M. Green, Biochemistry, 16, 4478-4482 (1977).
365. Y. Niiyama, K. Igarashi, K. Tsukamoto, T. Kurokawa, and Y. Sugino, J. Virol., 16, 621-623 (1975).
366. B. Öberg, J. Saborio, T. Persson, E. Everitt, and L. Philipson, J. Virol., 15, 199-207 (1975).
367. U. Pettersson, C. Tibbetts, and L. Philipson, J. Mol. Biol., 101, 479-501 (1976).
368. H. Polasa and M. Green, Virology, 25, 68-79 (1965).
369. K. Raska, Jr., L. M. Sehulster, and F. Varricchio, Biochem. Biophys. Res. Commun., 69, 79-84 (1976).

370. B. Rosenwirth, C. Anderson, and A. J. Levine, Virology, 69, 617-625 (1976).
371. J. L. Saborio and B. Öberg, J. Virol., 17, 865-875 (1976).
372. K. Shiroki, H. Shimojo, K. Sekikawa, K. Fujinaga, J. Rabek, and A. J. Levine, Virology, 69, 431-437 (1976).
373. H. Söderlung, U. Pettersson, B. Vennström, L. Philipson, and M. B. Mathews, Cell, 7, 585-593 (1976).
374. S. Sommer, M. Salditt-Georgieff, S. Bachenheimer, J. E. Darnell, Y. Furuichi, M. Morgan, and A. J. Shatkin, Nuc. Acids Res., 3, 749-765 (1976).
375. K. Sugawara, Z. Gilead, W. S. M. Wold, and M. Green, J. Virol., 22, 527-529 (1977).
376. J. S. Sussenbach and M. G. Kuijk, Virology, 77, 149-157 (1977).
377. J. S. Sussenbach, A. Tolun, and U. Pettersson, J. Virol., 20, 532-534 (1976).
378. P. C. van der Vliet, J. Zandberg, and H. S. Jansz, Virology, 80, 98-110 (1977).
379. B. Weingartner, E. L. Winnacker, A. Tolun, and U. Pettersson, Cell, 9, 259-268 (1976).
380. R. Weinmann, T. G. Brendler, H. J. Raskas, and R. G. Roeder, Cell, 7, 557-566 (1976).
381. M. Wu, R. J. Roberts, and N. Davidson, J. Virol., 21, 766-777 (1977).
382. S. G. Zimmer and H. J. Raskas, Virology, 70, 118-126 (1976).

Chapter 13

The Herpesviruses

Bernard Roizman

Departments of Microbiology and Biophysics/Theoretical Biology
University of Chicago
Chicago, Illinois

I. Introduction 770

A. The Object 770
B. The Objectives 771
C. The Boundaries 775

II. The Herpesvirion 775

A. Architectural Components 775
B. Herpesvirus DNA 779
C. Structural Proteins 792
D. Other Constituents 795
E. The Distribution of the Chemical Components in the Herpesvirion 796

III. The Replication of Herpesviruses in Permissive Cells 797

A. Initiation of Infection 797
B. Characteristics of the Reproductive Cycle 798
C. Transcription 799
D. The Synthesis and Function of Proteins Specified by Herpesviruses 802
E. Herpesvirus DNA Synthesis 817
F. Lipids 819

G. Morphogenesis and Egress of the Herpesvirion from Infected Cells 819

IV. Alterations of Cell Function and Structure during Productive Infection 824

A. Introduction 824
B. Structural Alterations 825
C. Effect of Viral Infection on Host Macromolecular Synthesis 827
D. Alteration in the Structure and Function of Cellular Membranes 831

Acknowledgments 837

References 838

Recent Developments 899

I. Introduction

A. The Object

This chapter deals with the properties of the herpesvirion and with the function encoded within its DNA. This brief statement of objectives must be amplified.

Election to membership in the herpesvirus family is based solely on virus architecture. The criteria for election are the possession of a relatively large double stranded linear DNA, a core, an icosadeltahedral capsid with 162 capsomeres (which, incidentally, no one ever reported counting), and an evelope. The most frequent identification of a virus as a putative herpesvirus or, as a herpes-type virus, is based on the presence of particles approximately 100 nm in diameter budding through the inner nuclear membrane. Based on these criteria, herpesviruses have been reported in nearly every eukaryotic species examined in detail, from fungi [1] to man. The list includes oysters [2], fish [3], frogs, birds, and a variety of domestic and wild animals.

Are all enveloped DNA viruses budding through a nuclear membrane necessarily related? What should be the criteria for relatedness in the face of the fact that there is apparently little DNA-DNA homology among herpesviruses other than herpes simplex 1 and 2 and the possibility that the common antigens reported at length might well be host or serum contaminants? From the point of view of this chapter, the questions are academic, particularly since of the numerous herpesviruses reported to date only a handful, i.e., herpes simplex (HSV), pseudorabies, Epstein-Barr (EBV), equine abortion viruses (EAV), and some cytomegaloviruses (CMV), have been studied in some detail.

It should be noted however that such was the extraordinary insight of the designers of the current system of viral classification that so far all attempts to establish criteria for subdivision of herpesviruses into smaller groups have met with little success. The problem is best illustrated by the apparent inability of most investigators to differentiate among the various herpesviruses on the basis of microscopy of thin sections or of negatively stained virions. However, detailed studies have clearly documented that there are significant differences in the base sequences of the DNA and structural proteins among the strains of 1 herpesvirus [4, 5].

Inevitably, herpesviruses must be identified by a name. Regrettably herpesviruses are unique among viruses to have been named by their discoverers after the host in which they abound, after the common name of the disease they produce, after even the histological manifestation of the infected cells, and on occasion, unabashedly after themselves. Until the time when such a nomenclature can be instituted, the ICNV Herpesvirus Study Group [6] proposed a provisional system for the labeling of herpesviruses based on the following rules: (1) The label for each herpesvirus would be in an anglicized form; (2) each herpesvirus would be named after the taxonomic unit, the family, to which its primary natural host belongs; and (3) the herpesviruses within each group would be given arabic numbers. The number would not be preceded by the word "type." New herpesviruses would receive the next available numbers.

To facilitate reading of this chapter, the common names now in use rather than the proposed new designation will be used throughout the text. Table 13.1 however lists both the new and the common names, the few abbreviations used in the text, as well as the pertinent properties of viral DNAs reported to date. When necessary the name of the virus strain is in parentheses after the virus designation or its abbreviation.

B. The Objectives

The most extensive information on the structure of the herpesvirion is available for herpes simplex viruses, and although it is not the presumption of the author to suggest that all herpesviruses are alike, the herpes simplex virion is a useful model against which to pit the expanding knowledge of the herpesvirus universe.

Our understanding of the function expressed by the virus seems to be more broadly based, and in fact, differences among herpesviruses with respect to viral functins have been claimed repeatedly. It should be pointed out, however, that many herpesviruses that are highly infectious in nature multiply poorly or not at all in culture. Although the functions expressed by viruses

Table 13.1 Provisional Labels, Common Names, and Properties of the DNA of Herpesviruses

Provisional label[a]	Common name	Abbreviation	G + C (moles %)	Buoyant density in CsCl	Reference	MW ($\times 10^{-6}$	Reference
Human herpesvirus 1	Herpes simplex type 1	HSV-1	67	1.726	14, 56	97-99	56, 58, 62
Human herpesvirus 2	Herpes simplex type 2	HSV-2	69	1.728	14, 56	99	56
Human herpesvirus 3	Varicella-zoster virus		46	1.705	21		
Human herpesvirus 4	Epstein-Barr virus	EBV	59	1.718	25		
Human herpesvirus 5	Cytomegalovirus	CMV	56-57	1.716	43	100	43
Ceropithecid herpesvirus 1	B Virus		-	-			
Ceropithecid herpesvirus 2	SA 6		51	1.710	49		
Ceropithecid herpesvirus 3	SA 8		67	1.726	49		
Cebid herpesvirus 1	Herpesvirus tamarinus; herpesvirus platyrrhini; marmoset herpesvirus						
Cebid herpesvirus 2	Herpesvirus saimiri		50	1.209	52		
Cebid herpesvirus 3	Spider-monkey herpesvirus		72	1.731	49		
Callitrichid herpesvirus 1	Marmoset herpesvirus		-	-	-		

Tupaiid herpesvirus 1	Tree shrew herpesvirus		66	1.725	15		
Canine herpesvirus 1	Canine herpesvirus		33	1.692	44		
Feline herpesvirus 1	Feline rhinotracheitis		46	1.705	44		
Equid herpesvirus 1	Equine abortion virus, equine rhinopneu-monitis virus	EAV	57	1.716	45, 15	84-94	45
Equid herpesvirus 2	Slowly growing, cyto-megalo-type viruses		58	1.717	44, 46		
Equid herpesvirus 3	Coital-exanthema virus	EAV	66	1.725	47		
Bovid herpesvirus 1	Infectious bovine rhinotracheitis virus		71-72	1.731	44, 15		
Bovid herpesvirus 2	Bovine mammalitis virus		64	1.723	48	82	48
Bovid herpesvirus 3	Wildebeest herpesvirus, malignant cattarrhal fever virus						
Bovid herpesvirus 4	Herpesvirus from sheep pulmonary adenomatosis						
Pig herpesvirus 1	Pseudorabies virus	PSV	72	1.731	44, 15	87	119
Pig herpesvirus 2	Inclusion-body rhinitis virus, pig cyto-megalovirus						

Table 13.1 (continued)

Provisional label[a]	Common name	Abbreviation	G + C (moles %)	Buoyant density in CsCl	Reference	(X 10^{-6}	Reference
Murid herpesvirus 1	Mouse cytomegalovirus from *Mus*		59	1.718	51	132	51
Murid Herpesvirus 2	Rat cytomegalovirus from *Rattus*						
Sciruid herpesvirus 1	Cytomegalovirus from European ground squirrel						
Caviid herpesvirus 1	Guinea-pig cytomegalovirus		57	1.716	24		
Lagomorph herpesvirus 1	Rabbit herpesvirus						
Phasianid herpesvirus 1	Infectious laryngotracheitis virus		45	1.704	44		
Phasianid herpesvirus 2	Marek's disease virus	MDV	46	1.705	67	103	67
Turkey herpesvirus 1	Turkey herpesvirus		46	1.706	317		
Anatid herpesvirus 1	Duck-plague herpesvirus						
Pigeon herpesvirus 1	Pigeon herpesvirus						
Cormorant herpesvirus 1	Cormorant herpesvirus						
Iguana herpesvirus 1	Iguana herpesvirus				25		
Ranid herpesvirus 1	Lucke virus		44-45	1.703	25		
Ranid herpesvirus 2	Frog virus 4		56	1.716	307		
Catfish herpesvirus 1	Catfish herpesvirus		56	1.715	49		

in abortive infections of restrictive or nonpermissive cells might be very significant from the point of view of human health, cell function, and even understanding viral functions, ultimately virus transmitted horizontally depends on productive infection in permissive cells for its survival in nature. Therefore, by necessity more than by choice, this chapter will focus on viral functions expressed in productive infections yielding infectious progeny.

C. The Boundaries

With few exceptions, the review of the literature covers the period to January, 1976. Because of limitations in space the chapter does not cover abortive infections, latency, cell transformation, and the tenuous relation to cancer. Hopefully, the serious student of the herpesvirus literature will consult the many excellent reviews on this subject currently in press.

II. The Herpesvirion

A. Architectural Components

The virion consists of 4 major architectural components. The innermost, the core, is surrounded by 3 concentric structures, i.e., the capsid, the tegument, and the envelope (Fig. 13.1), which are discussed in details which follow.

1. The Core

In thin sections of extracellular virions or virions in infected cells, the core has the appearance of an electron-dense ring surrounding an electron-translucent center or of an electron-dense bar (Fig. 13.2). These 2 images are compatible with the hypothesis that the core is an electron-dense toroid with a less-dense plus or spool filling the hole. Indeed many, although not all pictures of the core could be accounted for in terms of the angle of the plane in which the section is cut [7, 8, 9]. Two lines of evidence have led to the conclusion that the core contains the DNA. First, numerous investigators followed the example of Epstein [10] in showing that the central region of the virion is sensitive to DNase. Although enzymatic digestions showed that the DNA is in the core, it was not immediately clear whether the DNA was contained in the electron-dense area, the electron-translucent area, or both areas. Second, Furlong et al. [7] took advantage of a technique developed by Bernhard [11] to show that the DNA is contained in the electron-dense region making up the toroid. The technique is based on the observation that EDTA treatment of

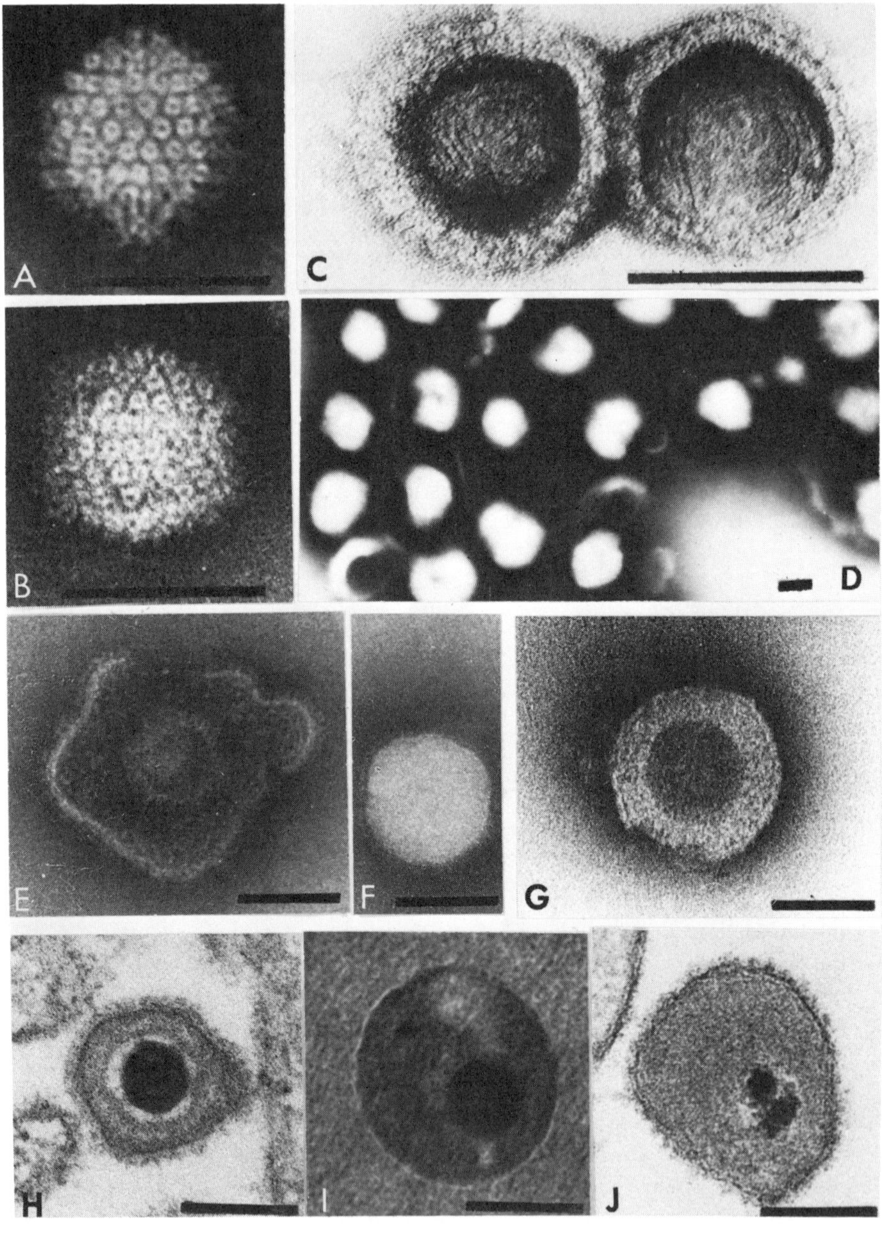
A
B
C
D
E
F
G
H
I
J

Fig. 13.1 The structure of the herpesvirus capsid and virion. (A, B, and C). Negatively stained preparations of HSV-1 capsids. (A and B) Capsids were stained with phosphotungstic acid and show 3- and 2-fold symmetry, respectively. (C) Capsids penetrated by uranyl acetate, bringing into relief threadlike structures showing periodic striations on the surface of the core. (D) Negative-stain preparation of purified HSV-1 virions. Intact virions are impermeable to negative stain and appear as white blobs with tails. The irregular shape and tailing probably result from stretching of the envelope during centrifugation. (E) Ruptured HSV-1 virion penetrated by negative stain which outlines the capsomeres, and the impermeable to negative stain shown at same magnification as E and G. (G) HSV-1 virion showing loss of membrane from 1 side. Note tegument adhering to capsid. (H, I and J) Thin section of HSV-1, MDV, and tree shrew herpesvirus virions, respectively. Note extensive tegument in I and J and densely staining membrane with spikes on the outer surface in H and J. Bar in this figure = 100 nm. From Roizman and Furlong [16].

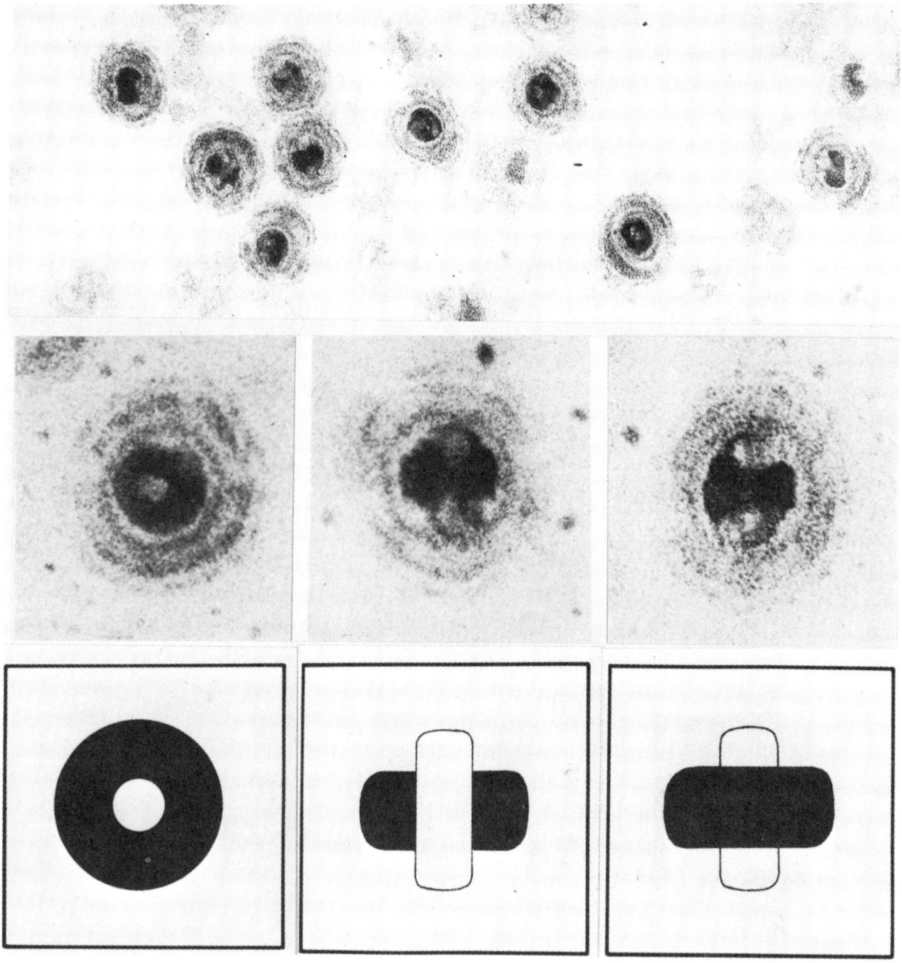

Fig. 13.2 Electron micrographs of thin sections of herpes simplex virions showing the core cut at various angles and drawings of a model cut in the same plan (left and center) or viewed from the same angle (right). The cylindrical structure passing through the toroid varies somewhat from virion to virion; X 200,000; figures are drawn on larger scale. From Furlong et al. [7].

thin sections, fixed with glutaraldehyde *only* and stained with uranyl acetate, selectively removed uranyl ions bound to DNA. The appearance of the plug was unaltered by this treatment. Nothing is known of the composition of the plug although structurally it appears contiguous with the inner layer of the capsid structure (Furlong, unpublished studies).

Strandberg and Carmichael [12] reported that the core of the canine herpesvirion had a helical appearance. HSV negatively stained with uranyl acetate showed coiled threadlike structures on the surface of the core. The threadlike structures were 4.0-5.0 nm wide and showed some indications of a periodic superstructure (coils?) along the thread. Furlong et al. [7] suggested that these structures were the spooled DNA.

2. *The Capsid*

In thin sections the capsid appears as a moderately electron-dense hexagon or ring separated from the core by an electron-translucent shell (Fig. 13.1). The outer diameter of the capsid has been reported to range from 85-110 nm. It is not clear whether the variability in the dimensions reported to date reflects artifacts of manipulation or inherent structural differences among herpesviruses. The morphologic subunits of the capsid, the capsomeres, are arranged to show 2-, 3-, and 5-fold symmetry. From the number of capsomeres (5) along the side of the triangular face and the axis of symmetry, Wildy et al. [13] concluded that the capsid is made of 162 capsomeres arranged in the form of an icosadeltahedron. As stated earlier, the number of capsomeres has not been confirmed by actual count. It would be expected that the predicted 12 pentameres and the 150 hexameres would differ with respect to structural subunits. However, the composition and molar ratios of HSV-1 capsid proteins [15] do not readily permit assignment of proteins to pentameric and hexameric structures. The hexameric capsomer appears to be 12.5 nm long. The end projecting outside the capsid has a diameter of 8.0-9.0 nm [13]. A hole, 4.0 nm in diameter, runs through the long axis of the capsomere. The negative stain completely fills the capsomeres broken off the capsid in a manner which suggests that the canal runs all the way through the hexamere. However, in the intact capsid, the hole fills only partly with negative stain, suggesting that in the intact capsid the hole is blocked at the proximal end.

3. *The Tegument*

The tegument is defined as the structure located between the capsid and the envelope [16]. A structure corresponding to the tegument was recognized by numerous authors [13, 17, 18] as a fibrous coat around the capsid. In some publications [19-22] it was designated as an inner membrane, a misnomer since the tegument does not have a trilaminar unit-membrane structure. In thin section it appears as a layer of amorphous material, and in negatively stained virions it has a fibrous appearance (Fig. 13.1). The material comprising the tegument seems to be present in most virions, but the amount seen is variable from virion to virion, even in the same cell [23]. The width of the layer varies considerably among the herpesviruses [1, 12, 24]; for example, in herpes simplex the layer is frequently interrupted [26], and in virons from Lucke adenocarcinoma the layer is generally complete [27]. In virions from Marek's disease virus [28] or from the herpesvirus of tree shrews [29], there is enough of this material to extend the diameter of enveloped particles to 250 or 260 nm in diameter. The extent of this layer may be determined by the virus strain. Thus, herpes simplex grown in tree shrew fibroblasts does not show enlarged virions. However, the tree shrew herpesvirus produces enlarged virions in both human embryonic fibroblasts and rabbit kidney cells, indicating that the structure of the tegument is controlled at least in part by the virus.

4. The Envelope

The outermost structure of the herpesvirion is the envelope [29-33], consisting of a trilaminar membrane with spikes (Fig. 13.1) projecting from its outer surface [13, 23, 34]. The envelope is impermeable to negative stain but becomes permeable if damaged [35]. Manipulation of the virion results in striking alterations of the envelope, and "tailing" frequently results [23, 24, 36]. The efficiency of envelopment, as well as the stability of the envelope, varies considerably among the herpesviruses [28, 37-40].

B. Herpesvirus DNA

Although it has been suspected for some time that herpesviruses contain DNA, definitive evidence did not emerge until some 13 years ago [41, 42]. The literature on herpesvirus DNA has steadily increased in recent years, with some notable and striking observations. We shall consider the herpesvirus DNA from the points of view of composition, structure, and genetic relatedness.

1. Composition

Cumulative catalogs of the base composition of herpesvirus DNAs have been published [16, 43] and a summary is presented in Table 13.1. There is basic agreement that herpesvirus DNAs vary in base composition from 33 to 74 G + C (guanine plus cytosine) moles percent, but for several reasons many of the measurements require confirmation. Most studies were done with labeled DNA dissolved in CsCl solutions and centrifuged to equilibrium in a preparative centrifuge. These often yield unreliable results. Although the results obtained from analytical centrifugations suitably corrected for conditions of centrifugation are far more reliable, here errors in sample preparation often render the results questionable.

Specifically:

1. Many determinations were done on DNA extracted from whole infected cells rather than from purified virus. This procedure will not

generate convincing data for poorly growing viruses and for DNAs whose density approaches host DNA.

2. The introduction of labeled precursors, into the DNA, in particular ^{14}C, may significantly increase its density due to the difference in atomic weights of the substituted isotopes [50].
3. Isopycnic centrifugation of some herpesvirus DNAs yielded more than one band differing in density. This was due in some instances to the banding of DNA fragments differing in base composition [51, 52] and in others to the presence of defective viral DNA with a different average base composition [53-55].
4. Determinations of base composition on the basis of buoyant density assume the absence of significant amounts of unusual bases which might throw such calculations off. In several instances [56, 57] the base composition calculated from buoyant density measurements was in accord with the results of analyses of the thermal denaturation of the DNA.

2. *The Size and Conformation of Herpesvirus DNA*

It is now well established that the DNA extracted from herpesvirions is linear and double-stranded, and fragments upon denaturation with heat or alkali. Several comments should be made concerning these conclusions.

The linearity of the DNA extracted from virions is firmly established; it is based on electron microscopic studies [58-62]. The conformation of the DNA in infected cells which have not been packaged is less certain; Adams and Lindahl [63] reported that EBV DNA in Raji cells is in a supercoiled form.

The size of herpesvirus DNAs has been reported to vary considerably. At least with respect to HSV-1, careful measurements of the counter length of the DNA [59-62], reassociation kinetics [64], sedimentation studies [56], and summations of the molecular weights of restriction enzyme fragments ([65]; Hayward, Buchman and Roizman, manuscript in preparation) all yield a molecular weight ranging from about 95 to 99 million. The differences in molecular weights observed among DNAs of different strains of HSV, however, are 1 to 2 million and cannot account for the significantly lower molecular weight of HSV-1 reported by Wagner et al. [66].

A significant property of HSV (Fig. 13.3), MDV, and EBV DNAs is their fragmentation upon denaturation with alkali [56, 67, 68], and it is likely that this is a general property of herpesvirus DNAs. In stock preparations of HSV-1, only 15% of DNA strands appear to be intact. Possible explanations for this observations are the presence of ribonucleotides, single strand nicks, or gaps in the DNA. Although ribonucleotides have been detected in the DNA [69-80], it

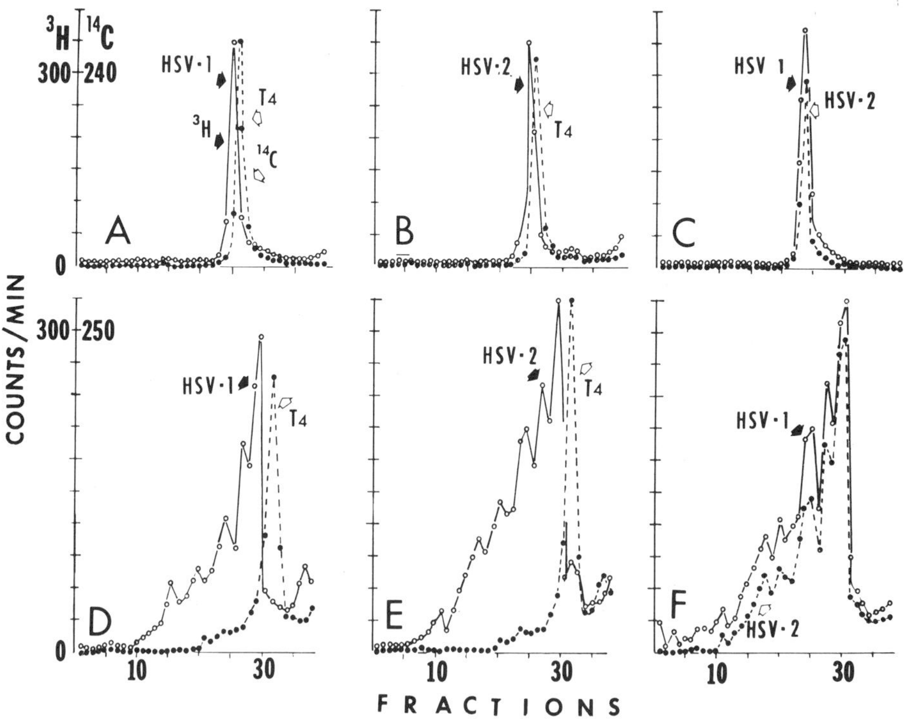

3H
14C
300
240
0
HSV-1
3H
T4
14C
A
HSV-2
T4
B
HSV 1
HSV-2
C
COUNTS/MIN
300
250
HSV-1
T4
D
10
30
HSV-2
T4
E
HSV-1
HSV-2
F
FRACTIONS

Fig. 13.3 Zone sedimentation of HSV DNA in neutral and alkaline sucrose density gradients. HSV-1 and HSV-2 DNAs were labeled with [^{3}H]- or [^{14}C] thymidine and cosedimented with each other (C and F) and with T4 DNA (A, B, D, and E). The DNAs were centrifuged for 3.5 hr in a SW41 rotor at 40,000 rpm and 20°C. Dashed line: ^{14}C-labeled DNA; solid line: ^{3}H-labeled DNA. Direction of sedimentation is to the right. Details of the preparation and denaturation of the DNA, and of the neutral and alkaline sucrose density gradients are given in Kieff et al. [56].

is not likely that they alone account for the fragmentation because the sedimentation profile in neutral sucrose solutions of DNA treated with alkali for 1-2 sec could not be differentiated from that of DNA exposed to alkali for much longer intervals or from that of DNA denatured with formamide (Wadsworth and Roizman, unpublished data). On the whole, the evidence supports the notion that there are gaps in the DNA for 2 reasons. First, attempts to "repair" HSV DNA by procedures which successfully ligate the nicks in T5 DNA have so far failed. Second, it has been observed that lambda exonuclease free of endonucleolytic activity digested internal single-strand stretches of HSV DNA, suggesting the presence of sizeable gaps (Wadsworth and Roizman, unpublished studies). The position of the gaps at this time is cloaked in uncertainty. The evidence available to date is that they are at nonrandom but probably not at unique sites [71-73].

3. Arrangement of the Sequences in the DNA

Recently it was reported that HSV DNA has 2 unusual features. First, several lines of evidence indicate that the DNA consists of 2 regions, L and S, containing 82 and 18% of the molecule, respectively. Both L and S regions contain inverted terminal repetitions. The terminally repetitive sequences of the L region designated as *ab* each comprise 6% of the total DNA, whereas the terminally repetitive sequence of the S region designated as *ac* each comprise 4.3% of the DNA. Only the sequences designated as *a*, each comprising no more than 0.5% of the DNA are shared by both L and S regions. The sequences in the entire molecule with the region L to the left may be represented by the sequence *ab-l-b'a'a'c'-s-ca* in which the prime letters represent an inversion of the sequences occuring at terminals. The evidence follows: (1) HSV-1 DNA digested with lambda phage exonuclease to about 0.2-0.5% of total will self-anneal (Fig. 13.4) to form circles [59]. The terminal repetitive sequence is limited to a maximum of 0.5% of total, i.e., to no more than 800 base-pairs [60, 62, 74]. (2) Evidence for the internal repetition of terminals is based on the finding that intact strands obtained upon denaturation of native DNA will self-anneal to form barbell structures consisting of a double-stranded stretch (*b'a'a'c'*) separating a large (L) and a single (S) stranded loop (Fig. 13.5) [61, 62, 75, 76]. The internal inverted repetition can also be demonstrated by digesting the DNA with lambda exonuclease beyond the S region, i.e., beyond the sequences *a'c'-s-ca,* and then letting the single-stranded terminal *ca* anneal to its inverted repeat *a'c'* [61, 62]. (3) Evidence that the bulk of the left (*ab*) and right (*ca*) terminal sequences repeated internally are not identical is based on partial denaturation studies [61, 62, 76]. These clearly differentiated the right and left termini and identified the left and right terminal sequences in the internal inverted repeat (Fig. 13.6).

The observation that digestion of the termini with lambda exonuclease beyond 0.5% of the DNA reduced the rate of circularization led to the demonstration that terminal sequence *a* itself contains inverted repetitions ([74]; Hyman, personal communication). The size of the sequence in *a* that is inverted is probably less than 100 bases long and is located 800-1600 bases from the terminus.

The second unusual feature of the DNA is that the arrangement of the sequences in populations of DNA molecules is not uniform. Specifically, recent studies indicate that HSV DNA populations consist of 4 kinds of

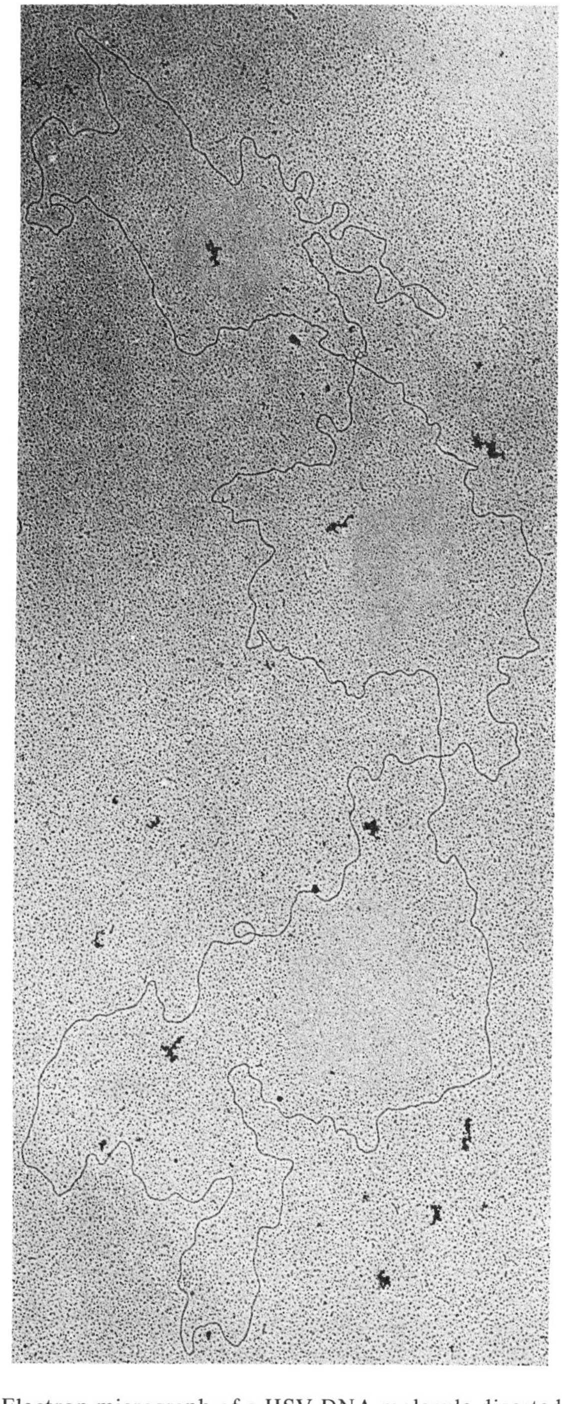

Fig. 13.4 Electron micrograph of a HSV DNA molecule digested with lambda exonuclease and allowed to self-anneal. The digestion was carried out until about 0.5% of the DNA was digested. From Wadsworth et al. [74].

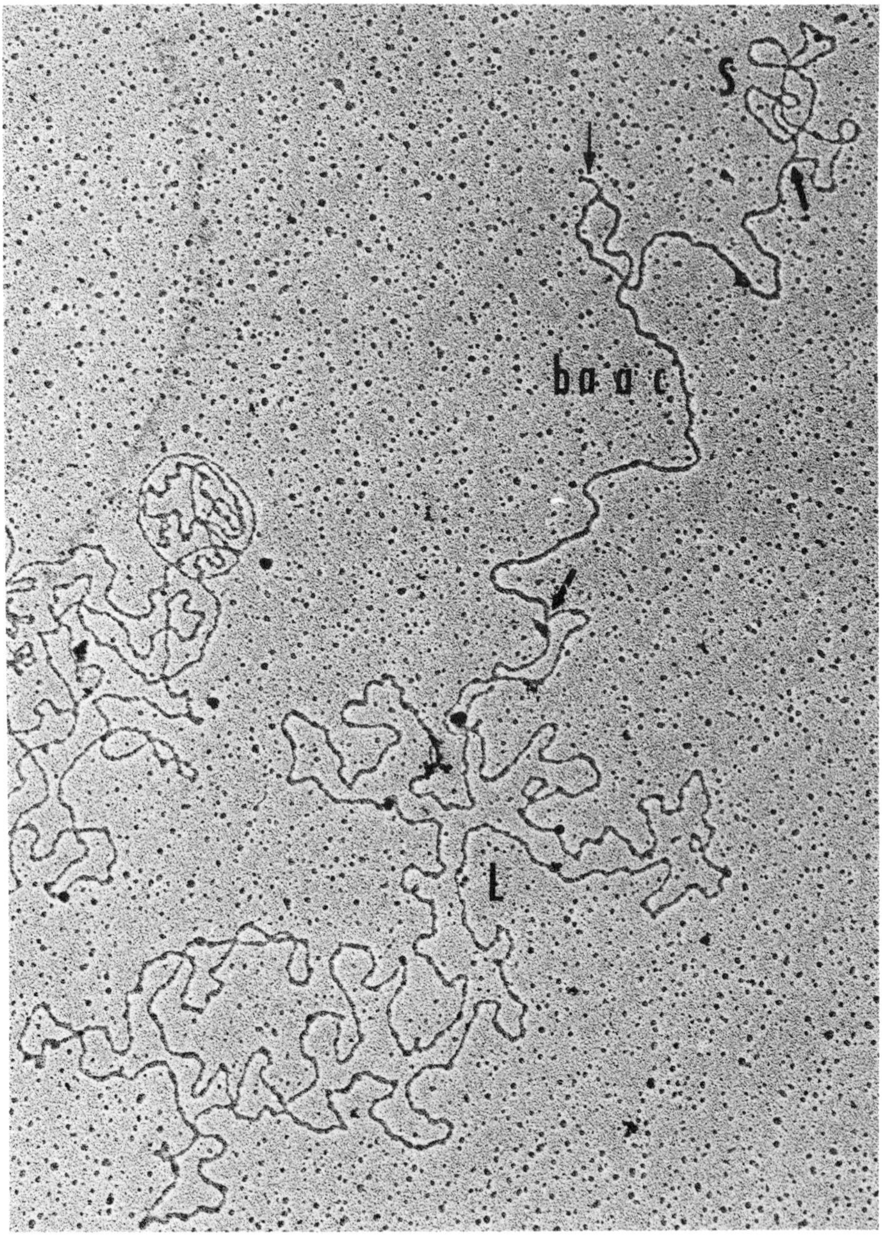
S
b'a'a'c'
L

Fig. 13.5 Electron microscopic demonstration of inverted internal repetitions in HSV-1 DNA. HSV-1 (F) DNA was denatured with alkali and sedimented in neutral sucrose density gradients. The DNA collected from the region containing intact strands was then self-annealed at room temperature in 66% v/v formamide, 0.01 M Tris, pH 8.0, 0.001 M ethylenediamine tetraacetic acid, then spread for electron microscopy. Note double-stranded region *b'a'a'c'* formed by annealing of the terminal sequences to the internal complementary sequences. The thick arrows point to the beginning of the single-stranded regions forming the small (S) and large (L) loops. The thin arrow points to a spoor, formed at the junction of the 2 termini. From Wadsworth et al. [62].

molecules differing solely in the orientation of L and S regions relative to each other [4, 61, 65, 76]. The evidence supporting this conclusion is as follows. If the L and S regions assort themselves at random relative to each other, it could be predicted that restriction endonucleases which do not cleave within the repeated sequences *ab* and *bc* would produce 3 sets of fragments. The first set should comprise 4 fragments arising from termini, and each of the 4 fragments should be present in 0.5 molar concentrations with respect to the molarity of the intact DNA. The second set should comprise 4 fragments spanning the junction of L and S regions; each of these fragments should be present in 0.5 molar concentration with respect to the molarity of the intact DNA. The third set should contain fragments arising from regions between the 0.5 and 0.25 molar fragments. Moreover, the fragments in the third set should be in 1 molar concentrations relative to that of the DNA. As illustrated in Fig. 13.7 and Fig. 13.8 these predictions were fulfilled [65]. In addition, analyses of partial denaturation of HSV DNA [65] and of heteroduplexes produced by renaturation of intact strands of HSV-1 DNA (Wadsworth and Roizman, unpublished studies) substantiated the conclusion that L and S regions were oriented at random relative to each other.

It should be noted that the mechanism responsible for the random orientation of the L and S regions is not known although the available evidence favors a high frequency legitimate recombination through the internal inverted repetition rather than independent replication and joining of the L and S regions (Hayward, Jacob, and Roizman, unpublished studies) (see Recent Developments at end of volume).

There is little information on sequence organization of the DNA of other herpesviruses. Presently available data suggest that EAV DNA may have L and S regions similar in size but differing from those of HSV in that only the left terminal region is demonstrably reiterated internally in an inverted form. The left terminus is either minimally or not reiterated (Sheldrick and Berthelot, personal communication) (see Recent Developments).

It is perhaps also worthy of note that these properties are shared by the DNAs of nearly 25 HSV-1 and HSV-2 strains [4, 65]. On the other hand, differences among strains of HSV-1 have been noted, and these were largely the absence of 1 or more restriction enzyme cleavage sites [4, 65]. The apparent random assortment of L and S regions in HSV DNA has numerous implications even though we do not know whether all 4 or only 1 of the resulting molecules is capable of independent replication. One immediate consequence is that the DNA populations contain 8 different kinds of single strands instead of the usual 2 strands. This structure therefore not only makes the task of mapping functions along the viral DNA more difficult but also predicts that genetic maps should show 2 linkage groups and that a considerable amount of symmetric transcription might reasonably be expected (see Recent Developments).

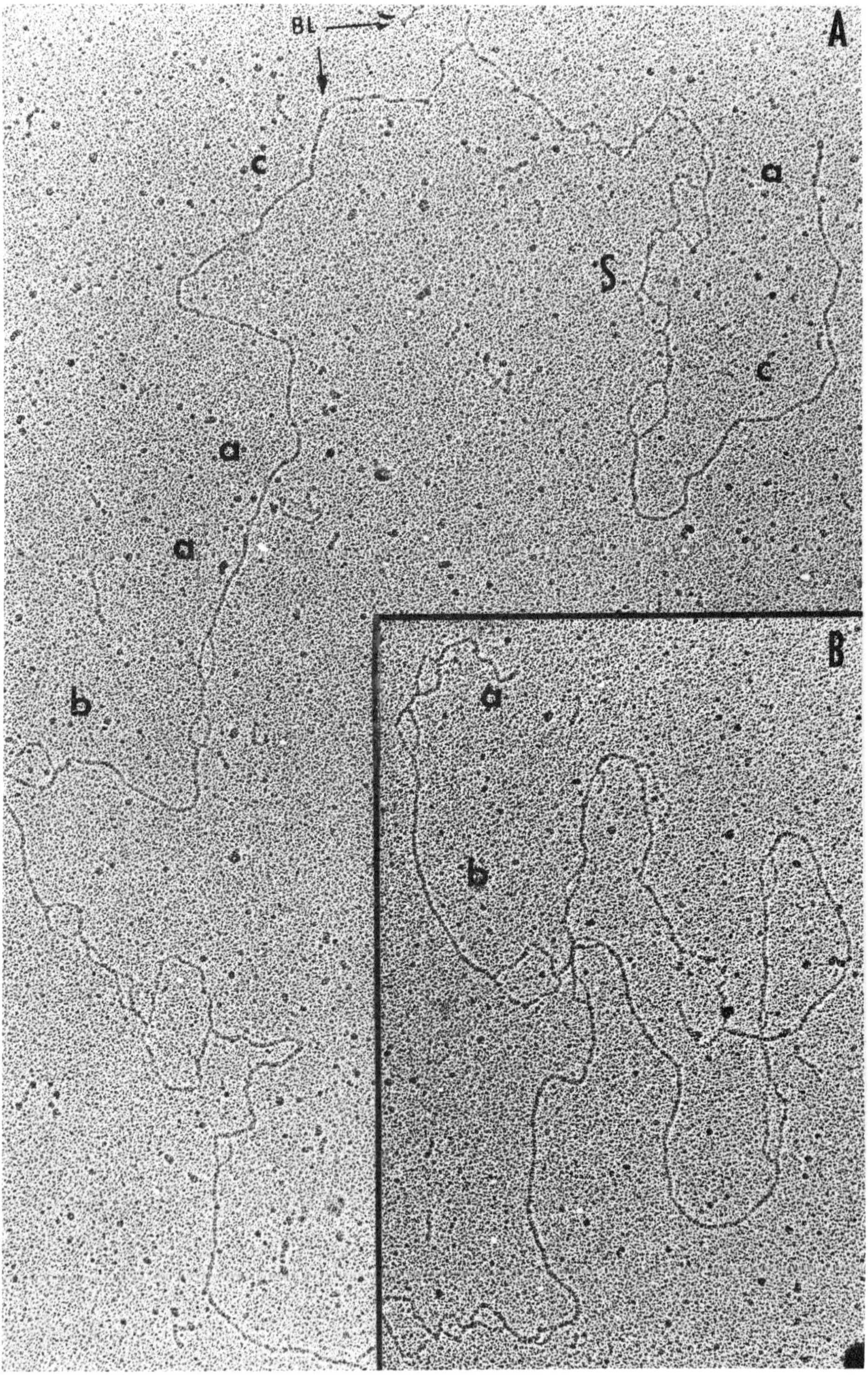
BL
A
c
a
S
c
a
a
b
B
a
b

4. *Genetic Relatedness Among Herpesviruses*

It has been known for many years that certain herpesviruses are immunologically related. Thus antisera against HSV-1 neutralize HSV-2 and vice versa, although the homologous reaction is favored [16, 77]. HSV has been reported to be neutralized by antisera to herpesvirus B [78], and to bovine mammalitis virus [79]. By tests other than neutralization, several other herpesviruses have been shown to share antigens [80-83]. However, the published evidence for a common antigen [84, 85] in several herpesviruses is rather weak.

Relatedness may be more precisely estimated from DNA-DNA and even DNA-RNA hybridizations, but this technique is useful only if there are sufficient amounts of homology between the DNAs, i.e., if the evolutionary divergence of these viruses is rather recent. Although demonstration of "relatedness" is simple, quantitation is not. The problem arises from 2 considerations. First, the DNA reassociation rate constant is greatly affected by the extent of matching of base-pairs and even minor mismatching of base-pairs greatly favors homologous rather than heterologous reassociation [86, 87]. In the light of this finding, simple hybridization of excess DNA from virus (A) with trace amounts of DNA from virus (B) is not very satisfactory because homologous (A) DNA in excess reassociates very rapidly and becomes unavailable for hybridization with the heterologous (B) DNA. In effect, "saturation" of heterologous (B) DNA requires repeated addition of denatured excess (A) DNA species. Ideally, therefore, one DNA in amounts so small that it cannot reassociate with itself during the allotted time should be hybridized to an excess amount of a second DNA fixed so that it is available for hybridization to the first DNA but not to itself. In our experience, immobilization of the second DNA to filters is not satisfactory because, for reasons poorly understood, the amount of DNA available for hybridization varies considerably from filter to filter, even though the amounts of DNA fixed to the filters are identical. Second, an entirely different set of problems arises when one of the DNAs is replaced with RNA. If the transcription is asymmetric, the use of RNA eliminates the problem of reassociation of at least 1 of the DNAs to itself. However, the RNA must be shown to contain all the sequences present in the DNA; all the RNA

Fig. 13.6 Electron micrographs of plaque purified HSV-1 (Justin) DNA partially denatured at pH 11.48 for 7 min at 25°C. (A) Right terminus showing sequences *ac* in the S region and the region *b′a′a′c′* containing the inverted terminal repetitions. (B) Left terminus of the same molecule. Note correspondence of the *ab* in the (B) with the internal inverted reiteration designated by the same letters and shown in (A). Data from Wadsworth et al. [62].

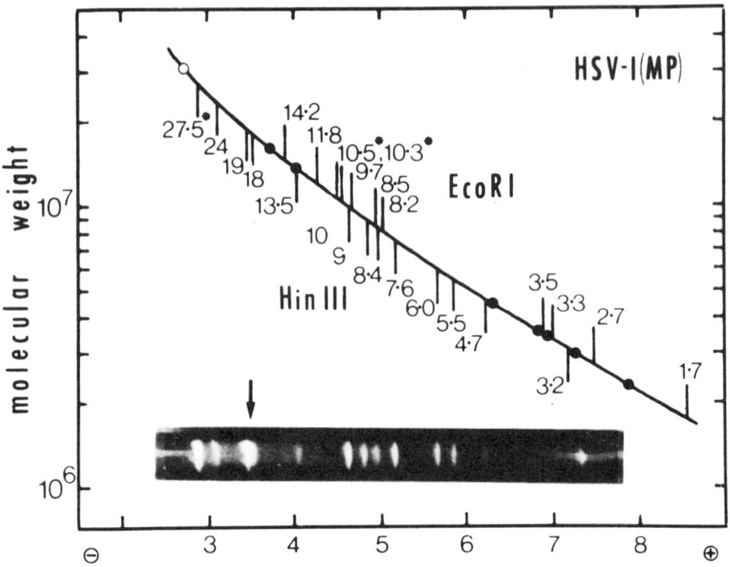
HSV-I(MP)
molecular weight
10^7
10^6
EcoRI
Hin III
27·5
24
19
18
14·2
13·5
11·8
10·5
10·3
9·7
8·5
8·2
10
9
8·4
7·6
6·0
5·5
4·7
3·5
3·3
2·7
3·2
1·7
⊖
3
4
5
6
7
8
⊕

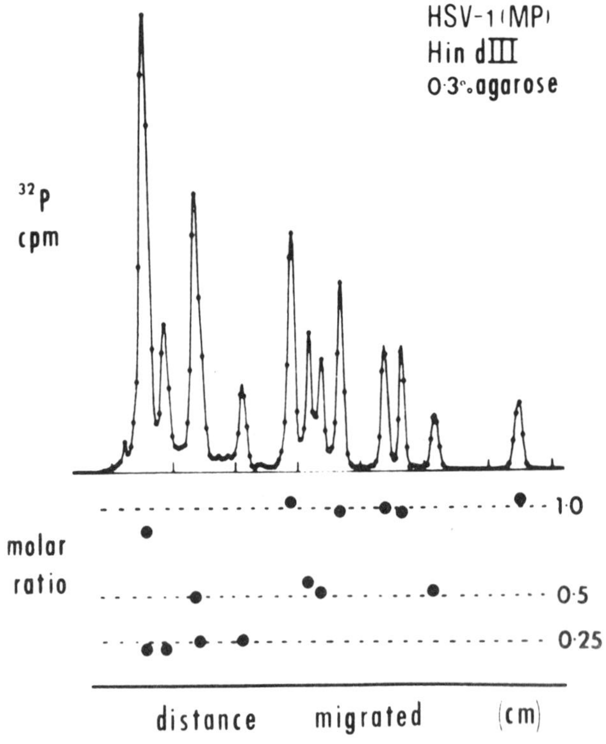
HSV-1(MP)
Hin dIII
0·3% agarose
^{32}P
cpm
molar
ratio
1·0
0·5
0·25
distance migrated (cm)

sequences must be present in equimolar amounts; and in addition, the kinetics of hybridization of RNA to the homologous DNA should be known.

There have been numerous publications dealing with the relatedness among herpesviruses. DNA-DNA hybridizations indicate that HSV-1 and HSV-2 DNAs show approximately 50% homology, with 85% matching of base-pairs of the homologous regions [88]. Trace amounts of homology between HSV and several other viruses [15, 79, 89-92] cannot be evaluated since the extent of base mismatch in the hybrid has not been determined. Thus, liquid-filter DNA-DNA hybridization between HSV-1 and HSV-2 DNA and the DNA of Marek's disease herpesvirus indicated less than 2% homology [89]. No homology was detected between HSV-1, HSV-2, Marek's disease herpesvirus, or CMV DNAs and EBV DNA labeled in vivo or in vitro [93, 94]. HSV-1, HSV-2, EBV, simian, and murine CMV DNAs had no effect on the kinetics of reassociation of in vitro labeled human CMV DNA of the AD 169 strain [94]. Lack of DNA homology does not exclude genetic relatedness because, in principle at least, the synthesis of identical polypeptides could result from mRNAs totally lacking in homology.

5. *Herpesvirus DNA: the Unanswered Questions*

One major endeavor, the mapping of viral genes along the DNA, is currently

---

Fig. 13.7 Size and molar ratios of the DNA fragments obtained by cleavage of HSV-1 (MP) DNA with Hin III restriction endonuclease. Top panel: Molecular weights of the Hin III and Eco R1 fragments. Mixtures of unlabeled fragments from HSV-1 (MP) DNA and reference Eco R1 fragments from lambda cl_{857} DNA were subjected to electrophoresis through 0.3% agarose gels (1 x 18 cm) and stained with ethidium bromide. The distances migrated by reference species were measured and used to construct the standard curve for molecular weight versus mobility. The reference species were intact linear lambda cl_{857} DNA 31 x 10^6 (open circle); Eco R1 fragments from lambda cl_{857} DNA A + F (15.8 x 10^6), A (13.7 x 10^6), B (4.7 x 10^6), C (3.7 x 10^6), D (3.5 x 10^6), and F (2.1 x 10^6) (closed circles). The asterisk indicates that 2 DNA fragments are present at each of these positions. Inset: ethidium bromide-stained Hin III fragments of HSV-1 (MP) DNA after electrophoretic separation in 0.3% agarose gels. Arrow indicates the double band of 18-19 x 10^6 molecular weight. Middle panel: ^{32}P-radioactivity profiles of electrophoretically separated fragments from limit digests of HSV-1 (MP) DNA with Hin III. Gel electrophoresis of digested ^{32}P-labeled HSV-1 (MP) DNA was carried out in 1 x 30 cm gels of 0.3% agarose at 1.6 v/cm and 4° in Tris-phosphate buffer for 36 hr. The radioactivity in each of the 300 fractions (1 mm gel slices) was measured in toluene scintillation fluid. Bottom panel: Relative molar ratios of the Hin III fragments calculated from the amount of DNA and the molecular weight in each band. Data from Hayward et al. [65].

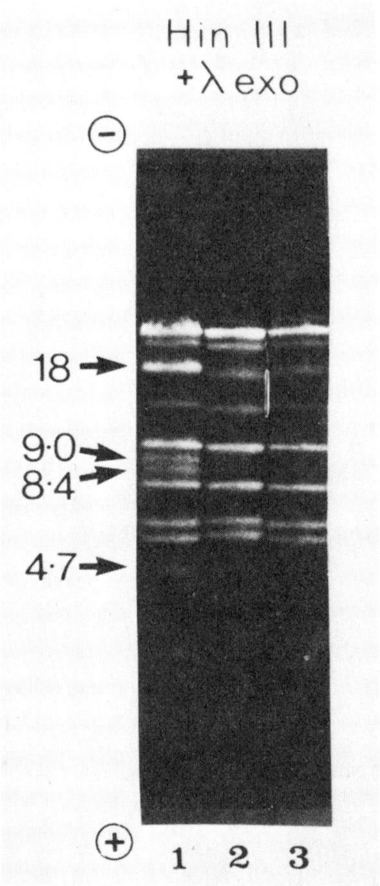
Hin III
+λ exo
18
9·0
8·4
4·7
1 2 3

Fig. 13.8 Identification of the restriction endonuclease fragments situated at the termini of HSV-1 DNA. HSV-1 (MP) DNA was digested with lambda exonuclease, then cleaved with Hin III endonuclease and subjected to electrophoresis on polyacrylamide gels. The electrophoretically separated fragments in the gel were stained with ethidium bromide. Gel 1, controls without exonuclease; gel 2, samples preincubated with lambda exonuclease for 15 min (approximately 6% digestion as measured by percent acid soluble ^{3}H-radioactivity); gel 3, 50 min incubations (14% digestion). Positions of terminal fragments are indicated by arrows. Data from Hayward et al. [65].

underway in several laboratories. It involves identification of the functions of viral polypeptides and mapping of the structural genes of viral polypeptides on DNA fragments generated with restriction endonucleases. Restriction endonuclease maps of HSV-1 (Fig. 13.9) and HSV-2 DNAs are already available (Hayward, Buchman, and Roizman, manuscript in preparation) and should be followed soon by maps of other herpesvirus DNAs (see Recent Developments).

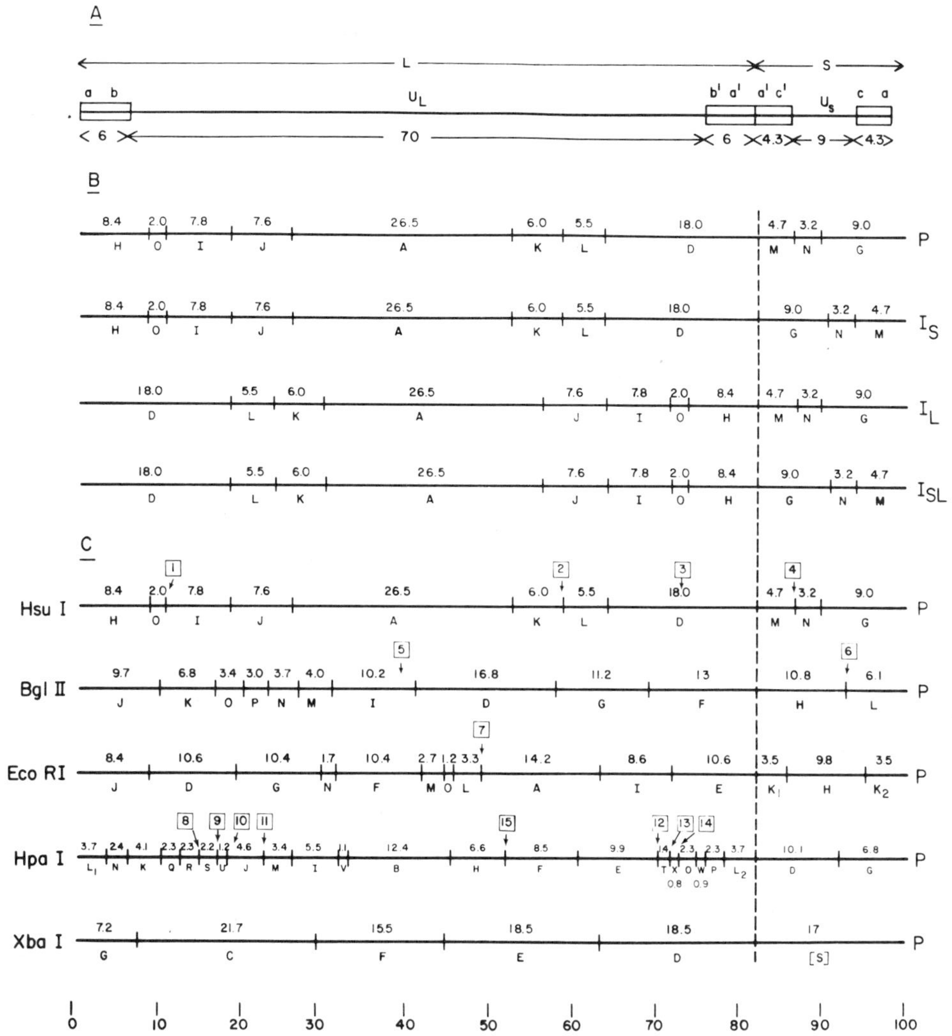

A
L
S
a b
UL
b' a' a' c'
Us
c a
< 6
70
6
4.3
9
4.3>
B
8.4 2.0 7.8 7.6 26.5 6.0 5.5 18.0 4.7 3.2 9.0
H O I J A K L D M N G
P
8.4 2.0 7.8 7.6 26.5 6.0 5.5 18.0 9.0 3.2 4.7
H O I J A K L D G N M
IS
18.0 5.5 6.0 26.5 7.6 7.8 2.0 8.4 4.7 3.2 9.0
D L K A J I O H M N G
IL
18.0 5.5 6.0 26.5 7.6 7.8 2.0 8.4 9.0 3.2 4.7
D L K A J I O H G N M
ISL
C
Hsu I
8.4 2.0 7.8 7.6 26.5 6.0 5.5 18.0 4.7 3.2 9.0
H O I J A K L D M N G
Bgl II
9.7 6.8 3.4 3.0 3.7 4.0 10.2 16.8 11.2 13 10.8 6.1
J K O P N M I D G F H L
Eco RI
8.4 10.6 10.4 1.7 10.4 2.7 1.2 3.3 14.2 8.6 10.6 3.5 9.8 3.5
J D G N F M O L A I E K1 H K2
Hpa I
3.7 2.4 4.1 2.3 2.3 2.2 1.2 4.6 3.4 5.5 1.1 12.4 6.6 8.5 9.9 1.4 2.3 2.3 3.7 10.1 6.8
L1 N K Q R S U J M I V B H F E T X O W P L2 D G
0.8 0.9
Xba I
7.2 21.7 15.5 18.5 18.5 17
G C F E D [S]
0 10 20 30 40 50 60 70 80 90 100

Fig. 13.9 Sequence arrangement and restriction endonuclease maps of HSV-1 DNA. (A) Sequence arrangement of the L and S components of HSV-1 DNA. The numbers below the line show the percent of total DNA contained within the reiterated sequences *ab* and *ca* and in the unique sequences U_L and U_S. (B) Linear arrangement of the restriction endonuclease fragments generated by digestion of HSV-1 DNA with Hsu I restriction endonuclease. P is the prototype arrangement of HSV-1 DNA determined on the basis of analysis of recombinants of HSV-1 × HSV-2 [320]. I_S, inversion of S component; I_L, inversion of L component; I_{SL}, inversion of L and S components. (C) Restriction endonuclease maps of the prototype arrangements of HSV-1 DNA generated by Hsu I, Bgl II, Eco RI, Hpa I, and Xpa I restriction endonucleases. Note that only the prototype arrangements are shown. The nomenclature employed in labeling restriction endonuclease maps of HSV DNA differs from that employed for other viruses in only one respect. The fragments are labeled alphabetically in order of decreasing size. The exceptions are the 0.25 M fragments which span the junction between L and S components. By convention these fragments are designated either by the letter of the alphabet reflecting their size or by two letters identifying the half molar constituents of these fragments. In the case of the digest generated by Hsu I, fragment B is a 0.25 M fragment spanning the L and S junction in the I_S arrangement of the DNA. Fragment B consists of the region D which in the I_L and I_{SL} arrangement is a terminal fragment and region G which in P or I_L arrangement is also a terminal fragment. Therefore, fragment B can also be designated as fragment DG. A single letter designation for 1 M and 0.5 M fragments and the two letter designation for the 0.25 M fragments were used exclusively in the maps presented in this figure. Numbers in boxes above restriction endonuclease maps refer to cleavage sites which may be absent in some virus strains and to fragments which in some strains differ in molecular weight.

There are, however, several major questions concerning the herpesvirus DNAs for which there are no solutions in sight as yet. Among the more interesting ones are the following.

We have no explanation for the considerable variation in the base composition of herpesvirus DNAs and even among viruses infecting a particular host species like humans. It is noteworthy that in humans the herpesviruses occupy distinct ecological niches, i.e., specialized cells and tissues usually at or near the site of entry to the body and that there is, by and large, relatively little overlap between the niches occupied by different herpesviruses. Could it be that the base compositions of herpesviruses reflect some peculiar characteristic of the differentiated cell in which they grow?

The arrangement of the terminal inverted repetitions in HSV-1 and HSV-2 invites 2 questions. The first arises from the fact that linear DNA molecules generally are expected to have terminal repetitions which are either tandem repeat, i.e., the *a* sequence in HSV whereas other viruses, notably the adenoviruses, have inverted terminal repeats. The first question therefore concerns the significance of the fact that HSV has both tandem and inverted repeats, i.e., that both L and S regions within the DNA molecules are each bound by inverted terminal repetitions. Could it be that herpesviruses have arisen by fusion of 2 genomes and that the junction between the 2 was conserved? Other obvious questions are whether the internal repetition of the terminal sequences has a function and whether the reiterated sequences were conserved in other herpesviruses and account for the trace amounts of homology that has been reported by several investigators. The answer to these questions are not known but could be most revealing! At least some answers could emerge from analyses of the other herpesvirus DNA for the presence of the internal inverted repetition and from studies on the replication of viral DNA (see Recent Developments).

C. Structural Proteins

Analyses of viral structural proteins require preparations of purified virions as well as high-resolution techniques for separation of virion polypeptides. Most of the problems associated with purification of virions stem from the lability of the viral envelope and from the similarity of the physicochemical properties of the virion to those of the membrane vesicles generated in the course of extraction. The membrane vesicle contains both host and viral proteins [94], whereas purified virions contain only variable, trace amounts of host proteins [36]. At least 2 techniques for purification of herpesvirions have been extensively documented [36, 95]. Both techniques require banding of the virions in dextran gradients, but whereas one [36] prescribes additional banding in either sucrose density gradients or potassium tartrate, the other [95] involves banding of the virus in colloidal silica prior to the centrifugation on dextran gradients.

To date, all separations of viral polypeptides have been done in polyacrylamide gels containing sodium dodecylsulfate. Earlier studies were based on a gel system with low resolving power; as a consequence, it was reported that herpesvirions contained 9-11 polypeptides. Current studies [36, 96, 97] based on high-resolution polyacrylamide gels indicate that the polypeptides in the purified virions form 33 bands and range in molecular weight from 25,000 to approximately 280,000 (Fig. 13.10). Virion polypeptides less than 25,000 in molecular weight have also been reported [98]. Analyses of the structural polypeptides of PSV, CMV, EAV, and EBV show that these herpesviruses are as complex as HSV [99-102].

Several points should be made concerning the available data on structural polypeptides of herpesviruses.

(1) A substantial number of virion polypeptides are not the primary products of translation. Thus HSV-1 and HSV-2 virions contain both glycosylated and phosphorylated polypeptides. Moreover, several virion polypeptides do not correspond in electrophoretic mobility to any of the infected cell polypeptides labeled during a short pulse [36, 96, 97, 103-104]. Pulse chase experiments [105], indicate that in only a few instances is the change in electrophoretic mobility due to cleavage of a larger precursor polypeptide. The difficulty in estimating the number of species of viral proteins arises chiefly from the real possibility that the virion contains not only the fully processed products but also the intermediate, partially processed, precursors.

(2) For any given virus strain, the molar ratio of virion polypeptides remained constant from preparation to preparation. The molar ratio of virion polypeptides to virions ranged from less than 20 (minor polypeptides) to well over 1000 (major polypeptides) [96]. Failure to detect some of the minor polypeptides by Watson and Powell [106] very likely reflects a lack of sensitivity of their system inasmuch as the conditions which allow optimal resolution of major polypeptides are not satisfactory for the detection of minor ones.

(3) Major differences in the amount and electrophoretic mobility of noncapsid polypeptides have emerged from comparisons among HSV-1 strains with a history of few passages outside the human host and those with an extensive history of extra human passages. The data demonstrated alteration in the number and electrophoretic mobility of the viral polypeptides upon extensive passaging outside the human host [96]. The unexpected finding, however, was the observation that HSV-1 isolates passaged less than 3-4 times outside the human host vary considerably in the electrophoretic mobility of several polypeptides [5]. Moreover, the apparent clustering of

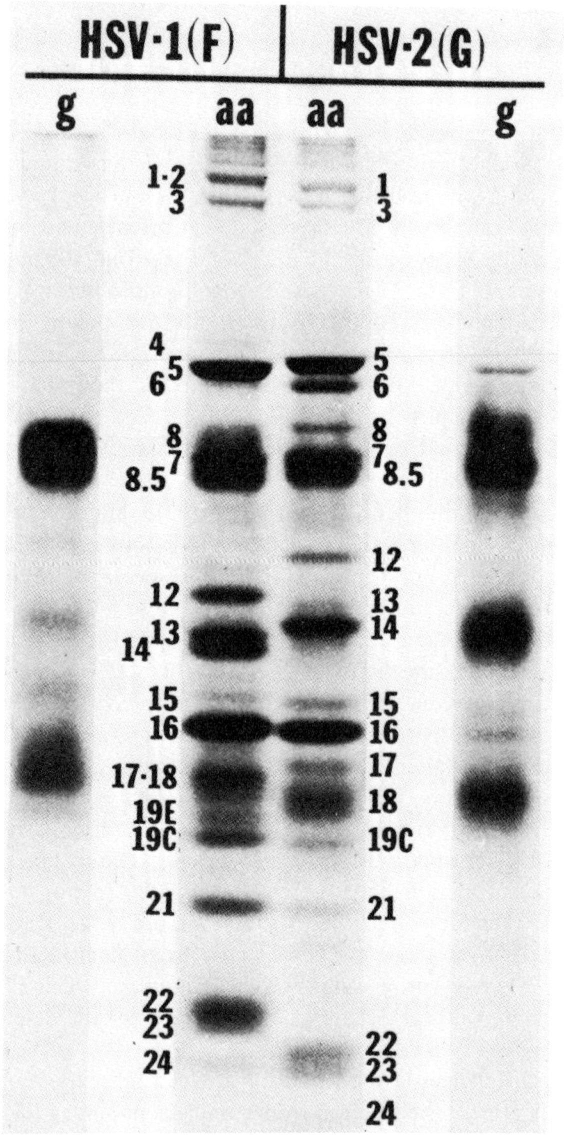
HSV-1(F)
HSV-2(G)
g
aa
aa
g
1·2
3
1
3
4
5
6
5
6
8
7
8.5
8
7
8.5
12
12
13
14
13
14
15
16
15
16
17·18
19E
19C
17
18
19C
21
21
22
23
24
22
23
24

Fig. 13-10 Structural polypeptides of HSV-1 and HSV-2. The figure is a print of an autoradiogram of sodium dodecyl-acrylamide gel slab showing the [^{32}S]-methionine-labeled (aa) or [^{14}C] glucosamine-labeled (g) polypeptides present in HSV-1 (F) and HSV-2 (G) virions. The polypeptides are numbered according to the nomenclature of Spear and Roizman [36]. Autoradiogram courtesy of P. G. Spear [97].

isolates from the same body site suggested the need for more extensive studies to determine whether there is in fact a correlation between virion structure and localization at a particular site on the human body. Studies on a small sample of HSV-2 strains [97] also revealed differences in the electrophoretic profiles of virion polypeptides. The variability of structural polypeptides upon continuous passaging in cell culture is especially worrysome since it can lead to inadvertent selection of laboratory strains significantly different from natural strains in the properties under study at that time.

D. Other Constituents

Herpesviruses have been shown to contain 2 other classes of chemical constituents: polyamines and lipids.

Highly purified preparations of herpes simplex virions were found to contain the polyamines spermidine and spermine in a nearly constant molar ratio of 1.6 ± 0.2. On the basis of polyamine-to-viral DNA ratios in these preparations, it was calculated that each virion contains approximately 40,000 molecules of spermine and about 70,000 of spermidine. The basic groups in spermine are adequate to neutralize over half of the viral DNA-phosphate groups [107, 108].

Polyamines seem to be specific structural components of the virion rather than nonspecifically bound host molecules. This conclusion is based on the results of an experiment in which an infected-cell lysate containing unlabeled virions was mixed with a similarly prepared lysate of uninfected cells containing labeled polyamines. Virions isolated from the mixture contained polyamines with specific activities which were less than 10% of those in the initial mixture, indicating that there had been little exchange between unlabeled polyamines bound to the virions and labeled extraneous polyamines.

Disruption of the viral envelope with a nonionic detergent and urea removed up to 95% of the viral spermidine, but left the amount of spermine essentially unchanged and in a nearly constant ratio to the viral DNA-phosphate. Although spermine was not shown to be specifically localized in the core, it is likely but remains unproven that it functions in this virus in a manner similar to that proposed for putrescine and spermidine found in the bacteriophage T4, namely, to neutralize DNA-phosphate [26, 109-119]. Since spermidine was removed in parallel with the envelope constituents following detergent treatment, it was suggested that this polyamine is specifically associated with the envelope [108].

Very little is known about the structural lipids of the herpesvirion. The problem arises from the fact that the purity of the virus analyzed for lipid content is uncertain. However,

1. The infectivity of the mature virion is destroyed by phospholipase C [111] suggesting that the lipids essential for infectivity are contained in the envelope.
2. Phospholipids made prior to infection cosediment with virions at least partially purified from extracellular fluid [112, 113].
3. The distribution of radiophosphorus in phospholipids of the virus was reported to be similar to that in phospholipids contained in the nuclei of infected cells, but the significance of this finding is not clear because the nuclei were prepared with the aid of detergents.
4. Studies on the buoyant density of the virus grown in different cells species suggests that lipid composition is determined at least in part by the cell in which the virus was grown [114].

E. The Distribution of the Chemical Components in the Herpesvirion

The distribution of the structural components in the virion is not yet well defined, but the general design of the virion is beginning to be discernible and elements of its topology are now available for 3 herpesviruses, i.e., HSV, EAV, and EBV.

Electron microscopic studies indicate that HSV DNA is contained in the core [7, 10]. The core probably contains proteins as well as polyamines, if the hypothesis concerning their functions proves to be correct. Comparison of the capsids lacking DNA with full capsids suggests that at least one polypeptide (NO. 21) functions as a core protein.

To date, 3 kinds of HSV [103, 104] and EAV [99] capsids have been described. In the case of HSV, the A and B capsids were isolated from nuclei of infected cells. The C capsids were obtained by stripping the envelope off the virion with nonionic detergents. The A capsids lack DNA, but contain 4 polypeptides, Nos. 5, 19, 23, and 24. The B capsids contain RNA and 2 additional polypeptides, Nos. 21 and 22A, as well as DNA. The C capsids contain DNA, the same polypeptides as those present in the empty A capsids, as well as polypeptides 21 and 1 and 2, and trace amounts of most glycoproteins, but lack polypeptide 22A. Polypeptide 22A is also absent from virions, which suggests that it is a precursor molecule that is cleaved or replaced at the time of envelopment and hence may be situated on the surface of the B capsid. The EAV capsids [99] were isolated by handing in Renografin gradients and superficially at least, the light (L), intermediate (I), and heavy (H) capsids appear to correspond to the A, B, and C capsids of HSV.

It is likely that all the glycosylated polypeptides are components of the virion envelope [16]. Apart from theoretical considerations involving the process of glycosylation of the polypeptides, it has been shown that the glycosylated polypeptides can be labeled in vitro by transfer of tritium from tritiated borohydrate during reduction of the Schiff's base in the presence of pyridoxal phosphate. In addition, the glycosylated polypeptides have been found in purified membranes extracted from infected cells [116, 120].

The localization of nonglycosylated polypeptides other than those accounted for in the capsid is not known. We suspect that at least some of these polypeptides are localized in the underside of the envelope and are thus not available for glycosylation whereas others are in the tegument.

III. The Replication of Herpesviruses in Permissive Cells

A. Initiation of Infection

1. The Infectious Unit

Analyses of infected cells show that they contain 2 kinds of particles, i.e., naked and enveloped capsids (virions). In addition, the nucleus of the infected cell contains excess viral DNA which does not become packaged. Of these products, the enveloped nucleocapsid is the epidemiologically significant unit. The HSV envelope need not be intact, although an intact envelope probably increases the stability of the particle [26, 110]. Naked nucleocapsids obtained from the nucleus before envelopment or from virions by stripping off the envelope by detergent treatment are not infectious [117-119, 121]. However, viral DNA rendered free of proteins is infectious [122, 123]. It is particularly interesting that both native and alkaline-denatured DNAs are infectious [123].

It is not known why naked nuclecapsids are not infectious. It is conceivable that the specific infectivity of the virus and its subunits may be determined by the mode of entry into the cell. The observation that HSV DNA is infectious indicates that the virion proteins are not essential for the multiplication of the virus. The infectivity of the virion is especially sensitive to detergent and lipid solvents, lipases and proteases, and protein-denaturing agents (reviewed in [16, 22, 124]). The conditions which best preserve the infectivity of the herpesviruses appear to be storage in distilled water at 4°C and storage in various other media at temperatures below −60°C.

Although particle infectivity ratios of 1:10 have been reported for HSV-1 [125], most preparations used in general laboratory work characteristically have a much higher ratio. For HSV-1 and HSV-2, standard virus stocks prepared at 34°C from Hep-2 cells infected at multiplicities less than

1 plaque forming units (PFU) per cell contain 1 infectious unit for every 40 (HSV-1) to 200 (HSV-2) particles.

2. *Adsorption, Penetration, and Uncoating*

Little is known of the receptors on the cell surfaces to which herpesviruses adsorb. Adsorption in general is slow and both volume and cation dependent. Sulfated polyanions, both natural (agar, mucopolysaccharides, heparin) and synthetic, prevent adsorption of HSV (see review in Roizman and Furlong, [16]). The events following adsorption are also unclear. On the basis of electron microscopic studies, 2 mechanisms have been proposed for the mode of entry of the virus into the cell. One [30, 126] suggests that herpesviruses are taken into the cell in phagocytic vesicles in which they are uncoated. The second [17, 127, 128] proposes that the entry is effected by fusion of the viral envelope with the plasma membrane followed by passage of the capsid through the fused membranes into the cytoplasm. Abodeely et al. [129] and Hummeler et al. [130] found particles in pinocytotic vesicles as well as particles whose envelope was fused with the plasma membrane. The critical experiment, i.e., demonstration of viral membrane proteins immediately following penetration either in the plasma membrane or in the cytoplasm, has not been done. It has been reported that the capsid disaggregates in the cytoplasm and that a DNA-protein complex is transported into the nucleus [131].

B. Characteristics of the Reproductive Cycle

1. *Duration and Temporal Landmarks*

The reproductive cycle of several herpesviruses, but specifically those of HSV-1, HSV-2, CMV, EAV, and pseudorabies virus has been characterized in considerable detail. Less is known about other herpesviruses, and especially about EBV because of the absence of a permissive cell line [132].

In general, the duration of the reproductive cycle, the time course of major events, and the yield of viral products are highly reproducible within a given cell host-virus system, but vary considerably from system to system depending on: (1) the degree to which the host restricts virus multiplication; (2) the multiplicity of infection; (3) the temperature; and (4) the nutritional properties of the medium [16, 22]. Under most optimal conditions, such as encountered in the multiplication of HSV-1 or HSV-2 in Hep-2 cells, the replicative cycle lasts 15-19 hr. The progeny virus is detected first by electron microscopy at about 4 hr after infection and then by infectivity measurements

1-2 hr later. Virus accumulates in the infected cell at exponential rates. Release of virus from infected cells varies from system to system, but is in general temperature dependent. Thus, at 34°C HSV-1 yield in human cells is optimal, but no more than 20% of the virus is released into the extracellular fluid [133]. The cycle is compressed slightly for pseudorabies virus, but is double or even triple in duration for CMV.

2. *The Sequence of Events*

The first event in the herpesvirus reproductive cycle is the transcription of viral DNA. It is followed by the synthesis of viral protein, viral DNA, and finally, assembly of viral progeny. Inherent in this sequence is the translocation of viral RNA from the nucleus into the cytoplasm, of viral protein from cytoplasm into the nucleus, and of viral progeny from the nucleus into the perinuclear space and cysternae of the endoplasmic reticulum. Concurrently the structure and metabolism of the infected cell become altered.

In this chapter the emphasis is on the most interesting of the problems which are currently amenable to experimental investigation, i.e., viral gene expression and its regulation.

It is convenient therefore to discuss the physical properties and the detection of viral RNA and proteins first, to be followed by the regulation of viral gene products.

C. Transcription

1. *Definition*

Viral RNA is the first viral product made in the infected cell. Since our information on viral RNA is based on fractionations dependent on its properties and on demonstrated homology to viral DNA, it is convenient to begin with a brief series of definitions.

The products of transcription of viral DNA are processed by cleavage, methylation, adenylation, and capping from the time it is synthesized in the nucleus and until it appears in the cytoplasm. Because we cannot block all processing, the initial product of transcription is virtually unknown. The product which becomes associated with polyribosomes has been characterized relatively well. It is convenient for the purpose of this discussion to define as mRNA the RNA present in polyribosomes and to define as "transcript" the products of transcription and all partially or fully processed RNAs other than those in polyribosomes.

The detection and estimation of viral RNA is based on its ability to hybridize to viral RNA. Two techniques have been used to identify viral RNA species. The technique described by Frenkel and Roizman [134] involves hybridization of trace amounts of labeled viral DNA to excess unlabeled viral RNA in solution. This technique measures the fraction of viral DNA complementary to the RNA sequences and the abundance of viral RNA sequences that have accumulated up to the time of extraction. It could, potentially, underestimate both the extent of transcription and abundance if both strands of the DNA are transcribed and give rise to stable transcripts. The technique described by Gillespie and Spiegelman [135], or variants thereof, involves binding of labeled viral RNA to DNA fixed to filters. Although the technique is susceptible to theoretical refinements, most often it measures only the RNA made during the labeling period and very frequently only if it is present in relatively high abundance. Current refinements of this technique involve hybridization of labeled RNA to restriction enzyme fragments fixed to filters as a rapid screening technique.

2. *Characterization of Viral RNA Sequences Accumulating in the Cytoplasm*

Viral RNA sequences accumulating at the time of viral maturation have been extensively investigated in HSV-1 infected cells. The data may be summarized as follows.

Viral RNA sequences accumulating in the cytoplasm of infected cells arise from approximately 43% of viral DNA and appear to form at least 2 groups differing in molar ratios [76, 136-138]. The abundant groups arise from approximately 21% of the DNA. The observation that self-annealing of the RNA scarcely affects the rate of hybridization or the amount of DNA driven into hybrid suggests that they are not self-complementary [139]. A perplexing finding for reasons that will be described later in the text is that even late in infection, when only few of the viral polypeptides are made, all of the RNA sequences found in the cytoplasm are also represented in polyribosomes [137, 138].

The sedimentation rate constant of viral RNA in the cytoplasm range to approximately 35S as determined by hybridization of RNA fractionated by sedimentation through a sucrose density gradient to DNA fixed to filters [140]. The 5′ end of the RNA is methylated and capped with 7-methylguanosine. Internal methylation may, however, be less frequent than in uninfected cell RNA (Bartkoski and Roizman, manuscript in preparation). Analyses of the RNA by poly(U) affinity chromatography indicate that all RNA sequences are adenylated and that the bulk of the viral RNA has poly(A) chains 150 nucleotides long (Millette, Silverstein, and Roizman, manuscript in preparation). An observation of considerable potential significance is that only a fraction of the viral RNA with full length poly(A) chains adheres to nitrocellulose filters and that this fraction contains only the abundant viral RNA. Scarce viral sequences that in the native state do not adhere to nitrocellulose filters [137] can be made to adhere by prior heat denaturation (Millette, Silverstein, and Roizman, manuscript in preparation). The data, therefore, suggest that the abundant and scarce sequences might differ in secondary structure (see Recent Developments).

3. *Characteristics of Viral RNA Transcripts Present in the Nucleus*

Nuclear viral RNA transcripts have not been thoroughly characterized but several observations are pertinent here.

Nuclear viral transcripts are complementary to more than 50% of the DNA, but the exact amount of DNA that gives rise to stable transcripts and, more specifically, the extent of symmetric transcription is not known [139]. Double-stranded viral RNA has been found to be complementary to at least 60% of the DNA, but this, it should be emphasized, is a minimal figure [141]. The data indicate, therefore, that transcription is at least in part symmetric. In view of the fact that the cytoplasmic transcripts are not self complementary, it is necessary to conclude that translocation of transcripts of their accumulation in the cytoplasm is regulated [76, 136-138].

The size range of nuclear viral transcripts is considerably larger than that of cytoplasmic transcripts. It has been estimated that transcripts as large as 20 million in molecular weight are present in nuclei of cells infected with HSV-1 [142]. Analyses of these high molecular weight RNAs show that they share sequences with polyribosomal mRNAs [140, 143]. This observation led to the suggestion that viral mRNA arises by cleavage of a high molecular weight precursor. A recent study noted that the high molecular weight RNA consisted of at least 2 classes differing 5000-fold in abundance. The abundant class consisted largely of symmetric transcripts [142].

4. *The Site of Transcription and the Enzymes Transcribing Viral DNA*

All of the available data are consistent with the hypothesis that the transcription of viral DNA takes place in the nucleus [140, 142]. Based on the fact that viral DNA is infectious, we may at least infer that a cellular polymerase can initiate the transcription of viral DNA and that this transcription can ultimately result in the biosynthesis of viral progeny. In the light of the observations that α-amanitine inhibits the synthesis of viral proteins it has been suggested that the transcription is at least initiated by host polymerase II [144].

D. The Synthesis and Function of Proteins Specified by Herpesviruses

1. *The Site and General Pattern of Protein Synthesis*

Protein synthesis in infected cells measured either by amino acid incorporation into peptides [105, 145-147] or by the amount of polyribosomes present in the infected cells [145, 146], takes place exclusively in the cytoplasm and follows a period of initial decline coinciding with disaggregation of polyribosomes specifying host proteins, a period of increased rates of synthesis coinciding with appearance of new, more rapidly sedimenting viral polyribosomes, and last, a period of slow irreversible decline. At least a portion of the polypeptides made on the cytoplasmic polyribosomes find their way into the nucleus [147-150, 318].

2. *Identification of Proteins Specified by the Virus*

Ideally, conclusive evidence that a protein is specified by a virus is its synthesis in an in vitro protein-synthesizing system directed by a viral mRNA. In view of the complexity of the herpesvirus genome, such data may not be available for the foreseeable future. Presently, classification of polypeptides as either viral or host specific done on HSV-1 and HSV-2 infected cells [105, 151, 152] was based on 3 observations. First, overall host protein synthesis declines rapidly after infection [145, 146]. Polypeptides synthesized at increasing rates at the time of decline of host polypeptide synthesis could therefore be expected to be virus-specific. Second, antisera prepared in rabbits against infected rabbit cells grown in rabbit serum precipitate a number of infected human or rabbit cell polypeptides but not uninfected cell polypeptides [153]. Polypeptides reactive with these antisera could be considered virus-specific if they were synthesized at increasing rates after infection. Last, it has been observed in cells infected with variants of HSV-1 or with HSV-2 that some polypeptides are either overproduced, underproduced, or replaced with polypeptides differing in electrophoretic mobilities [105]. The polypeptides whose production is under the genetic control of the virus could also be considered virus-specific provided that they were absent from uninfected cells and were synthesized at increasing rates after infection. On the basis of these criteria, approximately 50 polypeptides separated in high-resolution polyacrylamide gels from HSV-1 and HSV-2 infected cells have been classified as virus specific. They consist of both structural components of the virion and polypeptides for which a counterpart in the virion has not been found and are therefore nonstructural [105, 136, 152].

An assessment of the fraction of total DNA required to specify these polypeptides requires some evidence that these polypeptides are primary products of translation.

The results of several kinds of experiments involving the following: (1) comparison of polypeptides labeled during long and short intervals; (2) comparison of polypeptides extracted from infected cells after a short pulse and after a prolonged chase; (3) comparison of polypeptides labeled in the presence and absence of inhibitors of chymotryptic and tryptic proteases all led to the conclusion that the bulk of the polypeptide does not undergo rapid post translational cleavage even though cleavages of some polypeptides related to virus assembly were readily demonstrable [105, 136]. Based on the conclusion that approximately 50 polypeptides are primary gene products, the size of viral DNA, and the fraction of viral DNA represented in polyribosomes, it was estimated that the polypeptides account for at least 80% of the maximum amount of genetic information that could be encoded in viral DNA [76, 136] (see Recent Developments).

3. *Regulation of Synthesis of Viral Proteins*

The regulation of herpesvirus protein synthesis has been investigated most extensively in Hep-2 cells infected with HSV-1. The fundamental conclusions are that the synthesis of viral polypeptides is coordinately regulated and sequentially ordered. It is convenient to enumerate first the evidence supporting this conclusion and then the mechanisms responsible for this type of regulation.

a. Evidence for Coordinate and Sequential Synthesis of Viral Proteins. The evidence that herpesvirus proteins form several groups whose synthesis is coordinately regulated and sequentially ordered is based on 2 series of experiments. First, analyses of the rates of synthesis of viral polypeptides at different times after infection [76, 105, 136] revealed the presence of at least 5 groups differing in the kinetics of their synthesis (Fig. 13.11 and Fig. 13.12). Class E contained polypeptides which could not be identified as virus-specific and which were synthesized at decreasing rates throughout infection. Polypeptides in classes C and D reached maximum rates of synthesis early in infection and declined afterward. The 2 classes differed solely in the time at which maximal rates were attained; however, class C contained only structural polypeptides whereas class D contained both minor structural and nonstructural polypeptides. The rates of synthesis of class B polypeptides reached a maximum level early in infection and remained constant thereafter. The polypeptides in class A were identified as the major structural components of the virion and were synthesized at ever increasing rates until at least 15 hr

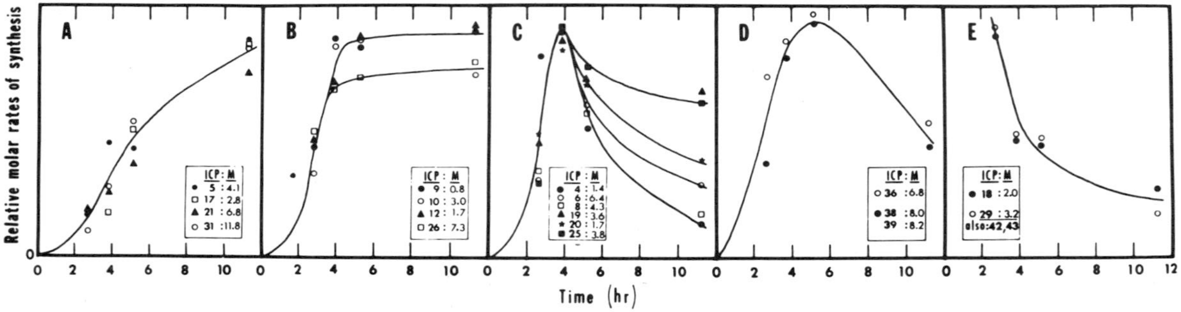
Relative molar rates of synthesis
A
ICP : M
5 : 4.1
17 : 2.8
21 : 6.8
31 : 11.8
B
ICP : M
9 : 0.8
10 : 3.0
12 : 1.7
26 : 7.3
C
ICP : M
4 : 1.4
6 : 6.4
8 : 4.3
19 : 3.6
20 : 1.7
25 : 3.8
D
ICP : M
36 : 6.8
38 : 8.0
39 : 8.2
E
ICP : M
18 : 2.0
29 : 3.2
also : 42,43
Time (hr)

Fig. 13.11 Different patterns of synthesis of infected cell polypeptides throughout the HSV-1 (F1) growth cycle. Curves A-E represent different patterns of synthesis of infected cell polypeptides throughout the HSV-1 (F1) growth cycle. Polypeptides labeled during short intervals at different times after a high-multiplicity infection were separated on polyacrylamide gel slabs and the amount of label in each polypeptide quantitated by computer-aided planimetry of autoradiogram absorbance tracings. The amount of each polypeptide made was then expressed on a molar scale and polypeptides conforming to a similar pattern of kinetic behavior were cluastered into the group shown. Infected cell polypeptides representative of each group and their maximal relative molar rates of synthesis are identified by symbols within the boxed legends. Polypeptides of class A, with gradually increasing rates of synthesis until late in infection, are mainly major structural polypeptides, whereas those of classes C and D, with rates of synthesis at first increasing and thereafter of class E, with gradually diminishing rates of synthesis throughout infection, appear to represent the declining synthesis of prominent host polypeptides. Data from Honess and Roizman [105].

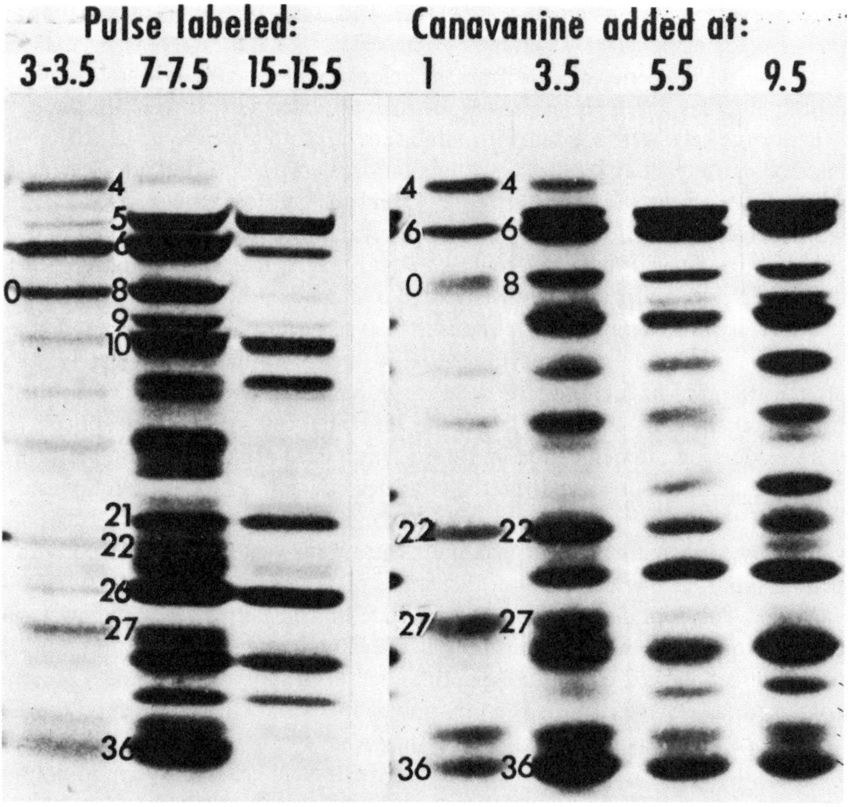
Pulse labeled:
3-3.5
7-7.5
15-15.5
Canavanine added at:
1
3.5
5.5
9.5
4
5
6
0
8
9
10
21
22
26
27
36
4
6
0
8
22
27
36
4
6
8
22
27
36

Fig. 13.12 HSV-1 polypeptide synthesis in infected Hep-2 cells. Left panel: autoradiogram of polyacrylamide gel slab containing electrophoretically separated polypeptides labeled in Hep-2 cells infected with HSV-1 (Justin), at intervals shown at the top. The ICP were designated according to the nomenclature reported previously [14]. All cultures were exposed to the same concentration of ^{14}C-labeled amino acids, and equivalent amounts of cell lysates were subjected to electrophoresis in each gel slab channel. Note shift is in polypeptide synthesis from α (Nos. 4 and 0) to β (Nos. 6, 8, etc.) to γ polypeptides made late in infection. Data from Frenkel et al. [54]. Right panel: autoradiogram of HSV-1 (F) infected cell polypeptides electrophoretically separated on polyacrylamide gels. The canavanine (2.8 mm) was added at time indicated above panel. The cells were labeled with [^{14}C]amino acids from 9.5-10 hr postinfection. Data from Honess and Roizman [154].

after infection. Analyses of the rates of polypeptide synthesis in the parasynchronously infected cells also indicated that although the temporal patterns of synthesis of the polypeptides in each class were identical, the absolute rates of synthesis varied [105, 136]. These observations suggested that a polypeptide abundance control was superimposed on the temporal control of synthesis.

Another interesting finding was that structural polypeptides could not be differentiated from nonstructural polypeptides with respect to their maximal or their average molar rate of synthesis [105, 136].

The second series of experiments demonstrated that the synthesis of the coordinately regulated groups was sequentially ordered [136, 151]. In these experiments cells were treated with inhibitors of protein synthesis, cycloheximide or puromycin at the time of infection or thereafter. After 1 to several hours of treatment the drug was removed and the cells were pulse labeled with [^{14}C] amino acids. Analyses of the polypeptides made after withdrawal of the drugs showed that they could be readily segregated into 3 groups designated α, β, and γ. Group α consisted of a minor structural polypeptide and several nonstructural ones. The synthesis of these polypeptides required no prior infected cell protein synthesis: i.e., they were made immediately after withdrawal of inhibitors of protein synthesis that had been added to the medium at the time of cell infection. In untreated cells, these polypeptides were made soon after infection and were synthesized at maximal rate between 2 and 4 hr postinfection and thereafter at decreasing rates. The α group also consisted of minor structural and nonstructural polypeptides. In untreated cells, they were synthesized at maximum rates somewhat later than α polypeptides, but, as the latter group, their rates of synthesis subsequently declined. Polypeptides of group β were made immediately after withdrawal of inhibitors of protein synthesis only if the inhibitors were added after 1-3 hr postinfection, i.e., after α polypeptides had been synthesized. The γ group contained largely the major structural polypeptides and corresponded to the A class; i.e., in untreated infected cells they were synthesized at ever increasing rates until at least 12 hr postinfection. In the treated cells, the γ group polypeptides were synthesized immediately after removal of inhibitors of protein synthesis only if the addition of the drug to the medium was delayed sufficiently long to permit both α and β polypeptides to be made [76, 136, 151].

b. Mechanisms and Requirements for Sequential Synthesis of Viral Gene Products. Most of the experimental studies (Fig. 13.12) have dealt with the transition from α to β to γ polypeptide synthesis. The data may be summarized as follows:

The initiation of synthesis of β polypeptides requires the presence in infected cells of functional α polypeptides and new RNA synthesis. Thus, only α polypeptides are made in cells treated with the amino acid analogs canavanine or azetidine from the time of infection; the transition to β and γ polypeptides does not ensue [76, 154]. Evidence that new RNA synthesis is required for the transition from α to β polypeptide synthesis emerged from experiments involving treatment of cells with cycloheximide from 0-7 hr postinfection. Cells pulse labeled immediately after removal of cycloheximide made only α polypeptides. With time α polypeptide synthesis declined and was replaced by β and subsequently by γ polypeptide synthesis. Addition of actinomycin D just prior to removal of cycloheximide reduced the rate of decay of α polypeptide synthesis and precluded the synthesis of β polypeptides [76, 136, 151].

The same series of experiments revealed that the shut off of α polypeptide synthesis required the synthesis of functional β polypeptides and new RNA synthesis. Specifically, addition of canavanine after onset of α polypeptide synthesis allowed the transition of α to β polypeptide synthesis. However, the bulk of polypeptides were made in the presence of canavanine and were therefore nonfunctional. Therefore, α polypeptide synthesis was not inhibited in such cells and in fact all of the polypeptides specified by the virus were made concurrently, a phenomenon not seen in untreated cells. Similarly, as noted earlier in the text, the cycloheximide-actinomycin D experiment cited above indicated that the shut off of α protein synthesis is related to appearance of β proteins which in turn required new RNA synthesis.

c. The Role of Transcriptional and Posttranscriptional Processes in the Regulation of Sequential Synthesis of Viral Proteins. Several lines of evidence indicate that posttranscriptional processing of viral RNA plays a major role in the regulation of synthesis of viral polypeptides. The evidence may be summarized as follows:

There is currently no compelling evidence to support the hypothesis that a temporal control of trancription exists, in spite of the published evidence. The problem arises from the fact that the quantities of viral RNA accumulating early in infection are too low to permit precise measurements of the amounts of viral DNA that are transcribed [136, 133]. If we equate "early in infection" with viral RNA synthesized in the absence of either viral DNA or protein synthesis, then the extent of DNA transcribed in infected cells treated with cycloheximide cannot be differentiated by current techniques from the amount of DNA transcribed in untreated infected cells [76, 136, 138].

The apparent relationship between the posttranscriptional processing and the ordering of viral polypeptide synthesis emerged from analyses of viral RNA sequences in the nucleus and polyribosomes of infected cells treated with inhibitors of protein synthesis and amino acid analogs. The major observations are that although the nuclei of the cells treated from the time of infection to the time of extraction contained sequences complementary to at least 40% of the viral DNA, the viral RNA sequences accumulating in the nucleus and cytoplasm were complementary to approximately 12% of the DNA [76, 136, 138]. Furthermore, even though the nuclei contained the sequences present in polyribosomes of untreated cells late in infection, new RNA synthesis in the presence of α polypeptides was required for the synthesis of subsequent groups. One conclusion that could be drawn from these data is that only functional mRNA sequences are transported to the cytoplasm and that transcripts which cannot be processed to function as mRNA are either retained in the nucleus or are rapidly degraded in the cytoplasm. Evidently, a posttranscriptional function discriminates between those transcripts which can be processed and those which cannot [76, 136, 138]. The obvious and very interesting question that is currently unanswered is: what is the difference between the transcript carrying β sequences made in the absence of α polypeptides which is not processed, and the transcript carrying β sequences made in the presence of α sequences and which is processed?

d. The Role of Viral DNA Synthesis in Effecting the Transition from Early to Late Protein Synthesis. With exceptions, numerous investigators have noted that inhibitors of viral DNA synthesis do not prevent the synthesis of structural polypeptides or assembly of empty capsids in EAV and HSV infected cells [155-158]. Analyses of proteins made during exposure of cells to inhibitors or in cells infected with temperature-sensitive mutants defective in DNA function and incubated at nonpermissive temperatures show that whereas α and β protein synthesis are normal, the α polypeptides are grossly reduced ([176, 136, 151, 158, 159]; Buchan, Honess, and Roizman, unpublished data). The observations that in cells treated with actinomycin D functional half-life of α and β mRNA is much longer than that of γ mRNA is consistent with the hypothesis that parental viral DNA is capable of supplying mRNA for all those groups of polypeptides but that for optimal rates of γ polypeptide synthesis the mRNA for γ polypeptides in untreated infected cells must be amplified by transcription of progeny DNA.

The notable exceptions are a study [160] in HSV-infected cells and several observations [161-163] in EBV-infected lymphoblastoid cell lines suggesting that viral DNA synthesis is required for the synthesis of "late", i.e., structural polypeptides. If the hypothesis just stated is correct, a possible explanation for these observations is that the amounts of structural proteins made in the absence of DNA was not sufficient to be detected in the systems employed in these studies.

A schematic diagram containing a model of the current concepts of regulation of HSV-1 gene products is shown in Fig. 13.13.

2. *Posttranslational Modification and Translocation of Viral Proteins*

At least 3 kinds of posttranslational modifications may occur: (1) rapid posttranslational cleavages such as those seen in picornavirus-infected cells [164]; (2) slow posttranslational cleavages occurring during intracellular translocation or during assembly of the virion; and (3) conjugation or addition of prosthetic groups, i.e., glycosylation, phosphorylation, amidation, acetylation, methylation, sulfation, etc.

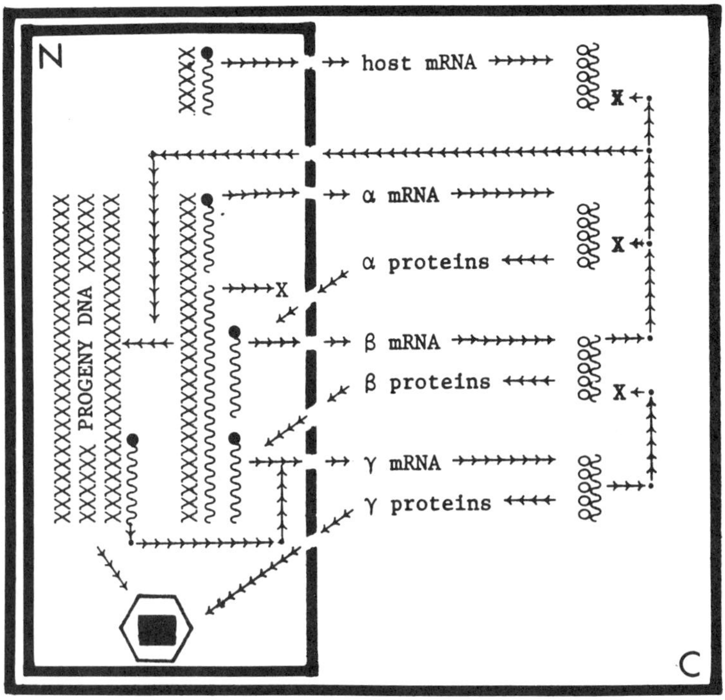
N
host mRNA
α mRNA
α proteins
PROGENY DNA
β mRNA
β proteins
γ mRNA
γ proteins
C

Fig. 13.13 The major features of regulation of synthesis of HSV-1 gene products in Hep-2 cells. Schematic diagram of the major features of regulation of synthesis of HSV-1 gene products in Hep-2 cells. N-nucleus, C-cytoplasm, Xed lines in nucleus represent parental and progeny DNA. Only a fraction of the transcripts made in the absence of prior infected cell protein synthesis is transported into cytoplasm. This includes α mRNA, which specifies α polypeptides. These enable new transcription generating β polypeptides that mediate: shut off of α and host polypeptide synthesis; the synthesis of viral DNA; and transcription of both parental and progeny DNA generating γ mRNA and γ polypeptides, the major structural polypeptides of the virus.

Earlier in the text it was noted that the bulk of the virus-specific polypeptides do not appear to undergo rapid posttranscriptional cleavages such as those observed in picornavirus-infected cells. Slow cleavages have been noted, and these will be dealt with in the section on assembly. Both phosphorylation and glycosylation of viral polypeptides have been observed.

Current studies indicate that all α, some β and some γ polypeptides are phosphorylated (Pereira et al. [318]). The phosphorylation of α polypeptides coincides with a decrease in their electrophoretic mobility and translocation in the nucleus. The electrophoretic mobility of phosphorylated β and γ polypeptides is not appreciably retarded relative to their nonphosphorylated counterparts. The possibility that phosphorylation renders α and β polypeptides "functional" and initiates self-assembly of γ polypeptides has not been adequately explored.

Glycosylated polypeptides have been found in all purified herpesvirions analyzed to date. Information concerning the glycosylation of viral polypeptides is fragmentary and restricted mostly to herpes simplex virus-infected cells, as follows.

1. The glycosylation of viral proteins occurs in the absence of any detectable glycosylation of host proteins. While we have no information on the stability of polysaccharide chains covalently bound to host proteins, there is no measurable exchange of sugars from host polysaccharides during infection [116, 165].
2. Viral glycoproteins present in the infected cell partition with membranes [166] and in fact, physical separation of infected and uninfected cell membranes based on the presence of antigenically distinct viral glycoprotein in the infected cell membranes has been reported [120, 167]. Pulse chase experiments show that the protein moiety binds to membranes after synthesis and is glycosylated in situ [166].

Analyses of the synthesis of HSV-1 glycoproteins show that the glycosylation of viral polypeptides occurs late in the virus growth cycle and that most of the precursors to major viral glycoproteins are polypeptides. Viral glycoproteins are formed by stepwise additions of heterosaccharide chains to completed precursor polypeptides. The precursor and the highly glycosylated product are separable by gel electrophoresis [168, 169]. Pulse chase experiments show that within 15 min of their synthesis, precursor polypeptides acquire heterosaccharide chains of about 2,000 molecular weight, which contain glucosamine but little or no fucose or sialic acid. Both precursor and product of this first stage of glycosylation accumulate in the cytoplasmic membranes but are either absent or in low concentrations in the virion or in the plasma membrane. The partially glycosylated product is then conjugated further in a slow, discontinuous process to form the mature glycoprotein of the virion and plasma membrane. These mature products bear large heterosaccharide units with molecular weights greater than 4000-5000; these contain fucose and sialic acid as well as glucosamine. It is noteworthy that the heterosaccharide chains from infected and uninfected Hep-2 cells are distributed among discrete size classes and that the smallest chains consist of multiple saccharide residues. The polysaccharide chains in the virion and plasma membranes of infected cells correspond in size to those present in the plasma membrane of uninfected cells and, on the average, are larger than those present in the cytoplasm.

It has now been well documented that in infected cells, at least some viral proteins are transported into the nucleus [148-150]. Radiolabeling experiments suggest that the equilibration between the cytoplasmic and nuclear pools of translocated polypeptides is rather slow [149]. If the movement of polypeptides from cytoplasm to nucleus is indeed slow, the observation by Roizman et al. [170] and by Ross et al. [171] of compartmentalization of herpesvirus antigens is significant; it implies that if the same proteins are present in both nucleus and cytoplasm, they must display different immunologic reactivities in the 2 compartments. Fujiwara and Kaplan [147] also reported that in pulse labeled cells infected with pseudorabies virus, radioactive peptides first appeared in the cytoplasm; they were then chased into the nucleus. Nuclear labeled peptides reacted with antiviral antibody whereas cytoplasmic labeled peptides did not. Mark and Kaplan [172, 173] reported that translocation of viral proteins from cytoplasm to nucleus requires arginine and is reduced or does not occur in deprived cells.

a. The Function of Herpesvirus-Specific Proteins. The polypeptides specified by herpesviruses may be classified in several ways. For example, structural and nonstructural categories classify polypeptides into those which are incorporated into the virion and those that are not. Another classification which ultimately might shed slightly more useful information is based on the ability of viral polypeptides to bind DNA. Indeed 16 HSV-1 infected cell polypeptides bind to calf thymus DNA [174] and presumably the remaining 34 odd polypeptides do not. In addition, many polypeptides have been identified as having a specific function other than components of capsid or envelope. A third classification is enumeration of function which is either known to be virus-specific or which by anology with other viruses, could be specified by viral polypeptides. We shall be concerned here mainly with these functions and the evidence that they are virus-specific.

b. DNA-Dependent RNA Polymerase (Transcriptase). The evidence supporting the hypothesis that viral DNA is at least initially transcribed by a host enzyme was marshalled earlier in the text (III.C.4). The data do not exclude the possibility that viral DNA transcription late in infection is by a host polymerase modified by viral factors or by a new, viral transcriptase. Keir [203], incidentally, reported as much as a 2-fold stimulation in the activity of this enzyme following infection.

c. Thymidine (TdR) and Deoxycytidine (CdR) Kinases. TdR and possibly CdR kinases are "scavenger" enzymes catalyzing the conversion of TdR and CdR to TdR-MP and CdR-MP, respectively, in cells grown in vitro. TdR kinases have been reported in cells infected with a number of herpesviruses [176-180]. In BHK-21 cells, activity increase as much as 20-fold 2-8 hr after infection with HSV [181]; thereafter activity falls off. An increase in activity was also reported in rabbit kidney cells infected with pseudorabies virus [182]. Stimulation of CdR kinase activity in BHK cells infected with HSV was also reported [183], but the function of the enzyme is not known since the product of the enzyme is not incorporated into DNA. In recent years a considerable amount of information has accumulated on the source of genetic information and properties of these enzymes. These may be summarized as follows:

In a rather elegant series of studies Kit and Dubbs [176, 177, 184] showed that TdR kinase activities are not essential for the growth of cells or for HSV replication in cells in culture. First, they obtained a BUdr-resistant strain of mouse fibroblasts lacking TdR kinase activity by growing the cells in media containing the analog. TdR kinase activity was induced in these cells by HSV and vaccinia, but prevented by puromycin and actinomycin D. Subsequently, they obtained TdR kinaseless mutants of HSV by growing wild strains in BUdR-resistant cells in the presence of analogs.

Several lines of evidence from many laboratories clearly established that the TdR kinase specified by different herpesviruses are readily differentiated among themselves and from host enzymes on the basis of their immunologic specificity. Thus, Klemperer et al. [181] showed that the thymidine kinase produced in cells infected with HSV was inhibited by antiserum prepared against rabbit kidney strain 13 (RK-13) cells infected with the same virus; the enzyme activity in infected cells was unaffected. Subsequently, Buchan and Watson [185] reported that thymidine kinase induced by pseudorabies virus in RK-13 cells was neutralized by homologous rabbit anti-infected-cell serum but not by serum produced against HSV-infected cells. Moreover, the enzyme induced by HSV in these cells was neutralized by homologous but not by heterologous antibody. In more recent studies, Thouless and associates [186, 187] reported that the thymidine kinases of HSV-1 and HSV-2 differ in their immunologic specificities. However, the study revealed several paradoxes which need clarification. First, common antigenic determinants were demonstrable in enzyme neutralization tests but not in the precipitin test. Second, some antisera to HSV-1 enzyme stabilized the TdR kinase specified by HSV-2 with only partial or no concommitant reduction in activity. The stabilized enzyme was still capable of being neutralized by homologous antiserum albeit at a reduced rate. One hypothesis which could conceivably explain these observations is that the enzyme is an oligomer which carries 2 sets of antigenic sites, a type specific antigenic site near the active center of the enzyme and a common antigenic site formed by the junction of monomers at some distance from the active center. The stabilization of HSV-2 TdR kinase by anti-HSV-1 serum could be visualized as cross-linking of the monomers in the dimeric form.

Hay et al. [183] postulated that both TdR and CdR kinase activities may be functions of the same polypeptides, but at different sites. This conclusion is based on the observation that HSV mutants selected for resistance to cytosine arabinoside or to BUdR are deficient in both CdR and TdR kinase activities, as is the BUdR-resistant mutant of Dubbs and Kit [184]. Genetic and biochemical studies by Jamieson and associates [188, 189] supported this conclusion and these authors concluded that the CdR and TdR kinases are "separate functions" albeit residing in the same molecule.

Compared with the enzyme from uninfected cells, the TdR kinase from infected BHK-21 cells was reported to have a low pH optimum, a low K_m, to be relatively stable at 40°C, and to be insensitive to inhibition by deoxythymidine triphosphate [181]. On the other hand, TdR kinase activity of African green monkey kidney cells (BSC-1) could not be differentiated from that of the same cells infected with HSV with respect to thermal stability and optimum temperature [190]. Ogino and Rapp [191] and Thouless and Skinner [192] reported that HSV-2 TdR kinase is more thermolabile than the HSV-1 enzyme and gives nonlinear dilution response. Munyon et al. [193] reported that the average mobility in polyacrylamide gels of HSV-1 TdR kinase was slower than that of the host enzyme by almost a factor of 2. The infected cells, and by inference, viral TdR kinase phosphorylates BCdR whereas the uninfected cell does not [194, 195].

Of special interest are a series of studies on the properties of the TdR kinaseless mutants of HSV. Not necessarily in chronological order, the data may be summarized as follows. Dubbs and Kit [196] reported on the properties of 3 kinaseless mutants, i.e., B2006, B2010 and B2015. B2006 produced extremely low levels of TdR kinase independent of the temperature of incubation. B2010 and B2015 produced no detectable TdR kinase when grown at 37°C irrespective of the temperature at which the enzyme was tested. At 31°C these mutants produced one-tenth the level of TdR kinase detected in extracts of cells tested at 31°C. The activity was independent of the temperature at which it was tested, and from autoradiographic assays it appears to have been distributed throughout most cells. Buchan et al. [197] found that extracts of cells infected with mutant B2006 did not block the serum neutralization of wild-type TdR kinase, and they concluded that mutant B2006 did not produce nonfunctional cross-reacting antigen of TdR kinase. The picture became more complex when it was found [198] that in cells doubly infected with wild-type and the TdR kinaseless mutant the enzyme activity was reduced 3- to 4-fold. Especially interesting was the observation that the TdR kinaseless mutant did not affect the production of thymidine kinase by vaccinia virus. On the other hand, thymidine kinaseless vaccinia virus is a potent inhibitor of herpesvirus TdR kinase in cells doubly infected with TdR kinaseless vaccinia and a wild strain of HSV. Munyon and Kit [198] excluded the possibility that vaccinia produces an inhibitor of TdR kinase activity or a repressor specific for the TdR kinase gene, but they left open the possibilities that TdR kinase might be an oligomeric protein which in doubly infected cells consists of functional and nonfunctional subunits, or that the parent and mutant viruses might compete for some cellular structure necessary for the expression of TdR kinase. Buchan et al. [199] confirmed the observation of Munyon and Kit [198] but failed to find a cross-reacting antigen or a viral antigen present in cells infected with B2006 mutant but absent in cells infected with the wild type. They concluded that it was unlikely that the inhibition of thymidine kinase production in cells infected simultaneously with the mutant and wild-type viruses can be explained by the interaction of functional and nonfunctional subunits. The data are puzzling; the hypothesis that the parental and mutant viruses compete for a common cellular site such as "ribosomes" or "polymerase" as suggested by Munyon and Kit [198] and by Buchan et al. [199] is not very appealing; it simply calls the black box by another name. Superficially, at least, the more appropriate line of investigation is to determine whether in cells infected with the TdR kinaseless mutants the translation of the TdR kinase mRNA is prematurely terminated, producing a product of limited immunogenicity but still capable of competing with the intact polypeptide in the assembly of the putative oligomeric enzyme. This hypothesis would explain the reduction in viral TdR kinase activity in cells infected with both parent and mutant viruses but not the inhibition of host enzyme activity in cells infected with the mutant virus alone [196] unless the host and viral enzymes are capable of interacting.

The availability of cells lacking TdR kinase activity has made possible extensive studies of viral thymidine kinase; in fact this enzyme is probably the most extensively studied viral function. The extensive treatment accorded this enzyme stems however from the fact that TdR kinaseless cells can acquire and perpetuate viral thymidine kinase in the course of abortive infection with HSV [175, 193, 200-202].

d. DNA Polymerase (DNA Nucleotidyl Transferase). The initial evidence that herpesvirus-infected cells contain a new DNA polymerase was summarized by Keir et al. [203]. These workers and subsequently Weissbach et al. [204] partially purified the HSV-1 enzyme from hamster and human infected cells. Boezi et al. [205] and Huang [206] purified the MDV and CMV enzymes, respectively.

The HSV-1 enzyme was reported to have a minimal molecular weight of 180,000, and required sulfhydryl groups for activity. The enzyme was maximally active at salt concentrations of 0.1-0.2 M K_2SO_4 or NH_4^+ which were inhibitory to host polymerases. The HSV-1 polymerase eluted from phosphocellulose columns at a lower salt concentration than the host polymerases and copied poly(dC)-oligo$(dG)_{12\text{-}18}$ faster than activated DNA. The CMV polymerase also eluted at a higher salt concentration than the host or HSV-1 polymerase, its optimum salt concentration was lower than that of HSV-1, and copied poly(dA)-oligo$(dT)_{12\text{-}18}$ as well as poly(dC)oligo$(dG)_{12\text{-}18}$. The MDV polymerase eluted from phosphocellulose columns at a lower salt concentration than host DNA polymerase, and could not use either deoxyribohomopolymer effectively.

Hay et al. [183] reported that the DNA polymerase activities in HSV-1 and HSV-2 infected cells differed in heat stability and NH_4^+ optimum. The PSV enzyme differs immunologically and in salt requirements from host as well as HSV-1 and HSV-2 polymerases [207].

It is noteworthy that all herpesvirus polymerases assayed to date appear to be inhibited by phosphonoacetic acid [208-211]. In an elegant study, Leinbach and coworkers [212] concluded that the drug binds to the polymerase at the pyrophosphate binding site and is a competitive inhibitor of pyrophosphate in the exchange reaction.

e. Deoxyribonuclease. An increase in DNAse activity (measured at pH 7.3) in BHK-21-C14 cells infected with herpes simplex virus was reported by Keir and Gold [213]. The increase in enzyme activity leveled off 7-9 hr after infection [213, 214]. Subsequently, McAuslan et al. [215] and Sauer et al. [216] reported an increase in activity of an "alkaline" DNAse in infected monkey kidney cells and L cells, respectively. An increase in "acid" DNAse has been observed in HSV-infected HeLa and L cells [217] but not in monkey kidney cells [215] or in KB cells [218].

The DNAse studied by Keir and Gold [213] and subsequently by Morrison and Keir [219] appeared to differ from the host DNase with respect to several properties. The induced DNase was readily inactivated at 45°C, adsorbed to DEAE-cellulose, preferred Mg^{2+} to Mn^{2+}, and required 50-60 mM Na^+ or K^+, whereas the uninfected cell enzyme was stable at 45°C, did not adsorb to DEAE-cellulose, did not differentiate between Mg^{2+} and Mn^{2+}, and was inhibited by Na^+ or K^+ at concentrations greater than 15 mM. Purified induced DNase emerged in the void volume during gel filtration through Sephadex G-200 whereas the uninfected-cell enzyme was retarded. The enzymes extracted from BHK-21 and Hep-2 cells infected with HSV or with pseudorabies were exonucleases capable of degrading both native and denatured DNA to deoxynucleoside-5′-monophosphates whereas the enzyme extracted from uninfected cells was an endonuclease effective primarily against denatured DNA. Rabbit antisera prepared against extracts of allotypic rabbit kidney cells infected with HSV neutralized the DNA endonuclease extracted from the uninfected cells [203].

f. Ribonucleotide Reductase. One of the functions of this enzyme is to catalyze the conversion of cytidylate to deoxycytidylate. In mammalian cells, dTTP acts as an allosteric inhibitor of this enzyme and the synchronization of mammalian cells by the addition and subsequent removal of excess thymidine is thought by many to be the result of depletion of dCTP from the cellular pool as a consequence of inhibition of this enzyme. Cohen [220] observed that excess thymidine blocked the synthesis of host DNA in uninfected KB cells as expected, but did not affect the synthesis of either viral or cellular DNA in infected cells. Based on the reasoning that blocked infected cells have a source of dCTP which is unavailable in blocked uninfected cells, Cohen tested the ribonucleotide reductase activity of infected and uninfected cells. The results showed that the activity of this enzyme was increased 2-fold at 3 hr postinfection and that the enzyme from infected cells retained 60% activity even in the presence of 2 mM dTTP, i.e., considerably more than required to inhibit the enzyme extracted from uninfected cells. There is as yet no genetic or immunochemical evidence that the enzyme is specified by the virus.

g. TdR-MP, CdR-MP, AdR-MP, and GdR-MP Kinases. Hamada et al. [182] reported that TdR-MP kinase activity increased in rabbit kidney cells infected with PSV whereas the activity of AdR-MP, GdR-MP, and CdR-MP kinase remained unaltered. The same laboratory previously [221] reported that TdR-MP kinase from PSV-infected rabbit kidney cells was more stable at 37°C than the corresponding enzyme from uninfected cells. Prusoff et al. [190] observed a similar increase in TdR-MP kinase in African green monkey kidney cells infected with herpes simplex, but they were unable to differentiate between the properties of the enzyme in extracts of infected and uninfected cells. AdR-MP kinase from infected and uninfected cells could not be differentiated in neutralization tests by antisera prepared against infected and uninfected cell extracts [182].

h. TdR-MP Synthetase. Frearson et al. [222] reported that TdR-MP synthetase activity did not increase in mouse fibroblasts, HeLa, and rabbit kidney cells infected with HSV.

i. CdR-MP Deaminase. An increase in CdR-MP deaminase in BHK-21 cells infected with HSV was reported by Keir [203]. The enzyme catalyzes one of the reactions concerned with de novo synthesis of TdR-MP. No additional information is available.

j. Protein Kinase. Randall et al. [223] reported on the presence of an "endogenous protein kinase" in EAV purified with respect to phosphorycholase activity. The kinase activity required Mg^{2+} and was enhanced by the addition of exogenous acceptors (protamine or arginine-rich histone) but not by cyclic

AMP, an indication that a regulatory subunit enhancing the activity found in uninfected cells was absent or inoperative. In an in vitro test, all 17 virion proteins detected in their polyacrylamide gels served as acceptors of ^{32}P, but many more proteins seemed to be phosphorylated in vitro than in vivo. The source of genetic information for this enzyme is not known. Rubenstein et al. [121] reported the presence of protein kinase in enveloped HSV particles and on the basis of the finding that capsids de-enveloped with a detergent did not possess protein kinase activity concluded that the enzyme is in the envelope of the virus. Although this is likely, since the kinase is normally associated with cellular membranes, the experiment itself does not offer compelling evidence on this point since the NP-40 extract containing the solubilized envelope proteins was not tested for the presence of protein kinase. This test is important as an indication that the detergents did not inactivate the enzyme. The authors also suggest that protein kinase is essential for infectivity and offer in support of this postulate the observation that heat treatment which reduced protein kinase activity also reduced infectivity. Quite obviously heat treatment denatures more than just protein kinase.

k. Proteases. In the course of concentration of highly purified HSV-1 polypeptides following their solubilization in SDS it was observed (Spear, Gibson and Roizman, unpublished data) that several proteins appeared to be degraded. The degradation was less severe or did not occur in the absence of SDS. The data suggested that the herpesvirions may contain a protease which is either activated or becomes accessible to the substrate in the presence of SDS. The source of this enzyme is not clear; it resembles superficially the enzyme reported to be present in other viruses [224] but could be a host contaminant.

E. Herpesvirus DNA Synthesis

1. Introduction

Most of the information on herpesvirus DNA synthesis comes from studies on PSV and HSV infected cells (reviewed in [225]) and on HSV-1 infected Hep-2 cells (reviewed in [16]). Both HSV and PSV contain DNA with high G + C content and these are readily separable from host DNA with a significantly lower G + C content. With the exception of studies on EAV DNA [226, 227], little has been reported on the synthesis of viral DNAs more closely approximating host DNA with respect to base composition. The PSV-infected rabbit kidney cells and HSV-infected human embryonic lung cells are particularly advantageous because only viral DNA is synthesized in rabbit infected confluent, contact inhibited cells; host DNA synthesis along with cell division remain inhibited throughout the reproductive cycle [228, 319].

Numerous studies involving cytochemical and biochemical determinations have shown that viral DNA is made in the nucleus [225]. The general pattern of viral DNA synthesis has been described in cells infected with PSV [225], EAV [227], and HSV [16]. In all these studies viral DNA synthesis replaces host DNA synthesis and ceases entirely as viral assembly procedes.

2. Requirements

Most of the early work [229, 230] involved the use of protein inhibitors to show that viral DNA synthesis required de novo protein synthesis. Once initiated, viral DNA synthesis continues in the absence of concomitant protein synthesis, but at a reduced rate. The reduction in rate is not readily interpretable for lack of information concerning thymidine pool size before and after addition of inhibitors of protein synthesis, integrity of replicating DNA in inhibitor-treated cells, and secondary effects of inhibitors on the intranuclear environment of the cell. There is, incidentally, no information concerning the stability in situ of the enzymes involved in the synthesis of viral DNA. The requirement for protein synthesis has become somewhat clarified by recent genetic studies by M. Timbury and P. Schaeffer (personal communication) which show that at least 9 functions must be expressed before viral DNA synthesis will commence.

A second possible requirement emerged in studies by Lawrence [226] showing that the onset of synthesis of EAV DNA in human (KB) cells varied as much as 5 hr, depending on the stage of the cell cycle. Thus synchronized KB cells infected in G1 or G2 replicated viral DNA at a time corresponding to the cellular S phase determined independently in mock-infected cells. Lawrence concluded that the initiation of EAV DNA synthesis was dependent upon some cellular functions which were related to the S phase of the KB cell. It is noteworthy that the S-phase cell did not fulfill all the requirements since viral DNA synthesis did not immediately follow infection of cells in S phase. The dependence on host cell functions does not appear to be universal for all herpesviruses or in all kinds of cells. Thus, Cohen et al. [231] demonstrated that HSV-1 DNA synthesis proceeds independently of the mitotic cycle in the same KB cell line, whereas O'Callaghan et al. [232] showed that EAV "adapted" to hamsters does not require host DNA synthesis for its own multiplication. It is conceivable that host DNA synthesis is required only in "heterologous" hosts for reasons that at present remain obscure.

d. Characteristics of the Replication of Viral DNA. Kaplan et al. [229] summarized studies reported between 1963 and 1967 showing that pseudorabies virus DNA replicates in a semi-conservative fashion and that less than one-half of the DNA not incorporated into virions and presumed available to function as template for replication is actually replicating. Current studies [319] indicate that only 2-5% of the input labeled DNA is actually replicated. The products made during a short thymidine pulse sediment as small fragments in alkaline sucrose density gradients and become larger as the labeling period increases [71, 72]. The data suggest that the small fragments are extended, repaired, and/or ligated and that only intact DNA defined as having only a minimum of gaps (see Section II.B.2) is incorporated into virions. As was noted earlier in the text (II.B.2), ribonucleotides have been reported in the DNA and it is conceivable that they represent residual primer RNA [69, 70].

The replicative form has not been described. Kaplan [233] reported the finding of PSV DNA molecules many times the unit length. Such molecules have not been found in HSV-infected cells although rapidly sedimenting pulse labeled HSV DNA can be "chased" on prolonged labeling to cosediment with marker HSV DNA (Jacob and Roizman, manuscript in preparation). In many ways, oligomeric forms of HSV DNA are an attractive explanation because "in phase" and "out of phase" cutting of the DNA at the junction of *ba* and *ac* regions would give rise to 2 of the 4 forms of the DNA [61, 76]. The remaining forms could originate from a recombinant which in turn was replicated by a mechanism giving rise to oligomeric forms. In HSV-infected cells, however, circular forms and linear forms with forks and loops, and lariats were the only forms present (Jacob and Roizman, manuscript in preparation) and the possible models are too numerous to be useful at this stage (see Recent Developments).

F. Lipids

Studies on the effect of viral infection on lipid metabolism are scarce and the results reported to date are inconsistent. Thus, selective increase in sphingolipid synthesis [113, 234], generalized decrease in phospholipid synthesis (Kieff and Roizman, unpublished studies), or no change at all [112] have been observed in herpesvirus-infected cells. No adequate studies of neutral lipid or glycolipids are available. Preliminary studies (Kieff and Roizman, unpublished data) suggested significant changes in the extent of glycosylation of glycolipids following infection.

G. Morphogenesis and Egress of the Herpesvirion from Infected Cells

1. Morphogenesis, Capsid, and Core

There is general agreement that the core and capsid assemble in the nucleus. Most of the information on the assembly of herpesviruses emerged from electron microscope studies and in fact a superficial count yielded over 200 publications dealing with 1 or more aspects of herpesvirus morphogenesis. These papers detail the major structures uniquely present in the nucleus of the infected cell and were summarized and reviewed elsewhere [16]. They draw attention to the fact that the nucleus of the infected cells contains the following.

1. "Arcs" of partially assembled capsids
2. Capsids varying in amount and shape of core, in the sharpness of the outline of the capsomeric layer
3. Filaments (DNA?) and other materials associated with the arcs and capsids
4. A profusion of granules and tubular structures dispersed throughout the cell

Interpretation of the data suffers from the fact that electron microscopic studies rely largely on a statistical approach to determine the sequential order of the intermediates seen in thin sections. This procedure is wrought with many dangers and could be misleading. With these reservations, the data [16] suggest that the herpesvirion capsid is first seen as an arc which is extended until it is complete. The data can be interpreted to suggest that a primitive core is assembled at the same time as the capsid or that the DNA enters a preformed capsid and that restructuring of the capsid and core takes place once the DNA is in place. It should be emphasized that this order could be erroneous and that the data could also be used to argue that all capsid forms other than that containing DNA are defective particles or artifacts of improper handling of infected cell material.

2. Envelopment

It is generally accepted that the envelope is acquired by the capsid budding through [40, 235, 236] a cellular membrane which has become suitably altered during the course of infection. Budding has been reported most consistently at the inner nuclear membrane. Other sites of budding are the Golgi apparatus, the endoplasmic reticulum, and the membranes lining the vacuoles in the cytoplasm. A reference guide to the literature on envelopment is given by Roizman and Furlong [16].

At the nuclear envelope the capsid buds through patches or macula wherein the inner lamella has been altered (Fig. 13.14). In the case of HSV, these patches can be seen as regions of increased density and altered curvature [237]. Dense, amorphous material is often present on the nuclear side of the membrane whereas spikes, morphologically resembling those on the surface of

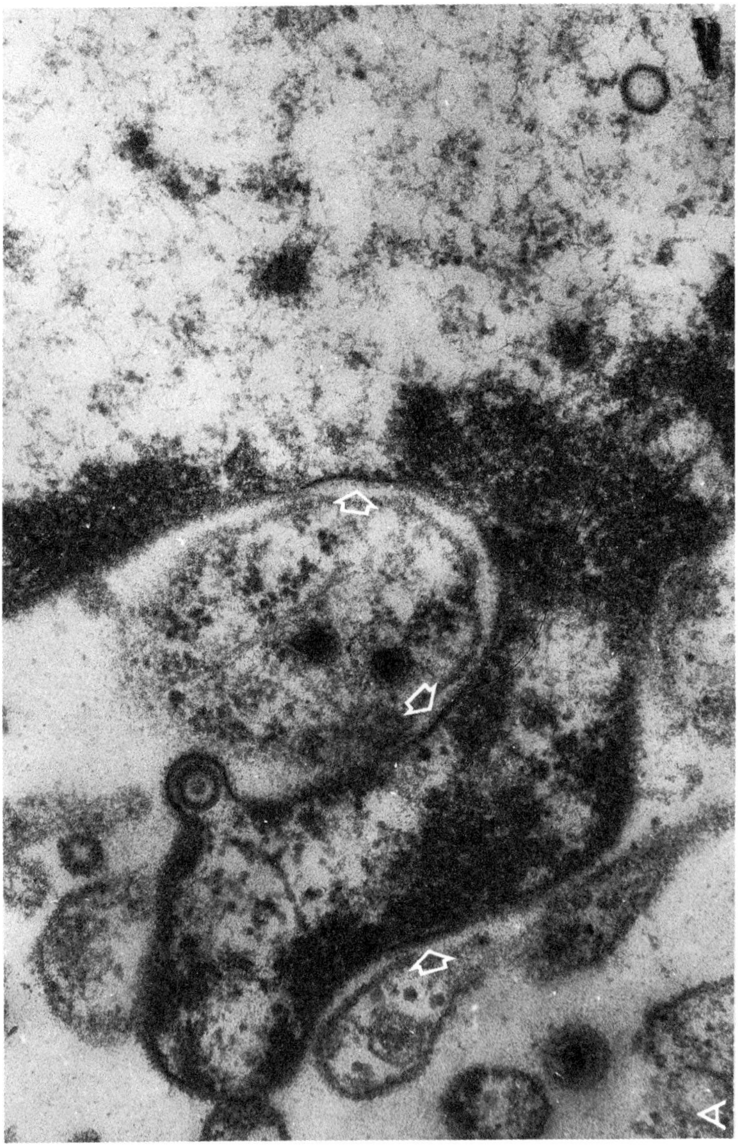

Fig. 13.14 Electron micrograph of thin section showing a capsid in the process of envelopment at one of the altered maculae, i.e., stretches of increased density and altered curvature on the inner lamella of the nuclear membrane (arrows). Note spike projections into the perinuclear space at the site of development. Any resemblance to a camel is purely coincidental.

the virion, are present on the cisternal side. The capsids budding through the inner nuclear membrane pick up the dense material that is present on the nuclear side of this lamella as well as the membrane itself. This material, by virtue of its location, would be defined as the tegument, or at least a part of it. This is surely not the only site at which the tegument is acquired since unenveloped capsids covered with a layer of some material can be seen in the nucleus [20, 24, 238, 239] and in the cytoplasm [1, 235, 240, 241]. It is not known whether the material seen around capsids not undergoing envelopment is the same as the material present in the tegument or whether the material seen on the cytoplasmic capsids is identical to, in addition to, or a replacement of the material seen around the capsids in the nucleus. The site of acquisition of the tegument varies. Thus the material defined as tegument is present in HSV-1 infected cells along the inner nuclear membrane as well as along the cytoplasmic membranes; this material could not be differentiated from the tegument of virions. In the guinea pig herpesvirus this material was not seen along the inner nuclear membrane or on capsids accumulated in the nucleus, and only those capsids that were exposed to the cytoplasmic sap and were enveloped at cytoplasmic membranes had a tegument [23].

In general, electron micrographs of capsids in apposition to or partially surrounded by altered membranes have been interpreted as showing capsids undergoing envelopment. This interpretation, particularly of structures found at the nuclear membrane, seems reasonable. The question arises, however, whether the same interpretation is valid for similar structures seen at cytoplasmic membranes, since particles in the process of being de-enveloped could give rise to similar images. The main objections to envelopment at the cytoplasmic membranes are 2-fold. A priori, it could be predicted that cytoplasmic membranes would differ structurally not only among themselves but also from nuclear membranes, and it could be expected that the structural requirements for envelopment would be somewhat rigid. The second objection stems from lack of information as to how unenveloped capsids arrive in the cytoplasm to become enveloped. These objections are not compelling since some herpesviruses do have a wide host range, implying that the requirement for a particular chemical composition of the membrane enveloping the capsid may not be rigid, or that the virus is able to alter it to suit its needs. Further, it is conceivable, as suggested by Stackpole [242] for the morphogenesis of the frog herpesvirus, that the capsid undergoes sequential envelopment and de-envelopment at each membrane barrier in its passage from the nucleus to extracellular space.

The foregoing discussion should not be interpreted to mean that the site of envelopment is either completely random or totally unknown. The impression gained from examination both of the literature and of infected cells in this laboratory [16, 18, 37, 243], is that the inner lamella of the nuclear membrane is the predominant site of envelopment and that the images interpretable as envelopment at the cytoplasmic membrane are seen more frequently with some herpesviruses than with others.

3. *Biochemical Aspects of Assembly and Envelopment*

A striking feature of viral morphogenesis is its inefficiency. Approximately 5- to 10-fold more viral DNA is produced than is utilized. The utilization of viral proteins is not known, but a similar order of inefficiency would not be too surprising.

There is very little biochemical information on assembly and envelopment inasmuch as intermediates in the assembly process have not been detected. A noteworthy observation is that unenveloped HSV-1 and HSV-2 capsids containing viral DNA differed from de-enveloped capsids [103, 104]. Thus neither the de-enveloped capsids nor the enveloped ones contain protein 22a present in capsids that have not undergone envelopment. Based on staining properties of polypeptides in polyacrylamide gels, efficiency of incorporation of in vivo labeled polypeptides into virions as a function of the time of addition of labeled amino acids into the medium, and in vivo phosphorylation, it has been deduced that polypeptide 22a is cleaved and that at least 1 product of this cleavage, polypeptide 22, is incorporated into the virion. Unenveloped capsids also lack polypeptides 1-3, which are quantitatively recovered in the de-enveloped capsids. Topologically, proteins 1-3 and 22a are in the tegument. These data suggest that the cleavage of polypeptide 22a and the addition of at least polypeptides 1-3 and possibly other tegument polypeptides occurs during envelopment.

4. *Egress*

Several pathways for the egress of virus from infected cells have been proposed in recent years. One envisions that throughout its journey from the perinuclear space to the extracellular fluid the virion is encased in some component of the cytoplasmic vericular system [26, 37, 40, 241, 244, 245]. It has been suggested that the virion remains in the endoplasmic reticulum which connects with the perinuclear space until it is released into the extracellular space via a direct connection of the endoplasmic reticulum [37, 245-247]. This is a continuous route. The other variation of this scheme is that the route is discontinuous, i.e., the virions are thought to travel in vacuoles and be secreted in a manner similar to other exported cellular products [248]. The virion would be nonetheless always within some membranous vesicle, protected from the cytoplasmic sap which initially served to release the core from the capsids in the inoculum [22]. In support of these release mechanisms are many pictures of virions in the cytoplasmic sap which appear to be disaggregating.

An alternative proposal is based on the assumption that the cytoplasmic sap is not an unsafe place for virions or capsids, and suggests that capsids may acquire their tegument there [1, 3, 23, 34, 127, 128, 236, 240, 249]. This hypothesis proposes that capsids exit by budding into and out of whatever membranes they encounter and that the virion envelope comes from the last membrane encountered. This requires that all membranes become identically altered at sites which serve for envelopment or that there be heterogeneity among virion envelopes. For almost every herpesvirus studied, virus particles have been seen inside and outside all membranous structures (excluding mitochondria), although the proportion of particles in each region varies a great deal from one infected cell system to another. A conservative evaluation of the available data is that, while no unique mode of egress from infected cells prevails, a particular mode of release may be more prevalent in some infected cell systems than in others.

IV. Alterations of Cell Function and Structure during Productive Infection

A. Introduction

All available evidence indicates that the consequence of productive infection with herpesvirus is cell death [157]. The causes of cell death are obscure. The symptoms, however, are abundant: the infected cell cannot divide, host DNA and protein synthesis are drastically reduced or cease altogether, host RNA synthesis is greatly altered, and the cell membrane acquires a new immunologic specificity. This is accompanied by gross alterations in cell morphology and in the structure of cellular organelles. Information concerning structural and functional modifications of the infected cell have come from direct visualization of the infected cell in the electron and light microscopes and biochemical studies on synchronously and parasynchronously infected cell populations. The reader should relate the 2 sets of observations with caution. Many of the electron microscopic examinations were done very late in infection, when all biochemical events ceased or were no longer of any significance. Further, the biochemical and electron microscopic measurements are not comparable since microscopic examinations involve very few cells, and by and large the photomicrographs tend to depict isolated instances of the more extreme rather than average situation.

B. Structural Alterations

1. The Nucleus

In addition to partially assembled, "mature" and "immature" capsids, and the various granules related to virus morphogenesis and mentioned earlier (see Section II.G), the infected cell nucleus exhibits gross changes in the structure of the nucleolus, chromatin, and the nuclear membrane. During infection the nucleolus becomes enlarged and is often displaced toward the nuclear membrane [18]. The nucleolar material appears to be altered as well. The amount of granular material, in comparison to the amorphous component of the nucleolus, increases [235, 250]. The presence of viral material in the nucleolus has been demonstrated with ferritin-labeled antiviral antibodies [251]. Late in infection the nucleolus appear to correlate with the inhibition processing of ribosomal precursor RNA. It is not known whether they are causally related.

It has been reported by Moretti et al. [252] that ribosomes contained in the nuclei of cells from chick chorioallantoic membranes infected with HSV-2 formed crystals in the presence of rifampicin or subsequent to hypothermia, whereas the ribosomes of uninfected cell nuclei do not. Lattice-like structures which resemble the crystallized ribosomes can sometimes be seen in Hep-2 cells infected with HSV-2 [18, 252].

A diagnostic feature of the infected cell is the displacement and condensation of the chromatin at the nuclear membrane. The displacement occurs quite early in infection and is, in fact, one of the very early signs that the cell is infected. Nothing is known of the mechanism of displacement of the chromatin although it has been suggested that it is not an invariant feature of cells infected with all herpesviruses [253]. Correlated with the displacement of chromatin are 2 phenomena. The first, erroneously labeled as amitotic nuclear division [254-256], is probably an extreme consequence of distortion and fragmentation of the nucleus. The fragments, frequently unequal in size, remain attached to each other. The second phenomenon is chromosome breakage described by numerous authors [257]. The chromosomal aberrations seen in herpesvirus-infected cultures cannot be differentiated with respect to site and nature from those occurring spontaneously or those induced by a variety of mutagenic agents [258-259]. A noteworthy observation by O'Neill and Rapp [260] is that the amount of chromosome breakages in cells infected with HSV-2 and treated with cytosine arabinoside was greater than in either uninfected cells treated with the drug or untreated infected cells. The authors concluded that some product made in the cell early after infection is responsible for the breakage of the chromosomes [261] since it was reduced in cells pretreated with interferon but not in those treated with cytosine arabinoside.

Parenthetically, many of the experiments designed to demonstrate chromosome damage are perplexing and difficult to understand. In typical experiments, chromosome analyses are done on cells arrested in mitosis a few hours after infection at multiplicities of 2.5 PFU per cell. The problem arises from the consideration of reports by Stoker [262], Stoker and Newton [263], and by Vantis and Wildy [264] that infection can both prevent and abort mitosis. If we take these observations at face value, and they are probably valid, chromosome studies are done on an unknown, but presumably small, fraction of cells which undergo mitosis either because the virus failed to inhibit mitosis and in that particular cell the infection aborted, or the cell did not become infected to begin with. The question arises, therefore, whether the infected cell undergoing mitosis is representative of the culture, i.e., whether the mitotic cell was truly infected and whether the mitotic infected and uninfected cells belong to the same or to entirely different populations. It could be, for example, that in the infected culture a small fraction of cells characterized by broken chromosomes is resistant to infection and goes into mitosis, whereas the bulk of the cells with intact chromosomes are infected and do not go into mitosis. Whatever the case, it could obviously be expected that interferon-treated cells would resemble the uninfected cell population rather than the infected one.

A characteristic of productively infected cells late in infection is the presence of long stretches of nuclear membranes folded upon themselves. Such stretches have been described by numerous authors [2, 19, 34, 37, 149, 235, 236, 265] as reduplicated nuclear membranes. The staining properties of these membranes superficially resemble those of nuclear membranes containing the "macula" at which the capsids become enveloped. The resemblance derives from the fact that both the macula and the folded membranes contain stretches of electron-opaque material. It remains to be shown that the folded membranes contain viral glycoproteins. Moreoever, it is not at all clear whether these membranes are synthesized de novo after infection or whether they represent modified stretches of the host nuclear membrane. In any event, it is probable that these refolded stretches represent an abberation of some cellular process necessary for the envelopment of the virus and that the multi-folded stretches of membranes play no functional role in the maturation of the virus.

2. *The Cytoplasm*

Other than the virions and capsids which abound late in infection, the cytoplasmic landscape contains no unique features characteristic of herpesvirus-infected cells. The most striking features are the increase in the size of polyribosomes and in the amount of membranous structures. The cytoplasm of infected cells often appears filled with polyribosomes. The polyribosomes appear to be much longer than those contained in uninfected cells; indeed, spirals containing 15-23 ribosomes have been seen and this finding agrees well with the increased sedimentation rate of infected cell polyribosomes reported by Sydiskis and Roizman [145, 146]. Both free and bound polyribosomes have been seen in infected cells. At present there is no information as to whether these polymerize the same or different classes of polypeptides.

Electron-opaque, "dense", unidentified bodies have been seen occasionally in cells infected with nearly all herpesviruses, but they are particularly prominent in cells infected with CMV [266] and MDV [267, 268]. The electron-opaque bodies appear to be membrane-bound, 200-500 nm in diameter, and to contain very dense, slightly granular material. Based on its reactivity with antiviral antibody it has been suggested that the dense body is virus-specific [266]. Sarov and Abady [101] reported the presence of viral structural proteins in purified CMV dense bodies. The dense bodies are formed by budding of granular material free in the cytoplasmic sap through membranes.

Plasma membranes from infected cells exhibit 3 kinds of alterations visible under the electron microscope. These are the appearance of patches of altered membranes corresponding in all likelihood to regions of the membranes containing viral glycoproteins described in the preceding section; a redistribution of receptors reacting with concanavalin A [265]; and an increase in the length and frequency of the microville [269]. Appearance of altered patches or plaques along the plasma membrane was shown very clearly by Nii et al. [270]. Following reaction of cells with ferritin-labeled antibody to HSV, the label was seen only along patches of the plasma membrane of infected cells which were denser and more curved than adjacent, unreacted areas. Filamentous material was often seen on the inside of the cell surface along these patches. These patches of altered membranes are very reminiscent of the macula seen on the nuclear membranes.

C. Effect of Viral Infection on Host Macromolecular Synthesis

1. General Description

Inhibition of host DNA synthesis has been observed in cells infected with PSV [228, 271], HSV-1 [22, 230, 272], EAV [227], and EBV [162, 273, 274]. A typical example of the decline of incorporation of thymidine into host DNA is shown in Fig. 13.15. The significance of the trace amounts of DNA synthesized late in infection is not clear; it could represent either asynchronously infected cells or a species of host DNA which escapes inhibition.

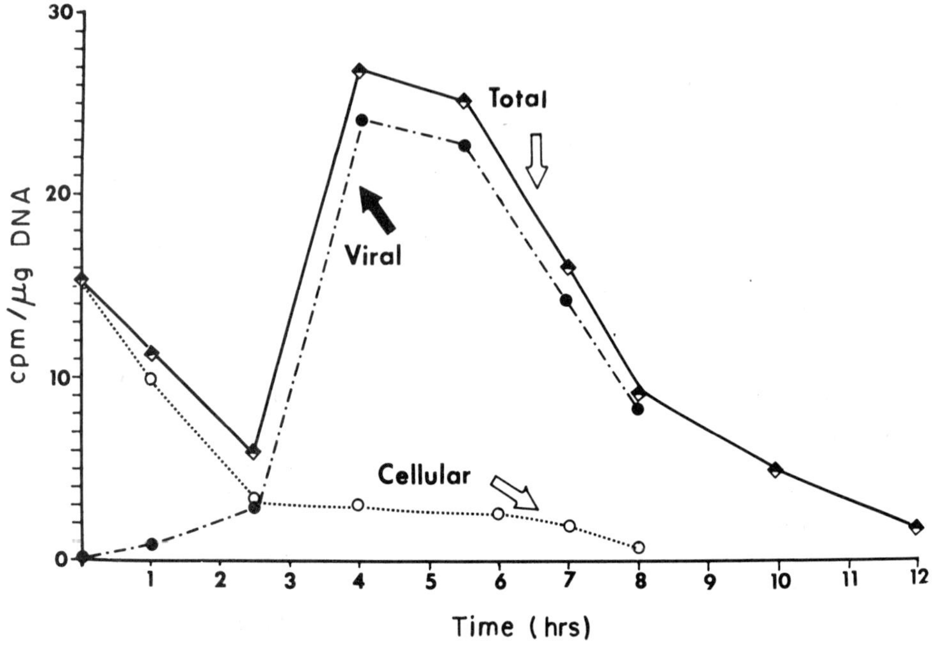

30
20
10
0
cpm / μg DNA
Total
Viral
Cellular
1
2
3
4
5
6
7
8
9
10
11
12
Time (hrs)

Fig. 13-15 The pattern of incorporation [^{3}H] thymidine into DNA of Hep-2 cells infected with HSV-1 (mP). The cells were pulse-labeled for 15 min at different times after infection. The DNA extracted after the pulse was centrifuged to equilibrium in CsCl density gradients [22].

Host protein synthesis also describes rapidly after infection. This may be deduced from analyses of autoradiograms of infected cell lysates shown in the Section B. Cessation of host protein synthesis is accompanied by a decrease in the number of polyribosomes and a concurrent drop in the amino acid incorporation rate soon after infection [105, 145]. Cessation of host protein synthesis is accompanied by a cessation of glycosylation of host proteins [116, 275].

The effects of viral infection on host RNA metabolism have been studied by numerous authors [183, 218, 272, 276]. The emerging picture is complex and poorly understood. Very briefly, there is an overall decrease in host RNA synthesis, but the decrease in the rate of synthesis is least for RNA greater than 28S and most pronounced for 4S RNA. However, the synthesis of small structural RNA species (5-7S) can no longer be detected (Millette, Silverstein and Roizman, unpublished observations). The appearance of new ribosomal RNA is nearly completely inhibited, but the site of action is not entirely at the level of synthesis of 45S precursor or its methylation, but rather at the level of subsequent processing. The nonribosomal RNA which continues to be made does not appear to be functional in the sense of being capable of directing host protein synthesis in the infected cell [140, 276, 277]. The report by Roizman et al. [140] that host RNA is transported from the nucleus to the cytoplasm without delay, as compared to the bulk of viral RNA which lingers in the nucleus some 10-20 min after it is made, indicates an alteration in processing of host RNA. Just why host RNA synthesis is not inhibited as completely as host DNA or host protein synthesis remains a mystery.

2. *Mechanisms of Inhibition of Host Macromolecular Synthesis*

The substantive observation is that the infected cell discriminates between host and viral macromolecular synthesis. The fundamental question raised here is the mechanism by which this discrimination takes place.

A major question is whether the inhibition of host macromolecular synthesis requires viral gene expression immediately following infection. In spite of a brief report by Newton [278] summarized in the proceedings of a meeting, there is no evidence that UV-irradiated virus inhibits host functions [146]. The data are not conclusive since rather high dosages of nonionizing radiation are required to substantially reduce the infectivity of virus preparations [279], and, therefore, UV-light irradiation might inactivate a sensitive component required for this function. There is no simple way to get around this objection. The hypothesis that the inhibitor is a structural component of the virus predicts that the rate of inhibition would be multiplicity-dependent. In fact, this is the case for RNA synthesis [277]. However, in contrast to adenovirus-infected cells, in which high multiplicity of infection results in cessation of both host and viral macromolecular synthesis [280, 281], increasing the multiplicity of herpesvirus infection does not have an adverse effect on virus yield and by extension, on viral macromolecular synthesis.

There is in fact suggestive evidence that viral gene expression at least accelerates and may even be necessary for the decay of host functions. Ben-Porat and Kaplan [282] demonstrated that the rate of inhibition of rabbit kidney cell DNA synthesis is less marked in PSV-infected cells treated with puromycin than in infected untreated cells. In another variation on this theme, Ben-Porat et al. [283] noted that in cells treated with cycloheximide for 5 hr from the time of infection, host polyribosomes were present and functional for 15 min after the drug was removed, then disappeared entirely. The authors concluded that the inhibition of host protein synthesis is mediated by a protein synthesized early after infection. The requirement for viral gene expression for the shut off of host proteins could also be deduced from studies on sequential shut off of host and viral polypeptide synthesis in HSV-1 infected Hep-2 cells [151, 154]. In these cells host protein synthesis is reduced to relatively low levels by 2-3 hr postinfection at the time of onset of β protein synthesis. A comparable, rapid reduction in host protein synthesis occurs following a relatively brief cycloheximide treatment with the onset of β protein synthesis. These observations have led to the conclusion that 1 or more β polypeptides shuts off host protein synthesis. However, it has also been noted that the rate of host protein synthesis immediately following withdrawal of cycloheximide added at the time of in fection is inversely proportional to the duration of the treatment. Thus the longer the treatment, the lower the level of host protein synthesis after reversal. Similarly, host protein synthesis decreases at a relatively slow rate in cells exposed to canavananine or to azetidine from the time of infection. We cannot exclude the possibility that a small amount of viral α proteins is made in the presence of cycloheximide and that this protein is responsible for the slow shut off of host protein synthesis. The alternative is that no new host mRNA is produced in infected cells treated with either cycloheximide or amino acid analogs for reasons that remain obscure and that the decrease in host protein synthesis results from normal decay of host mRNA.

The mechanism of inhibition of host DNA synthesis remains a puzzle. Kaplan [284] found most appealing the hypothesis that the inhibition of host DNA synthesis is due to inhibition of synthesis of host proteins required for replication of host DNA largely because the inhibition of host cell DNA synthesis is a relatively slow process. However the hypothesis does not explain the structural changes which accompany cessation of host DNA synthesis described in the preceding section. An alternate hypothesis, namely that herpesviruses specify histone-like proteins which bind to the host chromatin, has been exploited, and indeed in BHK-21 cells which were infected with pseudorabies virus, 2 viral acid-soluble polypeptides were found to be associated with isolated chromatin [285]. Although evidence that they play a role in the alteration of the structure and function of host DNA is lacking, the hypothesis that some viral product is responsible for the margination of the chromatin is attractive and will probably turn out to be correct.

There is yet no data on the mechanism of inhibition of host RNA synthesis and processing. Sasaki et al. [286] noted the presence of an inhibitor of nucleolar RNA polymerase in HSV-infected cells. The inhibitor has not been characterized as yet and its putative role in the modification of host RNA synthesis is unknown. Similarly, little is known concerning the mechanism by which the infected cells descriminate between host and viral mRNA or in fact, between the various classes of viral mRNAs. No doubt the answer to these questions will tell us much about the regulation of viral and eukaryotic cell metabolism.

Not to be overlooked in this discussion of putative viral inhibitors of host macromolecular metabolism is the fact that herpesviruses do contain genetic information for the inhibition of the host. One may wonder why viruses have acquired, retained, and expressed the capacity to inhibit the host. The only available data pertinent to this problem would seem to indicate that inhibition of host macromolecular synthesis is a prerequisite for virus multiplication. The conclusion is based on the observation that the HSV-1 (MP) multiplied and effectively inhibited cells of human derivation but did not produce infectious progeny and did not effectively inhibit macromolecular synthesis in dog kidney cells. The significant finding is that in dog kidney cells, viral DNA and proteins were synthesized only in cells infected at a multiplicity sufficiently high to inhibit host functions [272]. At low multiplicities of infection the cell made interferon only (Aurelian and Roizman 1965). These and other data would seem to indicate that: (1) host response to infection and inhibition of host macromolecular synthesis are competing processes initiated on infection, and (2) at low multiplicities of infection the cells attain the upper hand only because the amount of effective inhibitor specified by the virus is insufficient to inhibit the host in time to prevent it from inhibiting the virus [146, 272].

D. Alteration in the Structure and Function of Cellular Membranes

1. Introduction

The first inkling that herpesviruses alter the structure of the plasma membrane [287] was based on the numerous reports of isolation of mutants of variants of HSV, pseudorabies, and herpes B virus strains which differed from the "wild" or parental strains with respect to their effects on cells [257]. Whereas the parental strains caused the cells to round up and clump, the variants caused cells to fuse (Fig. 13.16). Ejercito et al. [288] classified HSV strains into 4 groups as follows.

1. Strains causing rounding of cells but no adhesion or fusion
2. Strains causing loose aggregation of rounded cells
3. Strains causing tight adhesion of cells
4. Strains causing polykaryocytosis.

The viruses comprising each group may differ in fine detail with respect to their effects on cells. Thus, polykaryocytes induced by various strains of HSV differ in size and morphology [289-291]. The term "social behavior of infected cells" was introduced [22, 288] to describe the different interactions

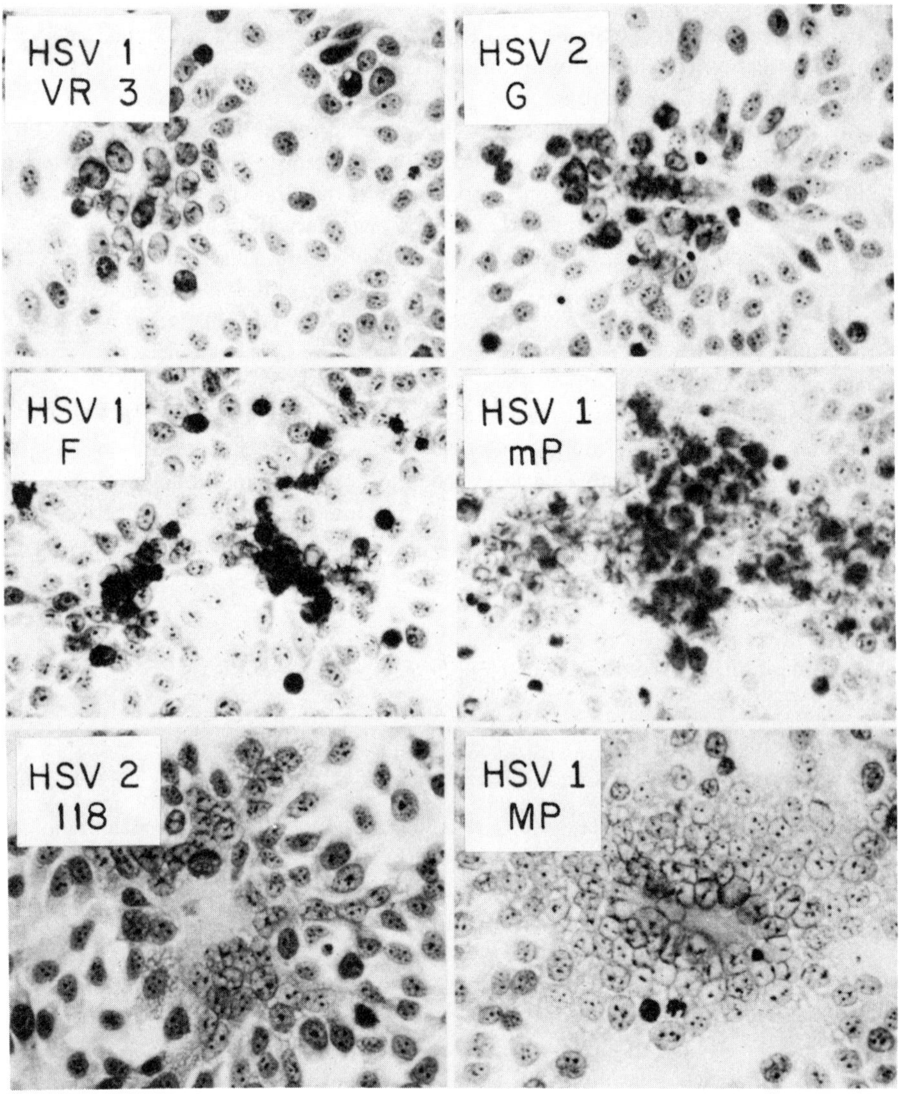
HSV 1
VR 3
HSV 2
G
HSV 1
F
HSV 1
mP
HSV 2
118
HSV 1
MP

Fig. 13.16 The social behavior of Hep-2 cells infected with HSV-1 and HSV-2. VR-3 virus: rounding but little or no clumping of cells; G virus: loose clumps of rounded cells; mP virus: tight large clumps of rounded cells; 118 virus: small polykaryocytes which tend to fragment; MP: large polykaryocytes. Hep-2 cells were infected at a multiplicity of 1 PFU/1000 cells and stained with Giemsa 24 hr after infection. Photomicrographs were taken and printed at the same magnification. The size of clumps, polykaryocytes, etc., is representative except for that produced by MP virus, which is the smallest found in that culture. Data from Keller et al. [297].

of cells among themselves. The alteration in the social behavior of infected cells is readily demonstrable in monolayer cultures of cells infected at very low multiplicities; in this instance the interaction of infected cells is readily apparent and easily differentiated from the surrounding lawn of uninfected cells. So striking is the difference between the appearance of the foci of infected cells that the technique was used not only for differentiation of strains differing in the effects on the social behavior of cells but also as a precise and reproducible plaque assay for measuring the infectivity of virus preparations [292, 293].

The reasoning that viruses alter the structure of the plasma membrane [287] was based on the necessary conclusion that the shape, adhesiveness, and social behavior of cells must reflect the structure of the plasma membrane and the observation that the social behavior of infected cells is genetically determined by the virus. All subsequent studies were directed toward elucidation of the nature of the structural function and immunologic properties of the infected cell membranes.

2. *Structural Changes in the Plasma Membrane*

Wilbanks and Campbell [269] differentiated between uninfected and HSV-infected cells by the number of microvilli. Changes in the reactivity of cells with concanavalin A following infection was also reported [265]. In this instance the cells became agglutinable by concanavalin A by 2 hr postinfection. Concomitantly, the infected cells leak macromolecules [277, 294, 295]. It is conceivable that the change in transmembrane potential observed in HSV-infected cells reflects the leakiness of the infected cell membranes [296].

3. *Incorporation of Viral Proteins into Plasma Membrane*

The presence of viral proteins in cellular membranes was reported in several publications [16, 116, 120, 275, 297]. The pertinent data may be summarized as follows.

Purified fractionated cytoplasmic membranes [275] and plasma membranes [116] extracted from HSV-1 (F)-infected cells were found to cosediment with proteins made after infection and absent from uninfected cells. In polyacrylamide gels, these proteins migrated with the major noncapsid virion proteins. Moreover, the glycosylation profile of the viral proteins associated with the plasma membrane was similar to that of the major noncapsid virion proteins. The studies on the plasma membranes of infected and uninfected cells did not reveal major changes in the composition of the host membrane proteins remaining associated with the plasma membrane after infection.

That the virus specific proteins cosedimenting with the plasma membrane throughout purification do not constitute either adventitious contaminants adhering to the plasma membranes or fragments of viral envelopes stripped off virions during the preparation of the infected-cell plasma membranes can be demonstrated as follows. Mixtures of infected and uninfected plasma membranes are readily separated by isopycnic centrifugation in sucrose gradients following reaction with antiviral antibody. The function of the antiviral antibody is to bind to viral proteins in the membranes, thereby augmenting the total protein mass relative to that of lipids. Consequently, membranes binding antiviral antibody band at a higher density than those lacking these antigens [167]. In a series of experiments, Heine and Roizman [120] have shown that whereas artificial mixtures of labeled, infected, and uninfected cell membranes are readily separable by isopycnic centrifugation following reaction with antiviral antibody, the host polypeptides labeled before infection band with the infected cell membrane and not with the host proteins contained in the membranes extracted from uninfected cells. Moreover, microcavitation employed for preparation of plasma membranes does not inactivate infectivity. The data indicated that the viral proteins are contained on the same membrane fragments as host proteins and that the binding or incorporation of viral proteins to membranes is sufficiently tenacious to withstand considerable hydrodynamic stress augmented by the presence of antibody bound to the polypeptides. The fact that not all viral glycoproteins appear in the plasma membrane, as discussed in a subsequent section, suggests that the association of viral polypeptides is not only tenancious but also specific.

The HSV glycoproteins in cell membranes are available for envelopment of RNA viruses, that is, viruses other than herpesviruses that mature by budding through cellular membranes [298].

4. *Alteration in Immunologic Specificity of the Plasma Membranes*

The evidence that infected cells acquire a new immunologic specificity was obtained with the aid of a test based on the observation that infected cells failed to produce plaques following injury by antibody and complement [299, 300]. The sensitivity of the test stemmed from the fact that very few cells were needed since nearly every infected cell produces a plaque. The assay was initially standardized with 2-hr infected cells and antibody made against uninfected cells [299]. The alteration of immunologic specificity after infection was demonstrated in in tests employing 20- to 24-hr infected cells and rabbit sera prepared against HSV-1 infected cells and rabbit sera prepared against HSV-1 infected cells [300]. The tests showed that complement and unabsorbed anti-infected-cell serum precluded the formation of plaques by 2-, 24-, and 48-hr infected cells. However, following absorption with uninfected cells, the serum and complement precluded the formation of plaques by 24- and 48-hr infected cells only; the absorbed serum was not effective against 2-hr infected cells, leading to the conclusion that 24- and 48-hr infected cells contained on their surface 1 or more antigens absent in uninfected cells.

The conclusion that the membranes of infected cells become altered with respect to structure and immunologic specificity was corroborated in a study by Watkins [301] showing that HeLa cells infected with the HFEM strain on herpes simplex acquire "stickiness" for sheep erythrocytes sensitized with rabbit anti-sheep-erythrocyte serum. The adhesion of sensitized erythrocytes to the infected cells could be abolished by exposing the infected HeLa cells to anti-viral serum. Normal rabbit serum and rabbit anti-sheep-erythrocyte serum failed to prevent the adhesion of the sensitized erythrocytes to infected HeLa cells. Virus-specific antigens have since been demonstrated by a variety of techniques, not only in cells infected with HSV [257], but also with EBV [302], MDV [303, 304], and herpes zoster [305]. Of particular interest are the membrane antigens associated with EBV infection of human lymphoblastoid cells in culture and in Burkitt lymphoma biopsy material. In the biopsy material the antigens appear to be associated with cells not engaged in virus production. In contrast lymphoblastoid cells carrying the EBV genome and growing in cell culture exhibit viral antigens on the cell surface only if they are on the path to virus production [306]. The reason for the difference in the apparent expression of membrane antigen in lymphoblastoid cell in vitro and in vivo is not known. More recent studies have differentiated between a set of "early" antigens present in nonproducer cells and a set of "late" antigens present on the surface of virions and virus producing cells [163, 307-309].

The weight of the evidence favors the hypothesis that the HSV-1 antigen is a structural component of the viral envelope. The evidence consists of the following findings.

Absorption of serum with partially "purified" HSV-1 virions removed both neutralizing and cytolytic antibody [300]. However, the significance of this observation is limited by the fact that most purified virus preparations available at that time were not free of host antigens. Similar evidence was also obtained for the antigens appearing on the surface of EBV-infected lymphocytes and the antigens on the surface of EB virions [310]. In this instance analyses of membrane immunofluorescent and virus neutralizing activities of sera indicated that the membrane antigens on the virion and lymphoblastoid antigens have common antigens but that the virion surface contains antigens absent from the plasma membrane [306, 308, 310, 311].

Assays of hyperimmune sera prepared against a variety of antigens extracted from infected permissive and nonpermissive cells showed excellent correlation between neutralizing and cytolytic titers [312].

Dog kidney (DK) cells abortively infected with HSV-1 (MP) virus

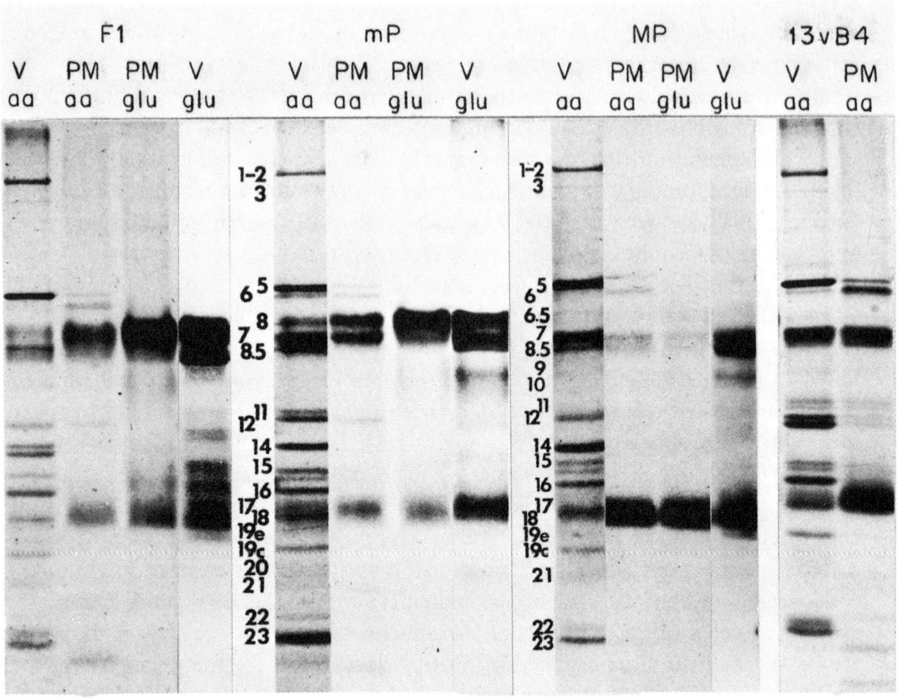
F1
mP
MP
13vB4
V aa
PM aa
PM glu
V glu
V aa
PM aa
PM glu
V glu
V aa
PM aa
PM glu
V glu
V aa
PM aa
1-2
3
6 5
7 8
8.5
12 11
14
15
16
17 18
19e
19c
20
21
22
23
1-2
3
6 5
6.5
7
8.5
9
10
12 11
14
15
16
18 17
19e
19c
21
22
23

Fig. 13.17 Comparison of polypeptides contained in virions and plasma membranes purified from Hep-2 cells with HSV-1 strains differing in their ability to cause cell fusion. The figure shows autoradiograms of polyacrylamide gels containing electrophoretically separated virions and plasma membrane polypeptides labeled in Hep-2 cells from a 4 hr postinfection in medium containing [^{14}C] leucine, isoleucine and valine or [^{14}C] glucosamine. The numbers refer to numerical designations given to structural viral polypeptides (see Fig. 13.10). Abbreviations are as follows: V, virions; PM, plasma membranes; aa, labeled with amino acids; glu, labeled with glucosamine. HSV-1 (MP) causes cell fusion and is a spontaneous mutant of HSV-1 (mP) which causes rounding and clumping of cells. 13vB4 is a polykaryocyte mutant of HSV-1 (13). Both HSV-1 (13) and HSV-1 (F) cause rounding and clumping of infected cells. Note the absence of the glycosylated viral polypeptide No. 8 from the virions and plasma membranes from cells infected with polykaryocyte inducing viruses. Heine and Roizman, unpublished data.

produced naked nucleocapsids only; envelopment did not take place [313, 314]. Rabbit hyperimmune sera produced against extracts of abortively infected DK cells lacked both neutralizing and cytolytic antibody [312].

In accordance with the hypothesis that the plasma membranes of infected cells contain virus-specific antigens, purified plasma membranes from HSV-1 infected cells were found to contain herpesvirus glycoproteins [116, 120]. The infected cell membranes readily react with anti-viral antibody as evidenced by the increase in the buoyant density of the infected cell membrane after exposure to immune serum [167]. Evidence that the antibody reactive with isolated membranes is capable of neutralizing virus emerged from the observation that addition of purified infected cell membrane to a mixture of virus and antibody competed with the infectious virus for a neutralizing antibody [176]. Last, direct evidence that the glycoproteins and not some other constituent of the membranes reacted with the antibody was furnished by the observation that glycoproteins were removed from solution during passage through immunoabsorbent columns prepared from immune sera [315].

5. Viral Membrane Proteins and the Social Behavior of Infected Cells

Since the studies of infected cell membranes were prompted by the hypothesis that structural alterations in the plasma membranes underlie the observed changes in the social behavior of cells following infection, it is of interest to review the data bearing on this hypothesis. Initial comparisons of smooth membranes from cells infected with HSV-1 (mP), HSV-2 (MP), and HSV-2 (G) revealed differences in the viral polypeptides bound to those membranes. In particular, the smooth membranes of cells infected with HSV-1 (MP), which causes cells to fuse, differed from the membranes of cells infected with HSV-1 (mP), which causes cells to clump, with respect to the number and electrophoretic mobility of viral proteins [297].

More recent analyses showed that HSV-1 (mP) and another HSV-1 strain causing cell fusion differed from the parental, nonpolykaryocyte forming strains in that they lacked VP8, a major virion glycoprotein (Fig. 13.17). However, not all polykaryocyte forming strains lack this polypeptide. Thus a variant of HSV-2 (G) which causes cell fusion cannot be differentiated from the parent strain in this regard [316] and it is conceivable that minor alterations in one or more membrane glycoproteins suffices to alter the social behavior of infected cells (see Recent Developments).

Acknowledgments

I am indebted to Norma Coleman for typing of the manuscript, to Betty Roizman for the bibliography, to Patricia Wiedner for the preparation of the illustrations, and to Elizabeth and Pampidou for help in sorting out papers. When not engaged in writing reviews, my work is supported by grants from American Cancer Society (VC 103K), National Institutes of Health, United States Public Health Service (CA 08494), University of Chicago Cancer Center (CA 14599) and Leukemia Foundation.

References

1. F. Y. Kazama and K. L. Schornstein, Science, 177, 696 (1972).
2. C. A. Farley, W. G. Banfield, G. Kasnic, Jr., and W. S. Foster, Science, 178, 759 (1972).
3. K. Wolf and R. W. Darlington, J. Virol., 8, 525 (1971).
4. G. S. Hayward, N. Frenkel, and B. Roizman, Proc. Natl. Acad. Sci. USA, 72, 1768 (1975).
5. L. Pereira, E. Cassai, R. W. Honess, B. Roizman, M. Terni, and A. Nahmias, Infect. Immun., 13, 211 (1976).
6. Herpesvirus Study Group, J. Gen. Virol., 20, 417 (1973).
7. D. Furlong, H. Swift, and B. Roizman, J. Virol., 10, 1071 (1972).
8. K. Nazerian, J. Virol., 13, 1148 (1974).
9. S. Nii and I. Yasuda, Biken J., 18, 41 (1975).
10. M. A. Epstein, J. Exp. Med., 115, 1 (1962).
11. W. Bernhard, J. Ultrastruct. Res., 27, 250 (1969).
12. J. D. Strandberg and L. E. Carmichael, J. Bacteriol., 90, 1790 (1965).
13. P. Wildy, W. C. Russell, and R. W. Horne, Virology, 12, 204 (1960).
14. C. R. Goodheart, G. Plummer, and J. L. Waner, Virology, 35, 473 (1968).
15. H. O. Ludwig, Med. Microbiol. Immunol., 157, 186 (1972).
16. B. Roizman and D. Furlong, in Comprehensive Virology, Vol. 3 (H. Fraenkel Conrat and R. R. Wagner, eds.). Plenum, New York, 1974, p. 299.
17. C. Morgan, H. M. Rose, and B. Mednis, J. Virol., 2, 507 (1968).
18. J. Schwartz and B. Roizman, J. Virol., 4, 879 (1969).
19. F. H. Shipkey, R. A. Erlandson, R. B. Bailey, V. I. Babcock, and C. M. Southam, Exp. Mol. Pathol., 6, 39 (1967).
20. U. Heine, D. V. Ablashi, and G. R. Armstrong, Cancer Res., 31, 542 (1971).
21. H. Ludwig, H. G. Haines, N. Biswal, and M. Benyesh-Melnick, J. Gen. Virol., 14, 111 (1972).
22. B. Roizman, in Current Topics in Microbiology and Immunology, 49 Springer-Verlag, Heidelberg. 1969, p. 1.
23. C. K. Y. Fong, R. B. Tenser, G. D. Hsiung, and P. A. Gross, Virology, 52, 468 (1973).
24. D. P. Nayak, J. Virol., 8, 579 (1971).
25. E. K. Wagner, B. Roizman, T. Savage, P. G. Spear, M. Mizell, F. E. Durr, and D. Sypowicz, Virology, 42, 257 (1970).
26. S. Nii, C. Morgan, and H. M. Rose, J. Virol., 2, 517 (1968).
27. C. W. Stackpole and M. Mizell, Virology, 36, 63 (1968).
28. K. Nazerian and R. L. Witter, J. Virol., 5, 388 (1970).
29. R. McCombs, J. P. Brunschwig, R. Mirkovic, and M. Benyesh-Melnick, Virology, 45, 816 (1971).
30. R. S. Siegert and D. Falke, Arch. Virusforsch., 19, 230 (1966).
31. D. Falke, R. S. Siegert, and W. Vogell, Arch. Virusforsch., 9, 484 (1959).

32. C. Morgan, S. A. Ellison, H. M. Rose, and D. H. Moore, J. Exp. Med., 100, 195 (1954).
33. H. C. Chopra, G. P. Shibley, and M. J. Walling, J. Microsc., 9, 167 (1970).
34. R. M. McCracken and J. K. Clarke, Arch. Virusforsch., 34, 189 (1971).
35. D. H. Watson, Symp. Soc. Gen. Microbiol., 18, 207 (1968).
36. P. G. Spear and B. Roizman, J. Virol., 9, 143 (1972).
37. J. Schwartz and B. Roizman, Virology, 38, 42 (1969).
38. P. A. Underwood, Microbios, 5, 231 (1972).
39. R. L. Witter, K. Nazerian, and J. J. Solomon, J. Natl. Cancer Inst., 49, 1121 (1972).
40. A. Gershon, L. Casio, and P. A. Brunell, J. Gen. Virol., 18, 21 (1973).
41. W. C. Russell, Virology, 16, 355 (1962).
42. T. Ben-Porat and A. S. Kaplan, Virology, 16, 261 (1962).
43. E. S. Huang, S.-T Chen, and J. S. Pagano, J. Virol., 12, 1473 (1973).
44. G. Plummer, C. R. Goodheart, D. Henson, and C. P. Bowling, Virology, 39, 134 (1969).
45. R. L. Soehner, G. A. Gentry, and C. C. Randall, Virology, 26, 394 (1965).
46. G. Plummer, C. R. Goodheart, and M. J. Studdert, Infect. Immun., 8, 621 (1973).
47. H. Ludwig, N. Biswal, J. T. Bryans, and R. M. McCombs, Virology, 45, 534 (1971).
48. W. B. Martin, D. Hay, L. V. Crawford, G. L. Le Bouvier, and E. M. Crawford, J. Gen. Microbiol., 45, 325 (1966).
49. C. Goodheart and G. Plummer, in Progress in Medical Virology, Vol. 19 (J. L. Melnick, ed.), Karger, Basel, 1975, p. 324.
50. E. Cassai and S. Bachenheimer, J. Virol., 11, 610 (1973).
51. T. R. Mosmann and J. B. Hudson, Virology, 54, 135 (1973).
52. B. Fleckenstein and H. Wolf, Virology, 58, 55 (1974).
53. D. L. Bronson, G. R. Dreesman, N. Biswal, and M. Benyesh-Melnick, Intervirology, 1, 141 (1973).
54. N. Frenkel, R. J. Jacob, R. W. Honess, G. S. Hayward, H. Locker, and B. Roizman, J. Virol., 16, 153 (1975).
55. B. Fleckenstein, G. W. Bornkamm, and H. Ludwig, J. Virol., 15, 398 (1975).
56. E. D. Kieff, S. L. Bachenheimer, and B. Roizman, J. Virol., 8, 125 (1971).
57. B. J. Graham, H. Ludwig, D. L. Bronson, M. Benyesh-Melnick, and N. Biswal, Biochim. Biophys. Acta, 259, 13 (1972).
58. Y. Becker, H. Dym, and I. Sarov, Virology, 36, 184 (1968).
59. R. H. Graftstrom, J. C. Alwine, W. L. Steinhart, and C. S. Hill, Cold Spring Harbor Symp. Quant. Biol., 39, 679 (1974).
60. R. H. Graftstrom, J. C. Alwine, W. L. Steinhart, C. W. Hill, and R. W. Hyman, Virology, 67, 144 (1975).
61. B. Roizman, G. Hayward, R. Jacob, S. C. Wadsworth, and R. W. Honess, Proc. XIth Int. Congress on Cancer, Vol. 2, Chemical and Viral

Carcinogenesis, International Congress Series No. 350, Excerpta Medica, Florence, 1974. p. 188.
62. S. C. Wadsworth, R. J. Jacob, and B. Roizman, J. Virol., 15, 1487 (1975).
63. A. Adams and T. Lindahl, Proc. Natl. Acad. Sci. USA, 72, 1477 (1975).
64. N. Frenkel and B. Roizman, J. Virol., 8, 591 (1971).
65. G. Hayward, R. J. Jacob, S. C. Wadsworth, and B. Roizman, Proc. Natl. Acad. Sci. USA, 72, 4243 (1975).
66. E. K. Wagner, K. K. Tewari, R. Kolodner, and R. C. Warner, Virology, 57, 436 (1974).
67. L. F. Lee, E. D. Kieff, S. L. Bachenheimer, B. Roizman, P. G. Spear, B. R. Burmester, and K. Nazerian, J. Virol., 7, 289 (1971).
68. M. Nonoyama and J. Pagano, Nature New Biol., 238, 169 (1972).
69. J. Hirsch and V. Vonka, J. Virol., 12, 1162 (1974).
70. N. Biswal, B. K. Murray, and M. Benyesh-Melnick, Virology, 61, 87 (1974).
71. N. Frenkel and B. Roizman, J. Virol., 10, 565 (1972).
72. N. M. Wilkie, J. Gen. Virol., 21, 453 (1973).
73. J. Reischig, I. Hirsch, and V. Vonka, Virology, 65, 506 (1975).
74. S. Wadsworth, G. S. Hayward, and B. Roizman, J. Virol., 17, 503 (1976).
75. P. Sheldrick and N. Berthelot, Cold Spring Harbor Symp. Quant. Biol., 39, 667 (1974).
76. B. Roizman, G. Hayward, R. Jacob, S. Wadsworth, N. Frenkel, R. W. Honess, and M. Kozak, in Oncogenesis and Herpesvirus, Vol. II (G. De The, M. A. Epstein, and H. Zur Hausen, eds.), IARC, Lyon. 1975, p. 3.
77. B. Roizman, P. G. Spear, and E. D. Kieff, in Perspectives in Virology, Vol. VIII. (M. Pollard, ed.), Academic, New York. 1973, p. 129.
78. G. L. Van Hoosier and J. L. Melnick, Texas Rep. Biol. Med., 19, 376 (1961).
79. H. Sterz, H. Ludwig, and R. Rott, Intervirology, 2, 1 (1974).
80. D. H. Watson, P. Wildy, B. A. M. Harvey, and W. I. H. Shedden, J. Gen. Virol., 1, 139 (1967).
81. G. Plummer, Br. J. Exp. Pathol., 45, 135 (1964).
82. D. L. Evans, J. W. Barnett, J. M. Bowen, and L. Dmochowski, J. Virol., 10, 277 (1972).
83. R. N. Hull, A. C. Dwyer, A. W. Holmes, E. Nowakowski, F. Deinhardt, E. H. Lennette, and R. W. Emmons, J. Natl. Cancer Inst. 49, 225 (1972).
84. J. Kirkwood, G. Geering, and L. J. Old, in Oncogenesis and Herpesviruses (P. M. Biggs, G. de The, and L. N. Payne, eds.), IARC, Lyon, 1972, p. 479.
85. L. J. N. Ross, J. A. Frazier, and P. M. Biggs, in Oncogenesis and Herpesviruses (P. M. Biggs, G. de The, and L. N. Payne, eds.), IARC, Lyon, 1972, p. 480.
86. E. M. Southern, Nature New Biol., 232, 82 (1971).
87. W. D. Sutton and M. McCallum, Nature New Biol., 232, 83 (1971).

88. E. D. Kieff, B. Hoyer, S. L. Bachenheimer, and B. Roizman, J. Virol., 9, 738 (1972).
89. S. L. Bachenheimer, E. D. Kieff, L. Lee, and B. Roizman, in Oncogenesis and Herpesviruses (P. M. Biggs, G. de The, and L. N. Payne, eds.), IARC, Lyon. 1972, p. 74.
90. D. L. Bronson, B. J. Graham, H. Ludwig, M. Benyesh-Melnick, and N. Biswal, Biochim. Biophys. Acta, 259, 24 (1972).
91. H. O. Ludwig, N. Biswal, and M. Benyesh-Melnick, Virology, 49, 95 (1972).
92. H. Zur Hausen, H. Schulte-Holthausen, G. Klein, W. Henle, P. Clifford, and L. Santesson, Nature, 228, 1056 (1970).
93. H. Zur Hausen and H. Schulte-Holthausen, Nature, 227, 245 (1970).
94. E.-S. Huang and J. S. Pagano, J. Virol., 13, 642 (1974).
95. A. G. Vahlne and J. Blomberg, J. Gen. Virol., 22, 297 (1974).
96. J. W. Heine, R. W. Honess, E. Cassai, and B. Roizman, J. Virol., 14, 640 (1974).
97. E. N. Cassai, M. Sarmeinto, and P. G. Spear, J. Virol., 16, 1327 (1975).
98. J. H. Subak-Sharpe, S. M. Brown, D. A. Ritchie, M. C. Timbury, J. C. M. Macnab, H. S. Marsden, and J. Hay, Cold Spring Harbor Symp. Quant. Biol., 39, 717 (1974).
99. M. L. Perdue, M. C. Kemp, C. C. Randall, and D. J. O'Callaghan, Virology, 59, 201 (1974).
100. M. Dolynink, R. Pritchett, and E. Kieff, J. Virol., 17, 935 (1976).
101. I. Sarov and I. Abady, Virology, 66, 464 (1975).
102. W. S. Stevely, J. Virol., 16, 944 (1975).
103. W. Gibson and B. Roizman, J. Virol., 13, 155 (1974).
104. W. Gibson and B. Roizman, J. Virol., 10, 1044 (1972).
105. R. W. Honess and B. Roizman, J. Virol., 12, 1347 (1973).
106. K. L. Powell and D. Watson, J. Gen. Virol., 29, 167 (1975).
107. W. Gibson and B. Roizman, Proc. Natl. Acad. Sci. USA, 68, 2818 (1971).
108. W. Gibson and B. Roizman, in Polyamines in Normal and Neoplastic Growth (D. H. Russell, ed.), Raven, New York. 1973, p. 123.
109. B. N. Ames, D. T. Dubin, and S. M. Rosenthal, Science, 127, 814 (1958).
110. B. N. Ames and D. T. Dubin, J. Biol. Chem., 235, 769 (1960).
111. S. B. Spring and B. Roizman, J. Virol., 2, 979 (1968).
112. Y. Asher, M. Heller, and Y. Becker, J. Gen. Virol., 4, 65 (1969).
113. T. Ben-Porat and A. S. Kaplan, Virology, 45, 252 (1971).
114. P. G. Spear and B. Roizman, Nature, 214, 713 (1967).
115. M. L. Perdue, J. C. Cohen, M. C. Kemp, C. C. Randall, and D. J. O'Callaghan, Virology, 64, 187 (1975).
116. J. W. Heine, P. G. Spear, and B. Roizman, J. Virol., 9, 431 (1972).
117. K. O. Smith, Proc. Soc. Exp. Biol. Med., 115, 814 (1964).
118. S. Stein, P. Todd, and J. Mahoney, Can. J. Microbiol., 16, 851 (1970).
119. A. S. Rubenstein and A. S. Kaplan, Virology, 66, 385 (1975).

120. J. W. Heine and B. Roizman, J. Virol., 11, 810 (1973).
121. A. S. Rubenstein, M. Gravell, and R. Darlington, Virology, 50, 287 (1972).
122. D. Lando and M-L. Ryhiner, C R Acad. Sci. Paris, 269, 527 (1969).
123. P. Sheldrick, M. Laithier, D. Lando, and M-L Ryhiner, Proc. Natl. Acad. Sci. USA, 70, 3621 (1973).
124. G. A. Gentry and C. C. Randall, in Herpesviruses (A. S. Kaplan, ed.), Academic, New York. 1973, p. 45.
125. D. H. Watson, P. Wildy, and W. C. Russell, Virology, 24, 523 (1964).
126. S. Dales and H. Silverberg, Virology, 37, 475 (1969).
127. Y. Iwasaki, T. Furakawa, S. Plotkin, and H. Koprowski, Arch. Virusforsch., 40, 311 (1973).
128. Y. C. Zee and L. Talens, J. Gen. Virol., 17, 333 (1972).
129. R. A. Abodeely, L. A. Lawson, and C. C. Randall, J. Virol., 5, 513 (1970).
130. K. Hummeler, N. Tomassian, and B. Zajac, J. Virol., 4, 67 (1969).
131. E. Hochberg and Y. Becker, J. Gen. Virol., 2, 231 (1968).
132. B. Roizman and E. D. Kieff, in Cancer: A Comprehensive Treatise, Vol. 2 (F. F. Becker, ed.), Plenum, New York. 1975, p. 241.
133. M. D. Hoggan and B. Roizman, Virology, 8, 508 (1959).
134. N. Frenkel and B. Roizman, Proc. Natl. Acad. Sci. USA, 69, 2654 (1972).
135. D. Gillespie and S. Spiegelman, J. Mol. Biol., 12, 829 (1965).
136. B. Roizman, M. Kozak, R. W. Honess, and G. Hayward, Cold Spring Harbor Symp. Quant. Biol., 39, 687 (1974).
137. S. Silverstein, S. L. Bachenheimer, N. Frenkel, and B. Roizman, Proc. Natl. Acad. Sci. USA, 70, 2101 (1973).
138. M. Kozak and B. Roizman, Proc. Natl. Acad. Sci. USA, 71, 4322 (1974).
139. M. Kozak and B. Roizman, J. Virol., 15, 36 (1975).
140. B. Roizman, S. L. Bachenheimer, E. K. Wagner, and T. Savage, Cold Spring Harbor Symp. Quant. Biol., 35, 753 (1970).
141. B. Jacquemont and B. Roizman, J. Virol., 15, 707 (1975).
142. B. Jacquemont and B. Roizman, J. Gen. Virol., 29, 155 (1975).
143. E. K. Wagner and B. Roizman, Proc. Natl. Acad. Sci. USA, 64, 626 (1969).
144. F. Costanzo, G. Campadelli-Finme, L. Foà-Tomasi, and E. Cassai, J. Virol., 21, 996 (1977).
145. R. J. Sydiskis and B. Roizman, Science, 153, 76 (1966).
146. R. J. Sydiskis and B. Roizman, Virology, 32, 678 (1967).
147. S. Fujiwara and A. S. Kaplan, Virology, 32, 60 (1967).
148. U. Olshevsky, J. Levitt, and Y. Becker, Virology, 33, 323 (1967).
149. P. G. Spear and B. Roizman, Virology, 36, 545 (1968).
150. T. Ben-Porat, H. Shimono, and A. S. Kaplan, Virology, 37, 56 (1969).
151. R. W. Honess and B. Roizman, J. Virol., 14, 8 (1974).

152. K. L. Powell and R. J. Courtney, Virology, 66, 217 (1975).
153. R. W. Honess and D. H. Watson, J. Gen. Virol., 22, 171 (1974).
154. R. W. Honess and B. Roizman, Proc. Natl. Acad. Sci. USA, 72, 1276 (1975).
155. D. J. O'Callaghan, J. M. Hyde, G. A. Gentry, and C. C. Randall, J. Virol., 2, 793 (1968).
156. S. Nii, H. S. Rosenkranz, C. Morgan, and H. M. Rose, J. Virol., 2, 1163 (1968).
157. B. Roizman, in Oncogenesis and Herpesviruses (P. M. Biggs, G. de The, and L. N. Payne, eds.), IARC, Lyon. 1972, p. 1.
158. K. L. Powell, D. J. M. Purifoy, and R. J. Courtney, Biochem. Biophys. Res. Commun., 66, 262 (1975).
159. D. R. Bone and R. J. Courtney, J. Gen. Virol., 24, 17 (1974).
160. R. L. Ward and J. G. Stevens, J. Virol., 15, 71 (1975).
161. L. Gergely, G. Klein, and I. Ernberg, Virology, 45, 10 (1971).
162. L. Gergely, G. Klein, and I. Ernberg, Int. J. Cancer, 7, 293 (1971).
163. I. Ernberg, G. Klein, F. M. Kourilsky, and D. Silvestre, J. Natl. Cancer Inst., 53, 61 (1974).
164. D. F. Summers, E. N. Shaw, M. L. Stewart, and J. V. Maizel, J. Virol., 10, 880 (1972).
165. B. Roizman and J. W. Heine, in Proceedings of the First California Membrane Conference (C. F. Fox, ed.), Academic, New York. 1972, p. 203.
166. P. G. Spear and B. Roizman, Proc. Natl. Acad. Sci. USA, 66, 730 (1970).
167. B. Roizman and P. G. Spear, Science, 171, 298 (1971).
168. R. W. Honess and B. Roizman, J. Virol., 16, 1308 (1975).
169. P. G. Spear, J. Virol., in press (1976).
170. B. Roizman, S. B. Spring, and P. R. Roane, Jr., J. Virol., 1, 181 (1967).
171. L. J. N. Ross, D. H. Watson, and P. Wildy, J. Gen. Virol., 2, 115 (1968).
172. G. E. Mark and A. S. Kaplan, Virology, 49, 102 (1972).
173. G. E. Mark and A. S. Kaplan, Virology, 45, 53 (1971).
174. G. J. Bayliss, H. S. Marsden, and J. Hay, Virology, 68, 124 (1975).
175. S. S. Lin and W. Munyon, J. Virol., 14, 1199 (1974).
176. S. Kit and D. R. Dubbs, Biochem. Biophys. Res. Commun., 11, 55 (1963).
177. S. Kit and D. R. Dubbs, Biochem. Biophys. Res. Commun., 13, 500 (1963).
178. S. Kit, W.-C. Leung, G. N. Jorgensen, and D. R. Dubbs, Int. J. Cancer, 14, 598 (1974).
179. S. Kit, G. N. Jorgensen, W.-C. Leung, D. Trkula, and D. R. Dubbs, Intervirology, 2, 299 (1973).
180. W.-C. Leung, D. R. Dubbs, D. Trkula, and S. Kit, J. Virol., 16, 486 (1975).
181. H. G. Klemperer, G. R. Haynes, W. I. H. Shedden, and D. H. Watson, Virology, 31, 120 (1967).
182. C. Hamada, T. Kamiya, and A. S. Kaplan, Virology, 28, 271 (1966).
183. J. Hay, P. A. J. Perera, J. M. Morrison, G. A. Gentry, and J. H. Subak-Sharpe, in Ciba Symposium on Strategy of the Viral Genome (G. E. W.

Wolstenholme and M. O'Connor, eds.), Churchill Livingstone, London. 1971.
184. D. R. Dubbs and S. Kit, Virology, 22, 493 (1964).
185. A. Buchan and D. H. Watson, J. Gen. Virol., 4, 461 (1969).
186. M. E. Thouless, J. Gen. Virol., 17, 307 (1972).
187. M. E. Thouless and P. Wildy, J. Gen. Virol., 26, 159 (1975).
188. A. T. Jamieson, G. A. Gentry, and J. H. Subak-Sharpe, J. Gen. Virol., 24, 465 (1974).
189. A. T. Jamieson and J. H. Subak-Sharpe, J. Gen. Virol., 24, 481 (1974).
190. W. H. Prusoff, Y. S. Bakhle, and L. Sekely, Ann. N.Y. Acad. Sci., 130, 135 (1965).
191. T. Ogino and F. Rapp, Proc. Soc. Exp. Biol. Med., 139, 783 (1972).
192. M. E. Thouless and G. R. B. Skinner, J. Gen. Virol., 12, 195 (1971).
193. W. Munyon, R. Buchsbaum, E. Paoletti, J. Mann, E. Kraiselburd, and D. Davis, Virology, 49, 683 (1972).
194. G. M. Cooper, Proc. Natl. Acad. Sci., USA, 70, 3788 (1973).
195. G. M. Cooper and S. Greer, Mol. Pharmacol., 9, 698 (1973).
196. D. R. Dubbs and S. Kit, Virology, 25, 256 (1965).
197. A. Buchan, D. H. Watson, D. R. Dubbs, and S. Kit, J. Virol., 5, 817 (1970).
198. W. Munyon and S. Kit, Virology, 26, 374 (1965).
199. A. Buchan, S. Luff, and C. Wallis, J. Gen. Virol., 9, 239 (1970).
200. W. Munyon, E. Kraiselburd, D. Davis, and J. Mann, J. Virol., 7, 813 (1971).
201. R. L. Davidson, S. J. Adelstein, and M. N. Oxman, Proc. Natl. Acad. Sci. USA, 70, 1912 (1973).
202. E. R. Kaufman and R. L. Davidson, Som. Cell Gen., 1, 153 (1975).
203. H. M. Keir, in Molecular Biology of Viruses, 18th Symposium Gen. of Microbiology, Cambridge University Press, Cambridge. 1968, p. 67.
204. A. Weissbach, S. Hong, J. Aucker, and R. Muller, J. Biol. Chem., 248, 6270 (1973).
205. J. A. Boezi, L. F. Lee, R. W. Blakesley, M. Koenig, and H. C. Towle, J. Virol., 14, 1209 (1974).
206. E. S. Huang, J. Virol., 16, 298 (1975).
207. I. W. Halliburton and J. C. Andrew, J. Gen. Virol., 30, 145 (1976).
208. J. C-H. Mao, E. E. Robishaw, and L. R. Overby, J. Virol., 15, 1281 (1975).
209. L. R. Overby, E. E. Robishaw, J. B. Schleicher, A. Rueter, N. L. Skipkowitz, and J. C.-H. Mao, Antimicrob. Ag. Achemother., 6, 360 (1974).
210. E.-S. Huang, J. Virol., 16, 298 (1975).
211. A. Bolden, J. Aucker, and A. Weissbach, J. Virol., 16, 1584 (1975).
212. S. S. Leinbach, J. M. Reno, L. F. Lee, A. F. Isbell, and J. A. Boezi, Biochemistry, 15, 426 (1976).
213. H. M. Keir and E. Gold, Biochim. Biophys. Acta, 72, 263 (1963).
214. W. C. Russell, E. Gold, H. M. Keir, H. Omura, D. H. Watson, and P. Wildy, Virology, 22, 103 (1964).

215. B. R. McAuslan, P. Herde, D. Pett, and J. Ross, Biochem. Biophys. Res. Commun., 20, 586 (1965).
216. G. Sauer, H. D. Orth, and K. Munk, Biochim. Biophys. Acta, 119, 331 (1966).
217. A. A. Newton, in Acid Nucleici e Lora Funzione Biologica, Istituto Lombardo Accademia di Scienze e Lettere, Convegno Antomio Baselli. 1964, p. 109.
218. J. F. Flanagan, J. Bacteriol., 91, 789 (1966).
219. J. M. Morrison and H. M. Keir, J. Gen. Virol., 3, 337 (1968).
220. G. H. Cohen, J. Virol., 9, 408 (1972).
221. M. Nohara and A. S. Kaplan, Biochem. Biophys. Res. Commun., 12, 189 (1963).
222. P. M. Frearson, S. Kit, and D. R. Dubbs, Cancer Res., 25, 737 (1965).
223. C. C. Randall, H. W. Rogers, D. N. Downer, and G. A. Gentry, J. Virol., 9, 216 (1972).
224. J. J. Holland, M. Doyle, J. Perrault, D. T. Kingsbury, and J. Etchison, Biochem. Biophys. Res. Commun., 46, 634 (1972).
225. T. Ben-Porat and A. S. Kaplan, in Herpesviruses (A. S. Kaplan, ed.), Academic, New York. 1973, p. 163.
226. W. C. Lawrence, J. Virol., 7, 736 (1971).
227. D. J. O'Callaghan, W. P. Cheevers, G. A. Gentry, and C. C. Randall, Virology, 36, 104 (1968).
228. A. S. Kaplan and T. Ben-Porat, Virology, 11, 12 (1960).
229. A. S. Kaplan, T. Ben-Porat, and C. Coto, in Molecular Biology of Viruses (J. Colter, ed.), Academic, New York. 1967, p. 527.
230. B. Roizman and P. R. Roane, Jr., Virology, 22, 262 (1964).
231. G. H. Cohen, R. K. Vaughn, and W. C. Lawrence, J. Virol., 6, 783 (1971).
232. R. O'Callaghan, C. C. Randall, and G. A. Gentry, Virology, 49, 784 (1972).
233. A. S. Kaplan, T. Ben-Porat, and A. S. Rubenstein, in Advances in Pathobiology, Cancer Biology II: Herpesviruses (C. Boreic and D. W. King, eds), Medical Book Corp., New York, 1976, p. 61.
234. T. Ben-Porat and A. S. Kaplan, Nature, 235, 165 (1972).
235. M. L. Cook and J. G. Stevens, J. Ultrastruct. Res., 32, 334 (1970).
236. J. E. Leestma, M. B. Bornstein, R. D. Sheppard, and L. A. Feldman, Lab. Invest., 20, 70 (1969).
237. S. H. Dillard, W. J. Cheatham, and H. L. Moses, Lab. Invest., 26, 391 (1972).
238. K. Nazerian, J. Natl. Cancer Inst., 47, 207 (1971).
239. U. Heine, D. V. Ablashi, and G. R. Armstrong, Cancer Res., 31, 1019 (1971).
240. J. E. Craighead, R. E. Kanich, and J. D. Almeida, J. Virol., 10, 766 (1972).
241. C. K. Y. Fong and G. D. Hsiung, Infect. Immun., 6, 865 (1972).
242. C. W. Stackpole, J. Virol., 4, 75 (1969).
243. B. Roizman and P. G. Spear, in Ultrastructure of Animal Viruses and Bacteriophages (A. J. Dalton and F. Haguenau, eds.), Academic, New York. 1973, p. 83.

244. T. J. Hill, H. J. Field, and A. P. C. Roome, J. Gen. Virol., 15, 253 (1972).
245. V. Jasty and P. W. Chang, J. Ultrastruct. Res., 38, 433 (1972).
246. A. Abraham and P. Tegtmeyer, J. Virol., 5, 617 (1970).
247. J. G. Strandberg and L. Aurelian, J. Virol., 4, 480 (1969).
248. C. Morgan, H. M. Rose, M. Holden, and E. P. Jones, J. Exp. Med., 110, 643 (1959).
249. K. Nazerian and B. R. Burmester, Cancer Res., 28, 2454 (1968).
250. C. Sirtori and M. Bosisio-Bestetti, Cancer Res., 27, 367 (1967).
251. K. Miyamoto, C. Morgan, K. C. Hsu, and B. Hampar, J. Natl. Cancer Inst., 46, 629 (1971).
252. G. F. Moretti, A. Zittelli, and A. Baroni, J. Submicrosc., Cytol., 4, 215 (1972).
253. J. D. Smith and E. de Harven, J. Virol., 12, 919 (1973).
254. M. Reissig and A. S. Kaplan, Virology, 11, 1 (1960).
255. S. Nii and J. Kamahora, Biken J., 6, 33. (1963).
256. T. F. McNair Scott, C. F. Burgoon, L. L. Coriell, and M. Blank, J. Immunol., 71, 385 (1953).
257. A list of references can be found in Ref. [15].
258. H. F. Stich, T. C. Hsu, and F. Rapp, Virology, 22, 439 (1964).
259. C. C. Huang, Chromosoma, 23, 162 (1967).
260. F. J. O'Neill and F. Rapp, J. Virol., 7, 692 (1971).
261. F. J. O'Neill and F. Rapp, Virology, 44, 544 (1971).
262. M. G. P. Stoker, Symp. Soc. Gen. Microbiol. (Cambridge), 9, 142 (1959).
263. M. G. P. Stoker and A. A. Newton, Ann. N.Y. Acad. Sci., 81, 129 (1959).
264. J. T. Vantis and P. Wildy, Virology, 17, 225 (1962).
265. S. S. Tevethia, S. Lowry, W. E. Rawls, J. L. Melnick, and V. McMillan, J. Gen. Virol., 15, 93 (1972).
266. R. E. Kanich and J. E. Craighead, Lab. Invest., 27, 273 (1972).
267. K. Nazerian, F. C. Lee, R. L. Witter, and B. R. Burmester, Virology, 43, 442 (1971).
268. S. Nii, I. Katsume, and K. Ono, Biken J., 16, 111 (1973).
269. G. D. Wilbanks and J. A. Campbell, Am. J. Obstet. Gynecol., 112, 924 (1972).
270. S. Nii, C. Morgan, H. M. Rose, and K. C. Hsu, J. Virol., 2, 1172 (1968).
271. A. S. Kaplan and T. Ben-Porat, Virology, 19, 205 (1963).
272. L. Aurelian and B. Roizman, J. Mol. Biol., 11, 539 (1965).
273. L. Gergely, G. Klein, and I Ernberg, Virology, 45, 22 (1971).
274. M. Nonoyama and J. S. Pagano, J. Virol., 9, 714 (1972).
275. P. G. Spear, J. M. Keller, and B. Roizman, J. Virol., 5, 123 (1970).
276. T. Rakusanova, T. Ben-Porat, and A. S. Kaplan, Virology, 49, 537 (1972).
277. E. K. Wagner and B. Roizman, J. Virol., 4, 36 (1969).
278. S. S. Cohen and W. K. Joklik, a report of A. A. Newton's paper, by International Virology, Vol. I (J. L. Melnick and S. Karger, eds.), Basel, New York. p. 65 and 253.

279. R. Duff and F. Rapp, Nature New Biol., 233, 48 (1971).
280. A. J. Levine and H. S. Ginsberg, J. Virol., 1, 747 (1967).
281. A. J. Levine and H. S. Ginsberg, J. Virol., 2, 430 (1968).
282. T. Ben-Porat and A. S. Kaplan, Virology, 25, 22 (1965).
283. T. Ben-Porat, T. Rakusanova, and A. S. Kaplan, Virology, 46, 890 (1971).
284. A. S. Kaplan, Cancer Res., 33, 1393 (1973).
285. J. K. Chandler and W. S. Stevely, J. Virol., 11, 815 (1973).
286. Y. Sasaki, R. Sasaki, G. H. Cohen, and L. I. Pizer, Intervirology, 3, 147 (1974).
287. B. Roizman, Cold Spring Harbor Symp. Quant. Biol., 27, 327 (1962).
288. P. M. Ejercito, E. D. Kieff, and B. Roizman, J. Gen. Virol., 3, 357 (1968).
289. B. Roizman and L. Aurelian, J. Mol. Biol., 11, 528 (1965).
290. H. Kohlhage, Arch. Ges. Virusforsch., 14, 358 (1964).
291. C. E. Wheeler, J. Immunol., 93, 749 (1964).
292. M. D. Hoggan, B. Roizman, and T. B. Turner, J. Immunol., 84, 152 (1960).
293. B. Roizman and P. R. Roane, Jr., Virology, 15, 75 (1961), Virology, 19, 198 (1963).
294. T. Kamiya, T. Ben-Porat, and A. S. Kaplan, Virology, 26, 577 (1965).
295. J. Zemla, C. Coto, and A. S. Kaplan, Virology, 31, 736 (1967).
296. M. E. Fritz and A. J. Nahmias, Proc. Soc. Exp. Biol. Med., 139, 1159 (1972).
297. J. M. Keller, P G. Spear, and B. Roizman, Proc. Natl. Acad. Sci. USA, 65, 865 (1970).
298. A. S. Huang, E. L. Palma, N. Hewlett, and B. Roizman, Nature, 252, 743 (1974).
299. B. Roizman and P. R. Roane, Jr., J. Immunol., 87, 714 (1961).
300. P. R. Roane, Jr., and B. Roizman, Virology, 22, 1 (1964).
301. J. F. Watkins, Nature, 202, 1364 (1964).
302. G. Klein, G. Pearson, J. S. Nadkarni, J. J. Nadkarni, E. Klein, G. Henle, W. Henle, and P. Clifford, J. Exp. Med., 128, 1011 (1968).
303. J. H. Chen and H. G. Purchase, Virology, 40, 410 (1970).
304. M. Ahmed and G. Schidlovsky, Cancer Res., 32, 187 (1972).
305. M. Ito and A. L. Barron, Infect. Immun., 8, 48 (1973).
306. G. Klein, in Oncogenesis and Herpesviruses (G. de The, A. M. Epstein, and H. Zur Hausen, eds.), IARC, Lyon. 1975, p. 293.
307. M. Gravell, Virology, 43, 730 (1971).
308. D. Silvestre, I. Ernberg, C. Neauport-Sautes, F. M. Kourilsky, and G. Klein, J. Natl. Cancer Inst., 53, 67 (1974).
309. A. Svedmyr, A. Demissie, G. Klein, and P. Clifford, J. Natl. Cancer Inst., 44, 595 (1970).
310. G. Pearson, F. Dewey, G. Klein, G. Henle, and W. Henle, J. Natl. Cancer Inst., 45, 989 (1970).
311. D. Silvestre, F. M. Kourilsky, G. Klein, Y. Yata, C. Neauport-Sautes, and J. P. Levy, Intl. J. Cancer, 8, 222 (1971).

312. B. Roizman and S. B. Spring, Proceedings of the Conference on Cross Reacting Antigens and Neoantigens (J. J. Trentin, ed.), Williams & Wilkins, Baltimore. 1967, p. 85.
313. S. B. Spring, B. Roizman, and J. Schwartz, J. Virol., 2, 384 (1968).
314. S. B. Spring and B. Roizman, J. Virol., 1, 294 (1967).
315. T. Savage, B. Roizman, and J. W. Heine, J. Gen. Virol., 17, 31 (1972).
316. E. Cassai, R. Manservigi, A. Corallini, and M. Terni, Intervirology, 6, 212 (1975-1976).
317. L. F. Lee, R. L. Armstrong, and K. Nazerian, Avian Dis., 16, 799 (1972).
318. L. Pereira, M. H. Wolff, M. Fenwick, and B. Roizman, Virology, 77, 733 (1977).
319. R. J. Jacob and B. Roizman, J. Virol., 24, in press (1977).
320. L. S. Morse, T. G. Buchman, B. R. Roizman, and P. Schaffer, J. Virol., in press (October, 1977).

Chapter 14

Poxviruses

Bernard Moss

Laboratory of Biology of Viruses
National Institute of Allergy and Infectious Diseases
National Institutes of Health
Bethesda, Maryland

I.	Introduction	850
II.	Classification	851
III.	Virion Structure	851
IV.	Chemical Composition of Virions	854
	A. General Considerations	854
	B. Deoxyribonucleic Acid	855
	C. Ribonucleic Acid	855
	D. Proteins	855
	E. Lipids	865
	F. Trace Substances	865
V.	Genetics	866
VI.	Growth Cycle	866
VII.	Entry into Cells and Uncoating	867
VIII.	Transcription	867
	A. Early Viral RNA	867
	B. Late Viral RNA	869
	C. Modification of Viral RNA	869
	D. Stability of Viral RNA	870
	E. Alterations of tRNA	870

IX. Protein Synthesis 870

A. Identification of Early and Late Viral Proteins 870
B. Virus-Induced Enzymes 872
C. Structural Proteins 873
D. Posttranslational Modification 874
E. Regulation of Early and Late Protein Synthesis 876

X. DNA Replication 876

XI. Virus Assemby and Maturation 877

XII. Dissemination of Poxviruses 879

XIII. Effects on Host Cells 879

A. Modification of Cell Membranes 879
B. Inhibition of Macromolecular Synthesis 880

XIV. Specific Inhibitors of Poxvirus Reproduction 881

XV. Summary 881

References 882

I. Introduction

Poxviruses are widespread, infect both vertebrate and invertebrate hosts, and may be responsible for a fulminating disease with a lethal outcome or a mild and more prolonged infection accompanied by the formation of benign tumors. The origins of 2 disciplines, virology and immunology, may be traced back to the use of vaccinia virus to prevent smallpox in the 18th century. Poxvirus particles were seen by light microscopy at about the turn of the 19th century and were identified as the infectious agents during the 1930s. The molecular biology of poxviruses may be considered to date from the time of their purification and detailed biochemical analysis in that same decade. Vaccinia virus was 1 of the first viruses grown in tissue culture and its ease of propagation in both primary and continuous cell lines has facilitated laboratory investigations. The poxviruses are among the very few groups of DNA viruses that replicate within the cytoplasm of the cell. This physical separation of host and viral genomes greatly simplified both electron microscopic and biochemical studies and permitted virologists to obtain an outline of the poxvirus growth cycle prior to that of other DNA viruses. The discovery in 1967 that purified infectious poxvirus particles contained an activity capable of synthesizing RNA in vitro had a great impact on the entire field of virology. Until then virions were considered to be little more than packaged genomes. At present a total of 10 enzymatic activities have been identified in poxvirus particles and it is unlikely that the list is complete. For the molecular biologist, poxviruses provide a unique system for studying transcription of DNA and modification of mRNA in vitro, the replication and expression of an independent genome within the cytoplasm of an animal cell, the apparent de novo synthesis of lipoprotein membranes, and complex morphogenetic processes. Since vaccinia virus and the closely related rabbitpox virus have been most often employed for such studies, their molecular biology will be emphasized in this chapter. The subject of poxvirus replication has previously been reviewed by Joklik [1, 2], McAuslan [3], Woodson [4], and Moss [5].

II. Classification

Viruses that have been grouped in the Poxviridae family share a similarity in size, morphology, molecular weight of DNA, antigenic properties, and replication within the cytoplasm of the host cell. The vertebrate poxivruses contain a related nucleoprotein antigen demonstrable by complement fixation or with fluorescent antibodies [6, 7]. The ability of poxviruses inactivated by heat to be rescued by other poxviruses, but not by viruses belonging to other families, which is referred to as nongenetic reactivation is another characteristic used for taxonomic purposes [8, 9]. The subclassification into 6 genera, recently approved by the International Committee on Taxonomy of Viruses, is presented in Table 14.1 [10]. Thus far nucleic acid hybridization, potentially the most useful method of further subclassification, has only been used to demonstrate relatedness within the orthopoxvirus group [11]. A detailed description of individual poxviruses is beyond the scope of this chapter but may be found elsewhere [12].

III. Virion Structure

Poxvirions have a distinctive morphology. Using electron microscopy, they appear as large oval or brick-shaped particles about 200 × 250-300 nm in diameter. Three principal architectural elements, a biconcave core, lateral bodies that fit into the concavities, and an envelope, have been recognized by negative staining of whole particles or in thin sections [13-15]. An electron micrograph of a thin section through the long axis of a poxvirus particle is shown in Fig. 14.1 (A).

Table 14.1 Poxviridae

Orthopoxvirus: Ether-resistant; close antigenic relationships among all members; agglutinate chicken red cells; members vary in virulence, host range, and pathogenicity for experimental animals; genetic reactivation observable among all strains; mass of vaccinia virus about 5.5×10^{-15} g; density of rabbitpox virus DNA 1.6949. Includes the viruses of vaccinia, variola, alastrim, cowpox, ectromelia, rabbitpox, monkeypox.

Avipoxvirus: Natural hosts: birds; ether-resistant; inclusion bodies type A contain lipids; antigenic relationships between members; molecular weight of viral DNA 200 to 240 million; density of fowlpox DNA 1.6945; virion dimensions 390 (220) $\times$ 240 (100) nm; molecular weight of virion 2,000 million. Includes the viruses of fowlpox, pigeonpox, turkeypox, canarypox, quailpox and lovebirdpox.

Capripoxvirus: Natural hosts: ungulates; ether-sensitive; antigenic relationships among all members; probably no hemagglutinating properties. Includes the viruses of sheeppox, goatpox, and lumpy skin disease.

Leporipoxvirus: Natural hosts: rodents; ether-sensitive; antigenic relationships among all members; possible mechanical transmission by arthropods. Includes the viruses of fibroma and myxoma.

Parapoxvirus: No multiplication in embryonated eggs or common experimental animals; probable immunological relationships among members of the genus but distinct from all other poxviruses. Mass of virion (bovine papular stomatitis): $2\text{-}3.7 \times 10^{-15}$ g. Virus-specific antigens: S antigen, probably LS antigen, NP antigen common to all members of the genus and probably to all other poxviruses, no hemagglutinating properties; nongenetic reactivation among all members and probably with other poxviruses; ether-sensitive. Includes the viruses of orf, ulcerative dermatosis of sheep, bovine papular stomatitis, milker's nodule.

Entomopoxvirus: Host range limited to arthropods; probably no multiplication in vertebrates.

When whole virus particles are examined by negative staining, electron transparent particles, termed M for mulberry, with a ridged and beaded surface (Fig. 14.1, part C) and more electron dense particles, termed C for capsule, exhibiting some internal structure (Fig. 14.1, part B) are generally seen [16, 17]. It appears that in the C particles the integrity of a lipoprotein layer at the surface of the particle has been destroyed, leading to penetration of the stain [18]. The surface projections of the M particles appear as loops of threadlike structures 9 nm wide which are themselves double-helices formed from coiled strands presumably made of protein or lipoprotein. Using the

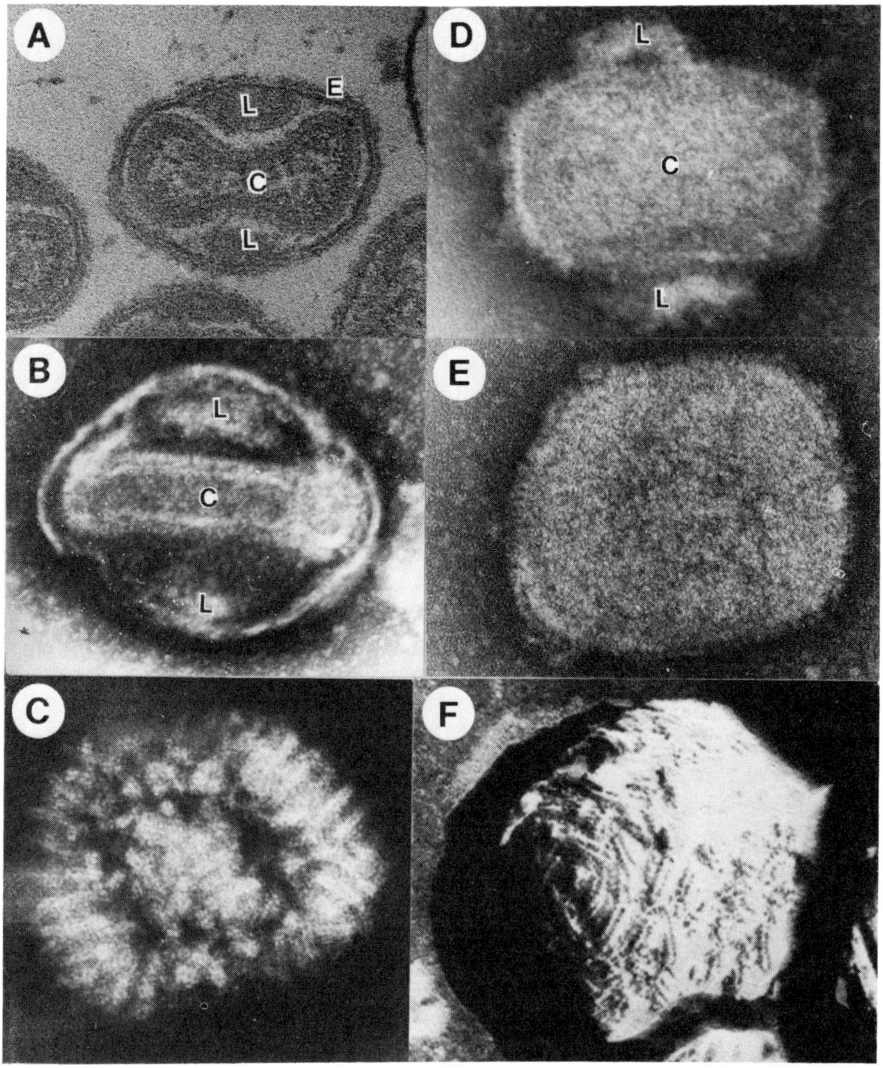
A
L
E
C
L
D
L
C
L
B
L
C
L
E
C
F

Fig. 14.1 Electron micrographs of vaccinia virus. Thin section (A); negatively stained whole particle, C form (B); M form (C); particle stripped of its envelope (D); core (E); freeze-etched particle revealing surface structure (F). E, envelope; L, lateral body; C, core.

technique of freeze-etching, which avoids some problems of fixation and dehydration, a smooth surface upon which double ridges composed of small globular units were randomly arranged has been observed ([19]; Fig. 14.1, part F). These ridges probably correspond to the threadlike structures previously revealed by negative staining (Fig. 14.1, part C).

Treatment of purified vaccinia virus particles with mercaptoethanol leads to a loosening of the envelope structure, perhaps due to breaking of protein disulfide bonds [20]. When a nonionic detergent is also added, the outer membranes are ruptured, leading to the release of the core with associated lateral bodies (Fig. 14.1, part D; [20, 21]). The lateral bodies sometimes appear to be composed of long filaments oriented in the direction of the long axis of the particle [20]. If the internal components of the virus, obtained by removal of the envelope, are treated briefly with a proteolytic enzyme such a trypsin, the lateral bodies become detached from the cores (Fig. 14.1, part E; [20, 21]). Negative staining of intact and partially degraded vaccinia virus particles indicated that the core wall is composed of cylindrical pegs that can be separated from an inner smooth layer [18, 20, 22]. A closely packed array of subunits on the external surface of the core has also been seen in freeze-etched preparations [19]. It is usually quite difficult to see fine structure within the central portion of the core or nucleoplasm, although after alkali treatment followed by negative staining, cylindrical structures sometimes assuming an "S" configuration embedded within a heavily contrasted matrix are revealed [23]. Using other chemical treatments, a more complex flower-like arrangement of tubules has been described in the nucleoplasm of fowlpox cores [24]. The manner in which the large DNA genome is folded and packed within the core, however, is unknown. Addition of sodium deoxycholate, dithiothreitol, and sodium chloride to a suspension of vaccinia virus cores leads to their rupture with the release of the DNA and about 20% of the protein, including many enzymes in a soluble form [25].

Other orthopoxviruses, avipoxviruses, and leporipoxviruses are quite similar in size and appearance to vaccinia virus, while the capripoxviruses are smaller [26] and the parapoxviruses, such as orf, are narrower and appear on negative staining to have a single long thread in the form of a left-handed spiral coiled around the outer surface [27]. An insect poxvirus isolated from *Melolantha* larvae is 400 × 250 nm, has spherical units on the surface, a kidney shaped core and a single lateral body [28]. Another insect poxvirus isolated from *Amsacta* larvae is smaller and quite different in appearance, with an electron dense core and material of intermediate density surrounding it, possibly in place of discrete lateral bodies [29].

IV. Chemical Composition of Virions

A. General Considerations

The rapid sedimentation rate of 5,000S enabled early investigators to obtain by differential centrifugation relatively pure preparations of vaccinia virus for chemical analysis [30, 31]. Presently, zonal sedimentation in sucrose density gradients [32-34] and equilibrium centrifugation in cesium chloride [35] or potassium tartrate [36] gradients are used to obtain highly purified particles composed of 92% protein (14.7% nitrogen), 3.2% DNA, 1.2% cholesterol, 2.1% phospholipid, 1.7% neutral fat, 0.2% nonnucleic acid carbohydrate, and trace amounts of copper, riboflavin, and biotin [37].

B. Deoxyribonucleic Acid

DNA comprises 3-4% of vaccinia virus [37] and the base composition and physical properties are consistent with a double-stranded structure [36, 38]. Fowlpox virus DNA appears to have a molecular weight of about 200 million [39, 40] while vaccinia virus DNA may be as small as 122 million [41-44]. The 2 strands of vaccinia virus DNA are cross-linked and cannot be separated by denaturing agents [44, 45]. The native DNA appears by electron microscopy as a linear duplex molecule but upon partial denaturation hair-pin loops are visible at each end (Fig. 14.2, part A). Upon complete denaturation, the molecule opens up to form a single-stranded circle with no detectable free ends (Fig. 14.2, part B). Since the cross-links are resistant to agents expected to remove protein, RNA, or lipids, it is possible that the vaccinia virus DNA molecule is a continuous polynucleotide chain. An internal cross-link within 50 nucleotides of the ends, however, has not been ruled out because of the limits of resolution.

C. Ribonucleic Acid

Although it is generally accepted that viruses contain DNA or RNA but not both, evidence for the presence of very small amounts of RNA in purified vaccinia virions has been obtained [35, 46]. The RNA is not covalently linked to the DNA and it is unclear whether its occurrence within the virion is accidental resulting from either inclusion during virus assembly or synthesis within the virion by the core-associated RNA polymerase. There is no evidence that this RNA is required for infectivity.

D. Proteins

1. *Structural Proteins*

The structural proteins of most viruses are difficult to solubilize and those of poxviruses are no exception. The only satisfactory methods of dissolving all of the proteins involve the use of reducing agents to break disulfide linkages combined with strong denaturing agents such as high concentrations of

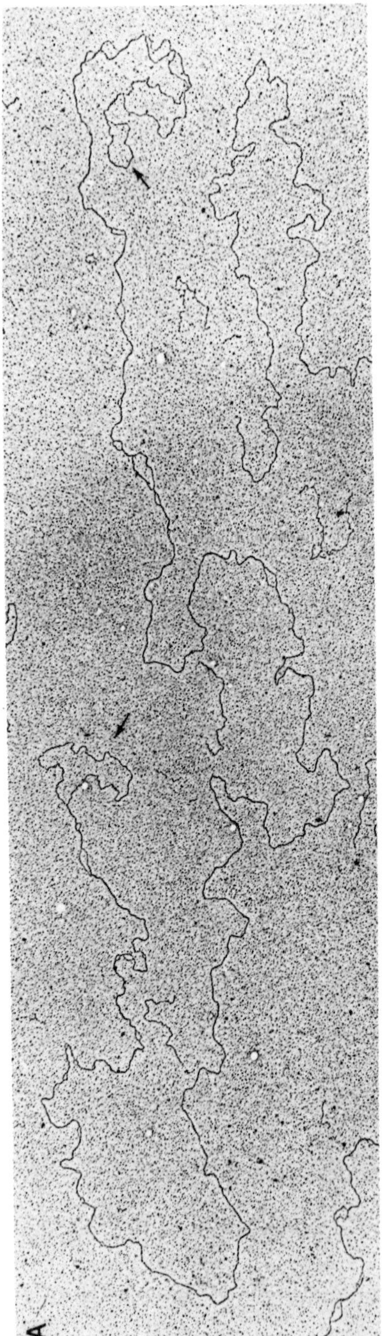
A

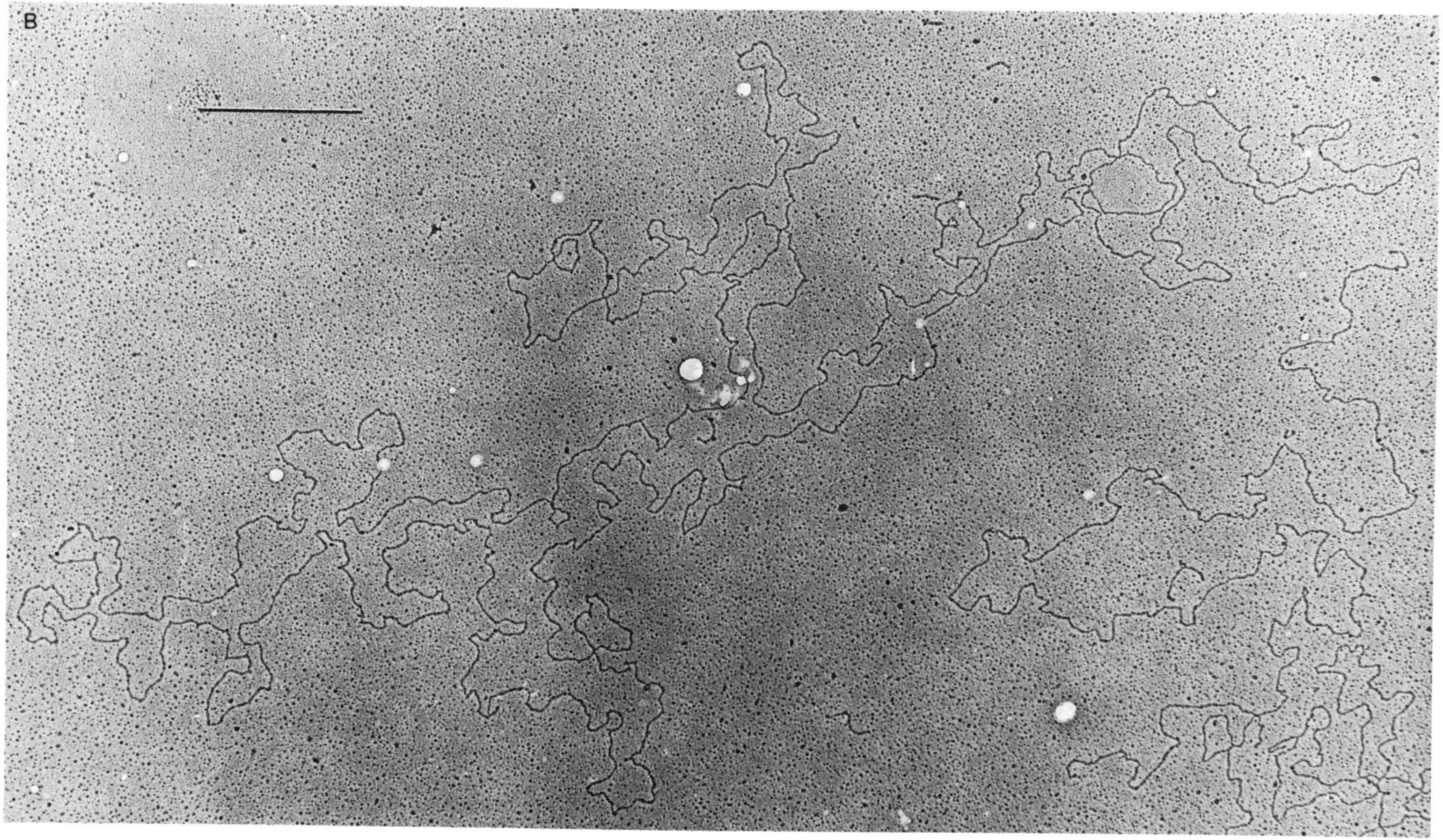
B

Fig. 14.2 Electron micrographs of vaccinia virus DNA. A, partially denatured with "bubbles" at ends; B, completely denatured with no visible free ends. Reprinted with permission from ref. [45].

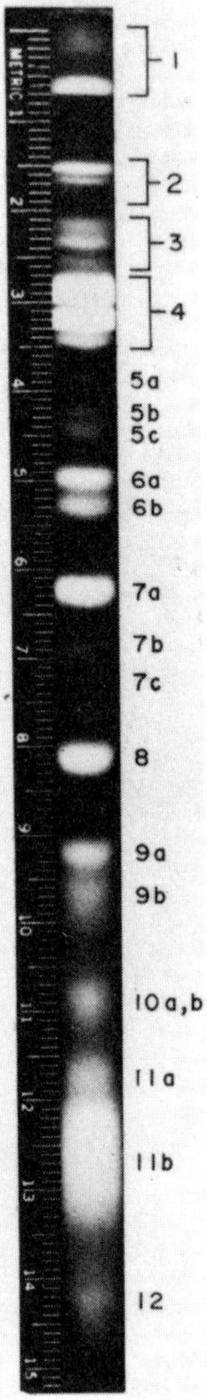

METRIC 1
1
2
3
4
5a
5b
5c
6a
6b
7a
7b
7c
8
9a
9b
10a,b
11a
11b
12

Polypeptide	MW	Percent
1a	200,000–	~0.5
1b	250,000	
1c	152,500	2.4
1d	145,000	
2a	94,000	2.2
2b	90,000	
2c	84,000	0.3
3a	78,500	2.0
3b	73,500	
3c	70,000	0.7
4a	63,000	13.8
4b	58,500	11.1
4c	56,000	2.4
4d	54,000	
5a	51,000	
5b	47,500	3.4
5c	46,000	
5P	46,750	<0.5
6a	41,000	9.1
6b	39,000	
7P	31,500	<0.1
7a	31,000	6.5
7b	27,000	1.0
7c	26,000	
8	23,000	7.0
9a	18,500	7.3
9b	17,000	
9P	17,000	<0.1
10a	16,000	10.5
10b	14,500	
11a	~11,800	1.5
11b	~11,000	11.4
12	~8,000	6.6
Total 30	$\sim 2 \times 10^6$	99.0

Fig. 14.3 Coomassie Brilliant Blue-stained vaccinia virus polypeptides separated by electrophoresis in polyacrylamide gels containing sodium dodecylsulfate. Molecular weights and approximate percentages of the total protein are indicated. Reprinted with permission from ref. [49].

guanidine hydrochloride or the detergent sodium dodecylsulfate. Polyacrylamide gel electrophoresis in dodecylsulfate separates polypeptides by molecular weight and is the single method providing the highest resolution. Depending on the percentage of acrylamide, the length of the gel, and the buffers, anywhere from 15-30 stained bands may be visualized from vaccinia [47-50] and fowlpox viruses [51]. The major polypeptides of vaccinia virus along with their estimated molecular weights and abundance are shown in Fig. 14.3. The molecular weights vary from approximately 200,000 to 10,000 or less. Assuming that all have unique primary sequences encoded by the viral genome, then about 20% of the coding capacity would be needed for their synthesis. Polypeptides 4a and 4b comprise about 25% of the total protein mass. Peptide analysis of these 2 polypeptides and of 3 polypeptides of group 6 resolved by dodecylsulfate-hydroxyapatite chromatography indicated that each has a unique amino acid sequence [50, 52].

2. *Modified Structural Proteins*

a. Glycoproteins. At least one [50, 53, 54] and possibly two [49] of the polypeptides of the 6 complex contain covalently bound sugar residues. The carbohydrate appears to be quite simple, containing only glucosamine [54].

b. Phosphoproteins. Several structural polypeptides are labeled when vaccinia virus is grown in the presence of ^{32}Pi [49, 55]. The major phosphoprotein has a molecular weight of 10,000-11,000, contains phosphoserine but no phosphothreonine [55], and is rich in arginine [56]. The site of phosphorylation is quite specific since only 1 tryptic phosphopeptide was detected [55].

3. *Localization of Structural Proteins*

As indicated in Fig. 14.4, at least 6 major polypeptides are released when vaccinia virions are treated with nonionic detergents and reducing agents [49, 50]. Some of these are on the exterior of the particle since they can react with specific surface labeling agents [49, 57] and are removed by digestion with proteolytic enzymes [49, 50], while others may lie beneath the envelope. The glycoprotein is susceptible to tryptic digestion [50] but does not react with some surface labeling agents [49], suggesting that it is just below the surface.

At least 12 major polypeptides are not released by treatment with nonionic detergents and reducing agents and remain associated with the core structure [49, 50]. These include the highest molecular weight polypeptides, the 2 most abundant polypeptides as well as the phosphoprotein (Fig. 14.4).

4. *Enzymes*

a. General Considerations. Ten enzymatic activities, at least 5 of which are clearly concerned with the synthesis and modification of RNA, have been located in vaccinia virus cores. They are listed in Table 14.2 and will be described in detail below.

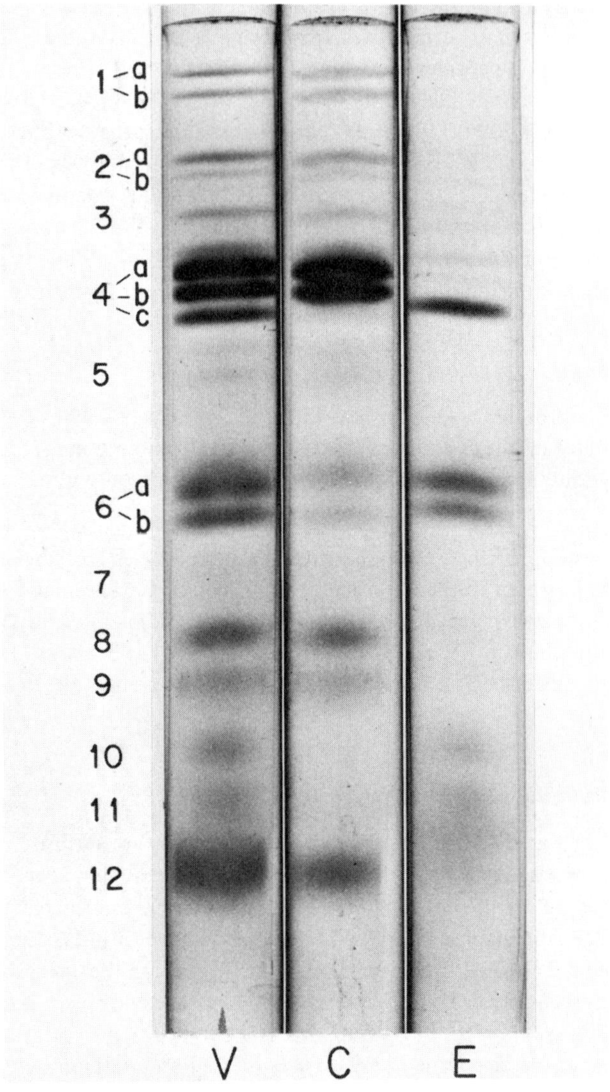
1 a
b
2 a
b
3
4 a
b
c
5
6 a
b
7
8
9
10
11
12
V
C
E

Fig. 14.4 Comparison of polypeptides from whole vaccinia virions and fractions obtained by detergent and mercaptoethanol treatment. Polypeptides were separated by polyacrylamide gel electrophoresis in sodium dodecylsulfate; V, virions; C, cores; E, envelope. Reprinted with permission from ref. [50].

Table 14.2 Vaccinia Virus Core Enzymes

Enzyme	In Vitro Reaction	MW[a]	Comment
RNA Polymerase	n NTP → RNA + n PPi		Not obtained in soluble or template-dependent form
Poly(A)Polymerase	n ATP + $N(-N)_m$ → $N(-N)_m(-A)_n$ + n PPi	80,000 (55,000; 30,000)	Specific for ATP donor; exhibits little acceptor specificity
RNA Guanylyltransferase	GTP + $(p)ppN(-N)_n$ → $GpppN(-N)_n$ + PPi		Specific for GTP donor; acceptor must have a diphosphate at 5′-terminus
RNA (Guanine-7-) methyltransferase	AdoMet + $GpppN(-N)_n$ → $m^7GpppN(-N)_n$ + AdoHcy	127,000 (95,000; 31,000)	Acceptor must be product of guanylyltransferase; appears to be associated with latter enzyme
RNA (Nucleoside-2′-) methyltransferase	AdoMet + $m^7GpppN(-N)_n$ → $m^7GpppN^m(-N)_n$ + AdoHcy		Not yet purified
Nucleoside triphosphate phosphohydrolase I	ATP + H_2O → ADP + Pi	61,000 (68,000)	Specific for ATP or dATP, requires DNA for activity
Nucleoside triphosphate phosphohydrolase II	NTP + H_2O → NDP + Pi	68,000 (68,000)	Not specific for ribo- or deoxyribonucleoside triphosphates; stimulated by DNA or RNA
Deoxyribonuclease (pH optimum 4.4)	DNA + H_2O → mono- and oligo-deoxyribonucleotides	105,000 (50,000)	Specific for single-stranded DNA; endo- and exonuclease activities
Deoxyribonuclease (pH optimum 7.8)	DNA + H_2O → oligodeoxyribonucleotides		Not yet purified; specific for single-stranded DNA
Protein kinase	ATP + protein → P-protein + ADP	62,000	Partially purified; phosphorylates serine or threonine residue of virion proteins; activated by protamine

[a]Obtained by gel filtration and/or sucrose gradient sedimentation; values in parenthesis obtained by polyacrylamide gel electrophoresis in sodium dodecyl sulfate.

b. RNA Polymerase. An RNA polymerase (RNA nucleotidyl transferase) has been detected in poxviruses, including those from rabbitpox [58], vaccinia [59], and Yaba [60] viruses, but not from any other family of DNA viruses. The enzyme is probably virus-specific since a sharp rise in RNA polymerase activity can be demonstrated late in infection [61, 62]. The polymerase activity can be detected simply by incubating purified virions with a reducing agent and nonionic detergents to disrupt the envelope, the 4 ribonucleoside triphosphates and Mg^{2+}. Under these conditions multiple rounds of synthesis occur and RNA with a sedimentation coefficient of 10-12S (approximately 1,000 nucleotides) is released from the virus particle. It has been estimated that there are about 40 RNA-chain growing points per particle and that the process of extrusion is ATP-dependent [63]. Only a portion of the viral genome corresponding to that expressed during the early part of the replicative cycle is transcribed in vitro. According to 1 report [63] this corresponds to 14% of the genome while according to another it corresponds to 50% [64]. The barrier preventing transcription of the entire genome is not known. One can imagine a situation in which the DNA is packaged within the virus core in such a manner that the portion corresponding to late genes is inaccessible to the RNA polymerase or one in which the sites for initiation of late RNA synthesis have sequences that are not recognized by the core-associated RNA polymerase. These possibilities cannot yet be evaluated since attempts to extract the viral RNA polymerase in a soluble form have been unsuccessful. Interestingly, when purified vaccinia virus DNA was used as a template for *Escherichia coli* RNA polymerase, synthesis was not restricted to the early sequences made by virus cores but late sequences were also transcribed [65]. The *E. coli* enzyme, however, transcribed the proper strand of the DNA suggesting that it recognized proper initiation signals.

The RNA synthesized in vitro by vaccinia virus cores has been used to stimulate protein synthesis in cell-free systems [64, 66, 67]. Although characterized in only a preliminary fashion, the products resemble early proteins made in infected cells.

c. Poly(A) Polymerase. RNA synthesized in vitro by vaccinia virus cores contains approximately 100 adenylate residues at the 3′ terminus [68, 69] attached either to cytosine or uridine residues [64]. The finding that there are no stretches of poly(dT) in vaccinia DNA long enough to act as a template for transcription of such poly(A) sequences [64, 70] and that the core-associated activities for RNA and poly(A) synthesis exhibit different characteristics [70, 71] suggest separate enzymes. Indeed, a poly(A) polymerase (polynucleotide adenylyltransferase or terminal adenylate transferase) with no apparent ability to synthesize RNA has been isolated from vaccinia virus cores [72]. The purified enzyme has a molecular weight of 80,000 and 2 polypeptides, possibly subunits, with molecular weights of 51,000 and 35,000 are detected by dodecylsulfate polyacrylamide gel electrophoresis [73]. The poly(A) polymerase catalyzes the addition of adenylate residues to the 3′ termini of short polyribo- or polydeoxyribonucleotide primers with little or no sequence specificity [73]. A rise in poly(A) polymerase activity has been detected in the cytoplasm of infected cells suggesting that the enzyme is virus-induced [74].

d. Guanylyl and Methyl Transferases. Very recently vaccinia virus mRNAs [75-77] as well as the mRNAs of other viruses and cells have been shown to contain 7-methylguanosine (m^7G) at the 5′ terminus. In the case of vaccinia virus, m^7G is connected from the 5′ position through a triphosphate to the 5′ position of an adjacent 2′-O-methyladenosine (A^m) or 2′-O-methylguanosine (G^m) residue [76]. This unusual structure is depicted in Fig. 14.5. Not only does this modification make the RNA resistant to the combined action of 5′-exonucleases and alkaline phosphatase [76], but in the case of reovirus and vesicular stomatitis virus, and presumably also vaccinia virus, it appears to be required for efficient translation of mRNA in vitro [78].

Three enzymatic activities responsible for the 5′ terminal modification have been isolated from vaccinia virus cores [79]. They include the following.

1. Guanylytransferase which transfers a GMP residue from GTP to the 5′ terminus of RNA containing a 5′ terminal di- or triphosphate
2. A methyltransferase, which transfers a methyl group from S-adenosylmethionine to the 7 position of the added guanine
3. A second methyltransferase which methylates the 2′ position of the ribose of the second nucleoside.

The first 2 activities appear to be associated in a single enzyme system with a molecular weight of 127,000 [80, 81]. Purified enzyme preparations contain 2 polypeptides with molecular weights of 95,000 and 31,400. The third activity is clearly a separate enzyme but has not yet been purified. Preliminary experiments (Boone, Ensinger, Martin and Moss, unpublished) suggest that the 5′ terminal modification enzymes are virus-induced.

e. Nucleoside triphosphate phosphohydrolases. An enzymatic activity that hydrolyzes nucleoside triphosphates to nucleoside diphosphates and inorganic phosphate has been demonstrated in several different poxvirions [60, 82-84]. Attempts to purify the enzyme were unsuccessful until it was discovered that nucleic acids were required for activity [85]. Two separate nucleic acid-dependent enzymes have been isolated [25, 86]. Phosphohydrolase I hydrolyzes ATP or dATP in the presence of DNA while II hydrolyzes

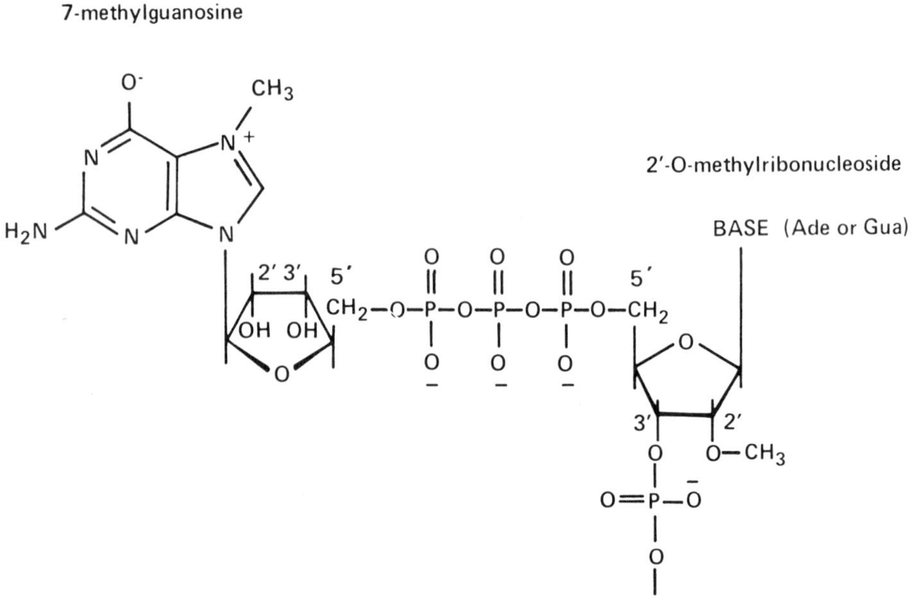
7-methylguanosine
O-
CH3
N
N+
H2N
N
N
2' 3'
5'
OH OH
O
O
O
O
CH2—O—P—O—P—O—P—O—CH2
O
O
O
5'
2'-O-methylribonucleoside
BASE (Ade or Gua)
O
3'
2'
O
O—CH3
O=P—O
O

Fig. 14.5 Modified 5′-terminal structures of vaccinia virus mRNA synthesized in vitro in the presence of S-adenosylmethionine.

all nucleoside triphosphates in the presence of DNA or RNA. Although both enzymes have molecular weights of about 68,000 they are immunologically unrelated [87]. Similar enzymes were not detected in uninfected cells but were induced after infection, suggesting that they are encoded by the virus genome [87, 88]. The functions of the viral phosphohydrolases are not known; some possibilities include the coupling of nucleotide hydrolysis with extrusion of RNA from the core, RNA chain termination, and virus assembly.

f. Deoxyribonucleases. Activities that specifically hydrolyze single-stranded DNA have been detected within the cores of several different poxviruses [21, 60, 84]. Two separate activities were suggested by biphasic pH optima with exonucleolytic activity at pH 5 and endonucleolytic activity at pH 7.8 [21, 89]. Thus far only 1 deoxyribonuclease has been extensively purified [90]. It is specific for single-stranded DNA, has a pH optimum of 4.4, and exhibits both endonucleolytic and exonucleolytic activities [91]. The native enzyme has a molecular weight of 105,000 and may be a dimer since a single polypeptide of 50,000 molecular weight was detected by dodecylsulfate polyacrylamide gel electrophoresis [90]. The deoxyribonuclease is able to cleave the hair-pin loops at the ends of vaccinia DNA but it is not known whether this is its biological role [45]. A possible function of the deoxyribonuclease active at neutral pH is to inhibit host DNA synthesis [92, 93]. Both the acid and neutral pH deoxyribonucleases are induced after infection suggesting that they are virus-specific [88, 94].

g. Protein Kinase. A protein kinase that transfers the γ-phosphate from ATP to serine or threonine residues of virion proteins has been detected in vaccinia virus cores [95, 96]. The kinase has been purified about 200-fold and has a novel feature in that protamine or other basic proteins are required as an activator for the phosphorylation of purified viral acceptor proteins [97, 98]. Cyclic nucleotides, which serve as activators for some cellular protein kinases, are ineffective. Although the protein kinase may be responsible for the phosphorylation of some virion proteins (Section IV, D and B), its biological function is obscure. The observation that protein phosphorylation precedes the linear synthesis of RNA when vaccinia virus cores are incubated in vitro suggests the intriguing possibility that there is a relationship between these events [95, 99]. No studies have yet been carried out on the induction of this enzyme after poxvirus infection.

E. Lipids

The poxviruses are lipid-containing enveloped viruses. In vaccinia virus, approximately 2.1% of the dry weight is phospholipid, 1.7% neutral fat, and 1.2% cholesterol [37]. The lipid composition is generally similar to that of the cell plasma membrane except for lower amounts of phosphatidylethanolamine and an increased amount of an unidentified material [100]. Much more striking differences are found in fowlpox virus in which approximately 34% of the dry weight is lipid [101, 102]. Cholesteryl ester, triglycerides, and free fatty acids make up a large proportion of the total, and squalene, an important intermediate in cholesterol biosynthesis present in only trace amounts in normal epithelial tissue, is abundant. An increase in the level of sterols in fowlpox-infected cells has been correlated with an elevation and alteration of the properties of 3-hydroxy, 3-methylglutaryl coenzyme A reductase, an enzyme considered to be rate-limiting in cholesterogenesis [103].

F. Trace Substances

Copper, riboflavin, and biotin are found in trace amounts in highly purified preparations of vaccinia virus [37]. The question of whether they are adsorbed impurities or cofactors of core-associated enzymes is unresolved.

V. Genetics

Conditional lethal temperature-sensitive mutants of rabbitpox virus and vaccinia virus have been obtained with the use of mutagens in several laboratories [104-107]. Twenty rabbitpox mutants that were unable to grow at 39.5°C all synthesized viral DNA but some exhibited impaired viral protein synthesis at the restrictive temperature [108]. The majority of these rabbitpox mutants appeared to be in different cistrons since complementation and recombination were observed. Of 15 vaccinia virus mutants unable to grow at 40°C, the temperature-sensitive stage occurred in the early phase of replication prior to DNA synthesis in 7 and at a later phase of replication in the remaining 8 [107]. By means of complementation tests, these mutants were divided into 3 early DNA^- and 2 late DNA^+ groups.

Host range rabbitpox mutants able to replicate in chick embryo fibroblasts but not in certain mammalian cell lines have been isolated [104, 109]. These so-called p mutants are blocked at various stages of replication but map very closely since they are unable to complement or recombine with each other [110]. The p mutants are actually a subclass of u mutants. The latter form white instead of red pocks on the chorioallantoic membrane and on the basis of pairwise crosses have been mapped in a linear order [111].

Additional poxvirus mutants that have been isolated include thymidine kinase negative (Section IX.B), rifampicin-resistant [112-114] and isatin β-thiosemicarbazone resistant [115-117] and dependent [117] types.

VI. Growth Cycle

Poxviruses may produce a rapidly lethal disease or a mild infection, the outcome depending on the virulence of the virus strain as well as the susceptability of the host. An account of the use of rabbitpox virus to control the rabbit population in Australia and the resultant mutual adaptation of both makes fascinating reading [118]. The ability of some poxviruses to produce benign tumors is particularly evident in the case of Shope fibroma [119, 120] and Yaba tumor [121, 122] viruses which infect rabbits and monkeys, respectively. Virus replication occurs in cells comprising the tumors which eventually regress by necrosis.

The growth cycle of poxviruses may be conveniently studied in a variety of tissue culture systems and a one-step growth curve exhibiting a definite eclipse phase can be demonstrated [125-126]. Under conditions employed for the majority of experiments in the following sections, vaccinia virus DNA replication begins in 1.5-2 hr and is nealry completed by 5 hr after infection; progeny virions are detected at 4-5 hr and a yield of about 10,000 particles per cell is achieved at 12-15 hr. Slower rates of development may occur with other viruses.

VII. Entry into Cells and Uncoating

Adsorption of vaccinia virus particles to the cell membrane is a rapid process and more than half of the inoculum may be attached within 15 min [127]. Breakdown of envelope phospholipid and dissociation of about 50% of the protein of the virus particle occurs without an appreciable lag [127]. Electron microscopy indicated that attachment of vaccinia virus occurs without specific orientation of the long and short axes and that particles are engulfed by phagocytosis [128]. The outer coat of the virion is then disrupted within the vacuole and the virus core lacking lateral bodies passes into the cytoplasmic matrix [128]. In Shope fibroma virus-infected cells, many particles are found in lysosomes [129]. Some investigators have presented evidence that at early times after vaccinia infection entry occurs by fusion of the envelope of the virus particle with the cell membrane, a mechanism that would lead to the direct transfer of the core to the cytoplasm, rather than by phagocytosis [130]. Penetration via membrane fusion has also been reported for an insect poxvirus [131].

Particles considered to be intermediate stages in the disassembly process have been isolated by sucrose gradient sedimentation [132-134]. The viral DNA is released from the core after a period of 0.5-2 hr [127, 135, 136] by an active process that is prevented by inhibitors of RNA and protein synthesis or by UV irradiation of the inoculum. Since uncoating occurs in infected enucleate cells [137-139], the requirement is presumably for viral RNA and protein synthesis. Although the existence of a specific uncoating protein is unresolved [140, 141], some electron microscopic images suggested that the DNA is released by rupture of the core wall [142, 143].

VIII. Transcription

A. Early Viral RNA

Since poxvirions contain enzymes for the synthesis, polyadenylation, and 5′-terminal modification of RNA (Section IV.D.4), functional mRNAs can be formed immediately after infection. Indeed, detection of the rapid onset of viral RNA synthesis even in the presence of inhibitors of protein synthesis [144-146] led to the postulation and eventual discovery of transcriptase activity within poxvirus cores. Two techniques are generally used to measure viral RNA synthesis in infected cells. The simplest takes advantage of the cytoplasmic site of poxvirus replication and the additional time required for the transfer of host mRNA and rRNA from the nucleus to the cytoplasm. By restricting the period of incubation with radioactively labeled ribonucleosides to 10 min and then carefully breaking the cell and removing nuclei, viral RNA significantly contaminated only with low molecular weight tRNA can be obtained [147]. The latter can be removed either by sucrose gradient sedimentation, by polyacrylamide gel electrophoresis or by adsorption of the polyadenylated RNA to poly(U). The early RNA sediments in sucrose gradients primarily as 10-12S species [148] and migrates on polyacrylamide gels as multiple partially resolved peaks [149]. The second method involves specific hybridization of RNA to viral DNA, either immobilized to filters [148] or free in solution [150]. Both the pulse labeling and hybridization methods provide evidence for a rapid burst of viral RNA synthesis during the first hour of infection. The amount of early viral mRNA made is dependent on the multiplicity of infection [146, 148] as expected if each virus particle is a separate transcriptional unit. In rabbitpox-infected chick embryo cells, RNA made early in infection or in the presence of cytosine arabinoside, an inhibitor of DNA synthesis, hybridizes with 17-20% of the viral DNA corresponding to 34-40% of the double-stranded genome [150]. There is some controversy concerning the percentage of the viral genome transcribed in the absence of protein synthesis. Competition hybridization experiments indicated that fewer vaccinia virus RNA sequences are made in the presence of an inhibitor of protein synthesis than in the presence of an inhibitor of DNA synthesis [58]. This study suggested that some early RNA species are made by cores before viral proteins are synthesized but that additional early RNA species are made following protein synthesis possibly after the core has been uncoated. In contrast, other investigators [150] found that RNA made in the presence of inhibitors of protein synthesis converted 28-33% of rabbitpox DNA to a hybrid form whereas 17-20% was converted using RNA made in the presence of an inhibitor of DNA synthesis. Perhaps more detailed studies on the relative abundance of different early mRNA species will resolve these apparent discrepencies.

Some experiments suggest that uncoating of the virus particle plays a role in the control of RNA synthesis. Under the usual conditions of infection, transcription appears biphasic with the relative rates of early and late RNA synthesis being dependent on the multiplicity of infection. The synthesis of early RNA proceeds in an uncontrolled manner for an extended period of time in the presence of an inhibitor of protein synthesis [144-145], interferon [151], or when the inoculum virus has been UV-irradiated [145, 146]. A common effect of all of these procedures is to prevent uncoating of the virus core, suggesting that dissociation of some element of the transcription complex normally accompanies this step. Both uncoating of virus cores and a decline in early RNA synthesis occur in the presence of inhibitors of DNA synthesis [146].

B. Late Viral RNA

Late RNA is made following the onset of DNA replication and is prevented by inhibitors of DNA synthesis. Competition hybridization experiments indicate that early vaccinia virus RNA competes with only half of the RNA made at late times whereas the late RNA competes with all of the early RNA [148]. Similarly, early rabbitpox virus RNA hybridized with 17-20% of the rabbitpox DNA, late RNA hybridized with 40-42% and no additional hybridization was obtained by mixing early and late species [150]. Additional experiments are needed to determine accurately the relative amounts of early sequences made at late times. Late RNA also appears to differ from early RNA in modal size. It sediments at 16-20S rather than 10-12S [148, 152] and migrates more slowly on polyacrylamide gels than early RNA [149, 153]. A small percentage of the RNA made at early and late times contains complementary sequences perhaps arising from overlapping transcription of opposite stands [154].

Late RNA sequences are presumably copied from progeny DNA, accounting for the requirement for DNA synthesis. DNA synthesis alone may not be sufficient, however, since DNA but not late RNA is made under certain conditions of arginine deprivation [155]. It is not known whether additional protein factors are required to modify the core-associated RNA polymerase or whether a completely different enzyme system is used for late transcription. The continued production of early RNA late in infection could take place in cores that have not been uncoated or on progeny DNA. Large aggregated complexes containing newly formed DNA and proteins have been isolated from vaccinia virus-infected cells [156]. These complexes synthesize both early and late RNA and should provide a useful system for further studies.

C. Modification of Viral RNA

RNA made at both early and late times after vaccinia virus infection contains about 100 adenylate residues at the 3′ terminus [64, 69] as does RNA made by virus cores in vitro (Section IV.D.4c). Presumably the poly(A) polymerase within the virus core is initially utilized for modification of RNA made immediately after infection. Additional poly(A) polymerase is induced during the first few hours [74, 157]. Preliminary experiments (Boone and Moss, unpublished) indicate that the mRNA made in infected cells has methylated 5′ termini as does RNA made by vaccinia virus cores in vitro (Section IV.D.4d).

D. Stability of Viral RNA

The stabilities of vaccinia and rabbitpox viral mRNAs have been measured following the addition of actinomycin D to stop further RNA synthesis. Assuming that this drug does not introduce additional complications, the half-life of early vaccinia virus mRNA as measured by hybridization to DNA is 2-3 hours in L cells and considerably longer in HeLa cells [148]. Using a functional assay, i.e., synthesis of vaccinia virus-induced antigens in actinomycin D-treated HeLa cells, a half-life of several hours was also obtained [158, 159]. However, the half-lives of mRNAs for thymidine kinase and early-induced deoxyribonucleases appear to differ [160].

Vaccinia virus mRNA in HeLa cells is less stable at late times than at early times. A value of 13 min was obtained by measuring the synthesis of viral antigens after addition of actinomycin D [160] and less than 60 min by DNA-RNA hybridization [148]. It is not clear whether the difference at early and late times reflects the intrinsic stabilities of the RNAs or a change in degradative activities within the cell.

E. Alterations of tRNA

RNA, the size of tRNA, continues to be made after vaccinia virus infection [147, 152]. Chromatographic analysis indicated an alteration in the elution of phenylalanine tRNA, quantitative changes in aspartyl tRNA [161], and increased methylation of tRNAs [162]. No evidence that the poxvirus genome codes for tRNAs has been presented.

IX. Protein Synthesis

A. Identification of Early and Late Viral Proteins

Immunodiffusion and immunoelectrophoresis with specific antisera have been used as analytical tools to resolve proteins labeled with radioactive amino acids at various times after infection and to distinguish viral proteins from those of the host [163, 164]. This procedure served to identify an early class of proteins that were synthesized exclusively during the first few hours of vaccinia virus infection or in the presence of inhibitors of DNA synthesis and a late class of proteins synthesized exclusively after the onset of DNA replication. Development of the technique of polyacrylamide gel electrophoresis in sodium

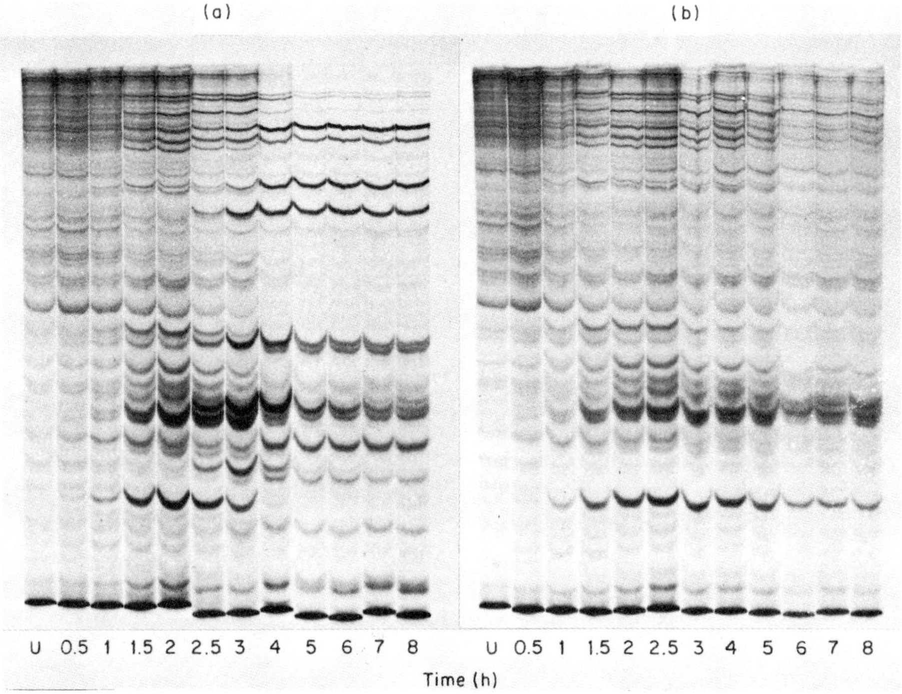
(a)
(b)
U 0.5 1 1.5 2 2.5 3 4 5 6 7 8
U 0.5 1 1.5 2 2.5 3 4 5 6 7 8
Time (h)

Fig. 14.6 Autoradiograms showing time of course of polypeptide synthesis in vaccinia virus-infected cells in the absence (left) and presence (right) of cytosine arab; noside, an inhibitor of DNA synthesis; U, uninfected cells. Data from ref. [166].

dodecylsulfate coupled with radioautographic analysis and the appreciation that host protein synthesis is rapidly turned off (Section XII.B) led to the enumeration of many more virus-induced proteins [164, 165]. With further modification of the electrophoresis buffer system, claims have been made for the reproducible detection of about 80 virus-induced polypeptides, of which 30 appeared before and 50 after the onset of DNA synthesis [166]. Radioautographs showing the induction of early and late proteins are presented in Fig. 14.6, (A). With inhibitors of DNA synthesis, the early labeling pattern is indistinguishable from that of untreated infected cells but no new polypeptides appear after 2 hr and the abrupt switch off of early polypeptide synthesis fails to occur (Fig. 14.6, part B). There is evidence for 3 classes of early proteins: 1 for which synthesis starts immediately after infection, continues for a period and then switches off; another for which the rate of synthesis builds up slowly; and a third which continues to be made for prolonged periods [165]. Two classes of late proteins have been described: 1 which begins at the time of maximal DNA synthesis, reaches a peak shortly thereafter, and then declines; and another which appears slightly later and continues throughout the period of virus growth [164, 166]. The induction of both early and late vaccinia virus polypeptides after infection of enucleated cells provides the best available evidence that they are virus coded [139].

B. Virus-Induced Enzymes

1. *Early*

Some of the enzymes induced early after poxvirus infection are involved in DNA synthesis. Large increases in thymidine kinase, an enzyme that catalyzes the transfer of a phosphate from ATP to thymidine, are detected within 1-2 hr after infection with vaccinia, rabbitpox, and cowpox viruses, and plateau at 4-6 hr [167-169]. The best evidence that thymidine kinase is virus-coded is its induction in thymidine kinase negative (TK^-) cells and the isolation of TK^- vaccinia virus [170, 171]. In addition, protein synthesis is required for kinase induction [172, 173] and host and virus-induced thymidine kinases have different enzymatic, immunological, and electrophoretic properties [174-176]. The vaccinia virus-induced thymidine kinase is able to use ATP but not UTP, CTP, or GTP as a phosphate donor [176]. The finding that the enzyme is feedback-inhibited by its distal product, deoxythymidine triphosphate, suggests that it may be a complex enzyme of the allosteric type [141]. Although the function of thymidine kinase is presumably to raise the thymidylic acid pool for viral DNA synthesis, TK^- vaccinia virus mutants grow well in TK^+ or TK^- tissue culture cells.

DNA polymerase (DNA nucleotidyl transferase) also increases during the early phase of vaccinia virus infection [172]. The induced enzyme differs from the DNA polymerase of uninfected cells with regard to chromatographic properties, molecular weight, heat stability, and immunological properties [177-179]. The partially purified DNA polymerase has a molecular weight of about 110,000, can only repair gaps in activated DNA primers, and may have an associated DNA exonuclease activity [179]. Although the DNA polymerase can bind adventitiously to vaccinia virus particles it is not an integral component [180]. Evidence for activities which utilize double- and single-stranded DNA primer-templates has been reported in Shope fibroma virus-infected cells [181].

Polynucleotide ligase, an enzyme that seals single-strand breaks in duplex DNA and is essential for DNA replication in prokaryotes, increases 10-fold in activity during the early period of vaccinia virus replication and in the presence of cytosine arabinoside [182]. The induced activity, which has the same ATP cofactor requirement as cellular ligase, has not yet been purified or extensively characterized. The absence of enzyme induction in the presence of puromycin suggests that the polynucleotide ligase is newly synthesized.

A deoxyribonuclease active at alkaline pH is induced early in infection [183-185]. The enzyme has a preference for double-stranded DNA and has been characterized as an exonuclease. Induction of the enzyme is not prevented by inhibitors of DNA synthesis but protein synthesis is required.

The ability of rabbitpox and vaccinia viruses to grow in a human citrullinemia cell line when citrulline was substituted for arginine as well as the enhanced utilization of citrulline in other cell lines after poxvirus infection suggested that enzymes involved in arginine biosynthesis might be induced [186]. Indeed, enhanced argininosuccinate synthetase and arginosuccinate lyase activities increase markedly after vaccinia virus infection [187]. Induction is not prevented by inhibitors of DNA synthesis, but both RNA and protein synthesis are required.

Poly(A) polymerase activity is maximally induced 3.5 hr after vaccinia virus infection and is not prevented by inhibitors of DNA synthesis [74]. The induced enzyme was isolated as an RNA-protein complex [157] and it is not yet certain that it is identical to the poly(A) polymerase extensively purified from vaccinia virus cores (Section IV.D.4c).

An activity that polymerizes ATP and UTP in the presence of poly(dAT) is detected after vaccinia virus infection only if DNA synthesis is prevented [62]. It is not known whether this activity is related to the DNA-dependent RNA polymerase.

2. *Late*

The enzymes that are induced late in infection include the acid [94] and neutral [88] pH optimum deoxyribonucleases and both nucleoside triphosphate phosphohydrolases [87, 88]. They are all components of the virus core and have been described in detail in Section IV.D.4. RNA polymerase activity also is detected late [61, 62] but since a template-free form has not been isolated the time of synthesis of the enzyme cannot be determined.

C. Structural Proteins

The time of synthesis of structural proteins has been estimated by pulse labeling infected cells at various times after infection and then either directly analyzing the labeled proteins by sodium dodecylsulfate polyacrylamide gel electrophoresis or by first washing the cells, resuspending them in media containing unlabeled amino acids, and isolating the virus particles at the end of the growth cycle. Both procedures provide evidence for a small number of early and a large number of late structural proteins [50, 163, 188, 189]. The early vaccinia virus structural polypeptides include 1a, 1b, and some components of the 6 complex (Fig. 14.7).

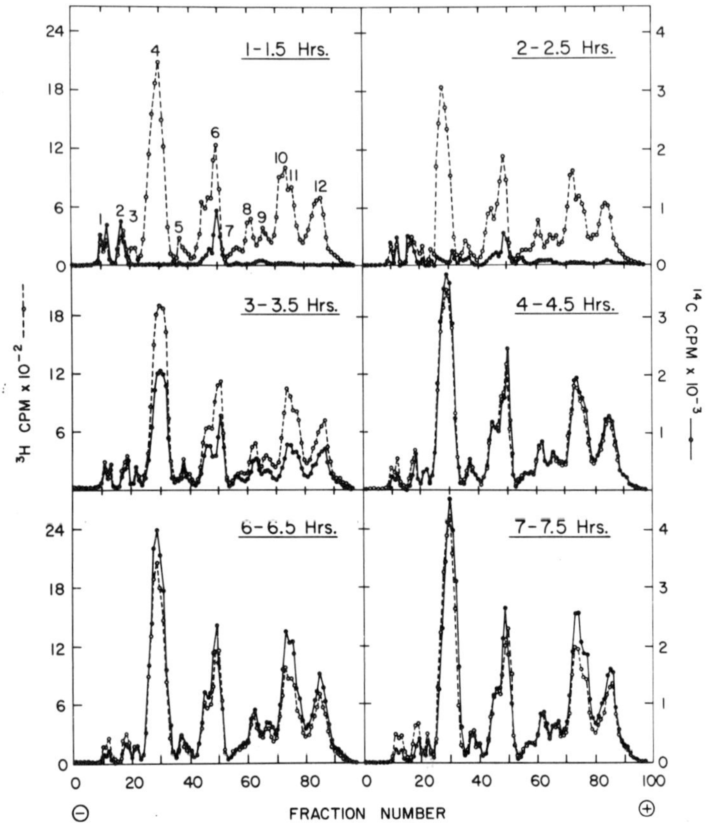
1-1.5 Hrs.
2-2.5 Hrs.
3-3.5 Hrs.
4-4.5 Hrs.
6-6.5 Hrs.
7-7.5 Hrs.
^{3}H CPM x 10^{-2}
^{14}C CPM x 10^{-3}
FRACTION NUMBER

Fig. 14.7 Electropherograms of early and late vaccinia virus structural proteins. Infected cells were labeled for the indicated 30 min intervals with [^{14}C] phenylalanine or for 7 hr with [^{3}H] phenylalanine. At the end of the growth cycle, each [^{14}C]labeled sample was mixed with [^{3}H] labeled cells and virus particles were purified and analyzed by polyacrylamide gel electrophoresis in sodium dodecylsulfate. Reprinted with permission from ref. [50].

D. Posttranslational Modification

1. *Cleavage*

At least 2 types of posttranslational processing are involved in the formation of poxvirus proteins: peptide cleavage and amino acid modification. Formation of the 2 major structural proteins from higher molecular weight precursors has been demonstrated by "pulse chase" type experiments [189, 190] as indicated in Fig. 14.8 (A). The precursor relationship of P4a (molecular weight 105,000) to structural polypeptide 4a (68,000), and of P4b (74,000)

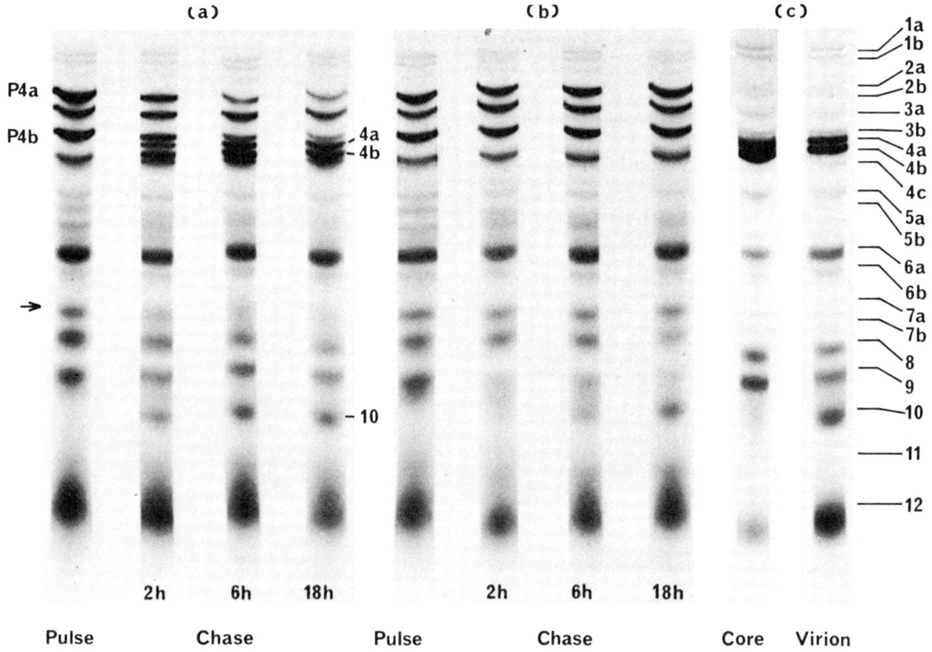
(a)
(b)
(c)
P4a
P4b
4a
4b
10
1a
1b
2a
2b
3a
3b
4a
4b
4c
5a
5b
6a
6b
7a
7b
8
9
10
11
12
2h
6h
18h
2h
6h
18h
Pulse
Chase
Pulse
Chase
Core
Virion

Fig. 14.8 Autoradiograms showing the formation of some vaccinia virus structural proteins from higher-molecular weight precursors. HeLa cells were infected in the absence (left) or presence (center) of rifampicin and incubated with [^{35}S]methionine for 20 min at 6 hr after infection. Cytoplasmic proteins obtained at the end of the labeling period and at intervals thereafter were analyzed by polyacrylamide gel electrophoresis. Reprinted with permission from ref. [52].

to 4b (62,000), has been established by tryptic peptide mapping [52]. The disappearance of an additional polypeptide indicated by an arrow and the appearance of polypeptide 10 during the chase are also evident in Fig. 14.8 (A). Evidence for several more precursor-product relationships has been obtained using a higher resolution electrophoretic system [166]. The precursor polypeptides P4a and P4b have a half-life of 1-2 hr and their cleavage appears to be coupled to assembly of virus cores [189, 190]. Evidence for the latter concept was obtained with the use of the drug rifampicin. Rifampicin interferes with the assembly of poxvirus particles (Sections XI and XIV) and under these conditions P4a and P4b are synthesized but not processed (Fig. 14.8, part B; [52, 189, 190]). When rifampicin is removed, the delay prior to cleavage is similar to the time required for virus particles to reach the core stage [191]. In contrast polypeptide 10, which is not associated with the virus core, is formed from a precursor in the presence of rifampicin (Fig. 14.8, part B).

2. *Amino Acid Modification*

Two types of amino acid modification, addition of sugar and phosphate residues, have been recognized. Evidence for the induction of several nonvirion glycoproteins that associate with cell membranes has been detected by labeling infected cells with radioactive glucosamine, fucose, and galactose [192, 193]. Induction requires viral protein synthesis but not viral DNA synthesis, indicating that at least some fall into the early category. An induced glycoprotein, tentatively identified as a component of the vaccinia hemagglutinin, is associated with the plasma membrane in the absence but not the presence of inhibitors of DNA synthesis [193]. The glycoprotein component of the virion has been described in Section IV.D.2a. There is no information regarding involvement of host or virus-induced enzymes in glycosylation.

Virion phosphoproteins are described in Section IV.D.2b.

E. Regulation of Early and Late Protein Synthesis

The existence of early and late classes of mRNA suggests that the onset of early and late protein synthesis is under transcriptional control. There is a possibility of 2 classes of early mRNAs (Section VIII.A), which could code for subclasses of early proteins, but no efforts as yet have been made to subclassify late mRNAs. The mechanisms affecting the switch off of early protein synthesis following DNA replication are not well understood. DNA-RNA hybridization experiments indicate that RNA with both early and late sequences are still present in polyribosomes at 8 hr after infection but that there is a significant reduction in the early sequences [148]. A possibility of translational regulation has come from studies which showed that the shut off of thymidine kinase formation is prevented by inhibitors of RNA and protein synthesis [194-196]. Further elucidation of these matters will require accurate measurements of the abundance of specific mRNAs at various times of infection and their translation in cell-free systems.

X. DNA Replication

In cells synchronously infected with vaccinia virus, DNA synthesis begins within 1.5-2 hr, peaks between 2-3 hr and is nearly completed by 5 hr [197, 198]. An early increase in several enzymatic activities, including thymidine kinase, DNA polymerase, and DNA ligase, has been discussed (Section IX.B.2). Presumably, additional proteins concerned with DNA replication remain to be identified. The abrupt cessation of poxvirus DNA synthesis at a time when progeny virus particles are just beginning to form is unique since DNA synthesis continues during the period of virus maturation with other groups of viruses. This regulation of poxvirus DNA synthesis may be related to the switch off of early protein synthesis since DNA synthesis can be interrupted by addition of puromycin [199] or to the accumulation of some late proteins.

Poxvirus DNA replication occurs within discrete cytoplasmic foci, the so-called factory areas [200, 201]. The cell nucleus appears not to be required since vaccinia virus DNA is replicated in enucleated cells [137-139]. Mature virus particles are not formed in enucleated cells, however, despite the synthesis of early and late RNA and proteins, albeit at reduced levels [139], leaving open the possibility of a nuclear function in poxvirus reproduction. Whether the latter function is related to recent reports [202, 203] of some poxvirus DNA replication in nuclei of infected cells remains to be determined. Virtually nothing is known about the biochemical mechanisms of poxvirus DNA formation, a fascinating subject in view of the unusual cross-linked structure of vaccinia DNA (Section IV.B).

XI. Virus Assembly and Maturation

The initial stages of virion formation occur in the circumscribed, granular, electron-dense factory areas of the cytoplasm. Detailed investigations by electron microscopy of thin sections of cells infected with vaccinia [15, 22, 204], variola [205], fowlpox [204, 206], Yaba tumor [207, 208], milker's nodule [209], and insectpox [29, 210] viruses suggest a common scheme of morphogenesis. Notable features of this process include the formation of viral envelopes with no obvious association with cellular membranes [211] and the differentiation of internal structures within the immature envelope. For convenience the developing particles have been grouped in 6 stages [207]. Type I particles are cupule-shaped segments appearing in thin section as short crescents or arcs composed of a typical membrane bilayer with a short spicule coat on the convex surface; type II are spheroidal; type III have electron-dense granular material or a distinct nucleoid within the cavity; type IV, which are very rare, have an elongated shape and some internal membrane structure; type V have recognizable cores and lateral bodies; and type VI appear mature.

The assembly process may be interrupted by addition of certain drugs. In the presence of rifampicin, vaccinia virus membranes appear at the usual time, but they appear irregular in contour and completely lack the external spicule coat [212-215]. Small segments of uncoated membrane are also found during Yaba virus infection in the absence of inhibitors [208] and after addition of actinomycin D to vaccinia virus-infected cells [211]. After rifampicin is removed from the medium, focal accretions of coat material

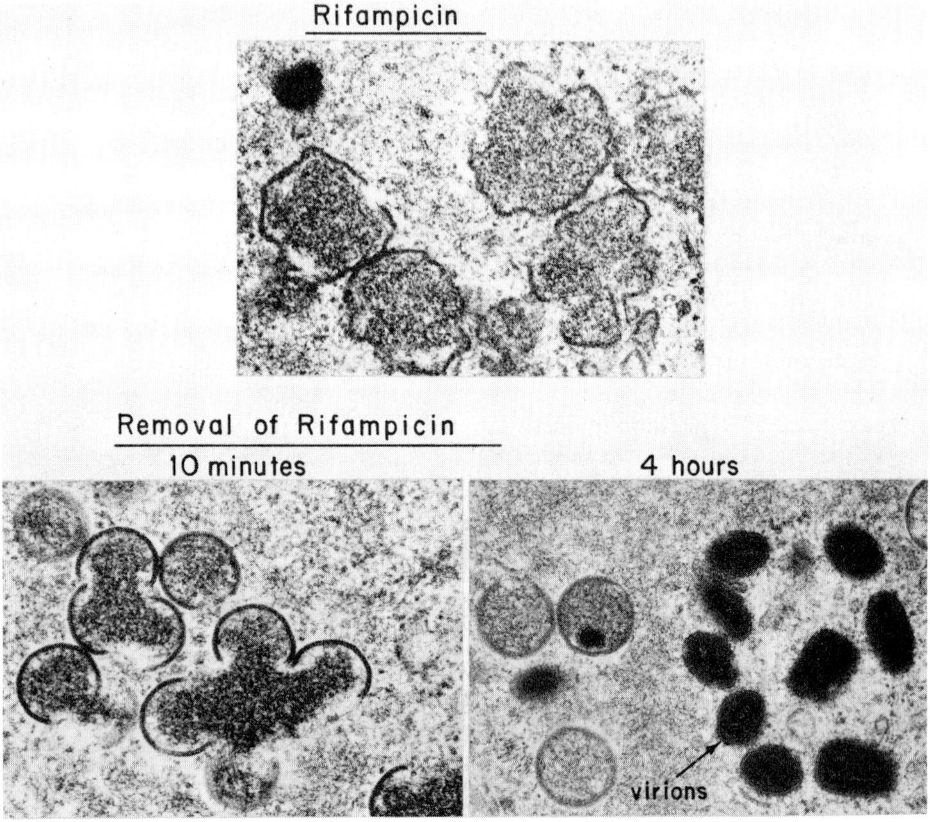

Rifampicin
Removal of Rifampicin
10 minutes
4 hours
virions

Fig. 14.9 Electron micrographs of uncoated vaccinia virus membranes formed in the presence of rifampicin and immature and mature virus particles formed after the removal of the drug. Reprinted with permission from ref. [212].

are detected on the surface of some membranes within 1 min most membranes within 2 min and virtually all within 5-10 min (Fig. 14.9; [213]). Coated segments of membrane always have a regular curvature, suggesting that the spicule layer affects the conformation. In the presence of inhibitors of DNA synthesis maturation is reversibly blocked at the type II particle stage [216-218].

Study of the assembly process in biochemical terms is not far advanced. Cytoplasmic aggregates containing newly replicated vaccinia virus DNA have been isolated [198, 219-222]. These structures contain 2 principal polypeptides with molecular weights of about 35,000 and 30,000. The latter is a phosphoprotein but neither correspond to a major virion structural protein. DNA associated with the virion phosphoprotein has been obtained by other procedures [56]. More mature forms containing some but not all of the structural proteins have also been identified [219, 221]. The possible occurrence of polypeptide cleavage during the assembly process has been mentioned (Section IX.D.1).

XII. Dissemination of Poxviruses

Results obtained in several cell culture systems have indicated that not more than 10% of the total yield of vaccinia virus is released into the medium [123-126]. Electron microscopic studies [15, 223] suggest that as mature virus particles migrate away from factory areas, some become enclosed within double-membrane cisternae apparently derived from Golgi membranes. The wrapped particles may then migrate toward the surface where the outer walls of the cisternae fuse with the plasma membrane and particles, either naked or enclosed by the inner cisternal membrane, are released. Other virus particles appear to migrate directly to the plasma membrane. The percentage of wrapped and unwrapped forms varies with the strain of vaccinia virus [223]. With some poxviruses, including cowpox, ectromelia, and fowlpox, there are dense proteinaceous masses in the infected cell cytoplasm called A-type inclusions [223-225] which contain mature virions and virus-induced antigens with molecular weights of 27,000 and 28,000 [226]. Integration of virions into the A-type inclusions is genetically controlled [227]. Presumably the inclusions are released as packets of virions after cell lysis.

Intercellular fusion [228, 229], which occurs after infection with some poxviruses, could lead to direct cell-to-cell spread. Such a mode of dissemination is consistent with the formation of virus plaques in cell monolayers in the presence of high titer antiserum [230].

Since most poxvirus particles remain intracellular at the end of a 1-step growth cycle, virus artificially released by cell disruption has been most frequently studied. Although naturally and artificially released particles are infectious, they differ in the nature of their surface antigens, apparently resulting from the presence of additional coat material acquired during passage of virus out of the cell [231-233]. This difference in antigencity has important implications for assessing immunity to poxvirus infection and for the development of inactivated vaccines.

XIII. Effects on Host Cells

A. Modification of Cell Membranes

Changes in cell surface antigens after poxvirus infection have been demonstrated by immune hemadsorption, fluorescent antibody, and ^{125}I-labeled antibody techniques [234-236], as well as by the incorporation of virus-induced glycoproteins into cell membranes [192, 237]. In addition, a hemagglutinin induced after infection with certain poxviruses appears to reside on the plasma membrane of infected cells [115, 238, 239]. Other evidence of cell surface alterations is the agglutination with concanavalin A [240, 241], fusion [120, 229, 242], density changes [243], and cytopathic alterations [244-248] of infected cells. Except for incorporation of the hemagglutinin and cell fusion, these changes occur early in infection and their biological significance is unknown. It would seem possible that surface changes might limit superinfection or be involved in the release of virus particles from the cell.

B. Inhibition of Macromolecular Synthesis

Host protein synthesis is inhibited after vaccinia virus infection [163-165]. The time required for inhibition is inversely related to the titer of the inoculum, occuring within 30 min at high multiplicities. The reduction in cell protein synthesis could be established in the presence of actinomycin D [249, 250] and cordycipin [165], both of which inhibit RNA synthesis, suggesting that a component of the virus particle might be responsible. Although infection with vaccinia virus in the presence of actinomycin D leads to a decrease in the size and amount of host polyribosomes, there is no change in the stability of the host mRNA [251]. The situation with actinomycin D may be anomalous, however, since very short, poly(A)-containing polyribonucleotides are made by the virion-associated enzymes [251]. An interesting feature of the inhibition of protein synthesis during a productive infection is its apparent selectivity for host mRNA. Such a result would be a natural outcome if the mechanism included competition of viral mRNA with host mRNA for limiting elements of the translation apparatus. Some inhibitor studies are also consistent with the synthesis of new translation factors after vaccinia virus infection [153].

Synthesis of high molecular weight nuclear RNA is reduced several hours after vaccinia virus infection and is preceded by a block in the transport of RNA to the cytoplasm [147, 152]. More detailed studies indicated that by 2 hr, the processing of precursor rRNA and the maturation of ribonucleoprotein particles were slowed [252]. The synthesis of ribosomal proteins was depressed prior to that of rRNA suggesting a possible cause and effect. It is of interest that synthesis of tRNA continues after vaccinia virus infection (Section VIII.E).

Inhibition of nuclear DNA synthesis also follows poxvirus infection [253-256]. Evidence suggesting that the inhibition results from degradation of nascent DNA by a single-strand specific deoxyribonuclease with a neutral pH optimum derived from the inoculum virus has been presented [92, 93]. Stimulation of host DNA synthesis and cell proliferation occur in cells persistently infected with Shope fibroma virus [257, 258] and transiently after fowlpox infection [259, 260].

XIV. Specific Inhibitors of Poxvirus Reproduction

Two substances, isatin β-thiosemicarbazone (IBT) and rifampicin (also referred to as rifampin), have been shown to be specific inhibitors of poxvirus replication, and a derivative of the former has been used for prophylaxis of smallpox.

IBT [261-263] is clearly virus-specific since both resistant [115-117] and dependent [117, 264, 265] mutants have been isolated. Although the structure-activity relationships of β-thiosemicarbazones have been extensively studied [266, 267], the mode of action of the drug is not understood. Early steps in replication are not affected and vaccinia virus DNA is synthesized [266, 268, 269]. Reduced synthesis of several late antigens [270] and a short half-life for late mRNA [271] suggest that late protein synthesis is the target of the drug. Other workers [268, 272], however, have detected synthesis of both early and late viral polypeptides and suggest that the drug inhibits maturation. The specific antipoxviral activity of rifampicin [112, 273] has also been demonstrated by the isolation of drug-resistant mutants [112-114] and structure-activity relationships have been demonstrated [274, 275]. The most striking effect of rifampicin is the interruption of viral morphogenesis at a unique stage (Section XI; Fig. 14.9). Some workers [212, 215, 276] have considered this to be the primary effect since DNA, early and late species of RNA [212, 277, 278], and all early and late proteins detectable by immunodiffusion and polyacrylamide gel electrophoresis [212, 276, 277] are synthesized. According to this interpretation, the failure of high molecular weight precursors of the major core polypeptides to be cleaved (Section IX.D.1) and the absence of core-associated RNA polymerase late in infection [279] are secondary to the assembly block. In contrast, other workers have presented evidence for an effect on late transcription [214, 280]. A more detailed discussion of the mode of action of rifampicin may be found in recent reviews [281, 282].

XV. Summary

Poxviruses are large enveloped viruses. The core contains a double-stranded DNA genome of 120-200 × 10^6 daltons and a variety of enzymes including those responsible for synthesis and modification of early mRNA. Investigations, particularly with vaccinia virus, indicate the following steps in virus reproduction (Fig. 14.10).

1. Entry of virions and release of the core in the cytoplasm
2. Synthesis of early mRNA by the virus cores
3. Translation of the early viral RNA into proteins, including some needed for DNA replication

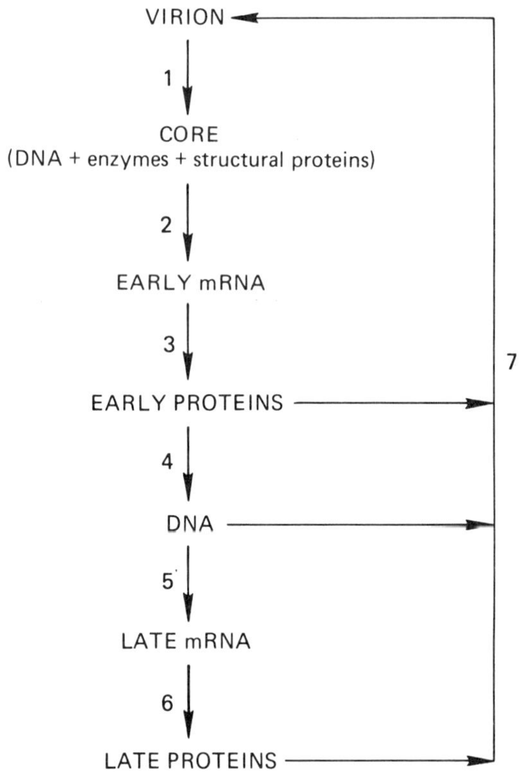
VIRION
1
CORE
(DNA + enzymes + structural proteins)
2
EARLY mRNA
3
EARLY PROTEINS
4
DNA
5
LATE mRNA
6
LATE PROTEINS
7

Fig. 14.10 Outline of the poxvirus growth cycle.

4. Replication of viral DNA
5. Synthesis of late viral mRNA
6. Translation of late viral mRNA
7. Assembly and dissemination of infectious particles

References

1. W. K. Joklik, Bacteriol. Rev., 30, 33 (1966).
2. W. K. Joklik, Ann. Rev. Microbiol., 22, 359 (1968).
3. B. R. McAuslan, in Virus Growth and Cell Metabolism (H. B. Levy, ed.), Dekker, New York. 1969, pp. 361-413.
4. B. Woodson, Bacteriol. Rev., 32, 127 (1968).
5. B. Moss, in Comprehensive Virology, Vol. 3 (H. Fraenkel-Conrat and R. R. Wagner, eds.), Plenum, New York. 1974, pp. 405-474.

6. M. Takahashi, S. Kameyama, S. Kato, and J. Kamahora, Biken J., 2, 27 (1959).
7. G. M. Woodroofe and F. Fenner, Virology, 16, 334 (1962).
8. H. Hanafusa, T. Hanafusa, and J. Kamahora, Biken J., 2, 85 (1959).
9. F. Fenner and G. M. Woodroofe, Virology, 11, 185 (1960).
10. F. Fenner, H. G. Pereira, J. D. Porterfield, W. K. Joklik, and A. W. Downie, Intervirology, 3, 193 (1974).
11. A. J. D. Bellett and F. Fenner, J. Virol., 2, 1374 (1968).
12. C. Andrewes and H. G. Pereira, Viruses of Vertebrates, Williams and Wilkins, Baltimore. 1972.
13. D. Peters, Nature (Lond.), 178, 1453 (1956).
14. M. A. Epstein, Nature (Lond.), 181, 784 (1958).
15. S. Dales and L. Siminovich, J. Biophys. Biochem. Cytol., 10, 475 (1961).
16. W. F. Noyes, Virology, 17, 282 (1962).
17. W. F. Noyes, Virology, 18, 511 (1962).
18. J. C. N. Westwood, W. J. Harris, H. T. Zwartouw, D. H. J. Titmuss, and G. Appleyard, J. Gen. Microbiol., 34, 67 (1964).
19. E. L. Medzon and H. Bauer, Virology, 40, 860 (1970).
20. K. B. Easterbrook, J. Ultrastruct. Res., 37, 132 (1966).
21. B. G. T. Pogo and S. Dales, Proc. Natl. Acad. Sci. USA, 63, 820 (1969).
22. S. Dales, J. Cell Biol., 18, 51 (1963).
23. D. Peters and G. Müller, Virology, 21, 266 (1963).
24. J. M. Hyde and D. Peters, J. Ultrastruct. Res., 35, 626 (1971).
25. E. Paoletti, H. Rosemond-Hornbeak, and B. Moss, J. Biol. Chem., 249 3273 (1974).
26. M. Abdussalam, Am. J. Vet. Res., 18, 614 (1957).
27. J. Nagington, A. A. Newton, and R. W. Horne, Virology, 23, 461 (1964).
28. M. Bergoin, G. Devauchelle, and C. Vago, Virology, 43, 453 (1971).
29. R. R. Granados and D. W. Roberts, Virology, 40, 230 (1970).
30. J. Craigie, Br. J. Exp. Pathol., 13, 259 (1932).
31. J. E. Smadel and C. L. Hoagland, Bacteriol. Rev., 6, 79 (1942).
32. W. K. Joklik, Virology, 18, 9 (1962).
33. W. K. Joklik, Biochim. Biophys. Acta, 61, 290 (1962).
34. H. T. Zwartouw, J. C. N. Westwood, and G. Appleyard, J. Gen. Microbiol., 29, 523 (1962).
35. D. N. C. Planterose, C. Nishimura, and N. P. Salzman, Virology, 18, 294 (1962).
36. C. J. Pfau and J. F. McCrea, Virology, 21, 425 (1963).
37. H. T. Zwartouw, J. Gen. Microbiol., 34, 115 (1964).
38. W. K. Joklik, J. Mol. Biol., 5, 265 (1962).
39. L. G. Gafford and C. C. Randall, Virology, 40, 298 (1970).
40. J. M. Hyde, L. G. Gafford, and C. C. Randall, Virology, 33, 112 (1967).
41. J. F. McCrea and M. B. Lipman, J. Virol., 1, 1037 (1967).
42. K. B. Easterbrook, J. Virol., 1, 643 (1967).
43. Y. Becker and I. Sarov, J. Mol. Biol., 34, 655 (1968).

44. K. I. Berns and C. Silverman, J. Virol., 5, 299 (1970).
45. P. Geshelin and K. I. Berns, J. Mol. Biol., 88, 785 (1974).
46. G. Roening and J. A. Holowczak, J. Virol., 14, 704 (1974).
47. J. A. Holowczak and W. K. Joklik, Virology, 33, 717 (1967).
48. B. Moss and N. P. Salzman, J. Virol., 2, 1016 (1968).
49. I. Sarov and W. K. Joklik, Virology, 50, 579 (1972).
50. B. Moss, E. N. Rosenblum, and C. F. Garon, Virology, 55, 143 (1973).
51. J. J. Obijeski, E. L. Palmer, L. G. Gafford, and C. G. Randall, Virology, 51, 512 (1973).
52. B. Moss and E. N. Rosenblum, J. Mol. Biol., 81, 267 (1973).
53. J. A. Holowczak, Virology, 42, 87 (1970).
54. C. F. Garon and B. Moss, Virology, 46, 233 (1971).
55. H. Rosemond and B. Moss, J. Virol., 11, 961 (1973).
56. B. G. T. Pogo, J. R. Katz, and S. Dales, Virology, 64, 531 (1975).
57. E. Katz and E. Margalith, J. Gen. Virol., 18, 381 (1973).
58. J. R. Kates and B. R. McAuslan, Proc. Natl. Acad. Sci. USA, 58, 134 (1967).
59. W. Munyon, E. Paoletti, and J. T. Grace, Jr., Proc. Natl. Acad. Sci. USA, 58, 2280 (1967).
60. J. Schwartz and S. Dales, Virology, 45, 797 (1971).
61. J. Kates, R. Dahl, and M. Mielke, J. Virol., 2, 894 (1968).
62. A. Pitkanen, B. McAuslan, J. Hedgpeth, and B. Woodson, J. Virol., 2, 1363 (1968).
63. J. Kates and J. Beeson, J. Mol. Biol., 50, 1 (1970).
64. J. R. Nevins and W. K. Joklik, Virology, 63, 1 (1975).
65. R. Dahl and J. R. Kates, Virology, 42, 453 (1970).
66. F. Fournier, D. R. Tovell, M. Esteban, D. H. Metz, L. A. Ball, and I. M. Kerr, FEBS Lett., 30, 268 (1973).
67. G. Jaureguiberry, F. Ben-Hamida, F. Chapeville, and G. Beaud, J. Virol., 15, 1467 (1975).
68. J. Kates and J. Beeson, J. Mol. Biol., 50, 19 (1970).
69. R. Sheldon, J. Kates, D. E. Kelley, and R. P. Perry, Biochemistry, 11, 3829 (1972).
70. R. Sheldon and J. Kates, J. Virol., 14, 214 (1974).
71. M. Brown, J. W. Dorson, and F. J. Bollum, J. Virol., 12, 203 (1973).
72. B. Moss, E. N. Rosenblum, and E. Paoletti, Nature, 254, 59 (1973).
73. B. Moss, E. N. Rosenblum, and A. Gershowitz, J. Biol. Chem., 250, 4722 (1975).
74. C. Brakel and J. Kates, J. Virol., 14, 715 (1974).
75. C. M. Wei and B. Moss, Proc. Natl. Acad. Sci. USA, 71, 3014 (1974).
76. C. M. Wei and B. Moss, Proc. Natl. Acad. Sci. USA, 72, 318 (1975).
77. T. Urishibara, Y. Furuichi, C. Nishimura, and K. Miura, FEBS Lett., 49, 385 (1975).
78. G. W. Both, A. K. Banerjee, and A. J. Shatkin, Proc. Natl. Acad. Sci. USA, 72, 1189 (1975).
79. M. J. Ensinger, S. A. Martin, E. Paoletti, and B. Moss, Proc. Natl. Acad. Sci. USA, 72, 2525 (1975).

80. S. A. Martin, E. Paoletti, and B. Moss, J. Biol. Chem., 250, 9322 (1975).
81. S. A. Martin and B. Moss, J. Biol. Chem., 250, 9330 (1975).
82. W. Munyon, E. Paoletti, J. Ospina, and J. T. Grace, Jr., J. Virol., 2, 167 (1968).
83. P. Gold and S. Dales, Proc. Natl. Acad. Sci. USA, 60, 845 (1968).
84. B. G. T. Pogo, S. Dales, M. Bergoin, and D. W. Roberts, Virology, 43, 306 (1971).
85. E. Paoletti and B. Moss, J. Virol., 10, 866 (1972).
86. E. Paoletti and B. Moss, J. Biol. Chem., 249, 3281 (1974).
87. E. Paoletti, N. Cooper, and B. Moss, J. Virol., 14, 578 (1974).
88. B. G. T. Pogo and S. Dales, Proc. Natl. Acad. Sci. USA, 63, 1297 (1969).
89. A. M. Aubertin and B. R. McAuslan, J. Virol., 9, 554 (1972).
90. H. Rosemond-Hornbeak, E. Paoletti, and B. Moss, J. Biol. Chem., 249, 3287 (1974).
91. H. Rosemond-Hornbeak and B. Moss, J. Biol. Chem., 249, 3292 (1974).
92. B. G. T. Pogo and S. Dales, Proc. Natl. Acad. Sci. USA, 70, 1726 (1973).
93. B. G. T. Pogo and S. Dales, Virology, 58, 377 (1974).
94. B. R. McAuslan and J. R. Kates, Virology, 33, 709 (1967).
95. E. Paoletti and B. Moss, J. Virol., 10, 417 (1972).
96. D. N. Downer, H. W. Rogers, and C. C. Randall, Virology, 52, 13 (1973).
97. J. H. Kleiman and B. Moss, J. Biol. Chem., 250, 2420 (1975).
98. J. H. Kleiman and B. Moss, J. Biol. Chem., 250, 2430 (1975).
99. Y. Watanabi, S. Sakuma, and S. Tanaka, FEBS Lett., 41, 331 (1974).
100. W. Stern and S. Dales, Virology, 62, 293 (1974).
101. C. C. Randall, L. G. Gafford, R. W. Darlington, and J. Hyde, J. Bacteriol., 87, 939 (1964).
102. H. B. White, Jr., S. Powell, L. G. Gafford, and C. C. Randall, J. Biol. Chem., 243, 4517 (1968).
103. D. S. Lyles, C. C. Randall, and H. B. White, Jr., Virology, 66, 106 (1975).
104. J. F. Sambrook, B. L. Padgett, and J. K. N. Tomkins, Virology, 28, 592 (1966).
105. A. Kirn, J. Braunwald, and R. Scherrer, Ann Inst. Pasteur, 108, 330 (1965).
106. C. Basilico and W. K. Joklik, Virology, 36, 668 (1968).
107. Y. Z. Ghendon, Prog. Med. Virol., 14, 68 (1972).
108. B. L. Padgett and J. K. N. Tomkins, Virology, 36, 161 (1968).
109. M. E. McClain, Aust. J. Exp. Biol. Med. Sci., 43, 31 (1965).
110. F. Fenner and J. Sambrook, Virology, 28, 600 (1966).
111. A. Gemmel and F. Fenner, Virology, 11, 219 (1960).
112. J. H. Subak-Sharpe, M. C. Timbury, and J. F. Williams, Nature, 222, 341 (1969).
113. B. Moss, E. N. Rosenblum, and P. M. Grimley, Virology, 45, 135 (1971).
114. Y. Ichihashi and S. Dales, Virology, 46, 533 (1971).
115. G. Appleyard and H. J. Way, Br. J. Exp. Pathol., 47, 144 (1966).
116. Y. Z. Ghendon and V. I. Chernos, Acta Virol., 16, 308 (1972).
117. E. Katz, B. Winer, E. Margalith, and N. Goldblum, J. Gen. Virol., 19, 161 (1973).

118. F. Fenner and F. N. Ratcliffe, Myxomatosis, Cambridge, London, 1965.
119. R. E. Shope, J. Exp. Med., 56, 793 (1932).
120. R. E. Shope, J. Exp. Med., 56, 803 (1932).
121. W. G. C. Bearcroft and M. F. Jamieson, Nature (Lond.), 182, 195 (1958).
122. C. H. Andrewes, A. C. Allison, J. A. Armstrong, G. Bearcroft, J. S. F. Niven, and H. G. Pereira, Acto. Unio. Int. Contra Concrum, 15, 760 (1959).
123. G. Furness and J. S. Younger, Virology, 9, 386 (1959).
124. R. Postlethwaite and H. B. Maitland, J. Hyg., 58, 133 (1960).
125. K. O. Smith and D. G. Sharp, Virology, 11, 519 (1960).
126. K. B. Easterbrook, Virology, 15, 404 (1961).
127. W. K. Joklik, J. Mol. Biol., 8, 263 (1964).
128. S. Dales, Prog. Med. Virol., 7, 1 (1965).
129. G. Ogier, Y. Chardonnet, and L. Gazzolo, J. Gen. Virol., 22, 249 (1974).
130. J. A. Armstrong, D. H. Metz, and M. R. Young, J. Gen. Virol., 21, 533 (1973).
131. R. R. Granados, Virology, 52, 305 (1973).
132. R. Dahl and J. R. Kates, Virology, 42, 453 (1970).
133. J. A. Holowczak, Virology, 50, 216 (1972).
134. I. Sarov and W. K. Joklik, Virology, 50, 593 (1972).
135. W. K. Joklik, J. Mol. Biol., 8, 277 (1964).
136. W. K. Joklik, Virology, 22, 620 (1964).
137. D. M. Prescott, J. Kates, and J. B. Kirckpatrick, J. Mol. Biol., 59, 505 (1971).
138. D. M. Prescott, D. Myerson, and J. Wallace, Exp. Cell Res., 71, 480, 1972.
139. T. H. Pennington and E. A. C. Follett, J. Virol., 13, 488 (1974).
140. P. M. Abel, Z. Vererbungsl., 94, 249 (1963).
141. B. R. McAuslan, Symposium on Enzyme regulation and Metabolic Control, Mexico City, 1966, National Cancer Institute Monograph 27, National Cancer Institute, Bethesda. 1967, p. 211-219.
142. S. Dales, J. Cell Biol., 18, 51 (1963).
143. S. Dales, Proc. Natl. Acad. Sci. USA, 54, 462 (1965).
144. W. H. Munyon and S. Kit, Virology, 29, 303 (1966).
145. J. R. Kates and B. R. McAuslan, Proc. Natl. Acad. Sci. USA, 57, 314 (1967).
146. B. Woodson, Biochem. Biophys. Res. Commun., 27, 169 (1967).
147. Y. Becker and W. K. Joklik, Proc. Natl. Acad. Sci. USA, 51, 577 (1964).
148. K. Oda and W. K. Joklik, J. Mol. Biol., 27, 395 (1967).
149. K. T. Atherton and G. Darby, J. Gen. Virol., 22, 215 (1974).
150. N. V. Kaverin, N. L. Varich, V. V. Surgay, and V. I. Chernos, Virology, 65, 112 (1975).
151. W. K. Joklik and T. C. Merigan, Proc. Natl. Acad. Sci. USA, 56, 558 (1966).
152. N. P. Salzman, A. J. Shatkin, and E. A. Sebring, J. Mol. Biol., 8, 405 (1964).

153. B. Moss and R. Filler, J. Virol., 5, 99 (1970).
154. C. Colby, C. Jurale, and J. R. Kates, J. Virol., 7, 71 (1971).
155. G. Obert, F. Tripier, and J. Guir, Biochem. Biophys. Res. Commun., 44, 362 (1971).
156. R. Dahl and J. R. Kates, Virology, 42, 463 (1975).
157. C. Brakel and J. R. Kates, J. Virol., 14, 724 (1974).
158. A. J. Shatkin, E. D. Sebring, and N. P. Salzman, Science, 148, 87 (1965).
159. E. D. Sebring and N. P. Salzman, J. Virol., 1, 550 (1967).
160. B. R. McAuslan, Biochem. Biophys. Res. Commun., 37, 289 (1969).
161. S. G. Clarkson and M. N. Runner, Biochim. Biophys. Acta, 238, 498 (1971).
162. M. Klagsbrun, Virology, 44, 153 (1971).
163. N. P. Salzman and E. D. Sebring, J. Virol., 1, 16 (1967).
164. B. Moss and N. P. Salzman, J. Virol., 2, 1016 (1968).
165. M. Esteban and D. H. Metz, J. Gen. Virol., 19, 201 (1973).
166. T. H. Pennington, J. Gen. Virol., 25, 433 (1974).
167. W. E. Magee, Virology, 17, 604 (1962).
168. S. Kit, D. R. Dubbs, and L. J. Piekarski, Biochem. Biophys. Res. Commun. 8, 72 (1962).
169. B. R. McAuslan and W. K. Joklik, Virology, 8, 486.
170. S. Kit, L. J. Piekarski, and D. R. Dubbs, J. Mol. Biol., 6, 22 (1963).
171. D. R. Dubbs and S. Kit, Virology, 22, 214 (1964).
172. C. Jungwirth and W. K. Joklik, Virology, 27, 80 (1965).
173. S. Kit, D. R. Dubbs, and L. J. Piekarski, Biochem. Biophys. Res. Commun., 11, 176 (1963).
174. B. R. McAuslan, Virology, 21, 383 (1963).
175. S. Kit and D. R. Dubbs, Virology, 26, 16 (1965).
176. S. Kit, W. C. Leung, D. Trkula, and G. Jorgenson, Int. J. Cancer, 13, 203 (1974).
177. W. E. Magee and O. V. Miller, Virology, 31, 64 (1967).
178. K. I. Berns, C. Silverman, and A. Weissbach, J. Virol., 4, 15 (1969).
179. R. V. Citarella, R. Muller, A. Schlabach, and A. Weissbach, J. Virol., 10, 721 (1972).
180. K. B. Tan and B. R. McAuslan, J. Virol., 9, 70 (1972).
181. L. M. S. Chang and M. E. Hodes, Virology, 32, 258 (1967).
182. J. Sambrook and A. J. Shatkin, J. Virol., 4, 719 (1969).
183. B. R. McAuslan, Biochem. Biophys. Res. Commun., 19, 15 (1965).
184. L. J. Eron and B. R. McAuslan, Biochem. Biophys. Res. Commun., 22, 518 (1966).
185. B. R. McAuslan and J. R. Kates, Virology, 33, 709 (1967).
186. B. C. Cooke and J. D. Williamson, J. Gen. Virol., 21, 339 (1973).
187. J. D. Williamson and B. C. Cooke, J. Gen. Virol., 21, 349 (1973).
188. J. A. Holowczak and W. K. Joklik, Virology, 33, 726 (1967).
189. E. Katz and B. Moss, J. Virol., 6, 717 (1970).
190. E. Katz and B. Moss, Proc. Natl. Acad. Sci. USA, 66, 677 (1970).
191. E. Katz, P. Grimley, and B. Moss, Nature, 227, 1050 (1970).

192. B. Moss, E. N. Rosenblum, and C. F. Garon, Virology, 46, 221 (1971).
193. S. Weintraub and S. Dales, Virology, 60, 96 (1974).
194. B. R. McAuslan, Virology, 20, 162 (1963).
195. B. R. McAuslan, Virology, 21, 383 (1963).
196. V. Zaslavsky and E. Yakobson, J. Virol., 16, 210 (1975).
197. N. P. Salzman, Virology, 10, 150 (1960).
198. W. K. Joklik and Y. Becker, J. Mol. Biol., 10, 452 (1964).
199. J. R. Kates and B. R. McAuslan, J. Virol., 1, 110 (1967).
200. J. Cairns, Virology, 11, 603 (1960).
201. S. Kato, S. Kameyama, and J. Kamahora, Biken J., 3, 135 (1960).
202. K. H. Walen, Proc. Natl. Acad. Sci. USA, 68, 165 (1971).
203. P. LaColla and A. Weissbach, J. Virol., 15, 305 (1975).
204. C. Morgan, S. A. Ellison, H. M. Rose, and D. H. Moore, J. Exp. Med., 100, 301 (1954).
205. A. A. Avakyan and A. F. Byckovsky, J. Cell Biol., 24, 337 (1964).
206. R. B. Arhleger and C. C. Randall, Virology, 22, 59 (1964).
207. E. deHarven and D. S. Yohn, Cancer Res., 26, 995 (1966).
208. T. Tsuruhara, J. Natl. Cancer Inst., 47, 549 (1971).
209. E. P. Cohen, S. S. Delaney, J. Sanders, and E. Mosovici, Virology, 23, 56 (1964).
210. M. Bergoin, G. Devauchelle, and C. Vago, Arch. Ges. Virusforsch, 28, 285 (1969).
211. S. Dales and E. H. Mosbach, Virology, 35, 584 (1968).
212. B. Moss, E. N. Rosenblum, E. N. Katz, and P. M. Grimley, Nature (Lond.), 224, 1280 (1969).
213. P. M. Grimly, E. N. Rosenblum, S. J. Mims, and B. Moss, J. Virol., 6, 519 (1970).
214. A. Nagayama, B. G. T. Pogo, and S. Dales, Virology, 40, 1039 (1970).
215. T. H. Pennington, E. A. Follett, and J. F. Szilagyi, J. Gen. Virol., 9, 225 (1970).
216. H. S. Rosenkranz, H. M. Rose, C. Morgan, K. C. Hsu, Virology, 28, 510 (1966).
217. B. G. T. Pogo and S. Dales, Virology, 43, 144 (1971).
218. W. Fil, J. A. Holowczak, L. Flores, and V. Thomas, Virology, 61, 376 (1974).
219. R. Dahl and J. R. Kates, Virology, 42, 453 (1970).
220. B. Polisky and J. R. Kates, Virology, 49, 168 (1972).
221. I. Sarov and W. K. Joklik, Virology, 52, 222 (1973).
222. B. Polisky and J. R. Kates, Virology, 66, 128 (1975).
223. Y. Ichihashi and S. Matsumoto, and S. Dales, Virology, 46, 507 (1971).
224. S. Kato, M. Hara, H. Miyamoto, and J. Kamahora, Biken J., 6, 233 (1963).
225. Y. Ichihashi and S. Matsumoto, Virology, 29, 264 (1966).
226. Y. Ichihashi and S. Dales, Virology, 51, 297 (1973).
227. Y. Ichihashi and S. Matsumoto, Virology, 36, 262 (1968).
228. G. Appleyard, J. C. N. Westwood, and H. T. Zwartouw, Virology, 18, 159 (1962).

229. H. Kaku and J. Kamahora, Biken J., 6, 299 (1964).
230. M. Nishmi and R. Keller, Virology, 18, 109 (1962).
231. G. Appleyard, A. J. Hapel, and E. A. Boulter, J. Gen. Virol., 13, 9 (1971).
232. G. S. Turner and E. J. Squires, J. Gen. Virol., 13, 19 (1971).
233. E. A. Boulter and G. Appleyard, Prog. Med. Virol., 16, 86 (1973).
234. Y. Ueda, M. Ito, and I. Tagaya, Virology, 38, 180 (1969).
235. H. Miyamoto and S. Kato, Biken J., 14, 311 (1971).
236. K. Hayashi, J. Rosenthal, and A. L. Notkins, Science, 176, 516 (1972).
237. S. Weintraub and S. Dales, Virology, 60, 96 (1974).
238. A. Shelokov, J. E. Vogel, and L. Chi, Proc. Soc. Exp. Biol., 97, 802 (1958).
239. K. E. Blackman and H. C. Bubel, J. Virol., 9, 290 (1972).
240. J. M. Zarling and S. S. Tevethia, Virology, 45, 313 (1971).
241. G. Bandlow, U. Koszinowski, and R. Thomssen, Arch. Ges. Virusforsch., 40, 63 (1973).
242. G. Appleyard, J. C. N. Westwood, and H. T. Zwartouw, Virology, 18, 159 (1962).
243. F. R. Ball and E. L. Medzon, J. Virol., 12, 588 (1973).
244. H. Bernkopf, M. Nishmi, and A. Rosin, J. Immunol., 83, 635 (1959).
245. A. Brown, S. A. Mayyasi, and J. E. Officer, J. Infect. Dis., 104, 193 (1959).
246. H. Hanafusa, Biken J., 3, 191 (1960).
247. R. Bablanian, J. Gen. Virol., 3, 51 (1968).
248. R. Bablanian, J. Gen. Virol., 6, 221 (1970).
249. A. J. Shatkin, Virology, 20, 292 (1963).
250. B. Moss, J. Virol., 2, 1028 (1968).
251. H. Rosemond-Hornbeak and B. Moss, J. Virology, 16, 34 (1975).
252. E. R. Jefferts and J. A. Holowczak, Virology, 46, 730 (1971).
253. T. Hanafusa, Biken J., 3, 313 (1960).
254. S. Kit, D. R. Dubbs, and T. C. Hsu, Virology, 19, 13 (1963).
255. S. Kato, M. Ogawa, and H. Miyamoto, Biken J., 7, 45 (1964).
256. C. Jungwirth and J. Launer, J. Virol., 2, 401 (1968).
257. H. C. Hinze and D. L. Walker, J. Bacteriol., 88, 1185 (1964).
258. H. C. Hinze and D. L. Walker, J. Virol., 7, 577 (1971).
259. W. Cheevers, D. J. O'Callaghan, and C. C. Randall, J. Virol., 2, 421 (1968).
260. L. G. Gafford, F. Sinclair, and C. C. Randall, Virology, 48, 567 (1972).
261. R. L. Thompson, J. Davis, P. B. Russell, and G. H. Hitchings, Proc. Soc. Exp. Biol. Med., 84, 496 (1953).
262. R. L. Thompson, S. A. Minton, J. E. Officer, and G. H. Hitchings, J. Immunol., 70, 229 (1953).
263. F. W. Sheffield, D. J. Bauer, and S. Stephenson, Br. J. Exp. Pathol., 41, 638 (1960).
264. E. Katz, E. Margalith, B. Winer, and A. Lazar, J. Gen. Virol., 21, 469 (1973).
265. E. Katz, E. Margalith, and B. Winer, J. Gen. Virol., 21, 477 (1973).

266. D. Bauer and P. Sadler, Br. J. Pharmacol., 15, 101 (1960).
267. E. Katz, E. Margalith, and B. Winer, J. Gen. Virol., 25, 239 (1974).
268. K. B. Easterbrook, Virology, 17, 245 (1962).
269. W. E. Magee and M. K. Bach, Ann. N.Y. Acad. Sci., 130, 80 (1965).
270. G. Appleyard, V. B. M. Hume, and J. C. N. Westwood, Ann. N.Y. Acad. Sci., 130, 92 (1965).
271. B. Woodson and W. K. Joklik, Proc. Natl. Acad. Sci. USA, 54, 946 (1965).
272. E. Katz, E. Margalith, B. Winer, and N. Goldblum, Antimicrob. Ag. Chemother., 4, 44 (1973).
273. E. Heller, M. Argaman, H. Levy, and N. Goldblum, Nature, 222, 273 (1969).
274. T. H. Pennington and E. A. Follett, J. Virol., 7, 821 (1971).
275. P. M. Grimley and B. Moss, J. Virol., 8, 225 (1971).
276. B. Moss, E. N., Rosenblum, and P. M. Grimley, Virology, 45, 123 (1971).
277. B. Moss, E. Katz, and E. N. Rosenblum, Biochem. Biophys. Res. Commun., 36, 858 (1969).
278. Z. Ben-Ishai, E. Heller, N. Goldblum, and Y. Becker, Nature, 224, 29 (1969).
279. B. R. McAuslan, Biochem. Biophys. Res. Commun., 37, 289 (1969).
280. B. G. T. Pogo, Virology, 44, 576 (1971).
281. B. Moss, in Selective Inhibitors of Viral Functions (W. A. Carter, ed.), CRC, Cleveland. 1973, pp. 313-328.
282. E. A. C. Follett and T. H. Pennington, Adv. Virus Res., 18, 105 (1973).

Recent Developments

Chapter 12

A. Biogenesis of Adenovirus mRNA

Since the original writing of this chapter, astonishing advances have been made in our understanding of Ad 2 mRNA biogenesis. These findings were presented at the XLII Cold Spring Harbor Symposium on Quantitative Biology (1977), and have been summarized by Sambrook [38]. J. E. Darnell and colleagues (R. Evans, N. Frazer, S. Goldberg, T. Weber, and N. Wilson) found that transcription of the 4 early gene blocks (see Fig. 12.11) begins at position 1.8 and 80 on *r* strand, and 98 and 72 on *l* strand, i.e. near the 3′ terminus of each gene block (determined by pulse-labeling RNA in isolated nuclei under conditions where RNA elongation but not initiation can occur, and annealing the shortest RNAs to sets of DNA restriction fragments). Primary transcripts are probably 4 polycistronic RNAs about the size of each gene block, as indicated by size analysis of Ad 2 early nuclear RNA under RNA denaturing conditions [2, 11, 12]. The primary nuclear transcripts then must be processed into smaller individual mRNAs that are found on polyribosomes [5, 11]. (Note: RNA polymerase II may read-through transcription terminators at the 5′ ends of the gene blocks, because low levels of RNA complementary to most or all of both strands are found in early nuclei.)

Late genes also are likely transcribed into polycistronic RNAs. Many workers have shown that the majority of pulse-labeled Ad 2 nuclear RNA is large, e.g., >30-45S. Most nuclear RNA is from *r* strand. Darnell and colleagues [45] showed that only very large pulse-labeled nuclear RNA annealed to DNA fragments at the right of the genome, whereas smaller RNAs annealed to fragments extending to the left of the genome. Ultraviolet irradiation of late infected cells resulted in a progressive decrease in synthesis of RNA from about position 16 to the extreme right of the genome [24]. These data provide strong (but not definitive) evidence that the majority of late nuclear RNA consists of a primary transcript initiated at ca. position 16 on *r* strand and extending to position 90-100.

Scientists at the Cold Spring Harbor Laboratories [8, 15, 27, 28] and at MIT [3] have made the amazing discovery that late Ad 2 mRNAs contain 5′ terminal "leader" sequences of 100-200 bases encoded by DNA not adjacent to the structural gene, but rather far upstream to the gene. The evidence for this is summarized below. (1) Gelinas and Roberts [22] first noted that $m^7G(5')ppp(5')A^mC^{(m)}U(C_4U_3)G$ was the predominant 5′ oligonucleotide produced by RNase T_1 digestion of late mRNA. (2) Klessig [27] showed that this oligonucleotide, plus an additional 100-150 bases, were removed after RNase digestion of purified 100K protein and fiber mRNAs hybridized to DNA fragments encoding the structural gene for these mRNAs (see Fig. 12.9). The 5′ oligonucleotide was shown to be coded within position 14.7 to 21, and the other 100-150 bases to be coded within positions 17-31.5. (3) Lewis et al. [28] found that many late mRNAs hybridize to their structural genes (determined by cell-free translation) but also to DNA within positions 16-32. (4) In electron microscopy studies, Chow et al. [8] found that mRNAs from genes that map between positions 36 and 92 hybridize simultaneously at their 5′ ends to DNA at positions 16.7, 19.7, and 26.7 (on *r* strand). Another late mRNA (encoding protein IX) mapping at 9.6-10.9, hybridized at its 5′ end to DNA at 4.9-6.0. (5) Also by electron microscopy, Berget et al. [3] found that hexon mRNA annealed to DNA at positions 51.7 to 61.3 (hexon structural gene) and simultaneously at its 5′ end to DNA at 17, 20, and 27. The above studies indicate that most or all of late mRNAs that map on *r* strand to the right of position 36 contain 100-200 bases, coded at position 17, 20, and 27, covalently linked at the 5′ termini of mRNA. Late mRNAs that map to the left of position 11 may be linked to a leader sequence coded at position 4.9-6.0. It seems likely that the late *l* strand-specific mRNAs may also contain a distally coded leader sequence.

In summary, these findings suggest the following model for mRNA biogenesis. Transcription of each early region is initiated at a specific 3′ terminal promotor, forming a polycistronic primary transcript corresponding to each early region, that is processed into discrete cytoplasmic mRNAs. The transcription of late genes is initiated at 2 "promotors" on *r* strand, one at positions 17-27 controlling the synthesis of polycistronic RNA coded between positions 36 to 91, and the other at position 4.9-6.0 controlling the synthesis of mRNA at positions 9-11. Discrete mRNAs are formed by cleavage of the primary transcript, and splicing of leader sequences to the 5′ end of the structural gene sequences. This model would provide for coordinate control of all mRNAs within a gene block by the promotor (leader sequences) of that gene block. It is unclear at present whether early mRNAs are ligated at their 5′ ends to leader sequences coded at the "promotor" sites. But if so, the promotor-leader sequences may define a eukaryotic operon, that is analogous to, but strikingly different from, a prokaryotic operon. Of course, the above model, which was suggested by Klessig, and by Sharp and colleagues, has not been proven as yet, and other models are possible.

B. Mapping Ad mRNA by Electron Microscopy of RNA-DNA Hybrids

Application of the "R-loop" procedure [44] to the Ad system has provided high resolution maps of early and late genes, and has resulted in the discovery of leader sequences linked to the 5′ termini of Ad mRNA. In 70% formamide, RNA-DNA hybrids are more stable than DNA-DNA hybrids; therefore, when hybridized, the RNA displaces the anticomplementary DNA strand to form an "R-loop." By this procedure, Chow et al. [9] obtained the following results: The transforming gene block extends from position 1.3 to 11.1 on *r* strand, and probably contains 2 genes. One gene, likely the transforming gene, 2.7 ± 0.3% of the genome, may map between 1.3 and 4.0. The second gene may extend from 5.0 to 11.1. There appears to be a "gap" (4.0-5.0) between the two genes. The exact dimensions of these genes is unclear, because the R-loops vary in length, some even spanning the entire gene block. The other *r* strand early gene block is located at position 78.6-86.1. A 15,500 dalton polypeptide has been observed after translation of this mRNA [29]. The *l* strand-specific early genes map at positions 62.4-67.9 (gene for 73,000 DNA binding protein) and 91.7-96.8. The latter block encodes 19K and 11K, as suggested by cell-free translation [29]. Neuwald et al. [33] have mapped Ad 2 early genes at 1.1-10.6, 61.6-68.1, 76.7-83.7, and 91.5-96.9.

Chow et al. [9] have also attempted to localize late genes. One mRNA mapped at 9.6-10.9 on *r* strand. This mRNA codes protein IX, a late virus structural protein [28]; M. Mathews and U. Pettersson, personal communication). Most remarkable, this late gene is located at the 5′ end of an apparent early gene(s). An mRNA (perhaps coding protein IVa_2) maps on *l* strand, with the 3′ terminus at 11.2 and heterogenous 5′ termini near 14.9. The mRNAs between positions 31 and 50 (coding proteins IIIa, V, III, and P-VII) are heterogeneous in length, with preferred 5′ termini at 45.5-47 and 48-50. Perhaps these mRNAs contain overlapping sequences. Hexon mRNA maps at 51.9-62.2 (52.7-61.3 [3]). The mRNA for P-VI probably maps to the left of the hexon gene. Long heterogeneous R-loops are seen between positions 68 and 78, regions where the 100K and P-VII proteins map. Fiber mRNA maps at 86.3-91.5.

Meyer et al. [32] have also mapped late mRNA using the R-loop procedure. Meissner et al. [31] examined late nuclear RNA, and detected R-loops spanning between position 30 and 90, possibly representing a primary transcript of the putative right-hand late gene block.

C. Ad 12 Transcription

Recent work by Smiley and Mak [42] and Doerfler and colleagues have shown that Ad 12 and Ad 2 (or Ad 5) are quite similar in terms of the organization and transcription of their genomes. Ad 12 specific early and late mRNA ranges from 0.3×10^6 to 2.3×10^6 daltons [34, 39]. The early genes are located approximately within positions 0-10 and 74-88 on *r* strand, and 63-74 and 89-100 on *l* strand. Late genes are located mainly on *r* strand in a manner remeniscent of Ad 2 [39]. Some, but not all, early genes are expressed as mRNA in Ad 12 transformed cells. Ad 12 transforming genes are located within position 0-10, because Ad 12 EcoRI-C fragment (0-16) can transform cells by transfection.

D. Polypeptide Linked to the 5′ Termini of Adenovirus DNA

Robinson et al. [37] originally reported that Ad DNA readily formed circles if the DNA was extracted in 4 M quanidinium hydrochloride, and without protease digestion. They suggested that a protein must be firmly associated with the DNA termini, and that this protein could play a role in DNA replication by circularizing the Ad DNA. Although it now seems that circular forms probably are not involved in Ad DNA replication (Section VII), it has become clear that the 5′ termini of Ad DNA is linked, probably covalently (although covalent linkage has not been proven chemically), to 5′ termini via a dC moiety. Evidence supporting this conclusion is given below. (1) DNA isolated without protease digestion readily forms circles of oligomeric structures, and the DNA molecules are always joined at the ends [26, 35, 37, 40]. Using procedures to vizualize proteins in the electron microscope, Keegsta et al. [26] and T. R. Broker and L. T. Chow (personal communication) have identified one "knob" in circular molecules, and two "knobs" in linear molecules, one at each end. Proteolysis destroyed the ability of the DNA to circularize, and removed the knobs. On the other hand, the ability of the DNA to circularize is resistant to treatment with 4 M guanidinium hydrochloride, 4 M urea, 2 M perchlorate, 50% formamide, 10% pyridine, 1% SDS, 1% 2-mercaptoethanol, chloroform-phenol, strongly suggesting that the protein is covalently linked to the DNA. (2) Ad protein-DNA complexes will not enter 0.3-1.4% agarose gels when subjected to electrophoresis, whereas protease-treated DNA readily enters gels. If the protein-DNA complex is digested with restriction endonucleases, and subjected to agarose gel electrophoresis, only end fragments remain at the top of the gels [4, 35, 40]; R. Padmanabhan, and G. Chinnadurai, personal communication). This indicates that a protein associated with Ad-terminal fragments prevents fragments from entering agarose gels. (3) Linear Ad DNA-protein complexes are resistant to digestion by a progressive nuclease (ATP-dependent DNase) that requires a free terminus for activity, whereas protease-treated DNA is susceptible to digestion [40]. (4) The majority of protease-treated DNAs cannot be labeled with (γ-^{32}P)ATP after treatment with alkaline phosphatase and polynucleotide kinase [6]. These DNAs are resistant to digestion with bacteriophage lambda exonuclease which hydrolyzes DNA starting from 5′ ends, whereas they can be digested with exonuclease III, which hydrolyzes DNA starting from 3-ends [6, 21, 36]; R. Padmanabhan, personal communication). This indicates that the 5′ but not 3′ termini of Ad DNA are blocked. (5) The 5′ terminal residue to which the protein is linked is believed to be dC [6, 35, 43]. (6) The protein component of the protein-DNA complex can be labeled with ^{125}I to yield a protein of about 55,000 daltons (following DNA digestion) [35].

The function of the 5′ terminal protein is not known (and it is not known whether it is an early or late viral protein). The protein could have a structural role in circularizing the DNA in the virion. It could also play a role in DNA replication, e.g., as an endonuclease in a "Cavalier-Smith" model (Section VII). Rekosh et al. [35] proposed that the protein could function to complete 5′ termini. They suggest that the protein-dC-complex may bind noncovalently to the 3′ end of the parental strand (perhaps at the inverted termini); the free 3′-OH on the protein-dC complex could then prime 5′ to 3′ DNA synthesis (or complete 5′ termini). This would yield a daughter strand covalently linked to the dC-protein complex at its 5′ terminus, as is found in virions.

Of significant practical interest, Ad protein-DNA complexes are about 100-fold more efficient in transfection-infectivity than protease-treated DNA ([40]; G. Chinnadurai, personal communication). This finding should facilitate in vitro genetic experiments and other studies that require efficient transfection.

E. Correlation Between Adenovirus-Transformed Cell Traits and Tumorigenicity

Virus-transformed cells display a variety of traits that differ from untransformed cells; these include changes in morphology, anchorage-independence, growth to increased saturation density, reduction in serum requirement, increased glucose uptake, alterations of surface components, loss of actin filament bundle, and induction of proteolytic enzymes (see refs. 7 and 19). It is important to establish which traits are correlated with cell tumorigenicity, considering that tumorigenicity may be most relevant to malignancy (i.e., cancer). Thus, a number of Ad 2-transformed rat embryo cells [17, 18] was examined for their ability to form tumors [7, 19, 25, 30]; tumors were produced in syngeneic rats only by T2C4 and 50A, in immunosuppressed newborn rats by T2C4, 50A, F4, and REM, and in nude mice by T2C4 (7.5 days), REM (8.5 days), 50A (18 days), and F4 (19 days). The nude mice tumors were locally invasive but did not metastasize. F19 cells produced benign tumors in nude mice. F17 and F18 cells did not produce any tumors. Most interesting, the only parameter that appeared to correlate with tumorigenicity was the appearance of a 250,000 dalton LETS (large external transformation sensitive) protein [7, 19]. For example, 100% of F17, F18, and F19 cells displayed LETS protein located at areas of cell-cell contact, whereas less than 1% of the tumorigenic lines displayed LETS protein. No correlation was observed between anchorage-independent growth and tumorigenicity, as has been reported for SV40 transformed cells [41]. The biological function of LETS protein is unknown, and it is unclear why Ad 2-transformed cells that produced LETS protein in areas of cell-cell contact are nontumorigenic, whereas cells without LETS protein are tumorigenic. It was suggested that LETS protein could play a role in cell-cell adhesion; if so, then adhesiveness between cells may be an important factor in tumorigenicity.

The Ad 2 DNA fragments present [20] and expressed as RNA [16] have been determined for most of these lines. All lines contain the left 14% of the genome, the transforming region, and express half of these sequences as mRNA. F17, F4, and T2C4 also produce candidate 53,000 and 15,000 Ad 2 transformation proteins ([23]; Wold and Green, unpublished data), that are apparently coded by the left 14% of the Ad 2 genome [23, 29]. Thus, the presence of Ad 2 putative transforming gene(s), and their expression as mRNA, and presumably protein, is not sufficient to render Ad 2-transformed cells tumorigenic. However, expression of transforming gene(s) probably is responsible for at least some of the transformed cell-specific properties of these cells. Conceivably, tumorigenicity of Ad 2-transformed cells could be related to the site of integration of Ad DNA, chromosome damage induced during the experimental transformation event, or even expression of other Ad 2-coded early proteins.

F. Nondefective Ad_2-SV40 Hybrids

Anderson et al. [1] recently identified 30K and 92K polypeptides specific to $Ad_2{}^+$ND1- and $Ad_2{}^+$ND4-infected cells, respectively. They also showed that the mRNAs for these polypeptides are encoded by both Ad 2 and SV40 sequences in the hybrids. Although they did not determine whether the polypeptides are encoded by the SV40 and/or Ad parts of the hybrid mRNA molecule, they pointed out that since the SV40 sequence in $Ad_2{}^+$ND1 can code only 16,000-20,000 daltons, the 30K must be at least partly coded by Ad sequences. Deppert and colleagues showed that SV40 T-serum (from

SV40-tumored hamsters) immunoprecipitated 42K and 56K from $Ad_2{}^+ND2$-infected cells [13], and 28K, 42K, 56K, 60K, 64K, 72K, 74K, and 95K from $Ad_2{}^+ND4$-infected cells [14]. Tryptic peptide maps indicate that $Ad_2{}^+ND2$-induced 42K and 56K share sequences [13]. The 42K, 56K, 60K, 72K, 74K, and 95K induced by $Ad_2{}^+ND4$, as well as the 42K and 56K induced by $Ad_2{}^+ND2$, apparently also have common peptides (unpublished data of K. Mann and G. Walter, cited in ref. 14). Therefore, it is likely that proteins induced by $Ad_2{}^+ND4$ and $Ad_2{}^+ND2$ are subspecies of the 95K. Further, it is likely that 95K shares sequences with SV40 T-antigen (i.e., both 95K and T-antigen are immunoprecipitated by SV40 T-serum). Since SV40 T-antigen has a molecular weight of 70,000-100,000 daltons, the 95K could even represent entire SV40 T-antigen. There is no direct evidence as yet that the 28K protein induced by $Ad_2{}^+ND1$ and $Ad_2{}^+ND4$ is encoded by the SV40 insertion.

References

1. C. W. Anderson, J. B. Lewis, P. R. Baum, and R. F. Gesteland, J. Virol., 18, 685-692 (1976).
2. S. L. Bachenheimer, J. Virol., 22, 577-582 (1977).
3. S. M. Berget, C. Moore, and P. A. Sharp, Proc. Natl. Acad. Sci. USA, in press (1977).
4. D. T. Brown, M. Westphal, B. T. Burlingham, U. Winterhoff, and W. Doerfler, J. Virol., 16, 366-387 (1975).
5. W. Büttner, Z. Veres-Molnár, and M. Green, J. Mol. Biol., 107, 93-114 (1976).
6. E. A. Carusi, Virology, 76, 380-394 (1977).
7. L. B. Chen, P. H. Gallimore, and J. K. McDougall, Proc. Natl. Acad. Sci. USA, 73, 3570-3574 (1976).
8. L. T. Chow, R. E. Gelinas, T. R. Broker, and R. J. Roberts, Cell, in press (1977).
9. L. T. Chow, J. M. Roberts, J. B. Lewis, and T. R. Broker, Cell, in press (1977).
10. J. Cordon, H. M. Engelking, and G. D. Pearson, Proc. Natl. Acad. Sci. USA, 73, 401-404 (1976).
11. E. A. Craig and H. J. Raskas, Cell, 8, 205-213 (1976).
12. E. A. Craig, M. Sayevedra, and H. J. Raskas, Virology, 77, 545-555 (1977).
13. W. Deppert and G. Walter, Proc. Natl. Acad. Sci. USA, 73, 2505-2509 (1976).
14. W. Deppert, G. Walker, and H. Linke, J. Virol., 21, 1170-1186 (1977).
15. A. R. Dunn and J. A. Hassell, Cell, in press (1977).
16. S. J. Flint, P. H. Gallimore, and P. A. Sharp, J. Mol. Biol., 96, 47-68 (1975).
17. P. H. Gallimore, J. Gen. Virol., 16, 99-102 (1972).

18. P. H. Gallimore, J. Gen. Virol., 25, 263-273 (1974).
19. P. H. Gallimore, J. K. McDougall, and L. B. Chen, Cell, 10, 669-678 (1977).
20. P. H. Gallimore, P. A. Sharp, and J. Sambrook, J. Mol. Biol., 89, 49-72 (1974).
21. C. F. Garon, K. W. Berry, and J. A. Rose, Proc. Natl. Acad. Sci. USA, 69, 2391-2395 (1972).
22. R. E. Gelinas and R. J. Roberts, Cell, 11, 533-544 (1977).
23. Z. Gilead, Y-H. Jeng, W. S. M. Wold, K. Sugawara, H. M. Rho, M. L. Harter, and M. Green, Nature, 264, 263-266 (1976).
24. S. Goldberg, J. Weber, and J. E. Darnell, Jr., Cell, 10, 617-621 (1977).
25. L. M. J. Harwood and P. H. Gallimore, Int. J. Cancer, 16, 498-508 (1975).
26. W. Keegstra, P. S. van Wielink, and J. S. Sussenbach, Virology, 76, 444-447 (1977).
27. D. F. Klessig, Cell, in press (1977).
28. J. B. Lewis, C. W. Anderson, and J. F. Atkins, Cell, in press (1977).
29. J. B. Lewis, J. F. Atkins, P. R. Baum, R. Solem, R. F. Gesteland, and C. W. Anderson, Cell, 7, 141-151 (1976).
30. J. K. McDougall, Prog. Med. Virol., 21, 118-132 (1975).
31. H. C. Meissner, J. Meyer, J. V. Maizel, Jr., and H. Westphal, Cell, 10, 225-235 (1977).
32. J. Meyer, P. D. Neuwald, S-P. Lai, J. V. Maizel, Jr., and H. Westphal, J. Virol., 21, 1010-1018 (1977).
33. P. D. Neuwald, J. Meyer, J. V. Maizel, Jr., and H. Westphal, J. Virol., 21, 1019-1030 (1977).
34. J. Ortin and W. Doerfler, J. Virol., 15, 27-35 (1975).
35. D. M. K. Rekosh, W. C. Russell, A. J. D. Bellett, and A. J. Robinson, Cell, 11, 283-295 (1977).
36. R. J. Roberts, J. R. Arrand, and W. Keller, Proc. Natl. Acad. Sci. USA, 71, 3829-3833 (1974).
37. A. J. Robinson, H. B. Younghusband, and A. J. D. Bellett, Virology, 56, 54-69 (1973).
38. J. Sambrook, Nature, 268, 101-104 (1977).
39. K-H. Scheidtmann and W. Doerfler, J. Virol., 22, 585-590 (1977).
40. P. A. Sharp, C. Moore, and J. L. Haverty, Virology, 75, 442-456 (1976).
41. S. Shin, V. H. Freedman, R. Risser, and R. Pollack, Proc. Natl. Acad. Sci. USA, 72, 4435-4439 (1975).
42. J. R. Smiley and S. Mak, J. Virol., 19, 36-42 (1976).
43. P. H. Steenbergh, J. S. Sussenbach, R. J. Roberts, and H. S. Jansz, J. Virol., 15, 268-272 (1975).
44. M. Thomas, R. L. White, and R. W. Davis, Proc. Natl. Acad. Sci. USA, 73, 2294-2298 (1976).
45. J. Weber, W. Jelinek, and J. E. Darnell, Jr., Cell, 10, 611-616 (1977).

Recent Developments

Chapter 13

This chapter was completed in the spring of 1975. Inasmuch as there has been considerable progress since that time, several sections are outdated and in some instances, the outdated conclusions have been replaced by data. The major developments that profoundly affected our understanding of herpesviruses may be summarized as follows:

1. Studies on the recombinants produced by crossing HSV-1 × HSV-1 have consistently indicated a single linkage map of markers in the L component relative to markers in the S component. Analyses of recombinants produced by crossing HSV-1 × HSV-2 indicated that at most 2 arrangements of HSV DNA are capable of being replicated (L. S. Morse, T. G. Buchman, B. Roizman and P. A. Schaffer, J. Virol., 24, 231–248, 1977). The data suggest, therefore, that the arrangements of HSV DNA are a consequence of some post-replicative event. Analysis of the replication of HSV DNA (Jacob and B. Roizman, August issue, J. Virol., 319; R. Jacob and B. Roizman, J. Virol., 23, 394–411, 1977) indicate that initiation of DNA synthesis is preceded by exonucleolytic digestion of one end (sequence *a*) and the folding of the single-stranded end and to form a hairpin. Replication involves the evolution of the hairpin into a lariat which ultimately gives rise to head-to-head concatemer. Processing of the concatemer to unit length DNA would necessarily require the regeneration of the digested sequence (*a*?) digested prior to replication. It is conceivable that this sequence is regenerated by (1) cleavage of concatemer, (2) annealing of the terminus formed by the cleavage to its inverted repeat following branch migration, (3) synthesis of the missing sequence using the inverted repeat as the template, (4) regeneration of the linear form. Depending on how the linear forms are regenerated, the concatemers would yield either the prototype form or the inversion of the DNA component whose end was repaired.

(2) Studies of HSV-1 × HSV-2 recombinants yielded considerable information concerning the functional anatomy of HSV DNA. It appears that HSV-1 and HSV-2 DNAs are largely if not entirely colinear. Recombinants formed by exchange of entire L and S components by exchange of regions such that they contain heterogeneous inverted repeats cannot be differentiated from the wild prototypes with respect to their ability to reproduce. It was

also of interest to note that preferred crossover sites exist and that most of the recombinants were formed by multiple crossover events. The recombinants also yielded information on the location of the templates for several polypeptides based on the fact that approximately half of the polypeptides of HSV-1 and HSV-2 differ in electrophoretic mobility on polyacrylamide gels.

3. Recent analyses of polypeptides specified by HSV-1 and HSV-2 (Pereira et al., 1977; Fenwick and Roizman, 1977) indicate that a very large number polypeptides specified by HSV-1 and HSV-2 are processed following synthesis. The immediate posttranslational modifications involve phosphorylation and possibly other modifications that result in an increased apparent molecular weight.

4. It has been known for several years that fusion of cells induced by HSV is recessive in the sense that cells doubly infected with polykaryocyte inducing and noninducing strains do not cause cell fusion. Recent studies by P. G. Spear and Associates (R. Manservigi, P. G. Spear, and A. Buchan, P.N.A.S., 74, 3913–3917, 1977) show that fusion is determined by two genes. The product of one designated as a fusion inhibitor precludes the fusion of cells. Mutations in the inhibitor genes produce perfectly viable progeny except that such virus strain mutants are capable of causing fusion. Mutations in the fusion promoting gene are of 2 kinds. One class of *ts* mutants (P. G. Spear, personal communication) do not fuse cells at the nonpermissive temperature. Moreover, the progeny made at this temperature is enveloped but is incapable of infecting cells. The second class of mutants specify fusion promoters that are no longer sensitive to the presence of functional fusion inhibiting gene product. Marker rescue experiments with restriction endonuclease fragments show that the two genes map in different regions of the DNA (D. Knipe, W. Ruyechan, and B. Roizman, manuscript in preparation).

Author Index

Numbers in brackets are reference numbers and indicate that an author's work is referred to although the name is not cited in the text. Italic numbers give the page on which the complete reference is listed.

Aaij, C., 603[141], *657*
Aaronson, S. A., 446[21], 474, *539, 540*
Aaslestad, H. G., 190[122, 127], 191[122], 203[195], 226[195], 227[195], 228[195], 229[195], 269[122], *275, 277*
Abady, I., 793[101], 827, *841*
Abdussalam, M., 854[26], *883*
Abel, M., 642[526], *669*
Abel, P. M., 867[140], *886*
Abelseth, M. K., 180[84], *274*
Abinanti, F. R., 351[1], *374,* 384[8], 386[8], *424*, 541[18], 543[18], *580*
Ablashi, D. V., 778[20], 822[20, 239], *838, 845*
Abodeely, R. A., 798, *842*
Abraham, A., 823[246], *846*
Abraham, G., 67[27], *106,* 229 [224], 230[224], 233[244, 225], 246[225], *278*, 374[272], *382*
Abrahams, J., 676[78], 686[78], 688[78], 732[78], 753[78], *759*
Abrahams, P. J., 638[497], 642 [497], *668*
Abrams, R., 628[368], 629[368], *664*
Acheson, N. H., 114[9], 115[26], 118[9], 123[26], *160*, 623[311], *662*
Ackerman, W. W., 614[248], *660*
Acs, G., 398[149], 400[156, 161], 401[165, 166], 402[149, 156], 403 [149, 181], 405[149, 156, 161], 406[149, 206], 407[207], 408[208, 211], 409[212], 410[219], 411[181, 212, 219], *429, 430, 431*
Action, J., 50[100], *59*
Ada, G. L., 302[97], 306[97], *344*
Adamova, V., 543[77], 550[77], 566[77], *581*
Adams, A., 780, *840*
Adams, J. M., 67[28], 367[2], *374, 378,* 719[1], *757*
Adams, W. R., 363[3], *374*
Adamson, R., 54[133], *60*
Adelstein, S. J., 814[201], *844*
Aden, D., 645[549], *670*

Volume 2 comprises pages 539-900.

Adler, R., 233[225], 246[225], *278*
Adler, S., 562, *584*
Adles, C., 143[139], *164*
Aebi, U., 20[1], *35*
Agabian-Keshishian, N., 20[77], 21[77], *38*
Agaphono, L. V., 368[29], *375*
Agol, V. I., 91[166], *110*
Aguet, M., 363[6], *374*
Ahl, R., 95[141], *109,* 147[153], *164*
Ahmed, M., 835[304], *847*
Ahne, W., 174[38, 40], 187[40], 220, *272*
Aitken, T. H. G., 170[14], *271*
Akami, M., 353[223], *381*
Albert, A. E., 596[64], *654*
Alberts, B. M., 727[2], *757*
Aldrich, C., 699[274], 700[274], *765*
Aldrich, C. D., 700[275], *765*
Aleksandrovskaya, I. M., 55[148], *61*
Alexander, D. J., 649[584], *671*
Alexander, V., 424[274], *433*
Alford, R. H., 286[18], *341*
Al-Janabi, J., 89[127], 90[127], *109*
Al-Lami, F., 549[63], 555[63], 559[63], 561[63], 568[197], 569[197], 570[197], *581, 585*
Allen, R., 180[82], *273, 274*
Allende, J. E., 69[53], *106*
Allison, A. C., 591[4], *652*, 744 [114], *760*, 866[122], *886*
Almeida, J. D., 7[3], 15[2], *35,* 169, *271*, 294[67, 236], 306, 333[236], *343, 348*, 591[9], *653*, 822[240], 824[240], *845*
Aloni, Y., 608, 616[275], 617 [275, 276, 287, 291], 620[276], 623[275, 291, 308, 309], 645 [276], 649[589], *659, 661, 662, 671,* 737[3], 752[3], *757*
Alving, C. R., 124[80], *162*
Alwine, J. C., 746[4], *757*, 780 [59, 60], 782[59, 60], *839*
Amano, Y., 358[4], *374*, 389[81], *427*
Amati, P., 613, 615[246], *660*
Amend, D. F., 186[107, 108], *274*
Ames, B. N., 795[109, 110], 797 [110], *841*
Amos, L. A., 4[105], *39*
Anderer, F. A., 15[4], *35*
Anderson, C., 634[447], *667*, 690 [370], 713[97], 728[370], 747 [97], 749[97], *760, 768*
Anderson, C. W., 630[387], 635 [387], *665*, 691[5], 695, 697[5], 706[158, 363], 711[363], 718 [363], 724[5, 157], 725[5, 6, 158], 738[363], 739, 752[6, 157, 158], 753[158, 363], *757, 761, 762, 768,* 891[28], 892[28], 893[28, 29], 896[29], *897, 898*
Anderson, D. M., 604[151], 606 [151], *657*
Anderson, M., 402[176], *430*
Anderson, S., 707[361], 717[361], 727[361], *768*
Anderson, T. F., 698[317], *766*
Ando, T., 604[165], *657*
Andrew, J. C., 815[207], *844*
Andrewes, C. H., 65, 66, 67[7], *105*, 283[14], *341*, 540[1], 551[1], *579*, 851[12], 866[122], *883, 886*
Anfinsen, C. B., 49[76], *59*
Anken, M., 612[243], 625[243], *660*
Ankerst, J., 738[255], *765*
Anschutz, W., 336[217], *347*
Anthony, D. D., 632[304], 638[304], *662*
Apostolov, K., 7[5, 41], *35, 36,* 293[60], 294[60], 300[60], *343*
Appleyard, G., 852[18], 854[18,

[Appleyard, G.]
34], 866[115], 879[228, 231, 233], 880[115, 242], 881[115, 270], *883, 885, 888, 889, 890*
Arber, W., 603[146], *657*
Archetti, I., 561[148], *584*
Archetti, J., 309[137], *345*
Arens, M., 715[243], 726[331, 332], 727[243, 331, 332], 728 [331, 332], 729[122, 332, 338], [121], *760, 764, 767, 768*
Arens, M. Q., 738[70], *759*
Argaman, M., 881[273], *890*
Arhleger, R. B., 877[206], *888*
Arif, B. M., 114[12], 117[12], *160*
Arkhipova, N. A., 55[148], *61*
Armstrong, G. R., 778[20], 822 [20, 239], *838, 845*
Armstrong, J. A., 46[43], *58,* 68 [35], 94[35], *106*, 309[136, 139], *345*, 578[229], *586,* 866[122], 867[130], *886*
Armstrong, R. L., 774[317], *848*
Arnott, S., 387[61], 392[61, 108], *426, 427*
Aronsen, B. E., 385[19], *425*
Arquedas, J. A., 546[41], *580*
Arrand, J., 638[498], 642[498], *668*
Arrand, J. R., 681[177], 683 [222], 687[177], 688[177], 736[177], 750[177], *762, 763,* 895[36], *898*
Arrobio, J. O., 351[128], *378*
Artsob, H., 177[62], *273*
Aruga, H., 422[256], *432*
Asai, J., 361[154], *379*
Asher, Y., 795[112], 796[112], 819[112], *841*
Asofsky, R., 54[125], 55[151], *60, 61*
Asso, J., 85[110], 89[110], *108*
Astell, C., 398[149], 402[149],
[Astell, C.]
403[149, 181], 405[149], 406 [149], 410[219], 411[181, 219], *429, 430, 431*
Astrin, S. M., 647[553, 554], *670*
Atabekov, J. G., 6[65], *37*
Atchison, R. W., 541[24], 552 [92], 554[118], 555[24], 567 [24, 173], 569[118], 570[209], 572[173], 573[173], 575[173], 578[118, 209, 229], *580, 582, 583, 584, 586*
Atger, P., 541[39], *580*
Atherton, K. T., 868[149], 869 [149], *886*
Atkins, G. J., 148[157], 149, *164*
Atkins, J. F., 706[158, 363], 711 [363], 718[363], 724[157], 725[6, 158], 738[363], 752[6, 157, 158], 753[158, 363], *757, 761, 762, 768,* 891[28], 892 [28], 893[28, 29], 896[29], *898*
Atkinson, P. H., 258[263], *279*
Atoyantan, T., 352[5, 105], *374, 377,* 595[54], *654*
Aubertin, A. M., 864[89], *885*
Aubertin, A. M., 864[89], *885*
Aucker, J., 814[204], 815[211], *844*
August, J. T., 202[184], *277*, 387 [52], 392[52], *426*, 451[26], 453[26], 454[26], 455[26], 462[26], 472, 473, 535[108], *539, 540, 542*
Aurelian, L., 823[247], 827[272], 828[272], 831[272, 289], *846, 847*
Austin, J. B., 385[13], *425*, 552 [90], 569[90], *582,* 679[55], 680[55], *758*
Austin, P. E., 701[323], 703[323], 705[323], 707[323], 708[323], 709[323], 712]323], 726[323], 754[323], *767*

Avakyan, A. A., 877[205], *888*
Avery, R. J., 313[158], 315[172], 318[172], *346*
Avila, J., 613[345], 625[345], 632 [345], 635[345, 463], 636[345], 637[345, 463], 638[345], 652 [345], *663, 667*
Aviv, A., 635[467], *667*
Aviv, H., 635[468], *667*
Axelrod, D., 610[218], 617[284, 285], *659, 661*
Axelrod, N., 624[319], *662*

Babcock, V. I., 778[19], 826[19], *838*
Bablanian, R., 723[7], *757,* 880 [247, 248], *889*
Bach, M. K., 881[269], *890*
Bachenheimer, S., 690[8], 735 [69, 339], [374], *757, 759, 768,* 780[50], *839*
Bachenheimer, S. L., 774[67], 780[56, 67], 781[56], 789 [88, 89], 800[137, 140], 801 [137, 140], 829[140], *839, 840, 842,* 891[2], *897*
Bächi, T., 293[61], *343,* 363[6], *374*
Bachmann, P. A., 174[38, 40], 187[40], 220, *272,* 540[2], 543[125], 555[125], 568[125], 569[125], *579, 583*
Bader, J. P., 248[247], *278,* 436 *538*
Baer, G. M., 43[10], 45[10, 41], *57, 58,* 179[67], 180[84], *273, 274*
Baer, P. N., 553[111], *582*
Baguley, B. C., 394[129], 395[129], *428*
Bailey, E. J., 390[90], *427*
Bailey, N. D., 458[36], *540*
Bailey, R. B., 778[19], 826[19], *838*
Bakhle, Y. S., 813[190], 816[190], *844*
Ball, F. R., 880[243], *889*
Ball, L. A., 51[113], *60,* 374[273], *382,* 862[66], *884*
Baltimore, D., 64[4], 67[4, 15, 18, 19], 68[36, 38, 43], 73[4], 73 [4], 75[4, 84], 76[4], 78[4], 79[4], 80[4, 18, 19, 102], 81 [18], 83[4], 85[19, 109, 110, 111], 86[115], 87[115], 88[4], 89[109, 110], 92[4, 131], 93 [4], 94[4, 36, 38, 43], 95[43], 96, 99[109, 151, 154], 100[84, 155, 156, 157], 101[151, 155, 158, 159, 160, 162], 102, 104, *105, 106, 107, 108, 109, 110,* 132, 139[201], *163, 166,* 190 [125], 201[172], 203, 226, 227 [217], 239[237, 238], 243[237], 246[238], 248[250], 257[238], 260, 264[270], 267[297], 268 [297], 269[250], 270[250], *275, 276, 277, 278, 279, 280,* 334[206], 335[206], *347,* 350 [7], 352[8], 366[107], 367[8, 107], 370[167], 373[106], *374, 377, 379,* 436[1], 454[29], 483 [1], 531[103], 532[103], 534 [106], *538, 539, 542*
Baluda, M., 353[208], *380*
Baluda, M. A., 302[104], 304[109, 112, 155], 317[109, 112, 115], 318[112], 319, 320[109, 115], *344*
Bancroft, J. B., 15[6], *35*
Banden, M. T., 54[136], *61*
Banerjee, A. K., 67[27, 31], *106,* 202[186, 187], 229[187, 224], 230[187, 224], 233[224, 225], 236[186, 226], 239[240], 246 [225, 240], 257[240], *277, 278,* 370[20], 374[272], *375, 382,* 394[117, 118, 119, 130], 395 [130, 134], 400[155, 160], 401[168], 404[193], 405[155, 160], 406[118, 119, 160, 204], 408[210], 410[168, 193], *428, 429, 430, 431,* 719[16], *757,*

[Banerjee, A. K.]
863[78], *884*
Banfield, W. G., 770[2], 826[2], *838*
Bang, F. B., 283[14], 290[41], 307[122], *341, 342, 345*
Baranska, W., 648[567], *670*
Barban, S., 47[49], *58*, 597[86, 89], 599[89], *655*
Barbanti, A., 596[67], *654*
Barbanti-Brodano, G., 614[252, 255], *660*
Barclay, M., 541[14], 543[103], 552[103], 553[14], 567[14], *579, 582*
Barclay, R., 560[145], 570[145], *584*
Barnett, J. W., 787[82], *840*
Baron, S., 41[16], 43[5, 9, 10], 45[10, 34, 35, 41], 47[49, 52], 54[131, 133, 135], 55[143], *56, 57, 58, 60, 61*
Baroni, A., 825[252], *846*
Barrett, A. N., 4[7], *35*
Barrington-Leigh, J., 4[7], *35*
Barron, A. L., 835[305], *847*
Barry, R. D., 282, 302[102, 108], 307[125], 313[152, 153], 317[108], *341, 344, 345*, 350[12], 353[11], 365[9, 10], 367[155], 369[11], 373[12], *374, 379*
Bartlett, N. M., 401[164], 405[164], *429*
Bartok, K., 604[167], 605[167], *657*
Barwise, A. H., 562[154], *584*
Baserga, R., 625[338], 641[524], *663, 669*
Basilico, C., 613, 614, 631[405], 636[482], 641[405], 642[529], *660, 665, 668, 669*, 713[184], *762, 866[106]*, 885
Bates, R. C., 543[79], 551[79], 552[79], 555[120], 568[120], 570[120], *581, 583*
Bates, S. R., 202[177], *276*
Bauer, D., 881[266], *890*
Bauer, D. J., 881[263], *889*
Bauer, H., 439[16], 533, *539, 542*, 853[19], 854[19], *883*
Bauer, S. P., 179[71, 72], 180[71], 181[71, 72], *273*
Bauer, W., 602[127, 131], 603[138], *656*
Baum, P. R., 691[5], 695[5], 697[5], 698[5], 706[158, 363], 711[363], 718[363], 724[5], 725[5, 158], 738[363], 752[158], 753[158, 363], *757, 762, 768*, 893[29], 896[1, 29], *897, 898*
Baum, S., 634[452, 457, 458], 635[452, 458], *667*
Baum, S. G., 402[174], 410[174], 420[174], *430*, 634[450, 451], *667*, 718[221], 726[112], 729[112], 738[9], 739[9], 744[228], *757, 760, 763, 764*
Bayer, M. E., 15[8], 20[8, 115], *35, 39*
Bayley, S. T., 631[410], 632[410], *665*
Bayliss, G. J., 811[174], *843*
Baz, T. I., 351[229], *381*
Beale, A. J., 351[13], *375*
Beals, T. F., 543[60], 549[60], 554[60], *581*
Bean, W. J., 314[164, 228], 315[173], 318, 336[242], *346, 348*
Bearcroft, G., 866[122], *886*
Bearcroft, W. G. C., 866[121], *886*
Beard, J. W., 592[13], *653*
Beard, P., 605[169], *658*
Beato, M., 135[122], *163*
Beatrice, S. T., 19[100], 34[100], 35[100], *38*
Beaud, G., 862[67], *884*
Beaudreau, G., 439[16], *539*
Becht, H., 293[57], 294[74], 299[92], 300[94, 95], 302[74], 308[57], 309[95], 329[204],

[Becht, H.] 332[204], *343, 344, 347,* 361 [220], 364[220], *381*
Becker, Y., 67[26], 75[77], 76 [77], 78[77], 80[26, 77], *106, 107,* 132[109], *163,* 772[58], 780[58], 795[112], 796[112], 798[131], 802[148], 811[148], 819[112], *839, 841, 842,* 855 [43], 868[147], 870[147], 876 [198], 878[198], 880[147], 881[278], *883, 886, 888, 890*
Beddow, T. G., 578[234, 235], *586*
Beemon, K., 459[37], 460, 502 [37], 504[37], *540*
Beerman, T. A., 605[172], *658*
Beeson, J., 862[63, 68], *884*
Begin, M., 712[10], *757*
Begin, M., 712[311], 713[311], *766*
Bell, J., 6[103], *38*
Bell, J. A., 387[67], *426*
Bell, P., 142[137], *164*
Bell, T. M., 385[12, 24, 25], 400 [12], *424, 425*
Bellamy, A. R., 386[47], 387[47, 52, 56, 64], 389[56], 391 [47], 392[52, 64], 394[127, 128, 129, 131, 132], 395[56, 129, 131, 132], 401[163, 164], 403[64], 404[64], 405[163, 164], 409[127], 410[131, 132], 423[266], *426, 428, 429, 432*
Bellett, A. J. D., [9], *35,* 148[158], 149[158], 151[158], *164,* 269, *280,* 681[225, 335, 336], 683 [224, 225], 728[12], 730[225], 733[11], *757, 764, 767,* 851 [11], *883,* 894[35, 37], 895 [35], *898*
Bello, L. J., 633[437], *666,* 716 [74], 720[74], 726[74], *759*
Belova, S., 634[441], *666*
Benda, R., 179[76], *273*
Ben-Hamida, F., 862[67], *884*
Ben-Ishai, Z., 617[286, 290], *661,* 881[278], *890*
Benjamin, T., 636[480, 481], 640[480, 481], *668*
Benjamin, T. L., 616[273], 631 [406], 634[440], 639[515], 640[440, 515], 641[406], *661, 665, 666, 669,* 731 *757*
Ben-Porat, T., 614[257], *660,* 779[42], 795[113], 797[113], 802[150], 811[150], 817[225], 818[225, 228, 229], 819[113, 233, 234], 827[228, 271], 828 [276], 829[276], 833[294], *839, 841, 842, 845, 846, 847*
Bentvelzen, P., 437, 439[16], *538, 539*
Benveniste, R. E., 463[47, 48], 488[48, 70], 509[88], 511, 514, 516[48], 536[90], *540, 541*
Benyesh-Melnick, M., 772[21], 778[21], 779[29], 780[53, 57, 70], 789[90, 91], 819[70], *838, 839, 840, 841*
Berckmans, R., 404[197], 407 [197], *430*
Bereczky, E., 309[137], *345,* 561 [148], *584*
Berezesky, I. K., 131[100, 101], *162,* 634[445, 446], *666,* 738 [217], *763*
Berg, K., 49[74, 77], *59*
Berg, P., 95[140], *109,* 604[154], 605[154, 169, 173], 607[154], 608[196, 198, 202], 610[196, 198], 611[196, 198, 202], 612 [202], 633[436], 637[491, 492], 638[491, 492], *657, 658, 659, 666, 668,* 713[175], 746 [176], 756[247], *762, 764*
Berg, V. V., 578[231, 232], *586*
Berge, T. O., 170[30], 174[30], 183[30], 215[30], *272*
Berget, S. M., 721[340], 728 [348], *768,* 891[3], 892,

[Berget, S. M.]
893[3], *897*
Bergoin, M., 854[28], 863[84], 864[84], 877[210], *883, 885, 888*
Bergold, G. H., 170[15], *271*
Berissi, H., 52[118], *60*
Berkaloff, A., 174[34], 185[34], 260[274], 267[292], *272, 279, 280*
Berkeley, W., 594[39], *653*
Berman, B. J., 49[80], 50[80], *59*
Bernard, J., 185[310], *280*, 593[34, 35], *653*
Bernard, J. P., 366[14], *375*
Bernardini, A., 616[274], 617[274], 623[274], *661*
Bernhard, W., 560[144], 636[478], *667*, 775, *838*
Bernkopf, H., 880[244], *889*
Berns, K. I., 559[150], 560[150], 562[152], 563[157], 564[142, 157], 565, 571[142], 576[225], *583, 584, 586,* 683[14], *757,* 855[44, 45], 857[45], 865[45], 872[178], *884, 887*
Berry, K., 611[225], *659*
Berry, K. W., 564[158], *584*, 683[67, 68], *759,* 895[21], *898*
Berthelot, N., 780[75], 782[75], *840*
Berthiaume, L., 350[120], *378*
Besmer, P., 267[297], 268[297], *280*
Bhaduri, S., 715[243], 727[243], 738[70], *759, 764*
Bhatt, P. N., 170[16], 181[89], *271, 274*
Bhayan, B. K., 43[7], 54[7], *56*
Bialy, H. S., 53, *60*
Bianchi, P. A., 625[333], *663*
Bibring, J. B., 20[92], *38*
Biggs, P. M., 787[85], *840*
Bijlenga, R., 20[1], *35*
Bikel, I., 360[15], *375*
Bilello, P., 69[61], *107*
Billheimer, F., 626[349], *663*
Billiau, A., 45[24], 47[55], 48[70], *57, 58, 59*
Bils, R. F., 18[10], 19[10], *35*
Binn, L. N., 352[16, 149], *375, 379,* 541[32], 543[32, 80], 551[32, 80], 552[80], 555[32, 80], 569[80], *580, 581*
Birdwell, C. R., 115[28], 124[77], 140[132, 134], 143[28, 132, 141], 144[203], 145[28], *160, 162, 163, 164, 166*
Bishop, D. H. L., 190[118, 119, 122, 123, 126], 191[118, 119, 122, 129, 131], 192[131], 193[129, 131], 194[135, 138], 195[123, 135, 138], 196[123, 135, 138], 197[118, 135, 142, 144], 198[150], 199[123, 138, 150], 200[142, 170], 201[138], 202[142, 183, 188], 203[119, 123, 129, 142, 195, 196, 197, 198, 200], 206[138], 218[138], 223[214, 215], 224[123, 144], 226[119, 195, 216], 227[123, 183, 188, 195, 200, 215, 219, 220], 228[123, 195, 196, 197, 198, 200, 216, 221, 222], 229[195, 196, 222], 230[197, 222, 314, 315], 231[126, 222], 232[196, 222], 233[197], 234[126], 237[197, 198], 238[142, 170, 200], 239[144, 170], 240[144, 197, 214, 215], 241[144, 197, 214], 242[215], 243[144, 214, 215], 244[215], 245[215], 250[144, 215], 251[144, 215], 252[197], 256[144, 215], 261[144], 262[144, 197, 215], 263[215], 264[144, 197, 215, 278], 265[144, 197, 215], 268[150], 269[118, 119, 122, 129, 198, 200], 270[309], *275, 276, 277, 278, 279, 280,* 302[99], 314[163, 164],

[Bishop, D. H. L.]
335[208], *344, 346, 347, 357, 380*
Bishop, J. M., 67[25], 80[25], 92, 94[134], *106, 109,* 459 [41], 496[77], 517[82], 519 [93], *540, 541, 542,* 733[298], *766*
Bishop, R. F., 385[26], *425*
Biswal, N., 772[21], 778[21], [47], 780[53, 57, 70], 789 [90, 91], 819[70], *838, 839, 840, 841*
Black, F. L., 365[35], *375*, 546 [49], *581*
Black, L. M., 422[257, 258], *432*
Black, M. M., 535[111], *542*
Black, P. H., 55[149], *61*, 593[33], 601[118, 124], 612[235, 241], 631[390, 395, 397], 648[390], 650[599, 600, 601, 603], 651 [608], *653, 656, 660, 665, 671, 672,* 679[55], 680[66], 744 [15], *757, 758*
Blacklow, N. R., 541[26], 545[26], 552[89, 90, 91], 555[26], 560 [139], 567[26, 139, 174, 175], 569[26, 90, 207], 570[175, 210], 572[175, 210], 573[139, 210], 574[139, 210], 575[174], 578 [26, 210, 237], *580, 581, 583, 584, 585, 586, 587*
Blackman, K. E., 880[239], *889*
Blackstein, M. E., 605[187], 610 [187], *658*
Blaese, R. M., 595[62], *654*
Blair, C. D., 353[17], 369[17], 370[17], 371[17], 372[18], *375*
Blair, G. E., 697[230], 710[230], 724[230], 727[230], *764*
Blakesley, R. W., 814[205], *844*
Blangy, D., 601[117, 120], *656*
Blank, M., 825[256], *846*
Blasi, F., 613[246], 615[246], *660*
Blaskovic, D., 45[23], *57*
Blatti, S., 313[161], 314[161], 320[161], *346*
Blödorn, J., 298[239], 326[239], *348*
Blomberg, J., 792[95], 793[95], *841*
Bloom, M., 546[47], *580*
Blough, H. A., 289[37], 290[40, 43], 292, 307[40], *342*, 357 [19], 360[19, 242], 361[243], 362[76], *375, 377, 381*
Bocharov, A. F., 15[8], 20[8], *35*
Bock, H. O., 20[92], *38*
Bock, S., 20[92], *38*
Bodo, G., 299[89], *344*
Boezi, J. A., 814, 815[212], *844*
Boggs, J. D., 541[34], 543[34], *580*
Bohl, E. H., 552[86], 553]86], *582*
Boime, I., 67[20], 80[20], 85[20], *105*
Boiron, M., 593[34, 35], *653*
Bolden, A., 815[211], *844*
Bollum, F. J., 862[71], *884*
Bolognesi, D., 439[16], *539*
Bolognesi, D. P., 451[26], 452 [28], 453[26], 454[26], 455 [26], 462[26], 473[28], 476 [63, 64, 65], *539, 541*
Bone, D. R., 808[159], *843*
Bonner, J., 599[94], *655*
Bootsma, R., 174[37], 187[37, 112], *272, 274*
Borecky, L. V., 45[23], *57*
Borgese, N. G., 594[38, 40], *653*
Borman, E., 595[48], *654*
Bornkamm, G. W., 780[55], *839*
Bornstein, M. B., 820[236], 824 [236], 826[236], *845*
Borsa, J., 391, 397[102, 103, 143], 400[154], 401[167], *427, 429*

Borst, P., 603[141], *657*
Bortolussi, R., 385[28], *425*
Bose, H. R., 117[41, 46], 132[46], 135[140], 143[140], 147[152], 152[140], *161, 164*
Bose, S., 47[58], 48[68], 49[76], *58, 59*
Bosisio-Bestetti, M., 825[250], *846*
Botchan, M., 643[532], 644[538, 539], 645[538], 648[572], 649[539, 572], 651[572], *669, 670,* 690[236], 732[236], 753 [236], *764*
Both, G. W., 67[31, 32], *106*, 236 [226], 239[240], 246[240], 257[240], *278*, 358[171], 367 [171], 370[20], *375, 379,* 395 [139], 398[139], 401[261], 404[139], 406[261], 407, 408 [209, 210, 236, 261], *428, 431, 432,* 719[16], *757,* 863[78], *884*
Boucher, D. W., 543[77], 550[77], 552[93], 566[77], 569[200, 201], *581, 582, 585*
Boulanger, P., 570[208], *586*
Boulger, L. R., 175[50], *272*
Boulter, E. A., 879[231, 233], *889*
Bouton, R. W., 114[13], 117[13], 119[13], *160*
Bourgaux, P., 614[250], 626[350, 354, 359], *660, 663, 664,* 728 [223], *763*
Bourgaux-Ramoisy, D., 626[350, 354, 359], *663, 664,* 728 [223], *763*
Bourgignon, M. F., 605[177], *658*
Bourne, G. H., 179[67], *273*
Bowen, J. M., 787[82], *840*
Bowling, C. P., 773[44], 774[44], *839*
Boxaca, M., 43[13], *57*
Boyd, V. A. L., 649[588], 650 [588], *671*
Boy de la Tour, E., 356[143, 144], 357[144], 358[143], *378*
Boyer, H., 620[303], 622[303], *662*
Boyer, H. W., 603[144], 609[210], *657, 659*
Bracha, M., 139[142], 152[204], 153[142], *164, 166*
Brackman, K. H., 718[364], *768*
Brackmann, K., 688[327], 717 [327], 719[103], 720[327], 749[327], 750[327], *760, 767*
Brackmann, K. H., 688[326], 715 [326], 717[326], 720[326], 723[326], 749[326], *767*
Bradley, D. E., 7[12], 15[11, 12, 64], *35, 37*
Bradshaw, R. A., 89[128], 90 [128], *109*
Brailovsky, C., 547, 555[58], 557 [58], 566, *581, 584*
Brakel, C., 863[74], 869[74, 157], 873[74, 157], *884, 887*
Brakke, M. K., 546[50], *581*
Brammer, K. W., 314[169], 315 [169], 316[169], *346*
Brand, C. M., 293[59], 294[236], 298[82], 306, 333[236], *343, 348*
Brandlow, G., 880[241], *889*
Brandner, G., 620[302], 622 [305], *662*
Brandt, C. D., 351[128], *378*, 385 [17], *425,* 552[91], *582*
Brandt, W. E., 114[18, 20], 119 [18, 20], 123[18], 124[18], 139 [20], 146[148], *160, 164*
Branton, D., 20[13], *35*
Bras-Herreng, F., 154[177], *165*
Bratt, M. A., 227[217], *278,* 353 [21], 360[89], 363[22, 23, 45, 77], 364[46, 89], 365[21], 366[107], 367[46, 89, 107, 258], 369[21, 46], 370[21, 24, 258], 371[21, 44], 373[263], *375,*

[Bratt, M. A.]
376, 377, 382
Braun, W., 55[141], *61*
Braunwald, J., 866[105], *885*
Bräutigam, A. R., 373[268], *382*
Brawner, T. A., 117[45], 132 [45], *161*
Brayton, C., 726[112], 729 [112], *760*
Breeden, J. H., 731[152], 749 [152], *761*
Breeze, S. S., Jr., 541[13], 555 [13], *579*
Bregliano, J. C., 174[34], 185 [34], *272*
Breindl, M., 74, *107,* 374[274], *382*
Breinig, M. K., 43[7], *57*
Breitburd, F., 592[20], *653*
Breitenfeld, P. M., 324[184], *346*
Breitmeyer, J., 604[158], 607 [158], *657*
Brendler, T. G., [380], *768*
Brenner, S., 16[55], *37*, 690[109], *760*
Bridger, J. C., 423[262, 263], *432*
Britten, R., 644[535], *669*
Britton, R. J., 719[35], *758*
Brockman, W. W., 605[192, 194], 608[194, 199], 609[192, 199], 610[194, 199], 611[192, 199], 628, 649[194], *658, 659*
Brodano, G., 596[67], 650[604], *654, 671*
Brodeur, B. R., 55[146], *61*
Broek, J. v. D., 20[1], *35*
Broek, R. v. D., 20[1], *35*
Broker, T. R., 891[8], 892[8], 893[9], *897*
Bronson, D. L., 780[53, 57], 789 [90], *839, 841*
Brooksby, J. B., 170[13], 206[13], *271*
Brotherus, J., 121[62], 158[62], *161*
Brouty-Boye, D., 54[127, 130, 136], *60, 61*
Brown, A., 147[159], *164*, 625 [340], 631[340], 641[340], *663*, 880[245], *889*
Brown, D. T., 118[52, 54], 123 [135], 130, 141[135], 143[52, 97], 146[144], 152[97], 154 [54], 156[196], 158, *161, 162, 163, 164, 165,* 697[17], 698 [17], 699, 716[41], *757, 758,* 894[4], *897*
Brown, F., 65[8, 9], 68[45, 46], 70[64], 71, 72[64], 91[165], 94[46], 99[152], *105, 106, 107, 110,* 122[71], *162*, 173[212], 191[128], 194[134], 195[210], 196[134], 197[145], 198[148], 199[148], 200[145, 167, 168, 171], 202[180], 204, 205, 206, 211, 212[171], 213, 214[205, 212], 221[210], 238[167], 248[249], 269[128, 203, 205, 301, 302], *275, 276, 277, 278, 280,* 351[245], *381*
Brown, G. E., 423[273], *433*
Brown, M., 612[232], 619[232], 620[232], 648[573], *660, 670,* 862[71], *884*
Brown, P., 351[25], *375*
Brown, S. M., 793[98], *841*
Brubaker, M. M., 390[86], *427*
Brugge, J., 638[504, 505], 639 [504, 505], *668*
Brugge, J. S., 705[18], 731[18], *757*
Brunell, P. A., 779[40], 820[40], 823[40], *839*
Brunschwig, J. P., 779[29], *838*
Bruschi, A., 356[142, 144], 357 [144, 145], *378*
Bruton, C. J., 359[198], *380*
Bryans, J. T., [47], *839*
Bryden, A. S., 385[27], 423[262, 263], *425, 432*
Bubel, H. C., 880[239], *889*

Buchan, A., 812[185], 813, 814 *844*
Bucher, D. J., 292[44], 298[44], *342*
Buchholz, M., 75[73], 80[73], *107*
Buchman, T. G., 791[320], *848*
Buchsbaum, R., 813[193], 814 [193], *844*
Buckler, C. E., 41[16], 43[10], 45 [10, 34], 47[49], *56, 57, 58*
Buckley, S. M., 177[60], 182[60], *273*
Buescher, E. L., 113[3], *159*
Buetti, E., 616[270], 617[270], 623[270, 311], 638[270], *661, 662*
Bukovic, J. A., 55[143], *61*
Bukrinskaya, A. G., 353[11], 359 [247, 265], 365[26, 27], 367 [28, 30], 368, 369[11], *374, 375, 381, 382*
Bulicheva, T. I., 55[148], *61*
Bullivant, S., 401[163, 164], 405 [163, 164], *429*
Bumgarner, S. J., 521, *542*
Burdon, M. G., 555[121], 559 [121], 561[121], 562[121], 564[121], *583*
Burdecea, O., 365[26], *375*
Burge, B. W., 114[7, 16], 115 [29], 118[7, 29, 47], 121[7, 16, 65, 66], 123[75], 124 [7], 134[7, 125], 137[7], 140 [133], 141[47], 143[75, 138], 146, 147, 148[149], 149, 150, 151[47, 162], 152[47, 165, 166], 153[169], *160, 161, 162, 163, 164, 165,* 198[153], *276,* 363 [31, 238], *375, 381*
Burgei, H., 716[41], *758*
Burger, M., 625[341], 634[341, 440], 639[341], 640[341, 440], *663, 666*
Burgess, G. W., 174[22], 184[22], 185[22], *271*
Burgoon, C. F., 825[256], *846*
Burgoyne, L. A., 599[104, 105], *655, 656*
Bürk, R. R., 592[27], *653*
Burke, D. C., 45[31, 32, 33, 36], 48[60], *57, 58,* 127[199], 135, *163, 165,* 310, *345,* 353[162], *379*
Burlingham, B. T., 697[17], 698 [17], 699[17], *757,* 894[4], *897*
Burmester, B. R., 774[67], 780 [67], 824[249], 827[267], *840, 846*
Burness, A. T. H., 69[60], *107*
Burnet, F. M., 283[14], 289[39], 307[129], 308[132], 336 [212], *341, 342, 345, 347*
Burnett, J. P., 690, *757*
Burns, W. H., 650[599, 600], *671*
Burroughs, J. N., 65[9], *105*
Burrows, R., 170[13], 206[13], *271*
Buss, J., 615[264], 616[264], *661*
Bussell, R. H., 351[254], 352[94], 358[32], 359[255], *375, 380, 382*
Bussereau, F., 174[98], 185[98, 101], *274*
Butel, J., 638[504, 505], 639[504, 505], *668*
Butel, J. S., 591[4], 592[18], 631 [393, 396, 399], 634[448], 648[565], 649[588], 650[588], *652, 653, 665, 667, 670, 671,* 705[18], 731[18], *757*
Butler, P. J. G., 5[15, 16], 17[14], 18[14], *35, 36*
Butterworth, B. E., 70[106], 83 [106], 84[106], 85[106, 114], 86[106, 118, 119], 87[119], 88[106], 91[119], *108, 109*
Büttner, W., 718, 736[341], 751, *757, 768,* 891[5], *897*

Byckovsky, A. F., 877[205], *888*
Bykovsky, A. F., 438[15], 457 [15], 480[15], *539*
Byrne, J. C., 611[224], 616[272], 618[272], 619[272], 620[272], 638[499], 646[499], *659, 661, 668*

Caffier, H., 677[88], 679[88], 688[88], 717[88], 718[88], 720[88], 732[88], 733[88], 734[88], 753[88], *759*
Caggiano, C. H., 170[12], *271*
Cairns, J., 626[353], *663*, 877 [200], *888*
Cairns, J. E., 261, *279*
Calberg-Bacq, C. M., 361[33], *375*
Caliguiri, L. A., 75[[74, 75], 80 [74, 75], 92[74, 75], 94[137], 97[146], 102, *107, 109, 110,* 292[49], 314[166, 167], 315 [166, 167], 316[166, 167], 323[166, 167], *342, 346,* 358 [168], 360[168, 213], 361 [213], 362[138], 364[213], 362[138], 364[213], 362[138], 364[213], *378, 379, 380*
Callahan, R., 463[47, 48], 476 [61], 486, 487[61], 488[48], 516[48], *540*, 752[316], *766*
Callan, E. A. O., 493[74], *541*
Callender, J., 636[487], 639[507], *668*
Camargo, E., 362[90], *377*
Came, P. E., 45[28], *57*
Campadelli-Finme, G., 801[144], *842*
Campbell, C. H., 288[32], 289[32], 307[32], 336[32], 339[226], *342, 347*
Campbell, J. A., 827[269], 833, *846*
Campbell, R. N., 7[52], *37*
Cancedda, R., 126, 135[119], 139 [129], 143[129], *162, 163*
Cann, B. W., 194[135], 195[135], 196[135], 197[135], *275*
Cant, B., 546[42], *580*
Cantell, K., 49[77], *59*, 350[34], 351[34], *375*
Cantor, H., 54[125], *60*
Carbon, J., 608[198, 202], 610 [198], 611[198, 202], 612 [202], *658, 659*
Cardiff, R. D., 114[20], 119[20], 139[20], *160*, 456[32], *539*
Carey, N. H., 68[44, 45, 46], 94 [46], *106*
Carl, G. Z., 147[152], *164*
Carmichael, L. E., 777, 778[12], *838*
Carncy, P. G., 634[443, 449], *666, 667*
Carnighan, J. R., 20[92], *38*
Caroline, L. N., 54[126], *60*
Carp, R. I., 617[279], 620[279], 622[279], 631[394, 404], *661, 665*
Carroll, R. B., 632[426], 633 [426], 648[426], *666*, 746 [22], *757*
Carski, R. T., 179[69], *273*
Cartas, M. A., 688[326], 715[326], 717[326], 720[326], 723[326], 749[326], *767*
Carter, B. J., 560[140], 564[140], 565[166], 566[166], 567[177, 194], 568[177], 575[222], 577 [140, 194], *583, 584, 585, 586,* 683[137], *761*
Carter, C., 365[35], *375*, 395[134], *428*
Carter, T. H., 720[342], 721[342], *768*
Carter, W. A., 47[57], 48[67, 69, 71], 49[57, 69, 71, 78], 50 [101, 51[111], *58, 59, 60*
Carter, W. B., 126[83], 127[83], *162*

Cartwright, B., 197[145], 198 [148], 199[148], 200[145, 167, 168, 171], 202[179, 180], 204[167, 201], 205 [171], 206[171], 211[171], 212[171], 214[205], 238 [167], 258[262], 269[205, 301, 302], *275, 276, 277, 279, 280*
Cartwright, K. L., 127[199], *165*
Cartwright, S. F., 543[95, 96], 552, 553[97], 568[96], *582*
Carusi, E. A., 895[6], *897*
Casals, J., 112[1, 2], 113[1], *159*
Casazza, A. M., 578[230], *586*
Casio, L., 779[40], 820[40], 823[40], *839*
Casjens, S., [23], *757*
Caspar, D. L. D., 2[19, 20, 21, 68], 3[18, 19, 68], 5[18], 8[21], 16[17, 19, 21, 68], 17[17], 19, 23[19], 24[19], *36, 37,* 118 [55], *161,* 357[36, 140], 358 [140], *375, 378*
Cassai, E., 771[5], 780[50], 793 [5, 96], 801[144], 837[316], *838, 839, 841, 842, 848*
Cassai, E. N., 793[97], 794[97], 795[97], *841*
Casto, B. C., 541[24], 554[118], 555[24], 567[24], 569[118], 570[209], 578[209], *580, 583, 586*
Cathala, F., 351[25], *375*
Caul, E. O., 546[43, 44], *580*
Causey, O. R., 175[51, 52, 54], 181[52], 182[54], 216[51, 54], 218[51, 54], *272, 273*
Cavalier-Smith, T., 730[24], *757*
Cellers, J., 286, *341*
Chakraverty, P., 287[24], *342*
Chamberlin, M., 95[140], *109*
Chamberlin, M. J., 45[40], *57*
Chambers, V. C., 186[107], *274*
Chambon, P., 599[101], 600[110], 602[110], *655, 656*
Champoux, J. J., 624[321], 626 [352], 627[352], 630[133], *656, 662, 663*
Chan, S. K., 69[52], *106*
Chandler, J. K., 830[285], *847*
Chandra, S., 541[14], 543[103], 546, 552[103], 553[14], 567 [14], *579, 581, 582*
Chaney, C., 541[13], 555[13], 566, *579, 584*
Chang, C., 560[144], *584*
Chang, G.-T., 402[180], 404[180], 405[180], *430*
Chang, K. S. S., 604[150], 606 [150], *657*
Chang, L. M. S., 872[181], *887*
Chang, P. W., 823[245], *846*
Chang, S. H., 203[196], 228[196], 229[196], 232[196], *277*
Chang, S. S., 304[232], 317[232], 320, *348*
Chaniotis, B. N., 176[59, 213], *273, 277*
Chanock, R. M., 283[16], [230], *341, 348,* 350[37], 351[1, 37, 128], 362[90], *374, 375, 377, 378*, 385[29], *425*, 546[40], 551 [40], 552[91], *580, 582*
Chapeville, F., 69[50, 51], *106*, 862[67], *884*
Chapman, J. D., 391[102, 103], 397[102, 103, 143], 401[167], *427, 429*
Chardonnet, Y., 677[257], 714 [25, 32], *757, 758, 765,* 867 [129], *886*
Charlwood, P. A., 298[82], *343*
Charman, H. P., 457[34], 465 [49], 466, 467[49], 468[49], 471[34], 480[34], 509[49], 537[116], *539, 540, 542*
Chattopadhyay, S. K., 437[11], 521[11], *539*

Cheatham, W. J., 820[237], *845*
Cheevers, W., 880[259], *889*
Cheevers, W. P., 617[282], *661,* 817[227], 818[227], 827[227], *845*
Chen, C., 361[38], 364[38], *375*
Chen, J. H., 835[303], *847*
Chen, L. B., 895[7, 19], 896[7, 19], *897, 898*
Chen, M., 605[176], *658*
Chen, M. C. Y., 604[150], 606 [150], *657*
Chen, S.-T., 772[43], 779[43], *839*
Chenault, S., 50[100], *59*
Cheong, L., 560, 570[145], *584*
Chernos, V. I., 866[116], 868 [150], 869[150], 881[116], *885, 886*
Cherry, J. D., 387[66], 390 [66], *426*
Chesebro, B., 546[47], *580*
Cheyne, I. M., 358[85], *377*
Chi, L., 880[238], *889*
Chibisova, V., 634[441], *666*
Chin, J. N., 402[172, 178], *429, 430*
Chinn, W. W., 723[343, 344], 725, *768*
Chinnadurai, G., 676[94], 677 [94], 678[94], 720[26], 723 [345], 732[94], 733[94], 735[26], 752, *757, 760, 768*
Cho, H. J., 546[48], *581*
Choppin, P. W., 6[22], 7[22], *36,* 198[149], 202[181], 267[149, 293], 268[149, 293], *276, 280,* 282[6], 289[36], 290[36], 292[56], 297[79], 298[78, 79, 240], 300[237], 302[105], 308[134], 310[78, 134], 312 [148], 317[105], 326[78, 79, 190, 240], 328[237], 329[190], 332[190], 335[105, 134], 336
[Choppin, P. W.] [105], *341, 342, 343, 344, 345, 346, 348,* 352[42, 84], 353[40, 52], 355, 358[50, 51, 54, 168], 359[39, 170, 215], 360[136, 137, 168, 213, 214], 361[38, 169, 213, 214, 215], 362[138, 139], 363[215], 364 [38, 170, 213, 214], 365[41], 368[170], 372[49], 373[40, 43, 53], *375, 376, 377, 378, 379, 380*
Chopra, H. C., 779[33], *839*
Chou, J. Y., 613[345, 417], 625 [345], 632[345, 417, 430], 635 [345, 430, 463], 636[345], 637 [345, 417, 463], 638[345, 506], 639[506], 652[345], *663, 666, 667, 668*
Chow, F. H., 189[117], *275*
Chow, J. Y., 705[168], *762*
Chow, L. T., 609[210], *659*, 891[8], 892, 893, *897*
Chow, N. L., 293[65], 313, 314 [65], *343*
Chow, T. L., 189[177], *275*
Christman, J., 406, 409[212], 410[219], 411[212, 219], *431*
Chu, L., 267[297], 268[297], *280*
Chubb, R. C., 508, *541*
Chumakov, K. M., 91[166], *110*
Citarella, R. V., 872[179], *887*
Claesen, M., 45[24], *57*
Clark, H. F., 178[66], 180[86, 87], 186[109], 190[122, 123], 191 [122, 131], 192[131], 193[131], 194[137, 138], 195[123, 138], 196[123, 137, 138], 199[123, 138], 201[137, 138], 202[137], 203[123, 195], 206[138], 218 [138], 224[123], 226[195], 227[123, 195], 228[123, 195], 229[195], 237[137], 258[137], 269[122], *273, 274, 275, 277,* 360[232], *381*

Clark, I. A., 174[26], *272*
Clarke, D. H., 112[1, 2], 113[1], *159*
Clarke, J. K., 779[34], 824[34], 826[34], *839*
Clarke, S. K., 546[43, 44], *580*
Clarkson, B., 641[524], *669*
Clarkson, S. G., 870[161], *887*
Clavell, L. A., 363[23, 45], 371 [44], *375, 376*
Clegg, J. C. S., 117[42], 132[42], 135[121], *161, 163*
Clifford, P., 789[92], 835[302, 309], *841, 847*
Clifford, R. L., 115[25], *160*
Cline, W. L., 385[29], *425*
Clinton, G. M., 556[137], 557, *583*
Cockburn, W. C., 283[16], *341*
Cocuzza, G., 571, *586*
Cohen, E. P., 877[209], *888*
Cohen, G. H., 258[263], *279*, 816, 818, 830[286], *845, 847*
Cohen, J. C., 795[115], *841*
Cohen, S. S., 829[278], *846*
Colby, C., 45[40], 50[92], 53, *57, 59, 60,* 869[154], *887*
Colby, D. S., 94[138], *109*
Colby, E., 552[98], 553[98], *582*
Colcher, D., 480[68], 535[68], *541*
Cole, C. N., 75[84], 100[84, 155, 156], 101[155, 158, 159], *107, 110*
Cole, G. A., 543[75], 550[75], 554[75], 567[191], 572, *581, 585*
Cole, R., 20[92], *38*
Coleman, D. V., 595[61], *654*
Coleman, M., 282[2], 287[2, 24], *341, 342*
Coleman, P. H., 174[32], 184[32], *272*, 568[198], *585*
Collins, B. S., 364[46], 367[46], [Collins, B. S.] 369[46], *376*
Collins, J. K., 631[403], 632[403], 638[403], *665*
Collins, M. J., Jr., 543[88], 199], 552[88], 568[88, 199], 573 [199], *582, 585*
Colonno, R. J., 360[48], 367[47, 48], *376*
Colter, J. S., 71[63], 83[105], *107, 108,* 610[212], *659*
Combard, A., 252[289], 265[285, 289], 266[289], 267[285], *280*
Comer, J. F., 120[60], *161*
Compans, R. W., 6[22], 7[22], *36*, 97[146], *109*, 122[68], *162*, 198[149], 267[149, 293], 268[149, 293], *276, 280,* 282 [6], 292[49, 55, 56], 293[62], 294[69], 297, 298[78, 83], 299[83, 91], 300[62, 83, 237], 302[69], 307[62], 309[135], 310[78], 314[166, 167], 315 [166, 167], 316[166, 167], 323 [166, 167], 326[78, 190, 191, 201], 328[201, 237], 329[190, 191], 331, 332[190, 191, 201], 333[62], 335[207], 336[62, 207], 337[83], *341, 342, 343, 344, 345, 346, 347, 348,* 352[50, 52], 353 [52], 358[50, 51, 54, 168], 359 [170], 360[168, 213], 361[38, 169, 213], 362[139, 184], 364 [38, 170, 213], 368[170], 372 [49], 373[43, 53], *375, 376, 378, 379, 380*
Conley, F. K., 595[57], *654*
Conrad, M. E., 541[34], 543[34], *580*
Consigli, R. A., 625[340], 631 [340], 641[340], *663*
Constantine, D. G., 180[80], 186 [78], *273*
Content, J., 52, *60*, 294[69], 302 [69], 304[100, 111], 305[116,

[Content, J.]
117], 306[100], *343, 344,* 360[55], *376*
Contreras, G., 78[91], *108*
Cook, G. M. W., 361[56], *376*
Cook, M. K., 351[1], *374*
Cook, M. L., 820[235], 822[235], 825[235], 826[235], *845*
Cooke, B. C., 873[186, 187], *887*
Coombes, J. D., 314[169], 315 [169], 316[169], *346*
Coon, H. G., 463[47], *540*
Cooper, G. M., 813[194, 195], *844*
Cooper, H. L., 55[137], *61*
Cooper, N., 864[87], 873[87], *885*
Cooper, P. D., 75[81, 82, 83], 88 [122, 123], 89, 93[122, 123, 124], *107, 109,* 146[150], 149 [150], *164,* 269, *280*
Cooperband, S. R., 55[134], *61*
Copeland, T., 469[52], 470[52], *540*
Coppey, A., 631[408], *665*
Copps, T. P., 391[103], 397[103, 143], *427, 429*
Corallini, A., 837[316], *848*
Cordell-Stewart, B., 75[86], *107*
Cordon, J., [10], *897*
Cords, C. E., 74, *107*
Corey, S., 719[1], *757*
Coriell, L. L., 825[256], *846*
Corley, L., 49[76], *59*
Cormack, D. V., 259[269], 260 [269, 276], 261[277], 262 [276], 264[276], 265[269, 282], *279*
Correa-Giron, E. P., 180[82], *274*
Cory, J., 155[191], *165*
Cory, S., 67[28], *106,* 367[2], *374*
Cossart, Y. E., 546[42], *580*
Costanzo, F., 801[144], *842*
Coto, C., 818[229], 833[295], *845, 847*
Cotton, W. E., 170[11], *271*
Courtney, R. J., 802[152], 808 [158, 159], *843*
Cowan, K., 632[304], 638[304], *662*
Cowie, A., 604[148], 606[148], 627[148], *657*
Cox, D. C., 411[230], *431*
Cox, J. H., 199[160], *276*
Cox, R. A., 45[38], *57*
Craig, E. A., 690[29], 718[29, 276], 720[30], 721[28], 723 [30], 736[27, 30], 748[30], 751[27, 28, 29, 30, 277], *757, 766,* 891[11, 12], *897*
Craig, R., 629[380], *664*
Craighead, J. E., 822[240], 824 [240], 827[266], *845, 846*
Craigie, J., 854[30], *883*
Cramer, R., 631[407], 641[407], *665*
Crawford, E. M., 624[317], *662,* 773[48], *839*
Crawford, L. V., 541[25], 543[25, 128], 549[25], 550[25], 555 [25, 121, 128], 559[25, 121, 128], 560[25, 128], 561[121, 128], 562[121], 564[121], 568 [121], 568[25], 570[189], *580, 583, 585,* 592[16, 22, 24, 26, 27, 28, 29], 597[26, 80], 598, 599[80], 601[116, 122, 124], 602[26, 134], 603[26, 122], 605[178], 614[249], 618[362], 624[317], 627[362], 628[366], 629[377], *653, 655, 656, 658, 660, 662, 664,* 773[48], *839*
Crick, F. H. C., 7, 16[23], *36,* 676[309], *766*
Crick, J. C., 173[212], 191[128], 213, 214, [205, 212], 269[128, 203, 205, 301, 302], *275, 277, 280*
Croce, C., 645[548], *670*

Croce, C. M., 645[544, 545, 546, 547, 549], 649[583], [585, 586], *670, 671*
Crocker, T. T., 80[100], *108*
Croissant, O., 592[20], 601[120], 605[174], *653, 656, 658*
Cross, R. K., 390[85], 395[138a], 396, 398[138a, 147], 399, 400 [153], 407, 413[85], 415[138b, 138c], 416, 417[138c, 147], 418[85], 420[138b, 147], 421 [138a, 147, 153], 424[153], *427, 428, 429*
Cross, S. S., 543[76, 88, 199]. 550 [76], 552[88], 568[88, 199], 573[199], *581, 582, 585*
Crouch, R., 629[382], *664*
Crouch, R. J., 756[315], *766*
Crowell, R. L., 19[100], 34 [100], 35[100], *38*, 74[67], *107*
Cruickshank, J. G., 6[116], *39*, 289 [38], *342*, 365[9], *374*, 543[72], 550[72], 550[72], *581*
Crumpacker, C. S., 633[434, 435], 634[434], *666*, 713[31, 106, 155], 746[154, 155], *757, 760, 761*
Cukor, G., 578[237], *587*
Culp, L. A., 651[608], *672*
Cuzin, F., 601[117, 120], 605 [174], *656*, *658*

Daesch, G. E., 678[85], 714[85], *759*
Dahl, R., 862[61, 65], 867[132], 869[156], 873[61], 878[219], *884, 886, 887, 888*
Dahlberg, J. E., 459, *540*
Dahmus, M., 599[94], *655*
Dales, S., 124[79], *162*, 199[162], 224[162], *276*, 312[148], *345*, 372[49], *376*, 387[58, 77], 392 [58], 402[58, 77, 177, 179], 403[179], 412[58], 423[177],
[Dales, S.] *426, 427, 430,* 714[25, 32], *757, 758,* 798[126], *842,* 851 [15], 854[21, 22], 859[56], 862[60], 863[60, 83, 84], 864 [21, 60, 84, 88], 865[88, 92, 93, 100], 866[114], 867[128, 142, 143], 873[88], 876[193], 877[15, 22, 211, 214], 878 [56, 217], 879[15, 223, 226, 237], 880[92, 93], 881[114, 214], *883, 884, 885, 886, 887, 889*
Dalgarno, L., 147[154], 154[181], 155[181], *164, 165*
Dalldorf, G., 541[14], 543[103], 552[103], 553[14], 567[14], *579, 582*
Dalrymple, J. M., 124[80], *162*
Dalton, A. J., 439, *539,* 546, *587*
D'Andrea, E., 320[244], 321[244], *348*
Daniel, R. W., 596[65], *654*
Daniel, W. A., 23[85], *38*, 636 [477], *667*
Danielescu, G., 51[109], *60*
Danna, K., 605[185], *658*
Danna, K. J., 604[161, 164], 607 [161], 619[298, 363], 620 [298], 622[298], 624[298], 627, 628, 631[411], *657, 662, 664, 665,* 728, 748[33], *758*
Darbushire, J. H., 386[42], *426*
Darby, G., 868[149], 869[149], *886*
Darlington, R., 797[121], 817 [121], *842*
Darlington, R. W., 186[110], *274,* 353[57], 358[132], 359[236], 360[236], 366[236], 367[132, 236], *376, 378, 381,* 770[3], 824[3], *838*, 865[101], *885*
Darnell, J. E., 67[16, 26], 75 [77], 76[77], 78[77], 80[26, 77], 92[131], *105, 106, 107, 109,* 132[109], *163*, 363[238],

[Darnell, J. E.]
381, 647[551, 552], *670,* 690 [8], 719[34, 65, 211], 720 [303], 733[304], 735[69, 304, 339], [374], *757, 758, 759, 766, 768*
Darnell, J. E., Jr., 16[78], *38,* 83[103, 104], *108,* 114[7, 16], 115[27, 29], 118[7, 27, 29], 119[27], 121[7, 16, 27], 123 [27], 124[7], 134[7], *160,* 737[108], *760,* 891[24, 45], *898*
Davenport, F. M., 283[16], 287 [23], *341, 342*
Davey, M. W., 47[57], 48[69], 49[57, 69, 78], *58, 59,* 147 [154], 154[181], 155[181], *164, 165*
David, A. E., 115[27], 118[27], 119[27], 120[59], 121[27, 59], 123[27], *160, 161,* 258 [264], *279*
David, D., 597[79], *655*
Davidson, G. P., 385[26], *425*
Davidson, H. E., 719[35], *758*
Davidson, N., 114[14], 117[14], 130[14], *160,* 644[534], *669,* [381], *768*
Davidson, R. L., 814[201, 202], *844*
Davies, A., 48[64], *58*
Davies, H., 385[27], 423[262, 263], *425, 432*
Davies, M. C., 15[58], *37*
Davies, P., 302[102, 108], 317 [108], *344*
Davis, D., 813[193], 814[193, 200], *844*
Davis, J., 467[50], 493[73], *540,* 541, *881*[261], *889*
Davis, R. W., 601[125], 608[198], 610[198], 611[198], 632[428], 633[428], *656, 658, 666,* 746 [220], *763,* 893[44], *898*
Davoli, D., 611[226], *659*
Dawson, P. S., 386[42], *426*
Dayan, A. D., 351[58], *376*
Dean, W. W., 605[171], *658*
Debbie, J. G., 180[84, 85], *274*
Debre, R., 286, *341*
DeClercq, E., 45[24], 48[70], 50 [95], *57, 59*
Defendi, V., 610[213], 612[233, 234, 240], 625[338, 339], 631 [339, 394], 632[420], 634 [444], 648[560, 561, 568], *659, 660, 663, 665, 666, 670,* 746[36], *758*
DeHaas, R. A., 170[19], 182[19], *271*
DeHarven, E., 825[253], *846,* 877 [207], *888*
Deichman, G. I., 634[441], *666*
Deinhardt, F., 787[83], *840*
Dekegel, D., 541[35], 545[35, 94], 550[35], 551[35], 552[35, 94], 555[94], 569[35], *580, 582*
DeKinkelin, P., 174[37], 187[37], 220[209], 221[209], *272, 277*
Delaney, S. S., 877[209], *888*
DeLavergne, E., 385[21], *425*
DeLeva, A. M., 23[86], *38,* 636 [476], *667*
Delius, H., 356[143], 358[143], *378,* 604[155], 605[155], 607 [155], 624[319], *657, 662,* 681[177, 178], 687[177, 203], 688[177], 732[207], 736[177], 749[178, 203], 750[177], *762, 763*
DeLorbe, W., 732[265], *765*
Del Villano, B. C., 632[420], *666,* 746[36], *758*
DeMaeyer, E., 49[73], 55[145], 55 [150], *59, 61*
DeMaeyer-Guignard, J., 49[73], 55 [145], 55[150], *59, 61*
Demissie, A., 835[309], *847*
DeMol, A. W., 617[288], *661*
Denhardt, D. T., 565[166], 566 [166], 577[166], *584,* 604

[Denhardt, D. T.]
[167], 605[167], *657*
Dennett, D. P., 154[181], 155 [181], *165*
Dennis, J., 390[260], *432*
DeNoronha, F., 473[56], *540*
Denys, P. H., Jr., 50[97], *59*
Deppert, W., 598, *655*, 897 [13, 14], *897*
DeRecondo, A. M., 605[168], *657*
DeRosier, D. J., 6[65], 22[24, 121], *36, 37, 39*
Derrick, J. M., 423[263], *432*
DeSimone, V., 625[333], *663*
Desmyter, J., 47[50], *58*
DeSomer, P., 45[24], 47[55], 48[70], 50[97], *57, 58, 59*
DeSousa, J., 174[21], 182[21], *271*
Desrosiers, R. C., 423[273], *433*, 719[37], *758*
Dessel, B. H., 595[60], *654*
Detjen, B. M., 67[17], 68[17], *105*
DeTorres, R. A., 612[236, 243], 625[236, 243], *660*
Deutsch, V., 265[286], 267[292], *280*
Dev, V. G., 73[66], *107*
Devauchelle, G., 854[28], 877 [210], *883, 888*
Devine, C., 688[326, 327], 715 326], 717[326, 327], 720 [326, 327], 723[326], 749 [326, 327], 750[327], *767*
Devinny, T., 603[139], *657*
DeVleesschauwer, L., 154[175, 183], *165*
DeVries, F. A. J., 638[497], 642 [497], *668*, 676[78], 686 [78], 688[78], 732[78], 753 [78], *759*
Dewey, F., 835[310], *847*
D'Halluin, J., 570[208], *586*
Dhar, R., 605[180, 181, 182, 183, 184], 619[180], 620[180], 624[184], *658*, 747[38], *758*
Diamandopoulos, G. T., 648[563, 564], *670*
Diamond, L. S., 19[84], *38*
Diche, R. J., 176[57], *273*
Dichfield, J., 351[59], *376*
Diefenthal, W., 75[73], 80[73], *107*
Diener, T. O., 3[25], *36*
Dierks, R. E., 178[65], 179[65], 216[207], 217[207], 269[207], *273, 277*
Dietzschold, B., 95[141], *109*, 199 [160], *276*
Dillard, S. H., 820[237], *845*
DiMayorca, G., 636[482, 487], 639 [507], 640[516], *668, 669*
DiMayorca, G., 596[66], 631[405], 641[405], *654, 665*
Dimmock, N. J., 293[62], 297[62], 300[62], 306[120], 307[62], 309[135, 140], 312, 313[158, 159], 315[172], 316[151], 318 [172], 329[203], 332[120], 333 [62], 336[62], [229], *343, 344, 345, 346, 347, 348*
Dinter, Z., 351[60], *376*
Dintzis, H. M., 627[364], *664*
Dion, A. S., 535[111], *542*
DiPorzio, U., 613[246], 615[246], *660*
Diskin, B., 89[126], 96[126], *109*
Dixon, C. B., 48[62], *58*
Dmochowski, L., 787[82], *840*
Doane, F. W., 154[178], *165*, 402 [176], *430*
Dobbertin, D., 94[136], *109*
Dobkin, M. D., 194[164], *276*
Dobos, P., 117[33], 126[85], *160, 162*
Dobson, P. R., 574[219], *586*
Doerfler, W., 681[39], 686, 697 [17], 698[17], 699[17], 701

[Doerfler, W.]
[77], 710[164], 716[41], *757, 758, 759, 762,* 894[4, 34, 39], *897, 898*
Dohan, C., 613[413], 632[413], 636[413], 637[413], *665*
Doherty, R. L., 174[26], 184[96], *272, 274*
Doi, Y., 352[117, 165], 364[117, 165], *378, 379*
Dolin, R., 567[175, 176], 570 [175], 572[175], 575[222], 577[222], *585, 586*
Dolynink, M., 793[100], *841*
Domoto, K., 541[33], 545[33], 562[33], *580*
Donaghue, T. P., 117[40], 132 [40], *161, 163*
Donelli, G., 7[26], 22[26], 24 [26], *36*
Donohoe, R. M., 55[147], *61*
Dorner, F., 47[56, 59], 48[59], *58*
Dorsch-Häsler, K., 96, *109*
Dorsett, P. H., 698, 755, *758*
Dorson, J. W., 862[71], *884*
Doughty, S. C., 117[106], 126 [103], 131[103, 106], *163*
Dourmashkin, R. R., 312[149], *345*
Dowdle, W. R., 282[2], 283 [16], 287[2, 24], *341, 342*
Dowling, H. F., 385[18], *425*
Downer, D. N., 816[223], *845,* 865[96], *885*
Downie, A. W., 851[10], *883*
Downie, J. C., 288[26], 307 [26], *342*
Doyle, M., 270[308], *280,* 313 [161], 314[161], 320[161], 326[197], 332[197], *346, 347,* 817[224], *845*
Drake, S., 574[217], *586*
Dreesman, G. R., 780[53], *839*
Dressler, D., 564[159], *584,* 626 [356], *664,* 683[329], *767*
Dressler, H. R., 55[147], *61*
Dressman, G. R., 592[18], *653*
Drost, S. D., 531[103], 532[103], *542*
Drouhet, V., 402[171], *429*
Drzeniek, R., 69[61], *107,* 292 [48], 307[127], 335[127], *342, 345*
Dubbs, D. R., 612[236, 243], 625 [236, 243, 323, 324, 326, 327, 328], 632[418], 636[418, 485], 637[418], 640[418], 648[579], 649[579, 580, 581], *660, 662, 666, 668, 671,* 808[176], 812 [176, 177, 178, 179, 180, 184], 813[197], 814[196], 816[222], 837[176], *843, 844, 845,* 872, [168, 170, 171, 173, 175], 880 [254], *887, 889*
Dubbs, S., 650[602], *671*
Dubin, D. T., 795[109, 110], 797 [110], *841*
Duesberg, P. H., 294[69, 71], 302 [69, 103, 106], 305[116, 117], 317[103, 106], 320, 335[106], *343, 344,* 353[61], 356[62], 360 [15, 55], *375, 376,* 459[37, 38], 460[37], 461[43], 502, 503[38, 80], 504[37], *540, 541*
Duff, R., 648[566], *670,* 829[279], *847*
Dulbecco, R., 2[20], *36,* 597[78], 598[78], 601[116, 121], 603 [136], 608[205], 616[268, 274], 617[274, 277, 281], 620[277, 301], 623[274], 625[322, 325, 332, 341, 342], 626[352, 354], 627[352], 630[133], 632[415, 416, 426], 633[426], 634[341], 635[416], 636[342, 415, 416], 637[415, 416], 639[341, 342], 640[341, 513], 641]522], 642 [530], 643[531], 644[540], 645[277, 540], 648[426, 571],

[Dulbecco, R.]
649[571], 650[571], 652[619], *655, 656, 659, 661, 662, 663, 665, 666, 669, 670, 672,* 676 [42], 746[22], *757, 758*
Dulbecoo, R., 597[77], 630[77], 635[77], *655*
Dumont, R., 260[270, 264[270], *279*
Duncan, I. B., 239[236], 265 [281], *278, 279*
Dunker, A. K., 6[28], 19[27], *36*, 69[56, 57], 71[56, 57], *107*
Dunn, A. R., 731[173], 733[43], *758, 762,* 891[15], *897*
Dunnebacke, T. H., 387[63], 392 [237], *426, 431*
Dupont, B., 351[188], 362[184], *380*
Duran-Reynals, M. L., 521[99], *542*
Duranton, H. M., 69[50, 51, 52], *106*
Durham, A. C. H., 5[69], *37*
Durr, F. E., 772[25], 774[25], *838*
Duscio, D., 571[212], *586*
Dutta, S. K., 541[30], 545[30], *580*
Dwyer, A. C., 787[83], *840*
Dym, H., 772[58], 780[58], *839*

Eason, R., 600[107], *656*
East, J. L., 353[63], 371[63], *376*
Easterbrook, K. B., 854[20], 855 [42], 866[125], 879[126], 881[268], *883, 886, 890*
Eaton, B. T., 117[40], 127[168], 132[40], 153[168], 132[40], 153[168], 154[168], 155 [168], *161, 163, 165*
Eckert, E. A., 292[46], 298[46],
[Eckert, E. A.]
300[46], *342*
Eckhart, W., 591[4], 625[341, 342, 343], 631[402], 632 [402], 634[341, 439], 636 [342], 639[341, 342, 343, 402, 508, 510, 512], 640[341, 513], *663, 665, 666, 668, 669*
Eckroade, R. I., 595[60], *654*
Eddy, B. E., 591[4], 594[38, 39, 40], *652, 653*
Eddy, G., 352[16], *375*
Eddy, G. A., 541[32], 543[32, 80], 551[32, 80], 552[80], 555[32, 80], 569[80], *580, 581*
Edenberg, H. J., 729, *758*
Edgel, M. H., 756[117], *760*
Edmonds, M., 68[35], 94[35], *106*
Edmonson, J. H., 385[23], *425*
Edvardson, B., 716[346], 725 [346], *768*
Efron, D., 359[64], *376*
Eggen, K. L., 67[21], 80[21], 85 [21], *105*
Eggers, H. J., 15[4], *35*, 390[88], *427*
Ehrenfeld, E., 75, 78[91], 89[125], 95[142], *107, 108, 109,* 198 [155], 236[227, 228], 237[227, 228], *276, 278*
Eikhom, T. C., 270[309], *280*
Eisenman, R., 497[78], *541*
Eiserling, F., 20[1], *35*
Ejercito, P. M., 831, *847*
El Dadah, A. H., 541[31], 543 [31], 550[31], *580*
Eliasson, R., 628[369], *664*
Ellem, K. O., 557[136], *583*
Ellens, D. J., 726[268], 728[267, 268], *765*
Ellis, D. A., 55[149], *61*
Ellis, R. J., 390[86], *427*
Ellison, S. A., 779[32], *839*, 877 [204], *888*
El Mishad, A. M., 545[73], 550

[El Mishad, A. M.]
[73], 569[202], *581, 585*
Elveback, L. R., 385[17], *425*
Emerson, S. V., 199[159], 200 [170], 201[173, 174], 238 [170, 173, 174, 232], 239 [170], 269[173], *276, 278,* 360[65, 66], *376*
Emmons, R. W., 180[80], *273,* 787[83], *840*
Enders, J. F., 350[88, 123], 351[88], 363[164], *377, 378, 379,* 631[398], 648 [563, 564], *665, 670*
Enders, J. T., 595[50], *654*
Enders-Ruckle, G., 351[248], *381*
Engelhardt, D. L., 53[121], *60,* 239[241], *278*
Engelking, H. M., [10], *897*
Engler, R., 386[50], 387[57], 390[95], *426, 427*
Engler, W. O., 552[111], *582*
Enomoto, C., 305[221], 336[221], 337[221], *347*
Ensinger, M., 704, 705[149], 707 [149], *761*
Ensinger, M. J., 701[46], 703, 704 [75], 705[46, 75, 295], 707 [46, 295], 708[46], 709[45, 75], 712[75], 726[75], 727 [75, 295], 738[295], *758, 759, 766,* 863[79], *884*
Ensminger, W. D., 79[98], *108,* 406[203], 411[203, 226, 227, 228], *431*
Enzmann, P.-J., 124[81], 156 [195], *162, 165,* 216[207], 217[207], 269[207], *277*
Ephrussi, B., 612[233], *660*
Epstein, L. B., 55[139], *61*
Epstein, M. A., 775, 796[10], *838,* 851[14], *883*
Erikson, R. L., 193[133], *275*
Erlandson, R. A., 778[19], 826 [19], *838*
Ernberg, I., 808[161, 162, 163], 827[162, 273], 835[163, 308], *843, 846, 847*
Eron, L., 634[454], *667,* 739, 752 [47, 316], *758, 766*
Eron, L. J., 872[184], *887*
Eskin, B., 603[145], *657*
Essex, M., 533[104], *542*
Esteban, M., 862[66], 871[165], 880[165], *884, 887*
Esteban, R. M., 51[105, 109, 113], 53, *59, 60,* 67[22], 80[22], 85 [22], *105*
Estes, M. K., 20[99a], *38,* 592[18], 597[84, 88], 599[84, 88], 600 [84], *653, 655*
Etchison, J., 326[197], 332[197], *347,* 817[224], *845*
Etchison, J. R., 198[154], *276,* 363[67], *376*
Etkind, P. R., 310[141], 320, 321, 323[238], *345, 346, 348*
Evans, C. A., 174[44], *272,* 592 [15], *653*
Evans, D. L., 787[82], *840*
Evans, H. E., 384[7], 385[7], *424*
Evans, J., 605[189], 610[189], *658*
Evans, M. J., 358[68], *376*
Everitt, E., 599[95], *655,* 691 [49, 50], 696, 697[49, 50], 699[49], 716[346], 720[366], 725[346], 727[366], *758, 768*
Evstigneeva, N. A., 298[84], *343*
Ezoe, H., 700[51], *758*

Fabiyi, A., 175[52], 180[88], 181 [52], *272, 274*
Fabrizio, D. P. A., 546[54], 555 [54], *581*
Falcoff, E., 51, 52, *60*
Falcoff, R., 51, 52, *60*
Falke, D., 779[30, 31], 798[30],

[Falke, D.]
838
Fambrough, D., 599[94], *655*
Familusi, J. B., 175[53], *273*
Fanning, E., 716[41], *758*
Fantes, K. H., 42[11], 47[54], 48 [54, 61], 49[54], 50[101], *57, 58, 59*
Faras, A. J., 459[41], *540*
Fareed, G. C., 609[207], 611[207, 224, 225, 226, 228], 627[361], 628[228, 367, 374], 630[384], *659, 664*
Farley, C. A., 770[2], 826[2], *838*
Farmilo, A. J., 605[187], 610 [187], *658*
Farnham, A. E., 289[34], *342,* 358[95], *377*
Farrell, J. A., 386[47], 387[47, 56], 389[56], 391, 395[56], *426*
Farringia, R., 649[582], *671*
Fasekas De St. Groth, S., 283 [16], *341*
Fauconnier, B., 49[73], *59*
Faulkner, P., 114[12], 117[12, 33, 40], 126[85], 127[168], 132[40], 153[168], 154[168], 155[168], *160, 161, 162, 163, 165*
Favre, M., 592[20], *653*
Fazekas De St. Groth, S., 310[144], 312[147], *345*
Federer, K. B., 170[13], 206[13], *271*
Feiz, G., 126[89], *162*
Feldman, L. A., 631[396], 634 [448], *665, 667,* 820[236], 824[236], 826[236], *845*
Fellner, P., 68[44, 45, 46], 94 [46], *106*
Felsher, B. F., 541[34], 543[34], *580*
Fennel, R., 97[148], 99[148], *109*
Fenner, F., 641[1], 74[1], 75[1], 80[1], 93[1], *105*, 351[69], *376,* 636[471], *667*, 851[7, 9, 10, 11], 866[110, 111, 118], *883, 885, 886*
Fenwick, M., 802[318], 809[318], *848*
Ferdinand, F. J., 752[316], *766*
Ferguson, J., 601[125], 632[428], 633[428], *656, 666,* 746[4, 220], *757, 763*
Ferguson, L. C., 552[86], 553[86], *582*
Fern, V. H., 553[110], *582*
Fernandes, M. V., 650[590], *671*
Fernandez, A. A., 170[12], *271*
Fernandez, S. M., 719[52], *758*
Fernandez-Munoz, R., 67[16], *105*
Fernandez-Tomas, C. B., 99[151, 154], 101[151], *110*
Ferris, D. H., 176[57], *273*
Fey, G., 598[93], 599[93], 600 [93], *655*
Field, A. K., 43[4], 45[4, 39], *56, 57*
Field, A. M., 546]42], *580*, 595 [61], *654*
Field, H. J., 823[244], *846*
Fields, B. N., 183[93], 199[158], 214[93, 158], 215[93], *274, 276,* 386[38, 40, 41], 390[85], 395[138a], 396, 398[138a, 147, 150], 399, 400[153], 402[174], 407, 410[174, 220], 413[85, 241, 242], 414[241, 242], 415[138b, 138c, 240, 244], 416, 417[138c, 147, 242], 418[85, 148], 420 [138b, 147, 174, 220], 421[138a, 147, 153], 424[153, 279, 280, 281], *425, 427, 428, 429, 430, 431, 432, 433*
Fields, H. A., 567[192], 568[192], 570[192], 572, *585*
Fiers, W., 604[164], 605[185], 638 [497], 642[497], *657, 658, 668*

Fife, K. H., 565[167], *584*
Fijan, N., 174[39], *272*
Fil, W., 878[218], *888*
Files, J. G., 597[87], 598[87], *655*
Filler, R., 869[153], 880[153], *887*
Finch, J. T., 2[35], 3[29, 35], 4[31, 34, 76], 5[38], 6[30, 33, 112], 15[32, 36, 37, 40, 71, 72], 16[32, 36, 37, 40, 70, 72, 73], 17[35, 39], 18 [35, 39, 75], 19[32], 20 [70], *36, 37, 39,* 358[70], [71], *376*, 591[1, 2], *652*
Finlay, M., 387[66], 390[66], *426*
Finnerty, V., 94[138], *109*
Finter, N. B., 50[94], *59*
Fischer, H., 595[59], 635[470], *654, 667*
Fischman, H. R., 179[68], 180 [83], *273, 274,* 351[72], *376*
Fishel, B. R., 604[151], 606 [151], *657*
Fiszman, M., 54[130], *60*
Fiume, L., 615[262], *661*
Fizman, M., 625[346], *663*
Flamand, A., 190[122], 191 [122, 131], 192[131], 193 [131], 197[144], 200[170], 203[197], 223[214, 215], 224 [144], 227[215], 228[197], 230[197], 233[197], 238[170], 239[144, 170], 240[144, 197, 214, 215], 241[144, 197, 214], 242[215], 243[144, 214, 215], 244[215], 245[215], 250[144, 215], 251[144, 215], 252[197], 256[144, 215], 259, 260[266, 271], 261[144], 262[144, 197, 215], 263[215], 264[144, 197, 215, 278], 265[144, 197, 215, 281], 269[122], *275, 276, 277, 278, 279*
Flanagan, J. F., 815[218], 828 [218], *845*
Fleckenstein, B., 772[52], 780 [52, 55], *839*
Fleissner, E., 451[26], 452[27], 453[26], 454[26], 455[26], 462[26], 473[27], [62], *539, 540*
Flewett, T. H., 7[5, 41], *35, 36,* 293[60], 294[60], 300[60], *343,* 385[27], 423[262, 263], *425, 432*
Flint, J., 619[300], 620[300], *622,* 676[347], 717[347], *768*
Flint, S. J., 688[53], 689[53], 690[53, 245], 707[245], 717 [349], 720[245, 349], 721 [340], 724[54], 728[348], 734 [53, 245], 735[53], 736[245], 737[53], 738[53], 747[54], 748[54], 750[53, 245], 751 [245], 753[53], *758, 764, 768,* 896[16], *897*
Flores, L., 878[218], *888*
Floyd, R. W., 405[200], *430*
Foà-Tomasi, L., 801[144], *842*
Fog, T., 351[188], 362[184], *380*
Fogel, M., 612[234], 645[543], 650[593, 594, 596, 597, 598], *660, 669, 671*
Fogh, J., 560[145], 570[145], *584*
Folk, W., 604[151], 606[151], 645[542], 650[542], *657, 669*
Follett, E. A. C., 188[312], 189 [116], 256[256], 269[116], *275, 279, 280,* 403[186], *430*, 543[128], 555[121, 128], 559[121, 128], 560[128], 561[121, 128], 562 [121], 564[121], *583*, 592[29], 605[178], *653, 658,* 867[139],

[Follett, E. A. C.] 871[139], 877[139, 215], 881[215, 274, 282], *886, 888, 890*
Fong, C. K. Y., 544[213], 571[213], 576[226], *586,* 778[23], 779 [23], 822[23, 241], 823[241], 824[23], *838, 845*
Ford, E. C., 350[91], *377*
Foreman, C., 469[53], *540*
Forrester, F. T., 117[32], 146[32], *160*
Foster, W. S., 770[2], 826[2], *838*
Fouad, M. T. A., 386[50], 387 [57], 390[95], *426, 427*
Fournier, F., 862[66], *884*
Fowler, A. K., 463[45], [113], *540, 542*
Fox, C. F., 364[211, 212], *380*
Fox, J. P., 385[17], *425*
Fox, R., 634[452], 635[452], *667*
Fox, R. I., 634[458], 635[458], *667,* 738[9], 739[9], *757*
Fox, S. M., 69[60], *107*
Francis, J., 174[28], *272*
Francis, T., Jr., 287[23], *342*
Francke, B., 639[508], *668*
Francki, R. I. B., 169[1], 173, 189, *271*
Frank, H., 15[4], *35,* 292[48], *342*
Franke, B., 628[370], 629[370, 376], *664*
Frankel-Conrat, H., 305[117], *344*
Franki, R. I. B., 7[42], *36*
Franklin, R. E., 16[73], *37*
Franklin, R. F., 697, *760*
Franklin, R. M., 310[143], 312, *345,* 353[208], 357[73], *376, 380,* 387[58], 391[100], 392 [58, 109], 402[58], 412[58], *426, 427*
Fraser, K. B., 351[79], 364[79], *377*
Fraser, M. J., 604[167], 605[167], *657*
Fraumeni, J. F., Jr., 595[55], *654*
Frazier, J. A., 787[85], *840*
Frazier, W. A., 89[128], 90[128], *109*
Frearson, P. M., 597[80], 599[80], 625[323, 324, 326, 327, 328], *655, 662, 663,* 816, *845*
Freedman, V. H., 896[41], *898*
Freeman, A. E., 534, *542,* 679[55], 680, *758*
Freeman, G., 601[121], *656*
Freeman, N. K., 307[130], *345*
Freifelder, D., 565[162], *584*
Frenkel, G., 704[149], 705[149], *761*
Frenkel, N., 609[206, 209], *659,* 771[4], 780[54, 64, 71, 76], 782[71, 76], 785[4, 76], 800 [76, 137], 801[76, 137], 803 [76], 805, 806[76], 807[76], 808[76], 819[71, 76], *838, 839, 840, 842*
Frey, L., 727[2], *757*
Frey, T. K., 117[91], 132]91], *162*
Friderici, K. H., 719[37, 200], *758, 763*
Fried, A. H., 624[316], *662*
Fried, M., 601[116], 604[148, 149, 152], 605[193], 606[148, 149, 152], 608[193, 197], 610 [193, 197], 611[152, 197, 229], 627[148], 628[152], 636[486], 639[509, 511], 649[193], *656, 657, 658, 660, 668*
Friedman, A., 265[285], 266[285], 267[285], *280*
Friedman, M. P., 634[455], *667*
Friedman, R. M., 50[86, 99, 101], 51, 55[137], *59, 60, 61,* 126 [83, 86], 127[83, 198], 131 [100, 101], 137[126], *162, 163, 165*
Friedman, T., 597[79], *655*

Friis, R. R., 446[22], 476[63], 501[22], 520[22], *539, 541*
Frisby, D., 68[45, 46], 94[46], *106*
Frisch-Niggemeyer, W., 302[98], 306[98], *344*
Fritz, M. E., 833[296], *847*
Frommhagen, L. H., 307[130], *345*
Fryer, J. L., 186[106], 221[106, 211], *274, 277*
Fuccillo, D. A., 287[22], *342*
Fujinaga, K., 676[56, 57, 58, 59], 677[56, 63, 88], 678[63], 679[88], 680, 683[372], 685 [62], 686[57], 688[88], 714 [60], 717[60, 61, 88], 718 [88], 720[88], 721[60], 730 [372], 731, 732[63, 88], 733 [57, 58, 59, 61, 62, 88, 91, 289], 734[59, 60, 88], 735[248, 289], 738[91], 751[60], 753[88], *758, 759, 764, 766, 768*
Fujiwara, S., 802[147], 811, *842*
Fukai, K., 117[44], 154[187], 155[187], *161, 165,* 364[102], *377*
Fukuda, A., 352[230], *381*
Fukumi, H., 283[16], 351[74], 352[74], *341, 376*
Fukunaga, T., 117[44], *161*
Fuller, W., 387[61], 392[61, 108], *426, 427*
Furakawa, T., 798[127], 824 [127], *842*
Furano, A., 605[179, 190], 610 [190, 223], 631[179, 223], *658, 659*
Furlong, D., 19[43], *36*, 773 [7], 775[7], 776, 777, 778 [16], 779[16], 787[16], 796 [16], 797[7, 16], 798, 817[16], 818[16], 820[16], 822[16], 833[16], *838*
Furman, P. A., 373[75], *377*
Furness, G., 879[123], *886*
Furuichi, Y., 67[29, 32], *106,* 203[191], 229[191, 223], 230 [191], *277, 278,* 358[171], 367[171], *379*, 394, 401[169], 406[169], 408[169, 209, 236], *428, 429, 431,* 719[64, 65], *[374],* 759, 768, *863*[77], *884*

Gaffney, B. J., 121[64], 142[64], *161*
Gafford, L. G., 855[39, 40], 859 [51], 865[101, 102], 880[260], *883, 884, 885, 889*
Gagė, Z., 733[304], 735[304], *766*
Gaidamovich, S. Ya., 154[176], *165*
Gajdusek, D. C., 351[25], *375,* 488[70], *541*
Galabov, A., 353[226], *381*
Galet, H., 236[231], 248[245], *278*
Galimard, B., 174[37], 187[37], *272*
Gallagher, R. E., 535[112], 537, *542*
Gallaher, W. R., 362[76], 363[22, 77], *375, 377*
Gallia, F., 175[48], 181[48], 216 [48], *272*
Gallimore, P., 690[236], 732[236], 753[236], *764*
Gallimore, P. H., 644[537, 539], 649[539], *669*, 677[66], 678 [66], 679[66], 688[53], 689 [53], 690[53, 66, 245], 707 [245], 720[245], 731[173], 732[66], 733[43], 734[53, 245], 735[53], 736[245], 737 [53], 738[53], 750[53, 245], 751[245], 753[53], *758, 759, 762,* 895[7, 17, 18, 19, 25], 896[7, 16, 19, 20], *897, 898*

Gallo, R., 439[16], *539*
Gallo, R. C., 535[66, 112], 537 [66], *541, 542*
Gallwitz, U., 4[54], *37*
Galton, M., 552, *582*
Gandhi, S. S., 298[80], 326[80, 198], 332[198], *343, 347*
Gardner, J., 571[214], *586*, 718[194], 720[194], *763*
Gardner, M. B., 437[13], 463[13], *539*
Gardner, S. D., 595[61], 596[71], *654, 655*
Garoff, H., 114[10], 119[57], 121 [10], 122[10, 57], 123[10, 57, 74], 133[10], 135[10], 142[57], 143[10], *160, 161, 162*
Garon, C., 611[227], 628[227], *659*
Garon, C. F., 560[140], 564[140, 158], 577[140], *583, 584,* 595 [62], 605[176], *654, 658,* 683 [67, 68, 137], *759, 761,* 859 [50, 54], 860[50], 873[50], 874[50], 876[192], 879[192], *884, 888,* 895[21], *898*
Garon, G. F., 627[361], *664*
Garrison, J., 575[222], 577[222], *586*
Garrison, M. S., 170[20], 174[20], 182[20], 184[20], 215[20], *271*
Gary, G. W., Jr., 117[32], 146[32], *160*
Gashmi, S., 73[66], *107*
Gauntt, C. J., 387[70], 392[111], 422[70], *426, 428*
Gautschi, M., 541[11], 543[11, 186, 190], 555[11], 556[11, 240], 559[11], 567[186, 190], 569[190], 570[190], 571[186], 573[186], 574[190], *579, 585, 587*
Gazdar, A. F., 54[135], 55[143], *61*
Gazzolo, L., 867[129], *886*
Geering, G., 787[84], *840*
Gehle, W. D., 546[55], 547[55], 554[117], 569[117], *581, 583,* 691[256], 765
Geissler, E., 88[122], 89[122], 93[122], *109*
Gelb, L., 643, 644[533], *669*
Gelb, L. D., 390[91], *427*, 609 [207], 611[207, 227], 628 [227], *659*
Gelboin, H., 43[8], 54[8], *56*
Gelinas, R. E., 719[350], *768,* 891[8], 892[8], *897, 898*
Gemmel, A., 866[111], *885*
Genevaux, M., 69[52], *106*
Gentry, G. A., 773[45], *839*, 797[124], 808[155], 812[183], 813[183, 188], 815[183], 816[223], 817[227], 818[227, 232], 827[227], 828[183], *842, 843, 844, 845*
Georgieff, M., 735[69], *759*
Gerard, G. F., 454[30], 497[30], *539*, 678[86], *759*
Gerber, P., 648[569], *670*
Gergely, L., 808[161, 162], 827 [162, 273], *843, 846*
Gerhard, W., 293[61], *343*
Gerin, J. L., 350[91], *377*, 546 [40], 551[40], *580*
Germanov, A. B., 298[84], *343*
Germond, J., 605[170], *658*
Germond, J. E., 600[110], 602 [110], *656*
Gerry, H. W., 563[157], 564 [157], *584*
Gershon, A., 779[40], 820[40], 823[40], *839*
Gershon, D., 612[238, 239], 625 [238, 239], *660*
Gershowitz, A., 719[312], *766,* 863[73], *884*
Geshelin, P., 855[45], 857[45],

[Geshelin, P.]
865[45], *884*
Gesteland, R., 597[85], 598 [85], *655*
Gesteland, R. F., 630[387], 635[387], *665*, 691[5], 695[5], 697[5], 698[5], 706[158, 363], 711[363], 718[363], 724[5, 157], 725[5, 6, 158], 738[363], 752[6, 157, 158], 753[158, 363], *757, 761, 762, 768,* 893[29], 896[1, 29], *897, 898*
Geuskens, M., 615[265], *661*
Ghamberg, C. G., 120[58], 122[69], *161, 162*
Gharpure, M., 701[320], *766*
Ghendon, Y. Z., 313[157], 336[224], *345, 347,* 866 [107, 116], 881[116], *885*
Ghittino, P., 185[104], *274*
Gibbs, A. J., 6[33, 44, 112], 15[93], 16[93], *36, 38,* 358[70], *376*
Gibbs, C. F., 43[9, 10], 45 [10, 41], *56, 57, 58*
Gibbs, C. J., Jr., 488[70], *541*
Gibson, W., 597[81], 598[81], *655*, 793[103, 104], 795 [107, 108], 796[103, 104], 823[103, 104], *841*
Gierthy, J. F., 556[135], 557 [135], 573[135], *583*
Gifford, G. E., 45[37], *57*
Gigstad, J., 385[36], *425*
Gilbert, P. F. C., 4[34, 45, 76], *36, 37*
Gilbert, W., 626[356], *664*
Gilboa, E., 635[467], *667*
Gilden, R., 473[58], 474[58], 475[58], 516[91], 519[91], *540, 542*
Gilden, R. V., 437[9, 13], 438
[Gilden, R. V.]
[15], 451[26], 453[26], 454 [26], 455[26], 456[31], 457 [15, 34], 462[26, 44], 463 [13, 44, 45, 46], 465[49], 466 [49], 467[49, 50], 468[49], 469[46, 52, 53], 470[52], 471 [34], 477[46], 480[34], 485 [69], 486[69], 490[69], 491 [72], 493[73, 74], 495[76], 508[9, 72], 509[9, 46, 49], 511 [46], 516[69], 518[92], 519 [72], 521[96], 537[116], *538, 539, 540, 541, 542,* 578[234, 235], *586,* 631[394, 404], *665,* 679[172], *762*
Gilead, Z., 715, 716[72], 721[71], 723[345], 725[263, 351], 727 [124, 263], 738[70, 351], [375], *759, 760, 765, 768,* 896[23], *898*
Gilles, S., 401[163, 164], 405[163, 164], *429*
Gillespie, D., 800, *842*
Gillmore, L. K., 676[229], *764*
Ginsberg, H., 716[299, 300], *766*
Ginsberg, H. S., 575[223, 224], 576 [224], *586*, 633[437], 634[453, 455], *666, 667*, 676[73, 352], 691[73], 698[317], 699[352], 701[46, 352], 703, 704[75], 705 [46, 75, 295], 707[46, 295], 708 [46], 709[75, 76], 712[75, 76], 716[74], 720[74, 163, 342], 721 [71, 163, 342], 726[74, 75], 727 [75, 295], 738[295], 751, 755, *758, 759, 762, 768,* 829[280, 281], *847*
Ginsburg, V. P., 313[157], 336[224], *345, 347*
Giordano, R., 636[487], *668*
Girard, M., 92[131], 95[143], *109,* 597[83], 605[168], 614[259], 622[306], 625[346], 636[475], *655, 657, 660, 662, 663, 667*
Girardi, A. J., 595[49], 634[444],

[Girardi, A. J.]
645[544, 546, 547], *654, 666, 670*
Gisler, R. H., 55[142], *61*
Glancy, T., 537[115], *542*
Glanville, N., 117[36, 39], 131 [36], 132[36], 134[118], 135[118, 120], 137[36, 118], 139[39], 153[118], *161, 163*
Glaser, R., 649[582], *671*
Glasgow, L. A., 55[138], *61*
Glazier, K., 374[270], *382*
Glickman, G., 719[211], *763*
Gliedman, J. B., 118[54], 154[54], 156, *161, 165*
Glover, D. M., 622[307], *662*
Gochenour, A. M., 594[38], *653*
Gökcen, M., 595[58], *654*
Gold, E., 595[55], *654*, 815 [213, 214], *844*
Gold, P., 863[83], *885*
Goldberg, A. R., 297[79], 298 [79], 326[79], *343*
Goldberg, M., 9[46], 29, *37*
Goldberg, S., 891[24], *898*
Goldblum, N., 866[117], 881 [117, 272, 273, 278], *885, 890*
Goldfarb, M. M., 45[27], *57*
Goldman, E., 636[481], 640[481], *668*
Goldman, R., 80[101], *108*
Goldstein, D., 617[280[, 622[280], *661*
Goldstein, D. A., 600[111], 602 [129], 626[358], *656, 664*
Goldstein, E. A., 294[73], 302[73], *343*
Goldwasser, R. A., 179[69], *273*
Golgher, R. R., 47[51], 49[80], 50[80], 54[129], *58, 59, 60*
Golini, F., 94[136], *109*
Gomatos, P. J., 114[15], 118[15], 126[15], 130, 131[96], *160, 162*, 386[49], 387[49, 58, 59, 60, 62, 77], 390[88, 94], 392 [49, 60], 395[136], 400[159],
[Gomatos, P. J.]
401[159], 402[58, 77], 403 [182], 404[159, 192], 406 [182], 411[182], 412[58], *426, 427, 428, 429, 430, 432*
Gonatas, L., 386[39], *425*
Goodenow, M., 385[23], *425*
Goodgal, S. H., 617[292], *661*
Goodheart, C., 772[49], 774[49], *839*
Goodheart, C. R., 772[14], 773 [44, 46], 774*[44]*, 805[14], *838, 839*
Goodman, H. M., 459[41], *540*, 609[210], 620[303], 622[303], *659, 662*
Goor, R. S., 597[86], *655*
Gordon, F. B., 55[147], *61*
Gordon, I., 50[100], *59*
Gorecki, M., 631[411], *665*
Goret, P., 351[113], *378*
Gorham, J. R., 543[203], 569 [203], *585*
Grace, J. T., Jr., 862[59], 863 [82], *884, 885*
Grace, M. A., 394[119], 406[119], *428*
Graessman, A., 613, *660*
Graessman, M., 613, *660*
Graffi, A., 439[16], *539*
Graftstrom, R. H., 780[59, 60], 782 [59, 60], *839*
Graham, A. F., 18[89], 18[89], *38*, 386[48], 387[70], 392 [110, 111], 394[48, 116, 121, 123, 126], 400[154], 403[126, 183, 185], 404[126, 185, 187], 406[187], 407[121, 197], 409 [216], 411[183], 412[121, 183, 185], 417[48, 245, 246], 418 [194], 421[121, 194], 422[70], 423[270], 424[276, 277, 278], *426, 428, 429, 430, 431, 432, 433*

Graham, B. J., 780[57], 789[90], *839, 841*
Graham, F., 701[101], *760*
Graham, F. L., 638[497], 642[497], *668*, 676[78], 686[78], 688 [78], 690, 732[78], 753[78], *759*
Granados, R. R., 854[29], 867 [131], 877[29], *883, 886*
Granboulan, N., 392[115], *428*, 636 [478], *667*
Granoff, A., 339[226], *347*, 363 [78], *377*
Grant, P. M., 537[115], *542*
Gravell, M., 774[307], 797[121], 817[121], 835[307], *842, 847*
Gray, H. B., Jr., 602[132], *656*
Graziadei, W. D., 397[142], 404 [190], 406[190], 407[190], *429, 430*
Green, H., 73[66], *107,* 613 [245], 614[245], 642[528], 651[605], *660, 669, 672*
Green, J. A., 55[134], *61*
Green, M., 454[30], 497[30], *539*, 541[29], 545[29], 555 [29], 559[29], 561[29], 571 [214], *580, 586,* 614[256], 635[256], *660,* 675[83], 676 [56, 57, 58, 59, 83, 194], 677 [56, 63, 83, 84, 88, 94, 353, 354, 355], 678[63, 83, 84, 85, 86, 92, 94, 212, 235], 679[83, 84, 88, 212, 213], 680[62, 84], 681[92, 133, 179], 682[92], 693[193], 685[62, 83, 138, 139, 140, 213], 686[57], 688 [88, 143, 326, 327], 714[60, 83, 85, 93, 195, 280, 281], 715 [72, 243, 326], 716[72, 81, 214, 216, 280], 717[60, 61, 88, 142, 143, 326, 327], 718[21, 88, 194, 195, 364], 719[103, 328], 720 [26, 88, 194, 326, 327], 721[60, 285], 723[102, 195, 326, 345], 725[263, 351], 726[214, 331,
[Green, M.]
332, 333], 727[124, 243, 263, 331, 332, 333], 728[332, 368], 729[122, 332, 338], 730[83, 193], 731[82], 732[63, 87, 88, 94], 733[57, 58, 59, 61, 62, 83, 88, 91, 94, 96, 289], 734[59, 60, 88], 735[26, 195, 289], 736[341], 737[95], 738[70, 91, 351], 749 [142, 326, 327], 750[327], 751 [60], 752[26], 753[88], [121, 161, 375], *757, 758, 759, 760, 761, 762, 763, 764, 765, 766, 767, 768,* 891[5], 896[23], *897, 898*
Green, M. H., 600[106], 615[264], 616[264], *656, 661*
Green, M. R., 676[94], 677[94], 678[94], 718[304], 732[94], 733[94, 96], 737[95], *760, 768*
Green, R. W., 476[65], *541*
Greene, E. L., 546[54], 549[65], 555[54], 581
Greer, S., 813[195], *844*
Gregg, M. B., 282[2], 287[2], *341*
Gregoriades, A., 296, 313[155], *345, 348*
Gresser, I., 54[123, 127, 130, 136], 55[142, 144], *60, 61*
Grieve, G. M., 384[5], *424*
Griffin, B. E., 604[148, 149, 152], 606[148, 149, 152], 608[197], 610[197], 611[152, 197], 627 [148], 628[152], *657, 658*
Griffith, J. D., 600[109], *656*
Grimley, P., 875[191], *887*
Grimley, P. M., 131[100, 101], *162,* 866[113], 877[212, 213], 878[212, 213], 881[212, 275, 276], *885, 888, 890*
Grodzicker, T., 634[447], *667*, 701 [238], 707[98, 170, 238, 321],

[Grodzicker, T.]
709[170, 238, 321], 713[98, 238], 747[97], 749[97, 98], 754[238, 321], 755[170], *760, 762, 764, 767*
Grogan, E. W., 565[167], *584*
Gromkova, R., 617[292], *661*
Gross, L., 438[14], 500[14], 538[14], *539*, 593[36, 37], 594[36, 37, 41], *653*
Gross, P. A., 778[23], 779[23], 822[23], 824[23], *838*
Gross-Bellard, M., 599[101], 600 [110], 602[110], *655, 656*
Grossberg, S. E., 50[98], *59*
Grosso, E., 625[333], *663*
Grover, J., 401[167], *429*
Grubbs, G., 594[38], *653*
Grubman, M. J., 198[155], *276*
Gruetter, M., 697, *760*
Guantt, C. J., 409[216], *431*
Guglielmi, F., 7[26], 22[26], 24[26], *36*
Guir, J., 869[155], *887*
Guntaka, R. V., 496, 517[82], *541*
Gupta, S. L., 51, 52, *60*
Gurari-Rotman, D., 48[63], 49 [76], *58, 59*
Gushchin, B. V., 359[247], *381*
Guskey, L. E., 314[165], 315 [165], *346*
Gussin, G. N., 732[265], *765*
Guttman, N., 67[19], 80[19], 85[19], 99[154], *105, 110*
György, E., 199[161], *276*

Haas, M., 643, *669*
Habel, K., 594[45], 631[400, 401], 648[400, 557, 558], *654, 665, 670*
Habermehl, K.-O., 75[73], 80[73], *107*
Hable, K., 631[391], *665*
Hackemack, B. A., 135[122], *163*
Hackett, A. J., 248[248], 269 [248, 303, 305], *278, 280*
Hackett, P. B., 373[269], *382*
Haddow, A. J., 385[12], 400[12], *424*
Hadlow, W., 546[47], *580*
Haenni, A. L., 69[51], *106*
Hagaki, A., 638[502], *668*
Hager, L., 632[426, 429], 633 [426, 429], 648[426], *666*, 746[22], *757*
Haguenau, F., 439[16], *539*
Hahn, R. G., 351[97], 352[97], *377*
Haines, H. G., 772[21], 778[21], *838*
Haire, M., 351[79], 364[79], *377*
Hall, C. E., 15[47], 18[10], 19 [10], *35, 37*
Hall, L., 99[150], *110*
Hall, M. R., 600[111], 602[129], *656*
Hall, T., 42[3], *56*
Hall, T. C., 69[54], *106*
Hall, W. W., 360[80], 364[81], *377*
Hallauer, C., 541[36], 543[36, 71, 123], 550[36, 71], 551[36], 554[36], 555[123], 567[36], 568[71], 569[71], *580, 581, 583*
Halliburton, I. W., 815[207], *844*
Halliwell, R. E. W., 552[101], *582*
Hallum, J. V., 48[65], *58*, 373 [75], *377*
Halonen, P., 390[87], *427*
Halonen, P. E., 199[158], 214 [158], *276*
Halperen, S., 69[59], *107*
Halpern, M. S., 476[63], *541*
Hama, S., 735[248], *764*
Hamada, C., 700[330], *767*, 812 [182], 816[182], *843*
Hamann, I., 179[75], *273*

Hamilton, R., 350[150], *379*, 385[28], *425*
Hammarskjöld, B., 353[82], *377*
Hämmerling, U., 299[92], *344*
Hammon, W. M., 113[3], *159*
Hammon, W. McD., 551[24], 554[118], 555[24], 567[24], 569[118], 570[209], 578 [118, 209, 229], *580, 583, 586*
Hampar, B., 825[251], *846*
Hampson, A. W., 298[77], 324 [188], 329[188], *343, 346*
Hampton, G. E., 567[187], 568 [187], 571[187], *585*
Hanafusa, H., 445[18], 447, 497 [18], 500[18], 502[79], 503 [80], *539, 541,* 851[8], 880 [246], *883, 889*
Hanafusa, T., 851[8], 880[253], *883, 889*
Hanawalt, P. C., 728[197], *763*
Hancock, R., 599[98], 600[112], 625[98], *655, 656*
Hand, R., 79[98, 99], *108*, 390, 398[92], 411]228, 229], *427, 431*
Hand, R. E., Jr., 567[183, 184], 568[183], 570[184], 571[184], *585*
Handa, H., 709[100], *760*
Hanna, I., 45[20], *57*
Hanna, M. G., Jr., 452[28], 473 [28], *539*
Hanson, M., 457[34], 471[34], 480[34], *539*
Hanson, R. P., 113[3], *159*, 169, 170[8], 176[8, 57], 177[8], 189[117], *271, 273, 275,* 351[83], *377*
Hapel, A. J., 879[231], *889*
Hara, M., 879[224], *888*
Harada, F., 459[40], *540*
Harboe, A., 308[131], *345*
Hardy, W., Jr., 452[27], 473 [27], *539*
Hare, R., 295, *343*
Hariri, E., 521[96], *542*
Härle, E., 597[78], 598[78], *655*
Harpaz, I., 422[259], *432*
Harrington, J. A., 690, *757*
Harris, J. I., 16[48], *37*
Harris, R. E., 568[198], *585*
Harris, T. J. R., 68[46], 91[165], 94[46], *106, 110*
Harris, W. J., 852[18], 854[18], *883*
Harrison, A. K., 117[32], 146[32], *160*, 175[51], 178[65], 179 [65, 71, 72], 180[71], 181[71, 72], 183[91], 216[51], 218[51], *272, 273, 274*
Harrison, B. D., 6[50, 51, 94], *37, 38*
Harrison, S. C., 20[49], *37*, 115 [27], 118[27, 53], 119, 121 [27], 123[27, 53], 143[53], *160, 161*
Harrison, T., 701[101], *760*
Harter, D. H., 352[84], *377*
Harter, M., 723[102], 727[124], *760*
Harter, M. L., 725[351], 738[351], *768*, 896[23], *898*
Hartley, J. T., 596[74], *655*
Hartley, J. W., 385[13], *425*, 446 [20], 449[23], 520[20, 23], 527[101], *539, 542,* 593[33], 594[43, 44], *653, 654*
Hartsough, G. R., 543[203], 569 [203], *585*
Hartsuck, J. A., 89[127], 90[127], *109*
Hartwell, L. H., 625[322, 325], *662, 663*
Hartzell, R. W., 361[177], 363 [176], *379*
Harvey, B. A. M., 780[80], 787 [80], *840*
Harvey, J. D., 386[47], 387[47,

[Harvey, J. D.]
56], 389[56], 391[47], 395 [56], 423[266], *426, 432*
Harwood, L. M. J., 895[25], *898*
Hasegawa, K., 650[595], *671*
Hashimoto, K., 145[202], *166*, 635[459], *667*, 739[104], *760*
Hashimoto, S., 719[103], *760*
Haslam, E. A., 292[47, 77], 300[47], *342, 343*, 358 [85], *377*
Hass, R., 625[334], *663*
Hassan, S. A., 385[35], *425*
Hassell, J. A., 891[15], *897*
Hastie, N. D., 309[139], 314 [168], 315[168], 316 [168], *345, 346*
Hatanaka, M., 437[13], 459[39], 462[44], 463[13, 44], 485 [69], 490[69], 491[72], 493[74], 494[75], 495[76], 508[72], 516[69, 91], 518 [92], 519[72, 91[*539, 540, 542*, 625[332], *663*
Hausen, P., 612[238], 625 [238], *660*
Hauser, R. E., 118[50, 51], *161*, 199[162], 224[162], *276*
Haverty, J. L., 894[40], 895 [40], *898*
Hay, A. J., 326[202], 332, *347*, 387[55], 389[55], 390[55], 397[55], 400[157, 162], 405 [157, 162], 408[162], 410 [162], *426, 429*
Hay, D., 773[48], *839*
Hay, J., 592[27], *653*, 793 [98], 811[174], 812[183], 813, 815, 828[183], *841, 843*
Hayashi, K., 691[231], 698 [231], *764*, 879[236], 889
Hayashi, M., 556[137], 557,
[Hayashi, M.]
583, 756[105], *760*
Haynes, G. R., 812[181], 816 [182], *843*
Hayry, P., 648[568], *670*
Hayward, C., 406[202], *430*
Hayward, G., 620, *662*, 780[61, 65, 76], 782[61, 76], 785[61, 65, 76], 789, 790, 800[76, 136], 801[76, 136], 802[136], 803 [76, 136], 805[136], 806[76, 136], 807[76, 136], 808[76, 136], 819[61, 76], *839, 840*
Hayward, G. S., 771[4], 780[54, 74], 782[74], 783[74], 785 [4], 805[54], *838, 839*
Hayward, W. S., 502[79], *541*
Haywood, A. M., 361[187], 364 [86], *377, 380*
Heberling, R. L., 352[122], *378*, [113], *542*
Hecht, T. T., 202[182], *277*, 353 [87], *377*
Hedgpeth, J., 620[303], 622[303], *662*, 862[62], 873[62], *884*
Hefti, E., 190[126], 191[129], 193[129], 202[188], 203[129, 196], 227[188], 228[196], 229 [196], 230[314, 315], 231[126], 232[196], 234[126], 269[129], *275, 277, 280*
Hehlmann, R., 535[110], *542*
Heijneker, H. L., 638[497], 642 [497], *668*, 676[78], 686[78], 688[78], 732[78], 753[78], *759*
Heine, J. W., 199[163], 224[163], 256[163], *276*, 793[96], 795 [116], 797[116, 120], 810[116, 120, 165], 828[116], 833[116, 120], 834, 836[116, 120], 837 [315], *841, 842, 843, 848*
Heine, U., 778[20], 822[20, 239], *838, 845*
Heine, U. I., 439[16], *539*
Heinemann, D. W., 170[19], 182

[Heinemann, D. W.]
[19], *271*
Heinemann, R., 463[48], 488 [48], 516[48], *540*
Hejna, C. J., 48[69], 49[69], *58*
Helleiner, C. W., 574[219], *586*
Heller, E., 43[14], 45[37], *57*, 881[273, 278], *890*
Heller, M., 795[112], 796[112], 819[112], *841*
Hellman, A., 463[45], [113], *540, 542*
Hellstrom, I., 648[559], *670*
Helmke, R. J., [113], *542*
Henderson, I. C., 632[421, 422], 648[422], *666*
Henderson, W. R., 632[423], *666*
Hendler, S., 600[106], *656*
Henle, G., 835[302, 310], *847*
Henle, W., 307[124], 312[145], *345*, 350[88], 351[88], 363 [78], 373[203], *377, 380*, 789[92], 835[302, 310], *841, 847*
Hennessy, A. V., 287[23], *342*
Henry, C., 310[143], 312, *345*
Henry, C. J., 567[178], 570[178], *585*, 634[456], *667*
Henry, P., 612[241], *660*
Henry, P. H., 633[434, 435], 634 [434], 648[563, 564], *666, 670*, 679[55], 680[55], 713 [31, 106, 155], 746[154, 155], *757, 758, 760, 761*
Henson, D., 773[44], 774[44], *839*
Herde, P., 815[215], *845*
Herman, A. C., 476[65], *541*
Hermann, F., 450[25], *539*
Hermodsson, L., 351[60], *376*
Hermodsson, S., 351[60], *376*
Herndon, R., 595[57], *654*
Herpesvirus Study Group, 771[6], *838*
Herzberg, M., 608[198], 610[198],
[Herzberg, M.]
624[320], *658, 662*
Heston, W., 439[16], *539*
Hetrick, F. M., 549[61], 552[83], 553[83], 555[61], 559[61], 560[61], 561[61], 564[61], *581, 582*
Heubner, R. J., 387[67], *426*
Heuschele, W. P., 184[95], *274*
Hewish, D. R., 599[104, 105], *655, 656*
Hewitt, D., 595[56], *654*
Hewlett, M. J., 67[15], 100[157], 101[160], *105, 110*
Hewlett, N., 834[298], *847*
Heydrick, F. P., 120[60], *161*
Heyduk, J., 154[189], 155[189], *165*
Heywood, P., 729[254], *765*
Hiatt, C. W., 391[4], *427*
Hightower, L. E., 360[89], 364 [89], 367[89], 370[24], *375, 377*
Hikita, M., 352[230], *381*
Hill, B. J., 195[210], 221[210], *277*
Hill, C. S., 780[59], 782[59], *839*
Hill, C. W., 780[60], 782[60], *839*
Hill, D. H., 175[47], 181[47], 216[47], *272*
Hill, M., 521, *542*
Hill, T. J., 823[244], *846*
Hilleman, M. R., 43[4], 45[4, 39], *56, 57,* 546[41], *580,* 594[46], 595[49], *654*, 676, *760*
Hillis, W. D., 324[185], *346*
Hillova, J., 521, *542*
Hills, G. J., 15[6, 104], 16[104], *35, 38*
Hills, V. M., 7[52], *37*
Hindley, J., 16[48], *37*
Hinze, H. C., 880[257, 258], *889*
Hirai, K., 610[213], *659*
Hirose, S., 628[371, 373], *664*

Hirsch, I., 780[73], 782[73], *840*
Hirsch, J., 780[69], 819[69], *840*
Hirsch, M. S., 55[149], *61*
Hirschberg, C. B., 121[63], *161*
Hirst, G. K., 293[66], 294[68, 70], 295, 302[107], 306[118], 307 [118, 123, 126], 317[107], 333[118, 123], 335[118], 336 [219], 339, *343, 344, 345, 347, 348*
Hirt, B., 597[85], 598[85, 93], 599[93], 600[93, 110], 602 [110], 605[170], 610[219], 625, 626[357], 627[357], 645, *655, 656, 658, 659, 663, 664*
Hirth, L., 6[53], *37*, 69[52], *106*
Hitchings, G. H., 881[261, 262], *889*
Hitotsumachi, S., 651[617], *672*
Ho, M., 43[17], 45[19], 46[43], 50[89, 96], *57, 58, 59*
Hoagland, C. L., 854[31], *883*
Hobbs, T. R., 385[34], *425*
Hobson, D., 336[218], *347*
Hochberg, E., 798[131], *842*
Hodap, M., 732[87], *759*
Hodes, D. S., 351[128], 362[90], *377, 378*
Hodes, M. E., 872[181], *887*
Hodge, L. D., 335[209], 336[209], *347*, 720[174], 729[254], *762, 765*
Hoffman, E. J., 350[91], *377*
Hoffman, M. N., 390[260], *432*
Hoffman, P. R., 737[108], *760*
Hoggan, M. D., 540[2, 4, 5], 541 [26], 543[5], 545[5, 26], 546, 547[5], 548[5], 549[5, 66, 69], 550[5], 551[5], 552[5, 8, 9, 90, 91], 553[113], 554[4], 555[26, 66, 69], 556[66, 69], 557[66], 559[150], 560[139, 141], 561
[Hoggan, M. D.] [141, 150], 565[163], 567[26, 139, 174, 175], 568[163], 569 [26, 90, 207], 570[175, 210], 572[175, 210], 573[139, 210], 574[139, 210], 575[174], 576 [225], 578[26, 210], *579, 580, 582, 583, 584, 585, 586,* 631 [395], *665*, 799[133], 807[133], 833[292], *842, 847*
Hoglund, S., 698, *763*
Höglund, S., 698[215], *763*
Hogue-Angeletti, R. A., 89[128], 90[128], *109*
Hohl, H. R., 387[76], 389[76], *426*
Hohn, B., 6[80], *38*
Holden, M., 823[248], *846*
Hole, L. V., 394[128, 129], 395 [129], *428*
Holland, J., 50[98], *59*, 155[194], *165*, 198[154], 202[185], 203 [199], 236[185], 248[246], 270 [308], *276, 277, 278, 280,* 292, 326[197], 332[197], *342, 347,* 363[67], 374[274], *376, 382,*
Holland, J. J., 817[224], *845*
Hollinshead, A. C., 744[114], *760*
Holloway, A. F., 259, 260[269, 276], 261[277], 262[276], 264 [276], 265[269, 282], *279*
Holmes, A. W., 787[83], *840*
Holmes, I. H., 184[96], *274*, 385 [26], 423[264], *425, 432*
Holmes, J., 352[16], *375*
Holmes, K. C., 2[35], 3[35], 4[7, 54], 17[35], 18[35], *35, 36, 37*
Holmes, K. V., 372[49], *376*
Holowczak, J. A., 855[46], 859 [47, 53], 867[133], 873[188], 878[218], 880[252], *884, 886, 887, 888, 889*
Holterman, O. A., 324[185], *346*
Holzel, F., 610[215], *659*
Homma, M., 363[93, 94, 116],

[Homma, M.]
373[239], *377, 378, 381*
Honess, R. W., 771[5], 780[54, 61, 76], 782[61, 76], 785[61, 76], 793[5, 96, 105], 800[76, 136], 801[76, 136], 802[105, 136, 151, 153], 803[76, 105, 136], 804, 805[54, 105, 136], 806 [76, 105, 136, 151, 154], 807 [76, 136, 151], 808[76, 136, 151], 810[168], 819[61, 76], 828[105], 830[151, 154], *838, 839, 840, 841, 842, 843*
Hong, S., 814[204], *844*
Hopkins, M. S., 549[63], 555 [63], 559[63], 561[63], 576 [226], *581, 586*
Hopkins, S., 566[170], 570[170], 578[170], *584*
Hopps, H. E., 45[21], *57*
Horisberger, M., 314[165], 315 [165], *346*
Horne, R. W., 2[79], 6[56], 15 [109, 117], 16[55, 117], *37, 38, 39,* 289[34, 35], *342,* 358[95], *377,* 547[59], *581,* 690, *760,* 778[13], 779[13], *838,* 854[27], *883*
Hornsleth, A., 350[96], *377*
Horsfall, F. L., Jr., 351[97], 352[97], *377*
Horst, J., 124[81], *162,* 305 [117], *344,* 459[38], 502 [38], 503[38], *540*
Horton, R. B., 720[26], 735[26], 752[26], *757*
Horwitz, M., 634[457], *667*
Horwitz, M. S., 690[356], 697, 708[148], 726[112], 728[110, 356], 729[112], *760, 761, 768*
Horzinek, M., 115[30], 118[30], 122[71], *160, 162,* 543[238], *587*
Hosaka, Y., 358[98, 99], 359[98], 364[100, 101, 102, 224], *377,*
[Hosaka, Y.]
381
Hoshino, M., 361[112], *378*
Hosty, T. S., 179[69], *273*
Houlditch, G. S., 578[236], *586*
Hovis, J. F., 387[67], *426*
Howard, H., 543[79], 551[79], 552[79], *581*
Howatson, A. F., 7[57], 15[2], *35, 37,* 169, 189, 206, 248 [251, 252], *271, 279,* 352 [172], 358[172], 359[172], *379,* 541[13], 555[13], *579,* 591[9], 592[16], *653*
Howatson, D., 15[58], *37*
Howe, C., 124[78], *162,* 352 [104], 363[6, 103, 163, 252], *347, 377, 379, 382*
Howe, C. C., 614[253], *660*
Howell, P. G., 422[255], *432*
Howk, R. S., 456[31], *539*
Howley, P., 612[232], 619[232], 620[232], *660,* 744[131], 746[131], 747[131], *761*
Hoyer, B., 789[88], *841*
Hoyle, L., 282[1], 289[35], 302 [98], 306[98], *341, 342, 344*
Hronovský, V., 179[76, 77], *273*
Hruska, J. F., 19[84], *38*
Hsiung, G. D., 352[5, 105], *374, 377,* 595[53, 54], *654,* 778 [23], 779[23], 822[23, 241], 823[241], 824[23], *838, 845*
Hsu, K. C., 332[205], *347,* 387 [77], 402[77], *427,* 825[251], 827[270], *846,* 878[216], *888*
Hsu, M.-T., 114[14], 117[14], 130, *160*
Hsu, T. C., 825[258], *846,* 880 [254], *889*
Huang, A. S., 85[111], 101[162], *108, 110,* 153, *164,* 189[115], 193[132], 198[153], 203, 214

[Huang, A. S.]
[115, 206], 226[194], 227 [217], 236[230], 239[237, 242], 243[237, 244], 248 [242, 250, 254], 251[255], 252[255], 254, 256[244], 257[257], 265[244, 288], 266[288], 267[297], 268 [297], 269[115, 132, 206, 242, 250], 270[206, 250, 307], *275, 276, 277, 278, 279, 280,* 334[206], 335 [206], *347,* 352[8], 363 [31], 366[107], 367[8, 107], 372[186, 233], 373 [106, 108], *374, 375, 377, 380, 381,* 834[298], *847*
Huang, C. C., 825[259], *846*
Huang, E.-S., 597[84], 599 [84], 600[84], 601, *655, 656,* 772[43], 779[43], 789[94], 792[94], 814, 815[210], *839, 841, 844*
Huang, J. W., 47[57], 48 [69], 49[57, 69, 78], *58, 59*
Huang, K., 55[147], 61
Huang, R. C. C., 599[94], *655*
Hubbs, J., 591[5], *652*
Huberman, E., 649[589], *671*
Huberman, J. A., 729, *758*
Hubert, E., 132[113], *163*
Huck, R. A., 543[95, 96], 552 [95, 96], 568[96], *582*
Hudson, B., 603[139], *657*
Hudson, J., 617[280], 622 [280], *661*
Hudson, J. B., 774[51], 780 [51], *839*
Huebner, K., 645[546, 547], 648[574], 649[574], 585, 586], *670, 671*
Huebner, R., 438]16], *539*
Huebner, R. J., 436, 437[7, 9, 13], 443[17], 463[13], 501
[Huebner, R. J.]
[5], 502[17], 505[87], 508 [87], 509[9], 534[107], *538, 539, 541, 542,* 578[234, 235], *586,* 593[33], 594[43, 44], 631 [390, 395], 648[390], *653, 654, 665,* 676[229], 677[115, 116], 679[55, 172], 680[55], 721[113], 731, 744[114], *758, 760, 762, 764*
Huez, G., 132[113], *163*
Huismans, H., 411[225], 423[271], *431, 432*
Hull, R., 7[59], *37,* 197[145], 200 [145], *275*
Hull, R. N., 351[110], 352[109], *377,* 384[9, 11], *424,* 594[47], *654,* 787[83], *840*
Hulme, B., 595[61], *654*
Hume, V. B. M., 881[270], *890*
Hummeler, K., 7[60, 61], *37,* 169, 201[176], *271, 276,* 614[254], *660,* 798, *842*
Hunsmann, G., 452[28], 472[55], 473[28, 56], *539, 540*
Hunt, D. M., 260, 261, *279*
Hunt, J. M., 53[121], *60,* 239 [241], *278*
Hunt, T., 75, *107*
Hunter, H. S., 114[5], 115[24], 118[5], 121[5], 142[24], *160*
Hunter, T., 628[370], 629[370], *664*
Huper, G., 452[28], 473[28], 476 [64], *539, 541*
Hurst, E. W., 591[10], 592[10], *653*
Hutchinson, C. A., III, 603[142], *657,* 756[117], *760*
Hutchinson, F., 387[61], 392 [61], *426*
Hutchinson, J. E., 366[111], 367 [155], *378, 379*
Huxley, H. E., 15[62], 16[62], *37*

Hyatt, D. F., 352[249], *381*
Hyde, J., 865[101], *885*
Hyde, J. M., 808[155], *843*, 854[24], 855[40], *883*
Hyman, R. W., 780[60], 782 [60], *839*
Hynes, R. O., 641[520], *669*

Iadarola, M., 43[10], 45[10, 41], *57*, *58*
Ichihashi, Y., 866[114], 879 [223, 225, 226, 227], 881 [114], *885*, *888*
Ideda, H., 452[27], 473[27], *539*
Igarashi, A., 117[44], 154 [187], 155[187], *161*, *165*
Igarashi, K., 719[365], *768*
Igel, H. J., 534[107], *542*
Iglewski, W. J., 392[109], *427*
Ihle, J. N., 452, 473[28], *539*
Iinuma, M., 361[112, 154], 373[228], *378*, *379*, *381*
Ikegami, N., 412, *432*
Illmensee, R., 461[42], *540*
Imagawa, D. T., 351[113, 114], *378*
Imblum, R. L., 202[189], 237 [189], 238[233], 258[189, 233], *277*, *278*, 360[115], *378*
Inaba, Y., 174[24], *272*
Ingram, D. G., 546[48], *581*
Inman, J. K., 555[129], 556 [129], 557[129], 568[129], *583*
Inui, S., 552[85], 553[85], *582*
Irisawa, J., 710[250], 711[250], *764*
Irlin, I. S., 634[442], *666*
Isaacs, A., 41, 45[30, 38], 46 [30, 46], 47[30], 48[60], *56*, *57*, *58*
Isbell, A. F., 815[212], *844*
Ishibashi, M., 545[179], 567 [179], *585*. 697[120], 698 [357], 712[118, 119], 716 [120], 723[357], 724[357], 725, *760*, *768*
Ishida, N., 358[4], 361[225, 244], 363[116], 373[239], *374*, *378*, *381*, 389[81], *427*
Ishikawa, A., 613[488], 637[488], 640[516], *668*, *669*
Ishitsuka, H., 635[460], *667*
Ishizaki, R., 476[64], *541*
Isoun, T. T., 180[88], *274*
Itagaki, A., 731[135], *761*
Ito, H., 174[24], *272*
Ito, K., [121], 729[122], *760*
Ito, M., 879[234], *889*
Ito, M., 545[179], 549[62], 550 [74], 554[119], 561[149], 567 [179, 181], 570[119], 572 [215], 575[181], *581*, *583*, *584*, *585*, *586*, 712[123], *760*, 835[305], *847*, 879[234], *889*
Ito, Y., 405[199], 409[217], 415, 418[199, 243], 420, 421, *430*, *431*, *432*, 592[12, 15], *653*, 735[248], *764*
Itoh, H., 352[117, 165], 364[117, 165], *378*, *379*
Ittensohn, O. L., 546[41], *580*
Ives, D. R., 365[9], *374*
Iwamoto, I., 174[25], *272*
Iwasaki, Y., 180[87], *274*, 798 [127], 824[127], *842*
Iwatsuki, N., 628[372], *664*
Izumida, A., 174[25], *272*

Jackson, A. H., 615[263], *661*
Jackson, G. G., 350[118, 119], 351[118, 119], *378*, 385[18, 31], 386[31], *425*, 677[358], *768*
Jacob, R., 780[61, 76], 782[61, 76], 785[61, 76], 800[76],

[Jacob, R.]
801[76], 803[76], 806[76], 807[76], 808[76], 819[61, 76], *839, 840*
Jacob, R. J., 772[62], 780[54, 62, 65], 782[62], 784[62], 785 [65], 787[62], 789[65], 790 [65], 805[54], 818[319], 819 [319], *839, 840, 848*
Jacobson, M. F., 83, 85[109, 110], 89[109, 110], 99[109], *108,* 139[201], *166*
Jacquemont, B., 801[141, 142], *842*
Jaenisch, R., 623[312], 626 [355], 627[355], 630[385, 386], *662, 664*
James, A. W., 605[175], *658*
James, C. G., 74[70], *107*
James, E. W., 629[377], *664*
Jamieson, A. T., 813, *844*
Jamieson, M. F., 866[121], *886*
Jamison, R. M., 386[53], 387 [53], 389[53], *426,* 545[53], 546[53], 547[53], 557[53], 561[147], *581, 584,* 636 [479], *668*
Jansz, H. S., 683[261], 726 [268], 728[267, 268], *765,* 895[43], *898*
Janz, H. S., [378], *768*
Janzen, H. G., 154[178], *165*
Jasty, V., 823[245], *846*
Jaureguiberry, G., 862[67], *884*
Jefferts, E. R., 880[252], *889*
Jelinek, W., 719[34, 65], *758, 759,* 891[45], *898*
Jeng, Y.-H., 723[345], 725[351], 727[124], 738[351], *760, 768,* 896[23], *898*
Jennings, A. R., 386[42], *426*
Jensen, A. B., 402[175], *430*
Jensen, F., 595[51], 625[339], 631[339], *654, 663*
Jensen, F. C., 595[52], 631[401],
[Jensen, F. C.]
648[570, 576, 577], 650[570], *654, 665, 670, 671*
Jensen, M. H., 174[42], 185[105], *272, 274*
Jenson, A. B., 184[94], *274*
Jerkofsky, M., 635[462, 464, 465], *667*
Jerofsky, M., 739[125, 126], *760*
Jersild, C., 351[188], 362[184], *380*
Johansson, K., 688[284], 749 [284], *765*
Johnson, F. B., 549[66, 69], 555 [66, 69], 556[66, 69], 557[66], 560[139], 567[139], 573, 574 [139], *581, 583*
Johnson, G. C., 385[18], *425*
Johnson, H. M., 55[143], *61*
Johnson, H. N., 170[17], 174[17], 177, 180, 183[91], *271, 273, 274*
Johnson, K. M., 176[59, 213], *273, 277*
Johnson, R. B., 410, 420[220], *431*
Johnson, R. H., 541[11, 22, 23], 543[11, 22, 23, 72, 78], 550 [72], 551[78], 552[22, 100, 101], 553[112], 555[11], 556 [11], 559[11], 567[100], 568 [22, 23, 100], 570[22, 100], *579, 580, 581, 582*
Johnson, R. T., 179[70], *273,* 595[57], *654*
Johnston, M. D., 70[153], 99[153], *110*
Johnston, R. E., 117[41, 46], 132 [46], *161*
Joklik, 829[278], *846*
Joklik, W. K., 50, 51, 52, *59, 60,* 385[30], 386[46], 387[46, 52, 55, 64], 389[55], 390[30, 46, 55], 392[30, 52, 64, 112, 113], 394[127, 131, 132], 395[30, 131, 132, 135, 137, 140, 141],

[Joklik, W. K.]
397[55, 135, 141], 398[113, 140], 400[112, 135, 157, 162], 401[112, 137], 402[135], 403 [30, 64], 404[30, 64, 188, 191], 405[112, 135, 157, 162, 199, 200], 406[113, 162], 407[113, 140, 141, 188, 189], 408[162], 409[127], 410[131, 132, 162, 222 222], 411[113, 224], 412[224], 413, 415[240], 417[247], 418 [199, 243, 248], 420[222], 421, 422[30, 137, 252, 253], 423 [271], *425, 426, 428, 429, 430, 431, 432, 433,* 678[241], 727 [243], *764,* 851[10], 854[32, 33], 855[38], 858[49], 859 [17, 49], 862[64], 866[106], 867[127, 134, 135, 136], 868 [147, 148, 151], 869[64, 148], 870[147, 148], 872[169, 172], 873[188], 876[148, 198], 878 [198, 221], 880[147], 881 [271], *882, 883, 884, 885, 886, 887, 888, 890*
Joncas, J., 350[120], *378*
Jones, E. P., 823[248], *846*
Jones, J. M., 423[262], *432*
Jones, K. J., 135[140], 143[140], 152[140], *164*
Jones, K. W., 733[43], *758*
Jones, T. C., 45[22], *57*
Jonkers, A. H., 170[14, 19], 177 [63], 182[19], *271, 273*
Jordan, L. E., 18[63], 19[63], *37,* 386[45, 53], 387[53, 75], 389 [53], 390[83], 400[83], 402 [45], *426, 427,* 545[53], 546 [53], 547[53], 549[62], 554 [119], 557[53], 561[147, 149], 570[119, 211], 574[217], *581, 583, 584, 586,* 636[479], *668*
Jorgensen, G. N., 812[178, 179], *843*
Jorgenson, G., 872[176], *887*
Jϕrgensen, P. E. V., 174[43], 186
[Jϕrgensen, P. E. V.]
[43], 221, *272*
Jori, L. A., 543[67], 549[67], 555[67], 559[67], 561[67], *581*
Jörnvall, H., 716[346], 725[346], *768*
Joske, R. A., 385[20, 33], 386 [37], *425*
Jumblatt, J., 115[27], 118[27], 119[27], 121[27], 123[27], *160*
Jungwirth, C., 872[172], 880[256], *887, 889*
Jurale, C., 869[154], *887*

Kääriäinen, L., 114[8, 15], 117[36, 39], 118[8, 15, 49], 120[58], 121[62], 123[72], 126[15], 131[36], 132[36], 133[72], 134[116, 118, 200], 135[72, 118, 120], 137[36, 118, 120, 128], 139[39], 148[128], 149, 151[116], 152, 153, 158[62], *160, 161, 162, 163, 166*
Kabigting, A., 437[13], 463[13], *539*
Kaesberg, P., 69[54], *106*
Kainuma, R., 628[372], *664*
Kajima, M., 541[32], 543[32, 80], 551[32, 80], 552[80], 555[32, 80], 569[80], *580, 581*
Kajioka, R., 623[313], *662*
Kakefuda, T., 493[74], *541,* 555 [122], 559[122], 561[122], 562[122], 564[122], *583,* 713 [31], *757*
Kaku, H., 879[229], 880[229], *889*
Kalica, A. R., 362[90], *377*
Kalter, G. V., 352[121], *378*
Kalter, S., 463[45], *540*
Kalter, S. S., 352[121, 122], *378,* [113], *542*
Kaluza, G., 317[178], 326[200], 332[200], *346, 347*

Kamahora, J., 825[255], *846*, 851 [6, 8], 877[201], 879[224, 229], 880[229], *883, 888, 889*
Kamen, R., 617[296], 618[296], 619[296], 620[296], 622[296], 623[296], 624[296], 644[296], 645[296], 646[296], *662*
Kamen, R. I., 92[132], *109*
Kameyama, S., 851[6], 877[201], *883, 888*
Kamiya, T., 624[318], *662*, 812 [182], 816[182], 833[294], *843, 847*
Kammer, K., 610[214], *659*
Kanarek, A. D., 6[116], *39*
Kang, C. Y., 214[204], 255, 257, 258[260], 269[204], *277, 279*
Kang, H. S., 626[349], *663*
Kanich, R. E., 822[240], 824 [240], 827[266], *845, 846*
Kapikian, A. Z., 385[29], *425*, 546[40], 551[40], 552[90], 569[90], *580, 582*
Kaplan, A. S., 614[257], 625 [337], *660, 663*, 773[119], 779[42], 795[113, 119], 797 [113, 119], 802[147, 150], 811[150], 812[182], 816 [182, 221], 817[225], 818 [225, 228, 229], 819[113, 234], 825[254], 827[228, 271], 828[276], 829[276, 283], 830, 833[294, 295], *839, 841, 842, 843, 845, 846, 847*
Kaplan, H. S., 508[84], *541*
Kaplan, J. C., 601[118], 650 [603], *656, 671*
Kapuler, A. M., 400[156, 158], 401[166], 402[156], 405 [156, 158, 198], 406[198, 202], *429, 430*
Kara, J., 625[329], *663*
Karabatsos, N., 170[20], 174 [20], 182[20], 184[20],
[Karabatsos, N.] 215[20], *271*
Karasaki, S., 547[57], 549[65], *581*
Karpas, A., 521, *542*
Kasel, J. A., 286[18], *341*
Kasnic, G., Jr., 770[2], 826[2], *838*
Kass, S. J., 592[19, 23], *653*
Kassanis, B., 6[44], *36*
Kasten, F. H., 560[144], *584*
Katagiri, S., 389[81], *427*
Kates, J., 714[305], *766*, 862 [61, 63, 68, 69, 70], 863[74], 867[137], 869[69, 74], 873 [61, 74], 877[137], *884, 886*
Kates, J. R., 862[58, 65], 867 [132, 145], 868[58, 145], 869 [154, 156, 157], 872[185], 873[157], 877[199], 878[219, 220, 222], *884, 886, 887, 888*
Kato, A. C., 604[167], 605[167], *657*
Kato, M., 605[186], 610[186], *658*
Kato, S., 851[6], 877[201], 879 [224, 235], 880[255], *883, 888, 889*
Katoh, T., 352[117], 364[117], *378*
Katsume, I., 827[268], *846*
Katz, E., 446[22], 501[22], 520 [22], *539*, 859[57], 866[117], 873[189], 874[189, 190], 875 [189, 190, 191], 881[117, 264, 265, 267, 272, 277], *884, 885, 887, 889, 890*
Katz, E. N., 877[212], 878[212], 881[212], *888*
Katz, J. R., 859[56], 878[56], *884*
Katz, S. L., 350[123], *378*
Kauffman, R. S., 704[75], 705 [75], 709[75, 76], 712[75, 76], 726[75], 727[75], *759*
Kaufman, E. R., 814[202], *844*

Kaverin, N. V., 369[126], 371[124], 372[125], *378*, 868[150], 869[150], *886*
Kawade, Y., 49[75], *59*
Kawai, S., 503[80], *541*
Kawakami, T., 516[91], 519[91], 537[116], *542*
Kawakami, Y., 552[85], 553 [85], *582*
Kawamura, H., 384[6], 385[6, 14], *424, 425*
Kay, D., 15[64], *37*
Kazama, F. Y., 770[1], 778[1], 822[1], 824[1], *838*
Ke, Y. H., 46[43], *58*
Keall, D. D., 385[20], *425*
Keegstra, K., 114[17], 121[17], *160*
Keegstra, W., 685[359], 690 [359], *768*, 894[26], *898*
Keihn, E. D., 292, 332, *342*
Keir, H. M., 592[27, 28], *653*, 811, 814, 815[203, 213, 214], 816, *844, 845*
Kellenberger, C., 20[1], *35*
Kellenberger, E., 20[1], *35*
Keller, J. M., 828[275], 832 [297], 833[275, 297], 837 [297], *846, 847*
Keller, R., 879[230], *889*
Keller, W., 602[128], 603[128], 616[267, 271], 618[271], 619 [271, 297], 620[271, 297], 621[297], 623[297], 624 [297], 629[382], 630[128, 383], 646[271], *656, 661, 662, 664*, 681[177], 683 [222], 687[177], 688[177], 736[177], 748[237], 750[177], *762, 763, 764*, 895[36], *898*
Kelley, D. C., 313[158, 159], 329 [203], *346, 347*
Kelley, D. E., 719[199, 200], *763*, 862[69], 869[69], *884*
Kelley, G. W., 287[19, 20, 21], *341, 342*
Kelley, J. M., 199[159], 200[166], *276*
Kelloff, G. J., 437[9], 508[9], 509[9], *538*
Kelly, F., 651, *672*
Kelly, T. J., 741, 742, 744, 746 [128], *761*
Kelly, T. J., Jr., 563[157], 564 [157], *584*, 626[351], 627 [351], 627[351], 633[433, 436], *663, 666*, 741, 742[145], 745 [129], 746[128, 145, 176], *761, 762*
Kelly, T. J. K., Jr., 683[14], *757*
Kelter, A., 385[17], *425*
Kemp, G. E., 175[51, 52, 54], 180, 181, 182[54], 218[51], *272, 273, 274*
Kemp, M. C., 793[99], 795[115], 796[99], *841*
Kendal, A. P., 7[5], *35*, 292[46], 293[60], 294[60], 298[46], 300[46, 60], *342, 343*
Kennedy, S. I. T., 117[42], 126 [88], 131[107], 132[42, 107], 135[121], 148[157], 149[157], *161, 162, 163, 164*
Kent, S. G., 287[22], *342*
Keränen, S., 123[72], 133[72], 134[116, 118], 135[72, 118], 137[118, 128], 148[128], 149, 151[116], 152, 153[118, 128], *162, 163*
Kern, J., 578[234, 235], *586*, 679 [172], *762*
Kerr, I., 67[22], 80[22], 85[22], *105*
Kerr, I. M., 51[113], *60*, 862[66], *884*
Kesteren, L. W., 681[293], *766*
Khakpour, M., 299[87], *344*
Khittoo, G., 712[311], 713[311], *766*
Khoury, G., 565[166], 566[166],

[Khoury, G.] 567[194], 577[166, 194], *584, 585,* 611[225, 228], 612[232], 616[272], 618 [272], 619[232, 272, 298], 620[232, 272, 298], 622 [298], 624[298], 628[228, 374], 634[454], 638[499], 646[499], 647[555], *659, 660, 661, 662, 664, 667, 668, 670,* 728[144], 739 [48], 741, 744[131, 146], 746[131], 747[131, 132], 748[132, 146], *758, 761*

Khutoretskaya, N. V., 154[176], *165*

Kibrick, S., 55[134], *61*, 578 [237], *587*

Kidwai, J. R., 617[278], 620 [278], 645[278], *661*

Kidwell, W. R., 601[119], *656*

Kieff, E., 793[100], *841*

Kieff, E. D., 774[67], 780[56, 67, 77], 781, 787[77], 789 [88, 89], 798[132], 831 [288], *839, 840, 841, 842, 847*

Kiehn, E. D., 617[295], 624 [295, 315], 630[388], 635 [388], *662, 665*

Kilbourne, E. D., 282[12, 231], 282[16], 288[33], 289[12, 23], 290[12], 292[44, 45], 298[44, 45], 299[87], 308 [133], 310[86], 324[186], 336[12, 214, 220], *341, 342, 344, 345, 346, 347, 348*

Kiley, M. P., 190[121], 248[121], 257, 258[261], 269[121], *275, 279*

Kilham, L., 363[127], *378*, 386 [39], *425*, 541[16], 543[12, 78], 546[241], 549, 551[12, 78], 552[12, 82, 87, 98, 104, 106], 553[12, 87, 98, 106,

[Kilham, L.] 107, 108, 110, 111, 112], 555 [12], 560[143], 567[12, 106], 568[82], 578[16, 106], *579, 581, 582, 584, 587,* 596[75], *655*

Kilhaus, S., 387[55], 389[55], 390[55], 397[55], *426*

Kim, C. S., 352[121], *378*

Kim, H. W., 351[128], *378*, 385 [29], *425*, 552[91], *582*

Kim, N., 465[49], 466[49], 467 [49], 468[49], 509[49], 537 [116], *540, 542*

Kimes, R., 678[92], 681[92, 133], 682[92], 686, *759, 761*

Kimes, R. L., 714[93], *759*

Kimura, G., 632[415, 416], 635 [416, 461], 636[415, 416, 461, 483], 637[415, 416], 638[502], *665, 667, 668,* 731 [135], 739[134], *761*

King, J., [23], *757*

Kingsbury, D. T., 155[194], *165*, 817[224], *845*

Kingsbury, D. W., 323, *346*, 353 [57, 63, 130], 356[192], 357 [134, 192], 358[68, 132], 359 [156, 234, 235, 236], 360[234, 235, 236], 361[195, 196], 362 [195, 196], 364[134, 196], 365 [129, 131], 366[130, 236], 367[132, 156, 157, 197, 236], 369[134], 371[63, 193, 195, 234, 237], 372[135], 373[133], 374[270, 271], *376, 378, 379, 380, 381, 382*

Kirckpatrick, J. B., 867[137], 877 [137], *886*

Kirkwood, J., 787[84], *840*

Kirn, A., 866[105], *885*

Kirschstein, R. L., 560[143], 578 [233], *584, 586*

Kiselev, N. A., 6[65], 23[66], *37*

Kissling, R. E., 179[69], *273*

Kit, S., 591[4], 612[236, 237, 243], 625[236, 243, 323, 324, 326, 327, 328], 632[418], 636[418, 485], 637[418], 640[418], 648[573, 579], 649[579, 580, 581], 650 [602], *652, 660, 662, 663, 666, 668, 670, 671,* 808 [176], 812[176, 177, 178, 179, 180, 181], 813[197, 198], 814[196], 816[222], 837[176], *843, 844, 845,* 867[144], 868[144], 872 [168, 170, 171, 173, 175, 176], 880[254], *886, 887, 889*
Kitahara, T., 612[243], 625[243], 631[393, 396], *660, 665*
Kitajima, E. W., 7[99], *38*
Kitamura, K., 552[85], 553[85], *582*
Kitayama, T. A., 351[74], 352 [74], *376*
Klagsbrun, M., 870[162], *887*
Klausner, R. D., 6[28], *36*
Kleiman, J. H., 865[97, 98], *885*
Klein, E., 835[302], *847*
Klein, G., 648[559], *670,* 789 [92], 808[161, 162, 163], 827[162, 273], 835[163, 302, 306, 308, 309, 310, 311], *841, 843, 846, 847*
Klein, R., 495[76], *541*
Kleinschmidt, A., 565[162], *584*
Kleinschmidt, A. K., 20[77], 21[77], *38*, 387[63], 392 [114, 237], *426, 428, 431,* 592[23], *653*, 681[39], 686, *758*
Kleinschmidt, W. J., 48[66], 50 [93], *58, 59*
Klement, V., 449[23], 520, *539*
Klemperer, H. G., 812[181], 816 [182], *843*
Klenk, H.-D., 202[181], *276*, 282 [9], 292[49], 293[57], 298 [239], 300[94, 95], 308[57], 309[95], 326[192, 193, 194, 195, 199, 239], 328, 332[192, 193, 194, 195, 199], 336[225], *341, 342, 343, 344, 345, 346, 347, 348,* 360[136, 137], 362 [138, 139], 373[267], *378, 382*
Klessig, D. F., 739, *761*, 891[27], 892, *898*
Klett, H., 400[156], 401[166], 402[156], 405[156], 406[206], 409[212], 410[219], 411[212, 219], *429, 431*
Kligman, A. M., 593[30], *653*
Klimenko, S. M., 359[247], *381*
Klinova, E. G., 55[148], *61*
Klontz, G. W., 188[113], *274*
Kluchareva, T. E., 634[441], *666*
Klug, A., 2[20, 68], 3[68], 4[7, 34, 45, 76, 105, 111], 5[16, 38, 69], 8[21], 15[36, 37, 40, 67, 71, 72], 16[21, 36, 37, 40, 68, 70, 72, 73, 74], 17[39], 18[39], 19[74], 20[13, 70], 22[24, 121], 23[66], *35, 36, 37, 39,* 118[55], *161,* 357[140], 358[140], *378,* 591[1, 2, 3], 592[1, 2], *652*
Knight, C. A., 307[130], *345*, 592 [19, 23], *653*
Knight, C. O., 402[172, 178], *429, 430*
Knight, V., 286[18], *341*
Knipe, D., 197[143], 239[239], 243[143], 246[239], 257[143], *275, 278,* 351[141], 370[141], *378*
Knippers, R., 574[218], *586*
Knowles, B., 650[604], *671*
Knowles, B. B., 648[576], *671*
Knox, A. W., 332[205], *347*
Knudson, D. L., 169, 173, 189, 206, *271*

Kobune, F., 352[230], *381*
Kocent, A., 268[299], *280*
Koch, G., 74, 76, 78[89, 90], 86 [90], *107, 108*
Koch, M. A., 15[4], *35*, 631[398], 648[562], *665, 670*
Kocylowski, B., 187[111], *274*
Koczot, F. J., 555[127], 559[127, 150], 560[140], 561[150], 562[127], 564, 565[163], 567 [195], 568[163, 195], 569 [207], 571[195], 572[195], 573[195], 575[195, 222], 577 [127, 140, 222], *583, 584, 585, 586*
Koczot, F. V., 683[137], *761*
Kodama, K., 174[25], *272*
Koenig, M., 814[205], *844*
Kogan, A., 385[17], *425*
Kohama, K., 358[178], *380*
Kohama, T., 361[225], *381*
Kohlhage, H., 831[290], *847*
Kohne, D., 516[91], 519[91], *542*
Kohne, D. E., 643[533], 644 [533, 535], *669*
Kohno, M., 45[21], *57*
Kohno, S., 45[21], *57*
Kokernot, R. H., 174[21], 182 [21], *271*
Kolakofsky, D., 92[133], *109*, 356[142, 143, 144], 357[144, 145], 358[143], 361[205], 370[206], *378, 379, 380*
Koldovsky, P., 648[567], *670*
Kolodner, R., 780[66], *840*
Komano, T., 574[218], *586*
Kongsvik, J. R., 555[130], 556 [130, 135], 557[135, 136], 573, *583*
Konigsberg, W., 397[142], *429*
Koprowski, H., 7[61], *37*, 188 [114], 194[140, 141], 195 [141], 196[141], 199[141, 161], 201[141], 203[195], 226[195], 227[195], 228
[Koprowski, H.] [195], 229[195], 256[114], *275, 276, 277*, 595[51, 52], 612[233], 614[252], 631[401], 645[544, 545, 546, 547, 548, 549], 648[567, 570, 574, 578], 649[574, 583], 650[570, 604], [585, 586], *654, 660, 665, 670, 671*, 798[127], 824[127], *842*
Korant, B. D., 69[59], 74[72], 85[112, 114], 90[112], *107, 108, 109*
Korbecki, M., 352[153], *379*
Kornberg, A., 726[239], *764*
Kornberg, R. D., 599[102], *655*
Kornilaeva, G. V., 298[84], *343*
Kort, J., 565[167], *584*
Kos, K. A., 123[76], 146[76], *162*
Koshelnyk, K. A., 147[155], 155 [155], *164*
Koszinowski, U., 880[241], *889*
Kourilsky, F. M., 808[163], 835 [163, 308, 311], *843, 847*
Kozak, M., 780[76], 782[76], 785[76], 800[76, 136, 138, 139], 801[76, 136, 138, 139], 802[136], 803[76, 136], 805 [136], 806[76, 136], 807[76, 136, 138], 808[76, 136, 138], 819[76], *840, 842*
Krainer, L., 385[19], *425*
Kraiselburd, E., 813[193], 814 [193, 200], *844*
Krempin, H. M., 521[96], *542*
Kronauer, G., 541[36], 543[36, 71], 550[36, 71], 551[36], 554[36], 567[36], 568[71], 569[71], *580, 581*
Kropachev, V. A., 45[27], *57*
Kruczinna, R., 336[225], *347*
Krueger, R. F., 45[29], *57*
Krug, R. M., 310[141], 320, 321, 323[238], 324[189], 336[222], 337[222], *345, 346, 347, 348*, 395[136], *428*

Krystal, G., 390[93], 398[93], 401[93], 423[270], 424[278], *427, 432, 433*
Kubes, V., 175[48], 181[48], 216 [48], *272*
Kuchino, T., 632[419, 427], 636 [427], 637[419], *666*
Kudo, H., 403[183, 185], 404 [185], 411[183], 412[183, 185], *430*
Kufe, D., 535[110], *542*
Kuijk, M. G., 728[267], [376], *765, 768*
Kundin, W. D., 385[36], *425*
Kung, H.-J., 114[14], 117[14], 130[14], *160*, 458[36], *540*
Kurokawa, T., 719[365], *768*
Kurstak, E., 541[38], 562[38], *580*
Kurth, R., 533[105], *542*
Kyner, D., 406[206], 407[207], 408[208], *431*

Lachmann, P. J., 351[146], *379*
Lachmi, B.-E., 134[118], 135, 137, 153[118], *163*
Lackey, D., 603[145], *657*
Lackland, H., 359[170], 364[170], 368[170], *379*
Lackovic, V., 45[23], *57*
LaColla, P., 877[203], *888*
Lacy, S., Sr., 685[138, 139, 140], *761*
Laemmli, U. K., 398[152], *429*
Lafay, F., 260[272, 273, 274], 265[272, 281], 267[272, 273], *279*
LaFiandra, A. K., 397[144], 406 [204], *429, 431*
Lagwinska, E., 143[139], *164*
Lai, C.-J., 608[200, 210], 610[200, 201], 611[200, 201], 632[412], 637[200, 490, 493], 638[200,
[Lai, C.-J.]
490, 493], *659, 665, 668*, 756 [141], *761*
Lai, M.-H., 411[224], 412[224], *431*
Lai, M. M. C., 459[38], 461[43], 502[38], 503[38], *540*
Lai, S.-P., 752[316], *766*, 893[32], *898*
Laine, R., 114[6], 121[6], *160*
Laipis, P. 602[126], 603[126], 629[378], *656, 664*
Laithier, M., 797[123], *842*
Lake, R. S., 597[89], 599[89], *655*
Lamb, R. A., 355[148], 356, 360 [147, 148], 361[147], 366[148], 367[148], 371[148], *379*
Lamont, P. H., 386[42], *426*
Lampson, C. P., 43[4], 45[4, 39], *56, 57*
Landgraf-Leurs, I., 677[88], 679 [88], 688[88], 717[88], 718 [88], 720[88], 732[88], 733 [88], 734[88], 753[88], *759*
Landgraf-Leurs, M., 686, 688[143], 717[142, 143], 749[142], *761*
Lando, D., 797[122, 123], *842*
Landsberger, F. R., 292, *342, 343*
Lane, W. T., 437[9], 508[9], 509 [9], *538*, 677[115, 116], 731 [116], *760*
Langridge, R., 387[60, 61], 392 [60, 61, 108], *426, 427*
Lantenberger, J. A., 603[145], *657*
LaPlaca, M., 596[67], 614[255], *654, 660*
Laskov, R., 398[148], 418, *429*
Latarjet, R., 631[407], 641[407], *665*
Lau, R. Y., 404[197], 407[197], *430*
Launer, J., 880[256], *889*
Laurence, G. D., 1[116], *39*

Lavelle, G., 728[144], *761*
Laver, W. G., 282[3, 8], 288[25, 26, 27, 28, 31], 289[31], 290, 292[45], 293[58], 295, 297, 298[45, 58, 81], 299[85, 86, 90], 307[25, 26, 27, 28, 31], 308[81], 310[86], 336[31], *341, 342, 343, 344*
Lavi, S., 395[139], 398[139], 404[139], 407[139], 408[139], *428*, 605[191], 609[191, 206, 208, 209, 211], 617[289], 643, *658, 659, 661*
Law, L. W., 43[6], 54[6], *56*, 594[43], *654*
Lawrence, C., 76[88], 77[88], 78[88, 92], 79[92], 90[130], 94[136], *108, 109*
Lawrence, W. C., 817[226], 818 [231], *845*
Lawson, D. E. M., 357[19], 360 [19], *375*
Lawson, L. A., 798[129], *842*
Layman, K. R., 567[183], 568 [183], *585*
Layton, J. E., 324[188], 329 [188], *346*
Lazar, A., 881[264], *889*
Lazar, E. C., 352[16, 149], *375, 379*, 541[32], 543[32, 80], 551[32, 80], 552[80], 555 [32, 80], 569[80], *580, 581*
Lazarides, E., 597[87], 598[87], *655*
Lazarowitz, S. G., 297, 298[78, 79, 240], 300[237], 310[78], 326[78, 79, 190, 240], 328 [237], 329[190], 332, *343, 346, 348*
Lazarus, H. M., 632[423], *666*
Lazdins, I., 292[47], 298[47], 300[47], *342*
Leak, P. J., 384[4, 5], 385[20, 33], 386[37], *424, 425*
Leamnson, R. N., 191[130], 243 [130], 269[130]
Leary, P., 55[144], *61*
Leavitt, R. W., 155[194], *165*
Leberman, R., 4[7, 34, 76], 16 [74], 17[39], 18[39, 75], 19 [74], *35, 36, 37*, 636[472], *667*
LeBerre, M., 220[209], 221[209], *277*
Lebleu, B., 51[114], 51[114, 118], *60*, 132[113], *163*, 423[273], *433*
LeBouvier, G. L., 773[48], *839*
Lebowitz, J., 592[25], 602[25, 126, 130], 603[126], 605[171, 172, 176], *653, 656, 658*
Lebowitz, P., 605[183], *658*, 741, 742, 744[146], 746[145], 748 [146], *761*
Leclerq, M., 132[113], *163*
Leder, P., 67[20], 80[20], 85[20], *105*
Ledinko, N., 543[171], 544[213], 549[63], 553[115], 555[63, 124], 559[63], 561[63], 566 [115, 169, 170], 569[115], 570 [169, 170], 571[213], 578[170], *581, 583, 584, 586*, 710[147], *761*
Lee, C. W., 595[54], *654*
Lee, F. C., 827[267], *846*
Lee, J. C., 117[45], 132[45], *161*
Lee, L., 789[89], *841*
Lee, L. F., 774[67, 317], 780[67], 814[205], 815[212], *840, 844, 848*
Lee, S. L., 385[23], *425*
Lee, T. N. H., 605[192], 608[199], 609[192, 199], 610[199], 611 [192, 199, 225, 227], 619[298], 620[298], 622[298], 624[298], 628[199, 227], 628[199, 227], *658, 659, 662*, 742[145], 746 [145], *762*
Lee, V. H., 175[54], 182[54], 216[54], 218[54], *273*

Lee, Y. F., 67[14, 17], 68[17], *105*
Leers, W. D., 385[16], *425*
Leestma, J. E., 820[236], 824 [236], 826[236], *845*
Leibowitz, J., 697, 708[148], *761*
Leibowitz, R., 75[78], 76[78], 78[78], 80[78], *107*
Leinbach, S. S., 815, *844*
LeMoyne, M. T., 385[21], *425*
Lenard, J., 202[178], *276*, 292 [55, 56], 335[207], 336[207], *342, 343, 347*
Lenches, E. M., 148[156], 149[156 [156], 150[156], 151[156], *164*
Lengyel, P., 51[115], 52[115, 119], *60*, 397[142], 404[190], 406 [190], 407[190], 423[273], *429, 430, 433*
Lennette, E. H., 390[260], 391 [96, 99], 402[172, 178], *427, 429, 430, 432,* 787[83], *840*
Lenoir, G., 220[209], 221[209], *277*
Leonard, K. R., 20[77], 21[77], *38*
Leone, A., 139[142], 153[142], *164*
Leong, J. K. L., 117[106], 126 [103], 131[103, 106], *163*
Lepow, M. L., 595[55], *654*
Leppla, S. H., 304[100], 306 [100], *344*
Lerner, A. M., 385[22], 387[66], 390[66, 89, 90, 91], *425, 426, 427*
Lerner, R. A., 335[209], 336 [209], *347*
Leung, W.-C., 812[178, 179, 180], *843*, 872[176], *887*
Levin, D. H., 386[51], 398[149], 400[156, 161], 401[165], 402 [51, 149, 156], 403[149, 181], 404[51], 405[149, 156, 161],
[Levin, D. H.] 406[149], 407[207], 408[51, 208, 211], 409[212], 411 [181, 212], *426, 429, 430, 431*
Levin, J. G., 126[86], 131[100, 101], *162*
Levin, M. J., 53, *60*, 633[434, 435], 634[434], 648[563, 564], *666, 670,* 713[31, 155], 746[150, 154, 155], 747[191], *757, 761, 762*
Levina, N. V., 55[148], *61*
Levine, A. J., 591[4], 600[112], 610[217], 625[335, 336], 626 [349, 360], 629[378], *652, 656, 659, 663, 664,* 683[372], 690[370], 704, 705[149, 295], 707[149, 294, 295, 361], 711 [226], 715[294, 362], 716 [74], 717[361], 720[74], 723 [362], 725[360], 726[74], 727[294, 295, 361, 362], 728 [370], 730[372], 738[295], 753[362], *759, 761, 764, 766, 768,* 829[280, 281], *847*
Levine, A. S., 619[300], 620[300], 630[385, 386], 633[435], 648 [563, 564], *622, 664, 666, 670,* 713[106, 155], 724[54], 731 [152, 153], 741[129, 130], 742[129, 130, 253], 745[129, 130, 253], 746[129, 130, 150, 155, 196], 747[54, 132], 748 [54, 129, 130, 132], 749[152, 153, 196], *758, 760, 761, 765*
Levine, M., 239[243], *278*
Levine, S., 50[87], *59*, 350[150], *379*
Levinson, A., 707[361], 717[361], 725[360], 727[361], 738, *768*
Levinson, A. D., 715[362], 723 [362], 727[362], 753[362], *768*
Levinson, W. E., 459[41], *540*

Levintow, L., 64[5], 67[5, 25], 73[5], 74[5], 75[5], 79[5], 80[5, 25], 88[5], 92, 94, [5, 134], *105, 106, 109*
Levisohn, R., 270[309], *280*
Levitan, D. B., 362[76], *377*
Levitt, J., 802[148], 811[148], *842*
Levitt, N. H., 74[67], *107*
Levy, A. H., 352[249], *381*
Levy, H., 126[83], 127[83], *162*, 881[273], *890*
Levy, H. B., 41[16], 43[6, 8, 9, 10], 45[10, 41], 50[90, 101], 51[111], 52[117a], 54[6, 8, 124, 125, 132, 133], 55[141, 151], *56, 57, 58, 59, 60, 61*
Levy, J., 437[12], 446[12], 509 [12], *539*
Levy, J. P., 593[34, 35], *653*
Levy-Koenig, R. E., 47[51], *58*
Lewandowski, L. J., 304[100], 306[100], *344*, 397[145], *429*
Lewin, B., 719[151], *761*
Lewis, A. M., 633[432, 433, 434, 435, 436], 634[434], 648 [432], *666*, 746[176], *762*
Lewis, A. M., Jr., 605[181, 183], *658*, 713[31, 106, 155], 731 [152, 153], 741[129, 130], 742[129, 130, 145, 353], 745 [129, 130, 253], 746[128, 129, 130, 145, 150, 154, 155, 196], 747[132, 191], 748[129, 130, 132, 156], 749[152, 153, 196], *757, 760, 761, 762, 765*
Lewis, J. B., 706[158, 363], 711 [363], 718[363], 724[157], 725[6, 158], 738, 752[6, 157], 753[158, 363], *757, 761, 762, 768,* 891[28], 892, 893[9, 28, 29], 896[1, 29], *897, 898*
L'Héritier, P., 174[33, 35, 36], 185[35, 36], *272*
Li, K. K., 306, 333[121], *345*
Lieber, M. M., 463[47], 476[61], 486[61], 487[61], 488[70], *540, 541*
Lieberman, M., 45[28], *57*, 508 [84], *541*
Lief, F. S., 299[85], *344*
Lilly, F., 521, 522[100], 525 [100], *542*
Lin, S. S., 814[175], *843*
Lind, P. E., 289[39], 336[212], *342, 347*
Lindahl, P., 55[142], *61*
Lindahl, T., 780, *840*
Lindahl-Magnusson, M. P., 55 [144], *61*
Lindberg, U., 647[551], *670*, 676[208, 209], 679[209], 690 [208, 209], 707[210], 714 [159], 716[159, 210], 718 [159, 210], 720[210], 739 [209], 749[210], 750[209, 210], 751[210], *762, 763*
Lindenmann, J., 41, 45[30], 46 [30], 47[30], 48[60], *56, 57, 58,* 293[61], *343*
Lindstrom, D. M., 617[296], 618[296], 619[296], 620 [296, 301], 622[296], 623 [296], 624[296], 644[296], 645[296], 646[296], *662*
Ling, H. P., 495[76], *541*
Link, H., 351[179], *380,* 897[14], *897*
Linn, S., 603[145], *657*
Linné, T., 101[161], *110*
Lipman, M. B., 170[20], 174 [20], 182[20], 184[20], 215[20], *271*, 855[41], *883*
Liss, A., 546[45, 46], *580*
Littauer, U. Z., 69[47], *106,* 132[113], *163*
Little, S. P., 251[255], 252 [255], 254[255], *279*, 372 [233], *381*

Littlefield, J. W., 592[21], *653*
Litvak, S., 69[53], *106*
Liu, C., 324[183], *346*, 385 [36], *425*
Liu, O. C., 307[124], *345*
Livingstone, D. M., 632[421, 422], 648[422], *666*
Lockart, R. Z., Jr., 46[45], 47[45], 50[88], *58, 59*, 126 [84], *162*
Locker, H., 623[309], *662*, 780[54], 805[54], *839*
Lodish, H., 139[129], 143[129], *163*
Lodish, H. F., 67[18, 19, 33], 80 [18, 19], 81[18], 85[19], *105, 106,* 239[238, 239], 246[238, 239], 257[238, 239], 257[238, 239], *278*, 351[141], 370[141, 167], 371[151], *378, 379,* 404[189], 407[189], *430*
Lodish, H. L., 635[466], *667*
Log, T., 449[24], *539*
Loh, P. C., 387[54, 76], 389[76], 392[54], 395[133], 402[173], 403[54], 411[223], 412[234], *426, 428, 430, 431*
Lomniczi, B., 45[32], *57*
Lonberg-Holm, K., 74[72], *107*, 698[215], 714[160], *762, 763*
London, W., 43[10], 45[10, 41], *57, 58*
London, W. T., 287[22], *342*
Long, C., 459[39], 493[73], *540, 541*
Long, C. W., 457[34], 471[34], 480 [34], *539*
Long, D. G., 391[102, 103], 397 [102, 103, 143], *427, 429*
Long, W. F., 310, *345*
Longley, W., 16[74], 19[74], *37*
Longley, W., 16[74], 19[74], *37*
Longworth, J. F., 540[7], 550[7], *579*
Loni, M. C., [161], *762*
Lopez-Reivilla, R., [162], *762*
Lou, T. Y., 385[32], 386[32], *425*
Louie, A. J., 600[108], *656*
Love, D. N., 267[298], 268[298], *280*
Lovinger, G. G., 495[76], *541*
Lowry, D. R., 437[11], 446[20], 520[20], 521[11], *539*
Lowry, S., 826[265], 827[265], 833[265], *846*
Lucas, J. J., 634[453], *667*, 720 [163], 721[163], 751, *762*
Lucas, M. H., 543[96], 552[96, 97], 553[97], 568[96], *582*
Lucas-Lenard, J., 94[138], *109*
Luchsinger, E., 541[35], 545[35, 94, 239], 550[35], 551[35], 552[35, 94], 555[94], 569 [35], *580, 582, 587*
Lucy, J. A., 364[152], *379*
Luczak, M., 352[153], *379*
Ludwig, H., 772[21], 778[21], [47], 780[55, 57, 79], 787 [79], 789[79, 90], *838, 839, 840, 841*
Ludwig, H. O., 773[15], 778[15], 789[15, 91], *838, 841*
Luff, S., 814[199], *844*
Luftig, R. B., 387[55], 389[55], 390, 397[55], *426*
Lule, M., 175[55], *273*
Lum, G. S., 541[19], 543[19, 84], 552[84], 578[19], *580, 582*
Lund, E., 608[197], 610[197], 611[197], *658*
Lundholm, U., 704[75], 705[75], 709[75], 710[164], 712[75], 726[75], 727[75], *759, 762*
Lundquist, R. E., 89[125], *109*
Luria, S. E., 16[78], *38*
Lust, S., 543[203], 569[203], *585*
Lutter, L., 691[49], 696[49],

[Lutter, L.]
697[49], 699[49], *758*
Luukkonen, A., 121[62], 158 [62], *161*
Lvova, A. I., 154[176], *165*
Lwoff, A., 2[20, 79], *36, 38*
Lyles, D. S., 865[103], *885*
Lyon, M., 591[6], *652*
Lyons, M. J., 154[189], 155[189], *165*, 634[455], *667*

Maass, G., 625[334], *663*
McAllister, P. E., 195[139], 196[139], 199[139], 201 [139], 221[211], *275, 277*
McAllister, R. M., 437[13], 458 [36], 463[13], 521[96], *539, 540, 542,* 679[172], 690[183], *762*
McAuslan, B., 862[62], 873[62], *884*
McAuslan, B. R., 64[1], 74[1], 75[1], 80[1], 93[1], *105,* 636 [471], *667*, 815, *845*, 851, 862 [58], 864[89], 865[94], 867 [141, 145], 868[58, 145], 870 [160], 872[141, 169, 174, 180, 183, 184], 873[94], 876[194, 195], 877[199], 881[279], *882, 884, 885, 886, 887, 890*
McCabe, C., 567[178], 570[178], *585*
McCain, B. B., 186[106], 221[106], *274*
McCallum, M., 787[87], *840*
McCarthy, B., 600, *656*
McCarthy, D., 6[88], *38*
McClain, M. E., 391[96, 99, 106], 417[106], *427*, 866[109], *885*
McClanahan, M. S., 567[174], 575 [174], *584*
McClelland, L., 295, *343*
McCombs, R., 779[29], *838*
McCombs, R. M., [47], *839*
McConnaughy, B. C., 624[321], *662*
McCormick, K. J., 545[73], 550 [73], 569[202], *581, 585*
McCracken, R. M., 779[34], 824 [34], 826[34], *839*
McCrae, A. D., 387[69], *426*
McCrea, J. F., 308[132], *345*, 854[36], 855[36, 41], *883*
McDonald, S., 701[320], *766*
McDougall, J. K., 731[173], 733 [43], *758, 762,* 895[7, 19, 30], 896[7, 19], *897, 898*
McDowell, M. J., 395[140], 398 [140], 404[188, 189], 407[140, 188], *428, 430*
McFalls, M. L., 194[138], 195[138], 196[138], 199[138], 201[138], 206[138], 218[138], *275*
McGeoch, D. J., 543[128], 555[121, 128], 559[121, 128], 560[128], 561[121, 128], 562[121], 564 [121], *583*
McGrath, C. M., 537[115], *542*
McGregor, S., 99, *110*
McGrogan, M., 690[29], 718[29], 736[27], 751[27, 29], *757*
McGuire, P. M., 720[174], *762*
MacHattie, L. A., 549[70], *581,* 678[92], 681[92, 282], 682 [92], *759, 765*
Macieira-Coelho, A., 54[127, 130], *60*
McIntosh, B. M., 174[21], 182 [21], *271*
Mackey, J. K., 676[94], 677[94, 354], 678[94], 718[364], 732 [94], 733[94, 96], 737[95], *760, 768*
McKenna, G., 643[532], *669*
Mackenzie, J. S., 336[223], [229], *347, 348*
McKerlie, L., 556[133, 134], 567 [193], 568[193], 570[193], 571[133, 134], 572[193], 574

[McKerlie, L.]
[193], 576[193], *583, 585*
McKerlie, M. L., 54[131], *60*, 630 [384], *664*
McKinney, R. W., 113[3], *159*
McLaren, L. C., 74[70], *107*
McLaughlin, B. C., 632[424], *666*
McLean, C., 69[62], *107*
MacLean, E. C., 15[47], *37*
MacLeod, R., 194[134], 196 [134], *275*
McMillan, V., 826[265], 827 [265], 833[265], *846*
MacNab, J. C. M., 793[98], *841*
McNair Scott, T. F., 825[256], *846*
MacPherson, I. A., 15[58], *37*, 641[517], 642[517], *669*
MacPherson, L. W., 351[59], *376*
McSharry, J. J., 198[149], 201 [175], 267[149], 268[149], *276*
Madeley, C. R., 595[63], *654*
Madin, S. H., 269[303], *280*
Madore, H. P., 190[123], 195 [123], 196[123], 199[123], 203[123], 224[123], 227 [123], 228[123], *275*
Maeda, A., 552[85], 553[85], *582*
Maeno, K., 324[186], *346*, 361 [112, 154], 363[158], *378*, *379*
Maess, J., 543[238], *587*
Magee, W. E., 872[167, 177], 881[269], *887, 890*
Magnusson, G., 628[368, 375], 629[368, 375, 380], *664*
Mahnel, H., 541[27], 543[27, 125], 555[125], 568[125], 569 [125], 570[27], *580, 583*
Mahoney, J., 795[118], 797[118], *841*
Mahy, B. W. J., 282, 309[138, 139], 314[168], 315[138, 168], 316[168], *341, 345, 346,* 350 [12], 355[148], 356, 360[148], 366[111, 148], 367[148, 155], 371[148], 373[12], *374, 378, 379,* 496[77], *541*
Maitland, H. B., 879[124], *886*
Maizel, J. V., 75[76], 78[91], 80 [76], 83, 85[113], 86[117], 92 [131], 95[142], *107, 108, 109,* 555[129], 556[129], 557[129], 568[129], *583,* 592[17], *653*
Maizel, J., Jr., 634[457], *667*, 691, 698[166], *762*
Maizel, J. V., [306], *766*, 808[164], *843*
Maizel, J. V., Jr., 20[77], 21[77], *38*, 83[103, 104], 89[125], 97 [147], *108, 109,* 633[438], *666,* 691, 697[120], 698[357], 716 [120], 723[343, 344, 357], 724 [357], 725, *760, 762, 768,* 893 [31, 32, 33], *898*
Major, E. D., 596[66], *654*
Mak, S., 700[51], 717[61], 733 [61, 91], 737[167], 738[91], *758, 759, 762,* 893, *898*
Mak, T. W., 71, *107*
Malíř, A., 268[299], *280*
Malmgren, R. A., 634[443, 449], *666, 667*
Maloney, J. B., 552[106], 553 [106], 567[106], 578[106], *582*
Mandeles, S., 305[117], *344*
Mandelkow, E., 4[54], *37*
Manders, E., 239[242], 248[242], 269[242], *278*
Mangel, W. F., 631[410], 632 [410], *665*
Maniloff, J., 546[45, 46], *580*
Manly, K. F., 85[111], *108*
Mann, J., 813[193], 814[193, 200], *844*
Manor, H., 645[543], *669*

Manservigi, R., 837[316], *848*
Manteuil, S., 614[259], 622[306], 625[346], 636[475], *660, 662, 663, 667*
Mao, J. C-H., 815[208, 209], *844*
Maraldi, N. M., 614[255], *660*
Marbaix, G., 132[113], *163*
Marchenko, A. T., 194[135], 195 [135], 196[135], 197[135], *275,* 336[224], *347*
Marcker, K. A., 87[121], *109*
Marcus, A., 359[64], *376*
Marcus, P. I., 50, 53, *59, 60,* 239[241], *278*
Margalith, E., 859[57], 866[117], 881[117, 264, 265, 267, 272], *884, 885, 889, 890*
Margolis, G., 386[39], *425*, 543 [78], 551[78], 552[87, 98], 553[87, 98, 107, 108, 112], *581, 582*
Marin, G., 636[487], 642[529], *668, 669*
Mark, G. E., 811, *843*
Markelova, T. A., 45[27], *57*
Markham, R., 15[6], *35*
Markhammer, M., 629[380], *664*
Markushin, S. G., 313[157], 336 [224], *345, 347*
Marquardt, H., 537[116], *542*
Marsden, H. S., 793[98], 811 [174], *841, 843*
Marsh, R. F., 596[64], *654*
Martin, E. M., 126[82], *162*
Martin, G. S., 530, *542*
Martin, H., 741, 742, 747[307], 748[307], *766*
Martin, M., 612[232], 619[232], 620[232], 643[533], 644[533], *660, 669,* 744[131], 746[131], 747[131], *761*
Martin, M. A., 609[207], 610[216], 611[207, 224, 225, 227, 228], 616[272], 617[284, 285], 618 [272], 619[272, 298], 620[272, 298], 622[298], 623[310], 624[298], 628[227, 228], 636 [484], 638[499], 646[499], 647[555], *659, 661, 662, 668, 670*
Martin, M. L., 423[265], *432*
Martin, R. G., 601[119], 613 [345, 417], 625[345], 632 [344, 345, 417, 430], 635 [344, 345, 430, 463], 636 [344, 345], 637[344, 345, 417, 463], 638[345, 506], 639[506], 652[345], *656, 663, 666, 667, 668,* 705[168], *762*
Martin, S. A., 397[146], *429,* 863 [79, 80, 81], *884, 885*
Martin, S. J., 70[153], 99[153], *110*, 269[301], *280*, 360[80], 364[81], *377*
Martin, W. B., 773[48], *839*
Martinet, C., 252[289], 265[285, 287, 289], 266[285, 289], 267 [285], *280*
Marty, L., 597[83], *655*
Maru, M., 353[240], *381*
Marushige, K., 599[94], *655*
Marvin, D. A., 6[28, 80, 81, 82, 114], *36, 38, 39*
Marvin, F. J., 6[81], *38*
Marx, M., 625[346], *663*
Marx, P. A., Jr., 359[156, 234], 360[234], 361[195, 196], 362 [195, 196], 364[196], 367 [156, 157], 371[195, 234], *379, 380, 381*
Maryak, J. M., 534[107], *542*
Mascoli, C. C., 384[11], 385[34], *424, 425*
Masler, L., 45[23], *57*
Mason, W. S., 476[63], 504[81], 531[103], 532[103], *541, 542*
Massie, A., 385[24, 25], *425*
Massie, E. L., 386[43], *426*
Mastrota, F. M., 387[67], *426*

Mathews, M., 604[156, 160], 607 [156, 160], *657*
Mathews, M. B., 76[87], 87[121], *108, 109,* 718[169], 720[373], 736, 750, *762, 768*
Matsubara, K., 610[223], 631 [223], *659*
Matsubara, T., 174[25], *272*
Matsuhisa, T., 409[219], 410 [222], 420[222], *431*
Matsumoto, S., 169, *271*, 879 [223, 225, 227], *888*
Matsumoto, T., 361[112, 154], 363[158], *378, 379*
Matsumura, T., 118[56], 146 [147], *161, 164*
Matsuo, Y., 553[114], *583*
Matsuya, Y., 613[245], 614 [245], *660*
Mattern, C. F. T., 2[83], 3[83], 7[83], 13[83], 15[106], 18 [83], 19[84], 23[83, 85, 86, 106], *38, 39,* 636[476, 477], *667*
Matthaeus, W., 216[207], 217 [207], 269[207], *277*
Matthews, M. B., 359[198], *380*
Matthews, R. E. F., 16[87], *38*
Matumoto, M., 174[24], *272*, 353[223], *381*
Matushia, T., 417[247], *432*
Maugh, T. H., II, 67[12], *105*
Mautner, V., 707[170], 709, 755, *762*
May, E., 559[165], *584*, 612 [231, 244], 614[258], 615 [265], 616[258], 617[258], 623[258], 652[258], *660, 661*
May, P., 559[165], *584*, 612 [231, 244], 614[258], 616 [258], 617[258], 623[258], 652[258], *660*
Mayer, A., 626[360], *664*
Mayer, A. J., 704[75], 705[75], 709[75], 712[75], 726[75], 727[75], *759*
Mayer, G. D., 45[29], *57*
Mayor, H. D., 18[63], 19[63], *37,* 386[45, 53], 387[53, 75], 389, 390[83], 400[83], 402[45], *426, 427,* 541[28], 545[28, 53], 546[53], 547[56], 549[62], 550[74], 554[119], 557, 561 [56, 147, 149, 151], 567[180, 182], 569[200], 570[119, 211], 572[28, 215], 574, 577[151], 578[236], *580, 581, 583, 584, 585, 586,* 636[479], *668*, 712 [171], *762*
Mayr, A., 541[27], 543[27, 125], 555[125], 568[125], 569[125], 570[27], *580, 583*
Mayyasi, S. A., 880[245], *889*
Meager, A., 45[36], *57*
Mechali, M., 605[168], *657*
Mécs, E., 126[82], *162*
Medappa, K. C., 65[10], 69[62], *105, 107*
Medeiros, E., 519[93], *542*
Mednis, B., 778[17], 798[17], *838*
Medrano, L., 73[65, 66], *107*
Medzon, E. L., 853[19], 854[19], 880[243], *883, 889*
Meier, H., 437[9], 505[87], 508 [9, 87], 509[9], *538, 541*
Meier-Ewert, H., 293[62], 297[62], 300[62], 307[62], 326[201], 328[201], 331, 332[201], 333 [62], 336[62], *343, 347*
Meindl, P., 299[89], *344*
Meinke, W., 600[111], 602[129], 626[358], *656, 664*
Meissner, H. C., 893, *898*
Melby, E. C., 541[31], 543[31], 550[31], *580*
Mellors, R. C., 591[11], *653*
Melnick, J. L., 47[50], *58*, 184 [94], *274*, 283[15], *341*, 352

[Melnick, J. L.] [159], *379*, 385[35], 387 [71, 72, 74], 391[71, 74, 97, 98, 101], 402[175], *425, 426, 427, 430,* 439, *539,* 540 [2], 541[28, 29, 34], 543[34, 77], 545[28, 29, 53], 546 [53], 547[53, 56], 550[77], 552[93], 554[119], 555[29], 557[53], 559[29], 561[29, 56, 147, 151], 566[77, 172], 569[200, 201], 570[119], 572[28, 215], 577[151], 578 [230], *579, 580, 581, 582, 583, 584, 585, 586,* 591[4], 612[236, 243], 625[236, 243, 328], 631[393, 396, 399], 636[479], 648[565], *652, 660, 663, 665, 668, 670,* 780 [78], 787[78], 826[265], 827[265], 833[265], *840, 846*

Mendelsohn, N., 400[156], 401 [166], 402[156], 405[156], *429*

Mendelson, C. G., 593[30], *653*

Mercer, C. K., 43[12], *57*

Meredith, C. D., 175[49], 181 [49], 220[49], *272*

Merigan, T. C., 42[3], 45[20], 47[48], 48[62, 66], 50[95], 51, 54[131, 132], 55[139, 146], *57, 58, 59, 60, 61,* 868[151], *886*

Merkow, L., 634[456], *667*

Merkow, L. P., 567[178], 570 [178], *585*

Merlie, J., 292, *342*

Mertz, J. E., 608[196, 198], 610 [196, 198], 611[196, 198], *658*

Meselson, M., 603[143], *657*

Mesyanzhinov, V., 20[1], *35*

Metselaar, D., 174[29], 183 [29, 92], 215[92], *272, 274*

Metz, D. H., 51[113], 53, *59, 60,*

[Metz, D. H.] 862[66], 867[130], 871[165], 880[165], *884, 886, 887*

Meyer, J., 893[31, 33], *898*

Meynadier, G., 541[39], *580*

Michel, M. R., 130, 131[96], *162,* 610[219], 625[331], *659, 663*

Middleton, P., 385[28], *425*

Mielke, M., 862[61], 873[61], *884*

Miles, J. A. R., 591[4], *652*

Millar, J. H. D., 351[79], 364 [79], *377*

Miller, C. G., 438[15], 457[15], 480[15], *539*

Miller, D. A., 73[66], *107*

Miller, H., 313[161], 314[161], 320[161], *346*

Miller, H. I., 78, *108*, 600[106], *656*

Miller, J. M., 457[34], 471]34], 480[34], *539*

Miller, O. J., 73[66], *107*

Miller, O. V., 872[177], *887*

Miller, R. H., 351[161], *379*

Miller, R. L., 68[37], 94[37], *106*

Millian, S. J., 385[23], *425*

Milliken, S. A., 352[104], *377*

Mills, D. R., 270[309], *280*

Mills, R. F. N., 43[12], *57*

Millward, S., 18[89], 18[89], *38,* 390[93], 394[116, 120, 121], 398[93], 401[93], 404 [120, 121, 194, 195, 196], 407[121], 412[121], 417 [195, 196], 418[194], 421 [121, 194], *427, 428, 430*

Milstein, C., 521, *542*

Milstein, J. B., 609[207], 611 [207], *659*

Mims, C. A., 64[1], 74[1], 75[1], 80[1], 93[1], *105,* 182[90], 184[90], 185[90], 215[90], *274,* 636[471], *667*

Mims, S. J., 877[213], 878[213], *888*
Minner, J. R., 352[109], *377*, 384 [9, 11], *424*, 594[47], *654*
Minowada, J., 738[255], *765*
Minton, S. A., 881[262], *889*
Minuse, E., 287[23], *342*
Mirkovic, R., 779[29], *838*
Mirkovic, R. R., 543[77], 550 [77], 566[77], *581*
Mitchell, F. E., 351[161], *379*
Miura, K., 863[77], *884*
Miura, K.-I., 394, *428*
Mivule, A., 175[55], *273*
Miyamoto, H., 879[224, 235], 880[255], *888, 889*
Miyamoto, K., 825[251], *846*
Mizell, M., 772[25], 774[25], 778[27], *838*
Mizutani, S., 436[2], 483[2], *538*
Mobbs, J., 599[104], *655*
Mobraaten, L. E., 55[145, 150], *61*
Moennig, V., 472[55], 473[56], *540*
Moffat, M. A. J., 324[185], *346*
Mogensen, K. E., 49[77], *59*
Mohler, J. R., 176, *273*
Molibog, E. V., 298[84], *343*
Molloy, J., 719[34], *758*
Molteni, P., 625[333], *663*
Monaco, A. P., 55[149], *61*
Mongillo, C. A., 170[20], 174 [20], 182[20], 184[20], 215[20], *271*
Montagnier, L., 601[116], 631 [407], 641[407], 642[527], *656, 665, 669*
Montelaro, R. C., 456[33], *539*
Montes De Oca, H., 554[117], 569[117], *583*
Montjardino, J., 605[175], *658*
Moody, M. F., 20[90], 22[90, 91], *38*
Mooney, J. J., 124[80], *162*
Moore, A. E., 541[14, 17], 543 [103], 552[17, 103], 553[14], 555[17], 567[14], *579, 582*
Moore, C., 891[3], 892[3], 893 [3], 894[40], 895[40], *897, 898*
Moore, D. H., 535[111], *542,* 591[8], *652,* 779[32], *839,* 877[204], *888*
Moore, D. L., 175[51, 52, 53, 54], 180[88], 181[52], 182[54], 216[51, 54], 218[51, 54], *272, 273, 274*
Moore, N. F., 198[150], 199[150], 200[166], 268[150], *276,* 353 [162], *379*
Moore, R. D., 54[126], *60*
Moorhead, P., 595[51], *654*
Moorhead, P. S., 650[590], *671*
Morahan, D. S., 568[198], *585*
Moretti, G. F., 825, *846*
Morgan, C., 124[78], *162*, 312 [150], 332[205], *345, 347,* 363[103, 163], *377, 379,* 778 [17, 26], 779[32], 795[26], 797[26], 798[17], 808[156], 823[26, 248], 825[251], 827 [270], *838, 839, 843, 846,* 877[204], 878[216], *888*
Morgan, E. M., 410[221], 420, 423[268, 269], *431, 432*
Morgan, H. R., 363[164], *379*
Morgan, M., 67[29], *106*, 229 [223], *278*, 401[168, 169], 406[169], 408[169], 410 [168], *429*, 719[65], [374], *759, 768*
Morgan, M. J., 50[92], *59*
Mori, R., 614[248], *660*
Morimoto, Y., 352[117, 165], 364 [117, 165], *378, 379*
Moritsugu, Y., 711[270], *765*
Morley, D. C., 351[166], *379*
Morris, A. D., 175[47], 181[47], 216[47], *272*
Morrison, J. M., 592[27, 28], *653*,

[Morrison, J. M.]
812[183], 813[183], 815 [183], 828[183], *843, 845*
Morrison, T. G., 239[238], 246 [238], 257[238], *278,* 370 [24, 167], *375, 379*
Morrow, J. F., 604[154], 605 [154, 169, 173], 607[154], 633[436], *657, 658, 666,* 713[175], 746[176], *762*
Morse, L. S., 791[320], *848*
Morser, M. J., 135[120], 137[120], *163*
Mosbach, E. H., 877[211], *888*
Moses, H. L., 820[237], *845*
Mosig, G., 20[92, 115], *38, 39*
Mosmann, T. R., 774[51], 780 [51], *839*
Mosovici, E., 877[209], *888*
Moss, B., 67[30], *106,* 203[193], *277,* 719[312], *766,* 851, 854 [25], 859[48, 50, 52, 54, 55], 860[50], 862[72], 863[25, 73, 75, 76, 79, 80, 81, 85, 86], 864[87, 90, 91], 865[95, 97, 98], 866[113], 869[153], 870 [164], 871[164], 873[50, 87, 189], 874[50, 189, 190], 875 [52, 189, 190, 191], 876[192], 877[212, 213], 878[212, 213], 879[192], 880[164, 250, 251], 881[212, 275, 276, 277, 281], *882, 883, 884, 885, 887, 888, 889, 890*
Mosser, A. G., 75[74], 80[74], 92[74], *107,* 456, *539*
Mountcastle, W. E., 352[42], 358 [54, 168], 359[170], 360[168], 361[169], 363[42], 364[170], 368[170], *375, 376, 379*
Mowshowitz, D., 117[35], 131[35], 132[35], *161*
Moyer, S. A., 202[187, 190], 229 [187], 230[187], 233[225], 237 [190], 239[240], 246[225, 240], 257[240], *277, 278,* 370[20],
[Moyer, S. A.]
375
Mudd, J. A., 155[194], *165,* 190 [120], 198[147], 199[165], 200[147], 233[229], 236[229], 239[229], 257[120], 258[229], *275, 276, 278*
Mueda, M., 384[6], 385[6], *424*
Mueller, N., 620[302], 622[305], *662*
Mühlethaler, K., 293[61], *343*
Mujomba, E., 175[55], *273*
Mulder, C., 604[155], 605[155], 607[155], 610[222], 638[497], 642[497], *657, 659, 668,* 676 [78], 681[177, 178], 686[78], 687[177, 203], 688[78, 177], 732[78], 736[27], 749[178, 203], 750[177], 751[27], 753 [78], *757, 759, 762, 763*
Muldoon, R. L., 350[118, 119], 351[118, 119], *378,* 385 [18, 31], 386[31], *425,* 677[358], *768*
Mullarkey, M. F., 595[62], 596 [69], *654, 655*
Müller, G., 389, *427,* 854[23], *883*
Müller, L., 20[1], *35*
Muller, R., 814[204], *844,* 872 [179], *887*
Mulligan, R. C., 631[411], *665*
Mumford, D. M., 578[236], *586*
Munk, K., 815[216], *845*
Munns, T., 719[328], *767*
Munyon, W., 813[198], 814[175, 193, 200], *843, 844,* 862[59], 863[82], *884, 885*
Munyon, W. H., 867[144], 868 [144], *886*
Munz, K., 170[15], *271*
Murakami, W. T., 15[106], 23 [106], *39*
Murnane, T., 352[16], *375*
Murphy, B. R., [230], *348*
Murphy, F. A., 117[32], 146[32],

[Murphy, F. A.]
154[179], *160, 165,* 169, 174 [32], 175[51, 54], 176, 177 [3], 178[65], 179[3, 65, 71, 72], 180[71], 181[71, 72], 182 [54, 90], 183[91, 92, 93], 184 [32, 90], 185[90], 188[3], 194 [135], 195[135], 196[135], 197[135], 199[158], 214[93, 158], 215[90, 92, 93], 216 [51, 54], 218[51, 54], 221[3], *271, 272, 273, 274, 275, 276*
Murphy, H. M., 503[80], *541*
Murphy, J. S., 290[41], 308[133], *342, 345*
Murray, B. K., 780[70], 819[70], *840*
Murray, K., 603[147], *657*
Murray, P. R., 457[35], *539*
Murray, R. E., 681[179], *762*
Murray, R. F., 591[5], *652*
Mussgay, M., 115[30], 118[30], 124[81], *160, 162,* 176[58], *273*, 543[238], *587*
Mustoe, T. A., 424[279, 280], *433*
Mutere, F. A., 174[29], 183[29], *272*
Muthukrishnan, S., 67[29, 32], *106,* 229[223], *278*, 358[171], 367[171], *379*, 394[122, 125], 401[169], 406[169], 408[169, 209, 236], *428, 429, 431*
Myers, D. D., 437[9], 508[9], 509 [9], *538*
Myers, P., 604[158], 607[158], *657*
Myerson, D., 867[138], 877[138], *886*

Nadkarni, J. J., 835[302], *847*
Nadkarni, J. S., 835[302], *847*
Nagai, Y., 361[112, 154], 373 [267], *378, 379, 382*
Nagaki, D., 650[595], *671*
Nagayama, A., 877[214], 881[214], *888*
Nagington, J., 854[27], *883*
Nahmias, A., 771[5], 793[5], *838*
Nahmias, A. J., 833[296], *847*
Nair, C. N., 94[139], *109*
Nakai, T., 352[172], 358[172], 359[172], *379*
Nakajima, H., 353[173], *379*
Nakajima, K., 635[459, 460], 636 [485], *667, 668,* 739[104, 180]. *760, 762*
Nakajima, M., 352[117], 364[117], *378*
Nakazato, H., 68[35], 94[35], *106*
Narajani, O., 596[68], *655*
Narayan, O., 595[57], *654*
Narita, M., 552[85], 553[85], *582*
Nathans, D., 604[161], 605[192, [94], 607[161], 608[194, 199, 200, 201], 609[192, 199], 610 [194, 199, 200, 201], 611[192, 199, 200, 201, 225, 227], 619 298, 363], 620[298], 622[298], 624[298], 627, 628[199, 277], 632[412], 637[200, 490, 493], 638[200, 490, 493], 649[194], *657, 658, 659, 662, 664, 665, 668,* 728, 742[145], 744[131], 746[131, 145], 747[131], 748 [33], 749[181], 756[141], *758, 761, 762*
Nathanson, N., 543[75], 550[75], 554[75], 567[191], 572, *581, 585*
Nayak, D. P., 287[19, 20, 21], 302[104], 304[109, 112, 113, 114, 115], 313[154], 316, 317 [109, 112, 113, 114, 115], 318, 319, 320[109, 113, 114, 115], 321, 323, 326[174], 327, 332 [174], 336[210], *341, 342, 344, 345, 346, 347, 348,* 457[35], 519[94], *539, 542,* 774[24], 778[24], 779[24], 822[24], *838*

Nazerian, K., 774[67, 317], 775 [8], 779[28, 39], 780[67], 822[238], 824[249], 827 [267], *838, 839, 840, 845, 846, 848*
Neauport-Sautes, C., 835[308, 311], *847*
Neiman, P. E., 506[83], *541*
Nelson, D., 595[62], *654*
Names, M. M., 45[39], *57*
Nermut, M. V., 293[64], 300 [64], *343*, 450[25], *539*, 697, *762*
Neurath, A. R., 194[164], *276*, 361[174, 175, 177], 363 [176], *379*, 559[156], 562 [156], *584*, 698[260], *765*
Neuwald, P. D., 893[32], *898*
Nevins, J. R., 862[64], 869 [64], *884*
Newbold, J. E., 617[286, 290], *661*
Newcomb, E. W., 352[104], *377*
Newman, C., 705[232], 708[232], 709[232], *764*
Newman, J. F. E., 65, 68[45, 46], 94[46], *105, 106,* 248[249], *278*
Newton, A. A., 815[217], 826, *845, 846,* 854[27], *883*
Newton, C., 68[45, 46], 94 [46], *106*
Ng, M. H., 46[44], 49[81], *58, 59*
Ngan, J. S. C., 260, 262, 264 [276], *279*
Nichols, J. L., 394[131, 132], 395 [131, 132], 400[162], 405 [162], 406[162], 408[162], 410[131, 132, 162], *428, 429*
Nicholson, B. L., 567[192], 568 [192], 570[192], 572, *585*
Nicholson, M. O., 679[172], 690 [183], *762*
Nicklin, P. M., 628[365, 366], *664*
Nicolayeva, O. G., 368[264], *382*
Nicoletti, G., 571[212], *586*
Nicolson, M., 437[13], 463[13], *539*
Nicolson, M. O., 458[36], 521 [96], *540, 542*
Nii, S., 775[9], 778[26], 795 [26], 797[26], 823[26], 825 [255], 827[268], *838, 846*
Niiyama, Y., 719[365], *768*
Nikolaeva, O. G., 359[265], *382*
Nishikawa, F., 351[74], 352 [74], *376*
Nishimoto, T., 690[29], 713 [184], 718[29], 751[29], *757, 762*
Nishimura, C., 854[35], 855 [35], 863[77], *883, 884*
Nishmi, M., 879[230], 880[244], *889*
Niveleau, A., 392[115], 428
Niven, J. S. F., 866[122], *886*
Nixon, H. L., 6[44, 50, 51, 94], 15[93], 16[93], *36, 37, 38*
Noble-Harvey, J., 74[72], *107*
Nohara, M., 816[221], *845*
Noll, M., 599[103], *655*
Nomoto, A., 67[14, 17], 68 [17], *105*
Nonomura, Y., 358[178], *380*
Nonoyama, M., 386[48], 394 [48, 120, 123], 404[120, 196], 417[48, 123, 196], *426, 428, 430,* 600[114], 601[114], *656,* 780[68], 827[274], *840, 846*
Noon, M. C., 473[58], 474 [58, 60], 475[58], 478 [67], *540, 541*

Noonan, C., 638[505], 639 [505], *668*
Norrby, E., 15[95, 96], *38*, 351 [179, 180, 183, 210, 246], 353[82], *377, 380, 381,* 691 [185], 698[186], *762*
Northrop, R. L., 366[14], 373 [181], *375, 380*
Notkins, A. L., 879[236], *889*
Novak, A., 543[123], 555[123], *583*
Nowakowski, E., 787[83], *840*
Nowinski, R. C., 451[26], 453 [26], 454[26], 455[26], 462 [26], 519[93], *539, 542*
Noyes, W. F., 591[7, 11], 593 [32], *652, 653,* 852[16, 17], *883*
Nudel, U., 132[113], *163*
Nusinoff, S. R., [230], *348*
Nuss, D. L., 76[89], 78[89], *108*

Obara, J., 353[173], *379*
Öberg, B., 718[371], 720[366], 725, 727[366], *768*
Öberg, B. F., 67[23], 69[49], 80 [23], 85[23], 93, 101[161], *106, 109, 110,* 126[90], *162*
Obert, G., 869[155], *887*
Obijeski, J. F., 190[122], 191 [122, 131], 192[131], 193 [131], 194[135, 138], 195 [135, 138], 196[135, 138], 197[135], 198[150], 199 [138, 150], 201[138], 203 [196], 206[138], 218[138], 226[216], 228[196, 216], 229[196], 232[196], 264 [279, 280], 268[150], 269 [122], *275, 276, 277, 278, 279,* 302[99], 306[99], 314 [163], *344, 346*
Obijeski, J. J., 859[51], *884*
O'Callaghan, D. J., 793[99], 795 [115], 796[99], 808[155], 817[227], 818[227], 827[227], *841, 843, 845,* 880[259], *889*
O'Callaghan, R., 818, *845*
O'Conor, G. T., 634[445, 446], *666,* 738[217], *763*
Oda, K., 616[268], 617[277], 620 [277], 635[459, 460], 645[277], *661, 667,* 739[180], *762,* 868 [148], 869[148], 870[148], *876[148],* 886
Oda, K.-I., 739[104], *760*
Odelola, A., 175[52], 181[52], *272*
Officer, J. E., 880[245], 881[262], *889*
Offord, R. E., 6[97], *38*
Ogawa, M., 880[255], *889*
Ogburn, C. A., 49[74, 77], *59*
Ogier, G., 867[129], *886*
Ogino, T., 813, *844*
Ohanessian, A., 174[34], 185[34], *272*
Ohe, K., 718[187, 188, 189], *762*
Ohtani, S., 180[86], *274*
Ohuchi, M., 363[93], *377*
Oie, H. K., 402[173], 411[223], *430, 431*
Okabe, H., 485[69], 486[69], 490[69], 516[69, 91], 519 [91], *541, 542*
Okada, Y., 353[253], 363[182], *380, 382*
Okawa, S., 352[117], 364[117], *378*
Okazaki, R., 628[373], *664*
Okazaki, T., 628, *664*
Old, L. J., 787[84], *840*
Old, R. W., 603[147], 617[296], 618[296], 619[296], 620[296], 622[296], 623[296], 624[296], 644[296], 645[296], 646[296], *657, 662*
Olins, A. L., 599[100], *655*
Olins, D. E., 599[100], *655*

Olive, D., 385[21], *425*
Oliver, L. J., 541, 543[12], 549, 551[12], 552[12], 553[12], 555[12], 567[12], *579*
Olshevsky, U., 802[148], 811 [148], *842*
Olson, L. C., 154[188], 155 [188], *165*
Olsson, J. E., 351[179], *380*
Omar, A. R., 386[42], *426*
Omori, T., 174[24], *272*
Omura, H., 815[214], *844*
O'Neill, C. F., 48[61], *58*
O'Neill, F. J., 825[261], *846*
Ono, K., 827[268], *846*
Onuma, M., 545[81], 551 *582*
Opperman, H., 76[89], 78[89], *108*
Orlich, M., 298[239], 326 [239], *348*
Oron, L., 89[126], 96[126], *109*
Oroszlan, S., 437[9, 13], 456 [31], 462[44], 463[13, 44, 45], 467[50], 469[52, 53], 470[52], 493[73], 508[9], 509[9], *538, 539, 540, 541,* 593[31], *653*
Orth, G., 592[20], *653*
Orth, H. D., 815[216], *845*
Ortin, J., 716[41], *758,* 894 [34], *898*
Orvell, C., 351[183], *380*
Osborn, J. E., 596[70], *655*
Osborn, K., 628[366], *664*
Osborn, M., 632[425], 638[501], 639[425, 501], 641[521], 648 [425], *666, 668, 669,* 705 [190], 707[361], 717[361], 727[361], 731[190], *762, 768*
Osenholts, M., 609[206], *659*
Ospina, J., 863[82], *885*
Osterrieth, P. M., 122[67], *162*
Oudet, P., 599[101], 600[110], 602[110], *655, 656*
Overby, L. R., 815[208, 209], *844*
Owens, M. J., 94[139], *109*
Oxford, J. S., 287[24], *342,* 632 [424], *666*
Oxman, M., 612[241], *660*
Oxman, M. H., 633[434], 634 [434], *666*
Oxman, M. N., 53, *60,* 631 [402], 632[402], 639[402], 648[563, 565], *665, 670,* 746 [150], 747[132, 191], 748 [132], *761, 762,* 814[201], *844*
Ozanne, B., 638[500], 640[514], 644[500, 538, 539], 645[538, 550], 646[500], 648[572], 649[539, 572], 651[500, 572, 606, 607], *668, 669, 670, 672,* 690[236], 732[236], 748[192], 753[236], *762, 764*
Ozer, H., 614[251], *660*
Ozer, H. L., 549[66], 555[66], 556 [66], 557[66], *581,* 601[119], 613[414, 488], 632[414], 635 [466, 469], 636[414, 469, 473], 637[414, 488], 638[414], *656, 665, 667, 668*

Pace, N. R., 270[309], *280*
Padgett, B. L., 595[60], 596[64, 70, 72], *654, 655,* 866[104, 108], *885*
Padmanabhan, R., 683, 729[338], 730, *763, 768*
Pagano, J., 603[142], 631[401], *657, 665,* 780[68], *840*
Pagano, J. S., 3[98], 20[99a], *38,* 591[4], 595[52], 597[84, 88], 599[84, 88], 600[84, 114], 601[114], *652, 654, 655, 656,* 772[43], 779[43], 789[94], 792[94], 827[274], *839, 841, 846*
Page, M. G., 314[169], 315[169],

[Page, M. G.]
316[169], *346*
Pages, J., 614[259], 625[346], 636[475], *660, 663, 667*
Palese, P., 298[83], 299[83, 89, 91], 300[83], 304[241], 323, 336[222], 337[222], 339, 340, *343, 344, 347,* 362[184], *380*
Palma, E. L., 248[254], 270[307], *279, 280,* 834[298], *847*
Palmer, E. L., 423[265], *432,* 859[51], *884*
Pan, J., 604[163], 605[180, 181, 182, 184], 607[163], 619[180], 620[180], 624[184], *657, 658,* 747[38], *758*
Panelius, M., 351[210], *380*
Panigel, M., [113], *542*
Paoletti, E., 813[193], 814[193], *844,* 854[25], 862[59, 72], 863[25, 79, 80, 82, 85, 86], 864[87, 90], 865[95], 873 [87], *883, 884, 885*
Paoletti, L., 7[26], 22[26], 24[26], *36*
Papadimitriou, J. M., 389[79], *427*
Parasiuk, N. A., 298[84], *343*
Pardo, M., 567[178], 570[178], *585*
Pardoe, I. U., 69[60], *107*
Parfanovich, M. I., 373[266], *382*
Parisot, T. J., 188[113], *274*
Parker, J. C., 352[185], *380,* 543[76, 88, 199], 550[76], 552[88], 568[88, 199], 573 [199], *581, 582, 585*
Parks, J. H., 43[5], *56*
Parks, W., 516[91], 519[91], *542*
Parks, W. P., 438, 456[31], 457 [15], 467[51], 473[58], 474 [60], 475[58], 478[67], 480 [15], 489[71], 536[114],
[Parks, W. P.]
539, 540, 541, 542, 541[28, 29], 545[28, 29], 552[93], 555 [29], 559[29], 561, 566[172], 569[201], 572[28], 578, *580, 582, 584, 585, 586*
Parodi, A., 601[120], *656*
Parrott, H., 676[229], *764*
Parrott, R. H., 350[37], 351[37, 128], *375, 378,* 385[29], *425,* 552[91], *582*
Parsons, J. T., 571[214], *586,* 677 [88], 679[88], 688[88], 714 [195], 717[88], 718[88, 194, 195], 720[88, 194], 723[195], 732[88], 733[88], 734[88], 735[195], 753[88], *759, 763*
Parvin, J. R., 174[44], *272*
Pascale, A., 45[28], *57*
Pashova, V. A., 359[247], *381*
Pass, F., 592[17], *653*
Patch, C., 728[144], *761*
Patch, C. T., 742[253], 745[253], 746[196], 749[196], *763, 765*
Paterson, B. M., 635[468], *667*
Pattyn, S. R., 154[175, 183], *165*
Paucha, E., 83[105], *108*
Paucker, K., 43[13], 47[51], 49 [74, 77, 79, 80], 50[80, 82, 83, 84], 54[129], *57, 58, 59, 60*
Paul, F. J., 155[192], *165,* 634 [445, 446, 449], *666, 667,* 738 [217], *763*
Paul, J. R., 64[2], *105*
Paver, W. K., 546[43, 44], *580*
Pavilanis, V., 350[120], *378*
Paxton, J., 292[55], *342*
Payne, F. E., 541[20], 543[20, 60], 549[60], 554[20, 60], 555[20], *580*
Payne, L. N., 508, *541*
Payne, P., 591[5], *652*
Pearce, C. A., 202[179], *276*
Pearson, G., 835[302, 310], *847*
Pearson, G. D., 728[197], *763,* [10], *897*

Pedersen, C. E., Jr., 143[136], *163*
Peebles, P. T., 463[45], *540*
Peleg, J., 154[182], 155[193], *165*
Pelnar, J., 174[45], *272*
Penhoet, E., 313[161], 314[161], 320[161], 326[197], 332[197], *346, 347*
Penhoet, E. E., 78, *108*
Penman, S., 67[26], 75[77, 78, 79, 80], 76[77, 78, 79], 78[77, 78, 79], 80[77, 78, 79, 80], *106, 107,* 132[109], *163,* 714[216], 718[216], *763*
Penney, J. B., 596[68], *655*
Pennington, T. H., 188[312], *280,* 403[186], *430*, 867[139], 871 [139, 166], 875[166], 877[139, 215], 881[215, 274, 282], *886, 887, 888, 890*
Percy, D. H., 181[89], *274*
Perdue, M. L., 793[99], 795[115], 796[99], *841*
Pereira, H. G., 65, 66, 67[7], *105,* 288[29, 30], 289[30], 307[29, 30], 309[136], 336, *342, 345,* 540[2], *579,* 691[198, 231, 291], 692[291], 697[198], 698[231, 291], 744[114], *760, 763, 766,* 851[10, 12], 866 [122], *883, 886*
Pereira, L., 771[5], 793[5], 802 [318], 809, *838, 848*
Pereira, N. S., 287[24], *342*
Perera, P. A. J., 812[183], 813 [183], 815[183], 828[183], *843*
Perlman, S. M., 243[244], 248 [254], 256[244], 265[244, 288], 266[244, 288], *278, 279, 280,* 372[186], *380*
Perrault, J., 203[199], *277*, 423 [270], *432,* 817[224], *845*
Perret, D., 384[5], *424*
Perry, B. T., 302[97], 306[97], *344*
Perry, R. P., 719[199, 200, 227], *763, 764,* 862[69], 869[69], *884*
Person, T., 707[210], 716[210], 718[210], 720[210], 749[210], 750[210], 751[210], *763*
Persson, T., 714[159], 716[159], 718[159], 720[366], 727[366], *762, 768*
Peters, D., 7[99], *38*, 389[82], *427*, 851[13], 854[23, 24], *883*
Peters, E. A., 578[233], *586*
Peters, R. L., 437[9], 508[9], 509[9], *538*
Peterson, J. E., 543[99], 552 [99], *582*
Peterson, R. L., 270[309], *280*
Petitjean, A. M., 185[310], *280*
Petric, M., 257[258], 269[258], *279*
Petrinec, Z., 174[39], *272*
Pett, D., 815[215], *845*
Pett, D. M., 20[99a], *38*, 395 [141], 397[146], 407[141], *429,* 597[88], 599[88], *655*
Pettersson, U., 644[536, 539], 645[550], 649[539], *669, 670,* 676[209], 679[209], 681[177, 178], 687[177, 203], 688[177, 283, 284], 690[209, 236, 246, 286], 691[50], 697 [50], 698[215], 707[210], 716[210], 717[283], 718[205, 210], 720[204, 210, 373], 721, 728[201, 286, 367], 732[206, 207, 236, 246], 736[177], 739 [209], 746[204], 749[178, 203, 210, 284], 750[177, 209, 210, 283], 751[210], 753[236], [264, 377, 379], *758, 762, 763, 764, 765, 768*
Pettijohn, D., 624[318], *662*

Petzoldt, K., 543[238], *587*
Pfau, C. J., 854[36], 855[36], *883*
Pfefferkorn, E. R., 113, 114[5], 115[24, 25, 29], 118[5, 29, 47, 52], 121[5], 125, 130, 134[114, 117], 140[133], 141 [47], 142[24, 117], 143[52, 138], 146[143, 144, 151], 147, 148[149], 149, 150, 151[47, 162], 152[47, 95, 165, 166], *160, 161, 162, 163, 164*
Pfendt, E., 80[100], *108*
Philipson, L., 19[100], 34[100], 35[100], *38*, 69[49], 74, 93, 101[161], *106, 107, 109, 110*, 126[90], *162*, 599[95], *655*, 676[208, 209], 679[209], 688 [284], 690[208, 209], 691[49, 50], 696[49], 697[49, 50], 698[215], 699[49], 707[210], 714[159, 160], 716[159, 346], 717, 718[159, 205, 210], 719 [211], 720[204, 210, 303, 366, 373], 721, 725[346], 727[366], 728[367], 736[367], 739[209], 746[204], 749[210, 284], 750 [209, 210, 367], 751[264], *762, 763, 765, 766, 768*
Phillips, B. A., 68[35], 94[35], 97[147, 148, 149], 99[148, 149], *106, 109, 110*
Phillips, C. A., 402[175], *430*
Piekarski, L., 612[236], 625[236], *660*
Piekarski, L. J., 872[168, 170, 173], *887*
Pierce, J. S., 140[145], 146[145], *164*, 361[187], *380*
Pigiet, V., 628[368. 369], 629 [368], *664*
Pigram, W. J., 6[81, 114], *38, 39*
Pilcher, K. S., 186[106], 221[106, 211], *274, 277*
Pina, M., 541[29], 545[29], 555 [29], 559[29], 561[29], *580*,
[Pina, M.] 677[88], 678[92, 212], 679[88, 212, 213], 680[62], 681[92], 682[92], 683, 685[62, 213], 686, 688[88], 714[93], 716 [214], 717[88], 718[88], 720[88], 721[285], 726[214], 732[88], 733 [62, 88, 91], 734[88], 738[91], 753[88], *758, 759, 763, 765*
Pinck, M., 69[50, 51, 52], *106*
Pincus, T.;, 522[100], 525[100], 527[101], *542*
Pinkerton, T. C., 576[225], *586*
Pirazzia, A. J. G., 170, *271*
Pister, L., 472[55], *540*
Pitkanen, A., 862[62], 873[62], *884*
Pivec, L., 353[231], *381*
Pizer, L. I., 830[286], *847*
Plagemann, P. G. W., 68[37], 94 [37], *106*
Planteoin, G., 541[39], *580*
Planterose, D. N. C., 854[35], 855 [35], *883*
Platz, P., 351[188], 362[184], *380*
Plotkin, S., 798[127], 824[127], *842*
Plowright, W., 351[189], *380*
Plummer, G., 772[14, 49], 773 [44, 46], 774[44, 49], 787[81], 805[14], *838, 839, 840*
Pogo, B. G. T., 854[21], 859[56], 863[84], 864[21, 84, 88], 865 [88, 92, 93], 873[88], 877[214], 878[56, 217], 880[92, 93], 881 [214, 280], *883, 884, 885, 888, 890*
Polasa, H., 721[285], 728[368], *765, 768*
Polisky, B., 878[220, 222], *888*
Polisky, T., 600, *656*
Pollack, R., 80[101], *108*, 641 [519, 521], 651[609], *669, 672*, 896[41], *898*
Pollack, R. E., 651[605, 616], *672*

Polli, E., 625[333], *663*
Pomeroy, B. S., 541[30], 545 [30], *580*
Pong, S. S., 76[90], 78[90], 86[90], *108*
Pons, M. W., 282[10], 293[66], 294[68, 70, 72, 73], 302[73, 105, 107], 304[110], 306[118], 307[118], 317[105, 107, 110], 320, 321, 323[233], 333[118], 335[105, 118], 336[105], 339, *341, 343, 344, 346, 348*
Ponten, J., 595[52], *654*
Ponten, J. A., 595[51], *654*
Popa, L. M., 361[190], *380*
Pope, J. H., 631[392], *665*
Porebska, A., 309[136], *345*
Porter, A., 68[44, 45], *106*
Porterfield, J. D., 851[10], *883*
Porterfield, J. S., 175[47, 50], 181[47], 216[47], *272*
Portmann, R., 595[59], *654*
Portner, A., 352[194], 353[57], 355[191], 356[192], 357 [192], 359[156, 235], 360 [191], 361[195, 196], 362 [195, 196], 364[196], 366 [235], 367[156], 371[191, 193, 195], 374[271], *376, 379, 380, 381, 382*
Portocala, R., 361[190], *380*
Portolani, M., 596[67], 614 [255], *654, 660*
Poste, G., 649[584], *671*
Postel, E. H., 715[362], 723 [362], 727[362], 753[362]
Postlethwaite, R., 879[124], *886*
Potter, C. W., 632[424], *666*
Powell, K. L., 793, 802[152], 808[158], *841, 843*
Powell, S., 865[102], *885*
Prage, L., 698[215], 716[346], 725[346], *763, 768*
Preble, O. T., 386[44], *426*
Prescott, D. M., 867[137, 138], 877[137, 138], *886*
Preston, R. E., 543[60], 549[60], 554[60], *581*
Prevec, L., 177[61], 194[134], 196[134], 214[204], 236[231], 248[245], 255, 257[258], 258 [260], 269[204, 258], *273, 275, 277, 278, 279,* 392[111], 394[126], 403[126], 404[126, 187], 406[187], *428, 430*
Prezozisi, T. J., 595[57], *654*
Price, P. J., 534[107], *542*
Price, R., 714[216], 718[216], *763*
Pridgen, C., 367[157, 197], *379, 380*
Prince, A. M., 363[3], *374*
Pringle, C. R., 188[312], 189[116], 194[136], 195[136], 196[136], 227[218], 239[234, 235, 236], 256[256], 259[235, 313], 260 [218, 235], 262[218], 264, 265 [235, 281], 267[235, 290], 269 [116], *275, 278, 279, 280,* 403 [186], *430*
Printz, P., 185[99, 100], 215[99], 265[285], 266[285], 267[285, 291], *274, 280*
Printz-Ané, C., 252[289], 265[285, 287, 289], 266[285], 267[285], *280*
Prinzie, A., 50[97], *59*
Pritchett, R., 793[100], *841*
Prives, C. L., 635[467, 468], *667*
Provost, P. J., 546[41], *580*
Prunieras, M., 677[257], *765*
Prusoff, W. H., 813[190], 816 *844*
Puentes, M. J., 456[32], *539*
Pugh, W. E., 449[23], 520[23], *539*
Purchase, H. G., 835[303], *847*
Purifoy, D. J. M., 808[158], *843*
Pursell, A. R., 351[161], *379*

Quigley, J. P., 121[61], *161*

Quintrell, N., 519[93], *542*
Qureshi, A. A., 114[21, 23], 117 [105], 118[23], 119[21], 123 [21], 126[23], 131[23, 105], 139, *160, 163*

Rabek, J., 683[372], 704[149], 705[149], 707[149], 730[372], *761, 768*
Rabin, E. R., 184[94], *274*, 385 [35], 402[175], *425, 430*
Rabinowitz, Z., 645[541], 651 [610, 611, 612, 613, 617], *669, 672*
Rabson, A. S., 43[6], 54[6], *56*, 560[143], 567[176], *584, 585*, 595[62], 634[445, 446, 449], *654, 666, 667*, 731[153], 738 [217], 749[153], *761, 763*
Race, R., 546[47], *580*
Rachmeler, M., 632[418], 636 [418], 637[418], 640[418], *666*
Rada, B., 403[184], 404[184], 406[184], 412[235], *430, 431*
Radloff, R., 602[126], 603[126, 138], *656*
Raghow, R., 374[270], *382*
Raine, C. S., 386[40, 41], 402 [174], 410[174], 420[174], *425, 430*
Rakusanova, T., 828[276], 829 [276, 283], *846, 847*
Ralph, R. K., 16[87], *38*, 610 [212], *659*
Ramig, R. F., 398[150], 400[153], 421[153], 424[153, 279, 280, 281], *429, 433*
Rand, K., 493[73], *541*
Randall, C. C., 773[45], 793[99], 795[115], 796[99], 797[124], 798[129], 808[155], 816, 817 [227], 818[227, 232], 827 [227], *839, 841, 842, 843,*
[Randall, C. C.] *845*, 855[39, 40], 865[96, 101, 102, 103], 877[206], 880[259, 260], *883, 885, 888, 889*
Randall, C. G., 859[51], *884*
Rands, E., 489[71], *541*
Ranki, M., 134[200], *166*
Ransom, J. C., 536[114], 537 [114], *542*
Rapoza, N. P., 552[92], *582*
Rapp, F., 387[71], 391[71], *426*, 631[393, 396, 399], 634[448], 635[464, 465], 648[566], *665, 667, 670*, 739[125, 126, 218], *760*, 813, 825[258, 261], 829 [279], *844, 846, 847*
Rasheed, S., 437[13], 463[13], *539*
Raska, K., Jr., 707[369], *768*
Raskas, H., 721[28], 751[28], *757*
Raskas, H. J., 690[29], 713[184], 716[219], 718[29, 276, 313], 719[52], 720[30], 723[30], 736[27, 30], 748[30], 751[27, 29, 30, 277], [380, 382], *757, 758, 762, 763, 765, 766, 768*, 891[11, 12], *897*
Rasmussen, A. F., 313[154], *345*
Rasmussen, C. J., 185[102], *274*
Ratcliff, G. A., Jr., 352[249, *381*
Ratcliffe, F. N., 866[118], *886*
Ratner, J., 352[121], *378*, 712 [171], *762*
Ratner, J. D., 567[180, 182], *585*
Rauth, A. M., 391[105], *427*
Ravdin, R. G., 595[51, 52], *654*
Rawls, W. E., 47[50], *58*, 592 [18], *653*, 826[265], 827 [265], 833[265], *846*
Reczko, E., 269[304], *280*
Redler, B. H., 576, *586*
Redman, D. R., 552[86], 553

[Redman, D. R.]
[86], *582*
Reed, G., 679[172], *762*
Reed, S., 42[3], *56*
Reed, S. I., 632[428], 633[428], *666*, 746[4, 220], *757, 763*
Reeder, R. H., 78[96], 79[96], *108*, 615[266], *661*
Reese, D. R., 199[158], 214 [158], *276*
Reeve, P., 649[584], *671*
Regelson, W., 45[26], *57*
Reginster, M., 361[33], *375*
Reháček, J., 154[184], *165*
Reich, E., 121[61], *161*
Reich, P. R., 634[450, 451], *667*, 718[221], *763*
Reichard, P., 628[368, 369], 629[368, 380], *664*
Reichmann, M. E., 189[116], 191 [130], 243[130], 265[283], 269[116, 130], *275, 279*
Reischig, J., 780[73], 782[73], *840*
Reiss, B., 406[206], *431*
Reissig, M., 825[254], *846*
Rekosh, D. M., 67[18], 80[18], 81[18], 85[111], 86[115, 116], 87[115], 89[116], *105, 108*
Rekosh, D. M. K., 894[35], 895 [35], *898*
Renkonen, O., 114[6, 10], 120 [58], 121[6, 10, 10, 62], 122 [10], 123[10], 133[10], 134 [200], 135[10], 143[10], 158 [62, 197], *160, 161, 165, 166*
Reno, J. M., 815[212], *844*
Repanovici, R., 361[190], *380*
Repik, P., 190[118, 122], 191[118, 122, 129, 131], 192[131], 198 [129, 131], 197[118], 198[150], 199[150], 203[129], 223[214], 240[214], 241[214], 243[214], 268[150], 269[118, 122, 129], *275, 276, 278*, 335[208], *347*
Rettenmier, C., 260[270], 264
[Rettenmier, C.]
[270], *279*
Reuveni, Y., 617[287], *661*, 737 [3], 752[3], *757*
Revel, M., 51[114], 52[114, 118], *60*, 132[113], *163*, 635[467, 468], *667*
Revet, B. M. J., 602[135], *656*
Reznikoff, C., 613[413], 632 [413], 636[413], 637[413], *665*
Rhim, J. S., 386[45], 387[72], 391[97, 98], 402[45], *426, 427*
Rho, H. M., 720[26], 725[351], 735[26], 738[351], 752[26], *757, 768*, 896[23], *898*
Rhode, S. L., III, 549[68], 556 [135], 557[135], 568[68, 196], 570[196], 571[68, 196], 573 [135, 196], 574[68, 196, 216, 221], 575, *581, 583, 585, 586*
Rhodes, A. J., 154[178], *165*
Rhodes, C., 637[491, 492], 638 [491, 492], *668*, 756[247], *764*
Rhodes, D. P., 67[27], *106*, 202 [186, 187], 229[187, 224], 230[187, 224], 233[224], 236 [186], *277, 278*
Ricceri, G., 571[212], *586*
Riccio, A., 613[246], 615[246], *660*
Rice, J. M., 43[10], 45[10, 41], *58*
Rich, A., 631[411], *665*
Rich, M. A., 537[115], *542*, 593 [31], *653*
Richards, K. E., 20[118], *39*
Richardson, J. P., 624[317], *662*
Richman, D. D., [230], *348*
Rifkin, D. B., 121[61], *161*
Rifkind, R. A., 332[205], *347*
Rigby, P., 637[492], 638[492], *668*
Rigby, P. W. J., 637[491], 638

[Rigby, P. W. J.] [491], *668*, 756[247], *764*
Riggs, J. L., 699[274], 700[274, 275], *765*
Rigo, P., 361[33], *375*
Riley, F. L., 52[117a], 54[124], *60*
Ripper, L. W., 613[489], 637 [489], *668*
Risser, R., 610[222], 651[609], *659, 672,* 896[41], *898*
Ritchey, M. B., 339[243], 340, *348*
Ritchie, D. A., 793[98], *841*
Ritzi, E., 625[335], *663*
Roane, P. R., Jr., 811[170], 818 [230], 827[230], 833[293], 834[299, 300], *843, 845, 847*
Robb, J. A., 613[488], 632[344], 635[344], 636[344], 637[344, 488], 638[494], 652[496], *663, 668*
Robberson, D. L., 608[197], 610 [197], 611[197], 629, *658, 664*
Robbersson, D. L., 611[229], *660*
Robbins, A. K., 628[365, 366], *664*
Robbins, E., 599[97], 625[97], *655*
Robbins, P. W., 121[63], *161*
Roberts, B. E., 359[198], *380,* 631[411], 635[468], *665, 667*
Roberts, D. W., 854[29], 863[84], 864[84], 877[29], *883, 885*
Roberts, J. M., 893[9], *897*
Roberts, R. H., 176[57], *273*
Roberts, R. J., 604[158, 159], 607[158, 159], *657,* 681[177], 683[222, 261], 687[177], 688 [177], 719[350], 736[177], 750[177], [381], *762, 763, 765, 768,* 891[8], 892[8], 895[36, 43], *897, 898*
Robertson, H. D., 76[87], *108*
Robertson, S. M., 596[70], *655*
Robey, R. E., 552[83], 553[83], *582*
Robin, J., 728[223], *763*
Robin, M. S., 678[235], *764*
Robinson, A. J., 681[225], 683 [224, 225], 730, *764,* 894[35], 895[35], *898*
Robinson, D. M., 549[61], 555 [61], 559[61], 560[61], 561 [61], 564[61], *581*
Robinson, W. S., 302[103], 317 [103], *344,* 353[17, 21, 61], 356[199], 357[199, 200], 358 [32], 360[202], 365[21], 366 [202], 369[17, 21], 370[17, 21], 371[17, 21, 202], 372[18, 201], *375, 376, 380*
Robishaw, E. E., 815[208, 109], *844*
Roblin, R., 597[77, 78], 598[78], 630[77], 635[77], *655*
Rodin, I. M., 45[27], *57*
Rodrigues, F. M., 170[16], *271*
Rodriguez, A. R., 352[121], *378*
Rodriguez, J. E., 373[203], *380*
Rodriguez, W. J., 385[29], *425*
Roeder, R. G., 78[92, 94, 95, 96], 79[92, 95, 96, 97], *108,* 615 [260, 261, 266], *661,* 718[313], [380], *766, 768*
Roening, G., 855[46], *884*
Rogers, H. W., 816[223], *845,* 865[96], *885*
Rogiers, R., 605[185], *658*
Rohrschneider, L. R., 533[105], *542*
Roizman, B., 15[101], 19[43], *36, 38,* 771[4, 5], 772[25, 62], 773[7], 774[25, 67], 775[7], 776, 777[7], 778[7, 16, 18, 22], 779[16, 36, 37], 780[54, 56, 61, 62, 64, 65, 67, 71, 74, 76, 77], 781[56], 782[61, 62, 71, 74, 76], 783[74], 784[62], 785[4, 61, 65,

[Roizman, B.]
76], 787[16, 62, 77], 789[65, 88, 89], 790[65], 791[320], 792[36], 793[5, 36, 96, 103, 104, 105], 794, 795[107, 108, 111, 114, 116], 796[16, 103, 104, 111, 114], 797[7, 16, 22, 116, 120], 798[22, 132], 799 [133], 800[76, 136, 137, 138, 139, 140], 801[76, 136, 137, 138, 139, 140, 141, 142, 143], 802[105, 136, 145, 146, 149, 151, 318], 803[76, 105, 136], 804, 805[54, 105, 136], 806 [76, 105, 136, 151, 154], 807 [76, 133, 136, 138, 151], 808 [76, 136, 138, 151, 157], 809 [318], 810[116, 120, 165, 166, 167, 168], 811[149], 817[16], 818[16, 230, 319], 819[61, 71, 76, 319], 820[16], 822[16, 18, 37, 243], 823[37, 103, 104], 824[22, 157], 825[18], 826 [37, 149], 827[22, 230, 272], 828[22, 105, 116, 145, 272, 275], 829[140, 146, 277], 830 [151, 154], 831[22, 146, 272, 287, 288, 289], 833[16, 116, 120, 275, 277, 287, 292, 293, 297], 834[167, 298, 299, 300], 835[312], 836[116, 120, 167, 312, 313, 314], 837[297, 315], *838, 839, 840, 841, 842, 843, 845, 846, 847, 848*

Roman, A., 626[352], 627[352], *663*

Rongey, R., 541[28], 545[28], 572[28], *580*

Rongey, R. W., 437[13], 463[13], *539*

Roome, A. P. C., 823[244], *846*

Rosati-Valente, F., 309[137], *345*

Rose, H. M., 312[150], 332[205], *345*, *347*, 778[17, 26], 779[32], 795[26], 797[26], 798[17], 808[156], 823[26, 248], 827

[Rose, H. M.]
[270], *838, 839, 843, 846*, 877[204], 878[216], *888*

Rose, J., 728[144], *761*

Rose, J. A., 540[8], 549[8], 551 [8, 127, 129], 556[129], 557 [129], 559[8, 127, 150], 560 [140, 141], 561[150], 562[127, 152], 564[140, 158], 565[163], 567[8, 177, 194, 195], 568 [129, 163, 177, 195], 569[207], 571[195], 572[195], 573[195], 575[195, 222, 223, 224], 576 [224], 577[127, 140, 177, 194, 222, 228], *579, 583, 584, 585, 586*, 634[450], *667*, 683[67, 68, 137], 718[221], *759, 761, 763*, 895[21], *898*

Rose, J. K., 67[15, 33], *105, 106*, 197[143], 239[239], 243[143], 246[239], 257[143, 239], *275, 278*, 351[141], 370[141], *378*

Rosemond, H., 131[108], *163*, 859 [55], *884*

Rosemond-Hornbeak, H., 854[25], 863[25], 864[90, 91], 880 [251], *883, 885, 889*

Rosen, L., 384[2, 3, 7, 8], 385[2, 3, 7, 15], 386[8, 15], 387[67], 390[15], 391[15], 422[15], *424, 425, 426*

Rosenberg, H., 89[126], 96[126], *109*

Rosenberg, N., 534[106], *542*

Rosenbergová, M., 267[294], *280*

Rosenblatt, S., 609[208], 623 [314], 631[411], *659, 662, 665*

Rosenblum, E. N., 859[50, 52], 860[50], 862[72], 863[73], 866[113], 873[50], 874[50], 875[52], 876[192], 877[212, 213], 878[212, 213], 879 [192], 881[212, 276, 277], *884, 885, 888, 890*

Rosenkranz, H. S., 808[156], *843*, 878

[Rosenkranz, H. S.] [216], *888*
Rosenthal, J., 879[236], *889*
Rosenthal, S. M., 795[109], *841*
Rosenwirth, B., 690[370], 704 [149], 705[149], 707[149, 361], 711, 717[361], 727 [361], 728[370], *761, 764, 768*
Rosijn, Th. H., [301], *766*
Rosin, A., 880[244], *889*
Ross, A. J., 174[45], *272*
Ross, J., 815[215], *845*
Ross, L. J. N., 787[85], 811, *840, 843*
Ross, M. G. R., 385[24, 25], *425*
Ross, M. R., 595[48], *654*
Ross, S., 385[29], *425*
Rossouw, A. P., 175[49], 181 [49], 220[49], *272*
Rotem, Z., 45[38], *57*
Rothblat, G. H., 202[177], *276*
Rothschild, H., 650[601], *671*
Rott, R., 292[48], 293[57], 298[239], 299[88, 92], 300 [94, 95], 307[127, 128], 308 [57], 309[95], 313[156], 314 [170], 315[170], 316[170], 317[176, 177, 178], 326[192, 193, 194, 195, 200, 239], 328 [194], 332[192, 193, 194, 195, 200, 239], 328[194], 332[192, 193, 194, 195, 200], 335[127], 336[216, 217, 225], *342, 343,* 344 *345, 346, 347,* 361[220], 362 [219], 364[220], 365[204], 367[218], 373[267], *380, 381, 382,* 780[79], 787[79], 789 [79], *840*
Rottman, F., 719[227], *764*
Rottman, F. M., 719[37, 200], *758, 763*
Rouget, R., 601[117, 120], *656*
Roumiantzeff, M., 75[76], 80[76], *107*
Rous, P., 592[13], *653*
Roux, L., 361[205], 370[206], *380*
Rovell, D. R., 51[113], *60*
Rovera, G., 625[338], *663*
Rowe, W. P., 385[13], *425*, 437 [10, 11], 446[20], 449[23], 520[23], 521[99], 527[101], *539, 542,* 541[26], 543[88, 199], 545[26], 552[88, 89, 90], 555[26], 567[26], 568 [88, 199], 569[26, 90], 570 [210], 572[210], 573[199, 210], 574[210], 578[26, 210], *580, 582, 585, 586,* 593[33], 594[42, 43, 44], 596[75], 631 [390, 392, 395, 397], 633[434], 634[434, 450], 648[390], *653, 654, 655, 665, 666, 667,* 676, 677[115, 116], 713[31], 718 [221], 731[116], 739[259], 744[228], 748[156], *757, 760, 761, 763, 764, 765*
Rowlands, D. J., 65[8], 99[152], *105, 110*
Roy, D., 397, *429*
Roy, P., 190[119, 122], 191[119, 122, 129, 131], 192[131], 193 [129, 131], 196, 197[142], 200 [142], 202[142, 183, 188], 203 [119, 129, 142, 198, 200], 226 [119], 227[183, 188, 200, 220], 228[198, 200, 222], 229[222], 230[222], 231[222], 232[222], 237[198], 238[142, 200], 269 [119, 122, 129, 198, 200], *275, 277, 278,* 314[164], *346,* 357, *380*
Roy, R., 190[123], 195[123], 196 [123], 199[123], 203[123], 224[123], 227[123], 228[123], *275*
Royce, K. R., 385[16], *425*
Rozenblatt, S., 609[206], 635 [468], *659, 667*

Rozijn, Th. H., 729[302], *766*
Rubenstein, A. S., 773[119], 795 [119], 797[119, 121], 817, 819[233], *841, 842, 845*
Rubin, B. A., 194[164], *276*, 361 [177], 363[176], *379*, 559 [156], 562[156], *584*, 698 [260], *765*
Rubin, H., 353[208], 362[209], *380*
Rubinstein, L. G., 595[57], *654*
Ruck, B. J., 314[169], 315[169], 316[169], *346*, 385[26], *425*
Rucker, R. R., 174[44, 45], *272*
Rueckert, R. R., 19[102], *38*, 64 [6], 65[10], 67[6], 69[6, 56, 57, 62], 71[6], 73[6], 74[6], 86[118, 119], 87[119], 91[119], 99[150], *105, 107, 109, 110*, 456[33], *539*
Rueter, A., 815[209], *844*
Rundell, K., 631[403], 632[403, 412], 638[403], *665*
Runevski, N., 353[226], *381*
Runner, M. N., 870[161], *887*
Ruprecht, R., 227[219], *278*
Ruschmann, G. K., 625[331], *663*
Russell, G. E., 6[56, 103], *37, 38*
Russell, P. B., 881[261], *889*
Russell, P. K., 114[18, 20], 119[18, 20], 123[18, 76], 124[18, 80], 139[20], 146[76, 148], *160, 162, 164*
Russell, W. C., 15[117], 16[117], *39*, 691[231], 697[230], 698[231], 705[232, 234], 708[232, 234], 709[232, 234, 324], 710[230], 723[7], 724[230], 727[230], [233], *757, 764, 767*, 778[13], 779[13, 41], 797[125], 815 [214], *838, 839, 842, 844*, 894 [35], 895[35], *898*
Rustigian, R., 373[263], *382*
Rutter, W. J., 78[94, 95], 79[95], *108*, 615[260, 261], *661*
Ruzijn, H., 728[267], *765*
Ryder, L., 351[188], 362[184], *380*
Ryhiner, M.-L., 797[122, 123], *842*
Rytel, M. W., 45[22], *57*

Saber, S., 313[156], *345*
Sabin, A. B., 66, *105*, 384, 385 [1], 387[1], 389[1], *424*, 631 [398], *665*
Sabin, A. S., 648[562], *670*
Saborio, J., 720[366], 727[366], *768*
Saborio, J. L., 76[90], 78[90], 86[90], *108*, 718[371], 725, *768*
Sachs, L., 591[4], 608[203], 612 [238, 239], 617[276], 620 [276], 625[238, 239], 645 [276, 541, 543], 649[589], 650[593, 594], 651[610, 611, 612, 613, 617], *652, 659, 660, 661, 669, 671, 672*
Sack, G. H., Jr., 604[161], 607 [161], 611[227], 628[227], *657, 659*
Sadler, P., 881[266], *890*
Sagik, B. P., 143[136], 147[152], *163, 164*
Saito, M., 650[595], *671*
Sakabe, K., 628[372], *664*
Saksela, E., 595[51], *654*
Sakuma, S., 390[93], 398[93], 401[93], 409, 411[213], *427, 431*, 865[99], *885*
Salb, J. M., 50, *59*
Salditt-Georgieff, M., [374], *768*
Salk, J., 64[3], 66, *105*
Salmi, A. A., 351[210], *380*
Salomon, C., 614[258], 616[258], 617[258], 623[258], 652 [258], *660*

Salomon, R., 69[47], *106*, 132 [113], *163*
Salvin, S., 55[140], *61*
Salzberg, S., 678[235], *764*
Salzman, L. A., 543[67], 549[67], 555[67, 122, 131], 556[131, 132, 133, 134], 557[131], 559 [67, 122], 561[67, 122], 562, 564[122], 567[193], 568[193], 570[193], 571[132, 133, 134], 572[193], 574[193], 576[193], *581, 583, 585, 586*
Salzman, N. P., 597[89], 599[89], 604[150], 605[176], 606[150], 626[351], 627[351, 361], 628 [367, 374], 629[379], 630 [384], *655, 657, 658, 663, 664*, 854[35], 855[35], 859[48], 869[152], 870[152, 158, 159, 163, 164], 871[164], 873[163], 876[197], 880[152, 163, 164], *883, 884, 886, 887, 888*
Samaha, R. J., 633[435], *666*, 713[106, 155], 746[155], *760, 761*
Sambrook, J., 64[1], 74[1], 75[1], 80[1], 93[1], *105*, 604[156, 157], 607[156, 157], 616[271], 617[157], 618[271], 619[271, 297, 300], 620[271, 297, 300], 621[297], 623[297], 624[297], 625[330], 630[330], 634[447], 638[498, 500], 640[514], 641 [519], 642[498], 644[500, 536, 537, 538, 539, 540], 645[538, 540, 550], 646[271, 500], 648 [572], 649[539, 572], 651[500, 572, 606], *657, 661, 662, 663, 667, 668, 669, 670, 672*, 677 [66], 678[66], 679[66], 687 [244], 690[66, 236, 246], 701 [238], 707[98, 170, 238, 321], 709[170, 238, 321], 713[97, 98, 238], 724[54], 732[66, 206, 207, 246], 747[54, 97], 748[54, 192, 237], 749[97,
[Sambrook, J.]
98], 753[236], 754[98, 321], 755[98, 120], *758, 759, 760, 762, 763, 764, 767*, 866[110], 872[182], *885, 887*, 891, 896 [20], *898*
Sambrook, J. F., 148[158], 149 [158], 151[158], *164*, 636 [471], *667*, 866[104], *885*
Samson, A. C. R., 364[211, 212], *380*
Samuel, C. E., 52, *60*, 423, *433*
Samuel, I., 361[190], *380*
Samuels, J., 148[157], 149[157], *164*
Sanders, J., 877[209], *888*
Sanderson, P. J., 691[231], 698 [231], *764*
Sangar, D. V., 99[152], *110*
Sanmartin, C., 170[18], 215[18], *271*
Sanpe, T., 352[117], 364[117], *378*
Santesson, L., 789[92], *841*
Santoli, D., [586], *671*
Santos, G. W., 541[31], 543[31, 75], 550[31, 75], 554[75], *580, 581*
Saral, R., 601[119], 613[345], 625[345], 632[345], 635 [345], 636[345], 637[345], 638[345], 652[345], *656, 663*
Sargent, M. D., 391[102, 103], 397[102, 103, 143], *427, 429*
Sarkar, N. H., 519[93], *542*
Sarma, P. S., 437[9, 13], 443[17], 449, 463[13], 502[17], 508 [9], 509[9], *538, 539*
Sarmeinto, M., 793[97], 794 [97], 795[97], *841*
Sarov, I., 772[58], 780[58], 793 [101], 827, *839, 841*, 855[43], 858[49], 859[49], 867[134], 878[221], *883, 884, 886, 888*
Sasaki, R., 830, *847*

Sasaki, Y., 830, *847*
Sasao, F., 154[187], 155[187], *165*
Saski, N., 174[25], *272*
Sato, K., 174[24], *272*, 353 [240], *381*
Sato, N., 543[203], 569[203], *585*
Sato, T., 352[230], *381*
Sauer, G., 610[214], 612[240], 617[278, 279, 283], 620[278, 279], 622[279, 283], 635 [470], 645[278], *659, 660, 661, 667,* 815, *845*
Sauerbier, W., 373[268, 269], *382*
Saunders, E. L., 546[54], 555 [54], *581*
Savage, T., 772[25], 774[25], 800[140], 801[140], 829 [140], 837[315], *838, 842, 848*
Sawyer, R. C., 459[40, 41], *540*
Sayevedra, M., 891[12], *897*
Schachanina, K. L., 634[441], *666*
Schaeffer, B., 649[584], *671*
Schaeffer, M., 179[68], *273*
Schäfer, W., 324[184], *346*, 450 [25], 452[28], 472[55], 473 [28, 56], *539, 540*
Schaffer, F. L., 190, 248[248], 269[248, 303, 306], *275, 278,* 387[63, 73], *426*
Schaffer, P., 791[320], *848*
Schaller, G. B., 351[221], *381*
Schäperclaus, W., 174, *272*
Scharff, M. D., 398[148], 418 [148], *429*, 633[438], *666,* 691[165, 166], 698[166], *762*
Scheele, C. M., 130, 134[114, 117], 152[95], *162, 163*
Scheid, A., 352[42], 355, 359[215], 360[213, 214], 361[213, 214, 215], 363[42, 215], 364[213, 214], *375, 380*
Scheidtmann, K-H., 894[39], *898*
Schekman, R., 726[239], *764*
Scher, C. D., 534[106], *542*, 638 [494], 641[518], *668, 669*
Scherrer, K., 132[109], *163*, 623 [311], *662*
Scherrer, R., 866[105], *885*
Scherrer, S., 75[77], 76[77], 78 [77], 80[77], *107*
Schick, J., 716[41], *758*
Schidlovsky, G., 835[304], *847*
Schieble, J. H., 614[248], *660*
Schild, G. C., 283[16], 287[24], *341, 342*
Schilling, R., 690[240], 728[240], *764*
Schincarial, A. L., 248[251, 252], *279*
Schinkaryl, A. L., 678[241], 727 [241], *764*
Schlabach, A., 872[179], *887*
Schleicher, J. B., 815[209], *844*
Schlesinger, H. R., 201[176], *276*
Schlesinger, M. J., 121[65], 123 [73], 126[87], 133, 134[73, 115], 135[119], 139[129, 142], 143[73, 129], 152[204], 153 [169], *161, 162, 163, 164, 165, 166*
Schlesinger, R. W., 114[22], 117[22, 31], 118[22, 56], 126[22], 131 [22, 104], 132[22], 146[147], *160, 161, 163, 164,* 676[242], 721[242], *764*
Schlesinger, S., 121[65], 123[73], 133, 134[73, 115], 143[73, 139], 153[169, 171], 154[171], *161, 162, 163, 164, 165*
Schloemer, R. H., 198[151, 152], 199[151, 157], 204[157], *276*
Schlom, J., 480[68], 535[68], *541*

Schluederberg, A., 353[216], 365 [35, 217], *375, 380, 381*
Schlumberger, H. D., 15[4], *35,* 199[161], *276*
Schmidt, J. R., 175[55], *273*
Schmidt, N. J., 390[260], *432*
Schmidt, R. W., 541[20], 543[20], 554[20], 555[20], *580*
Schmir, M., 602[135], *656*
Schmookler, R. J., 615[264], 616, *661*
Schnagl, R. D., 423[264], *432*
Schnaitman, C. A., 199[163], 224 [163], 256[163], 257, 258 [261], *276, 279*
Schnaitman, T. C., 197[146], *276*
Schneider, C. C., 389[82], *427*
Schneider, L. G., 179[73, 74, 75], 199[160], 216[207], 217[207], 269[207], *273, 276, 277*
Schnipper, L. E., 713[106], 731 [152], 749[152], *760, 761*
Schnitzer, T. J., 362[90], *377*
Schofield, F. W., 540, *579*
Scholtissek, C., 294[74], 302[74], 307[127, 128], 313[156, 162], 314[162, 170], 315[170, 171], 316[170, 171], 317, 320[162], 326[193, 194, 195, 196, 200], 328[194], 332[193, 194, 195, 196, 200], 335[127], 336[216, 217, 218, 225], *343, 345, 346, 347,* 365[204], 367[218], *380, 381*
Schonberg, M., 386[51], 398[149], 400[156], 402[51, 149, 156], 403[149], 404[51], 405[149, 156], 406[149], 408[51], 409 [212, 411[212], *426, 429, 431*
Schonne, E., 45[24], 47[55], 48 [70], 50[97], *57, 58, 59*
Schornstein, K. L., 770[1], 778[1], 822[1], 824[1], *838*
Schriner, A. W., 541[19], 543[19], 578[19], *580*
Schuerch, A. R., 417, 418[248], 421, *432*
Schulman, J. L., 283[16], 299 [85, 86, 89], 300, 304[241], 310[86], 323[241], 339[241, 243], 340, *341, 344, 348*
Schulte-Holthausen, H., 789[92, 93], *841*
Schulze, I. T., 282[5], 289[5], 292[53], 294[68, 70], 297, *341, 342, 343*
Schwartz, J., 778[18], 779[37], 822[18, 37], 823[37], 825 [18], 826[37], 836[313], *838, 839, 848,* 862[60], 863[60], 864[60], *884*
Schwartz, L. B., 78, 79[92], *108*
Schwartz, M., 631[403], 632[403], 638[403], *665*
Schwarz, R. T., 313[162], 314 [162], 320[162], 326[199], 332[199], 336[216], *346, 347*
Schwöbel, W., 147[153], *164*
Scolnick, E. M., 438[15], 446[21], 456[31], 457[15], 467[51], 473[58], 474[58, 60], 475 [58], 480[15], 489, 516[91], 519[91], 521, 536[114], 537 [114], *539, 540, 541, 542*
Scotti, P. D., 88[122], 89[122], 93[122], *109*
Scraba, D. G., 69[58], 71[63], *107*
Scriba, M., 47[56], *58*
Scroggs, R. A., 361[196], 362 [196], 364[196], *380*
Seals, M. K., 358[32], *375*
Sebring, E., 575[224], 576[224], *586*, 605[176], *658*
Sebring, E. A., 869[152], 870 [152], 880[152], *886*
Sebring, E. D., 626[351], 627 [351], *663*, 870[158, 159, 163], 873[163], 880[163], *887*
Seebeck, T., 599[99], 602[99],

[Seebeck, T.]
625[99], *655*
Seehafer, J., 83[105], *108*
Seehafer, J. G., 597[82], 630[82], 635[82], *655*
Sefton, B. M., 114[17], 121[17, 64], 123[75], 134[125], 142 [64], 143[75], *160, 161, 162, 163*
Segal, S., 130, *162*
Sehulster, L. M., 707[369], *768*
Seifert, E., 472[55], *540*
Seiler, P., 626[350], *663*
Sekellick, M. J., 53[121], *60*, 239[241], *278*
Sekely, L., 813[190], 816[190], *844*
Sekikawa, K., 677[63], 678[63], 683[372], 730[372], 732 [63], 735[248], *758, 764, 768*
Sela, I., 69[55], *107*
Selimova, L. M., 359[265], 368 [264], *382*
Semba, T., 364[102], *377*
Sen, A., 600[112], *656*
Sen, G. C., 423, *433*
Sengbusch, P., 4[7], *35*
Sepence, L., 177[62], *273*
Sereno, M. S., 552[91], *582*
Seto, J., 299[88], 304[232], 306, 317[232], 320, 333[121], *344, 345, 348,* 361[220], 362[219], 364[220], *381*
Sever, J. A., 287[22], *342*
Shah, K., 595[57], *654*
Shah, K. V., 351[221], 352[222], *381,* 596[65, 73], *654, 655*
Shaila, S., 423[273], *433*
Shamugam, G., 723[102], *760*
Shand, F. L., 352[172], 358[172], 359[172], *379*
Shani, M., 617[287], 645[541], 649[589], *661, 669, 671*
Shanmugam, G., 715[72, 243], 716 [72], 727[243], 738[70], *759, 764*
Shani, M., 737[3], 752[3], *757*
Shanthaveerappa, T. R., 179[67], *273*
Shapiro, D., 113, 114[18, 20], 119[18, 20], 123[18, 76], 124 [18], 125, 139, 146[76, 148], *160, 162, 164*
Shapiro, L., 20[77], 21[77], *38*, 387[52], 392[52], *426*
Sharp, D. G., 866[125], 879[125], *886*
Sharp, P., 644[538, 539], 645 [538], 649[539], *669*, 707 [98, 321], 709[321], 713[98], 749[98], 754[98, 321], 755 [98], *760, 767*
Sharp, P. A., 604[157], 607[157], 608[195], 610[195], 611[195], 616[271], 617[157], 618[271], 619[271, 297, 300], 620[271, 297, 300], 621[297], 623[297], 624[297], 634[447], 638[500], 644[500, 536, 537], 645[550], 646[271, 500], 651[500], *657, 658, 661, 662, 667, 668, 669, 670,* 677[66], 678[66], 679[66], 681[177, 178], 687[177, 203, 244], 688[53, 177], 689[53], 690[53, 66, 236, 245, 246], 701 [238], 707[170, 238, 245], 709 [170, 238], 713[97, 238], 717 [344], 720[245, 349], 721[340], 724[54], 728[348], 732[66, 236, 246], 734[53, 245], 735 [53], 736[177], 737[53], 738 [53], 747[54, 97], 748[54, 192, 237], 749[97, 178, 203], 750 [53, 177, 245], 751, 753[53, 236], 754[238], 755[170], *758, 759, 760, 762, 763, 764, 768,* 891[3], 892[3], 893[3], 894 [40], 895[40], 896[16, 20], *897, 898*
Sharpe, A. S., 424[279, 280], *433*
Shatkin, A. J., 67[21, 23, 29, 31,

[Shatkin, A. J.]
32], 80[21, 23], 85[21, 23], 85[21, 23], *105, 106,* 203 [192], 229[192, 223], 230 [192], 236[226], *277, 278,* 282[11], *341,* 358[171], 367 [171], *379,* 387[54, 65], 390 [84], 392[54, 107], 394[65, 117, 118, 122, 124, 125], 395 [133, 134, 139], 397[84, 144], 398[139], 400[84, 155, 160], 401[168, 169, 261], 403[54, 184], 404[139, 184, 193], 405[84, 155, 160], 406[118, 160, 169, 184, 201, 204, 205, 261], 407[139], 408[139, 169, 209, 210, 236, 261], 409[65], 410[168, 193], 412[232, 233, 235], 422[232, 254], *426, 427, 428, 429, 430, 431, 432,* 555 [129], 556[129, 138], 557[129, 138], 560[138, 141], 561[141], 565[163], 568[129, 163], 569[207], 570[189], *583 584, 585,* 625[330], 630[330], *663*, 719[16, 165, 227], [374], *757, 759, 764, 768,* 863[78], 869]152], 870[152, 158], 872 [182], 880[152, 249], *884, 886, 887, 889*

Shatkin, A. S., 617[289], *661*

Shaw, E. D., 386[43], *426*

Shaw, E. N., 85[113], *108*, 808 [164], *843*

Shaw, J. E., 411[230], *431*

Sheaff, E. T., 45[36], *57*

Shechmeister, I. L., 190[124], *275*

Shedden, W. I. H., 780[80], 787 [80], 812[181], 813[181], *840, 843*

Shedlarski, J. G., Jr., 248[245], *278*

Sheer, C. J., 535[109], 537[109], *542*

Sheffield, F. W., 881[263], *889*

Sheffy, B. E., 543[125], 555[125],

[Sheffy, B. E.]
568[125], 569[125], *583*

Shein, H., 595[50], *654*

Shein, H. M., 631[398], *665*

Sheinin, R., 612[242], 617[282], 625[242], *660, 661*

Shelburne, C. E., 595[58], *654*

Sheldon, R., 862[69, 70], 869[69], *884*

Sheldrick, P., 780[75], 782[75], 797[123], *840, 842*

Shellabarger, C. J., 541[20], 543 [20], 554[20], 555[20], *580*

Shelokov, A., 880[238], *889*

Shenk, T. E., 127[172], 147[155], 153[170, 173], 154[172, 173, 174, 186], 155[155, 173, 174], *164, 165,* 608[202], 611[202], 612[202], 637[491, 492], 638 [491, 492], *659, 668,* 756[247], *764*

Sheppard, R. D., 820[236], 824 [236], 826[236], *845*

Sherr, C. J., 463[47], 488[70], 509[88], 537[117], *540, 541 542*

Shibley, G. P., 779[33], *839*

Shibuta, H., 353[223], *381*

Shih, T. Y., 647[555], *670*

Shimada, K., 735[248], *764*

Shimiza, F., 384[6], 385[6], *424*

Shimizu, K., 358[99], 361[225], 364[100, 224], *377, 381*

Shimizu, Y. K., 361[225], 364 [101, 224], *377, 381*

Shimojo, H., 631[409], 635[459], 642[525], 650[525], *665, 667, 669,* 683[372], 700[330], 701 [77], 709[100], 710[250, 251], 711[226, 250, 269, 270], 712 [251, 271], 726[249], 727 [251], 729[252], 730[372], 739[104], *759, 760, 764, 765, 767, 768*

Shimono, H., 625[337], *663*, 802

[Shimono, H.]
[150], 811[150], *842*
Shimonski, G., 45[28], *57*
Shin, D. S., 69[54], *106*
Shin, S., 896[41], *898*
Shindarov, L., 353[226], *381*
Shipkey, F. H., 778[19], 826 [19], *838*
Shiroki, K., 642[525], 650[525], *669,* 683[372], 709[100], 710 [250, 251], 711[226, 250], 712[251], 726[249], 727[251], 729[252], 730[372], *760, 764, 768*
Shope, R. E., 170[14], 175[51, 54], 181, 182[54], 183[91, 92], 205[202], 206[202], 208[202], 209[202], 215[92], 216[51, 54p, 218[202], 219 [202], 220, *271, 272, 273, 274, 277,* 591[8], *652*, 866[119, 120], 880[120], *886*
Shope, R. W., 591[10], 592[10], *653*
Showe, M., 20[1], *35*
Shur, P. H., 402[170], *429*
Shure, H., 617[296], 618[296], 619[296], 620[296], 622[296], 623[296], 624[296], 644[296], 645[296], 646[296], *662*
Siegel, S., 741[129, 130], 742[129, 130], 745[129, 130], 746[129, 130], 748[129, 130], *761*
Siegel, S. E., 742, 745[253], *765*
Siegert, R. S., 779[30, 31], 798 [30], *838*
Siegl, G., 541[11, 36], 543[11, 36, 71, 123, 125, 186, 190], 544[64], 549[64], 550[36, 71], 551[36], 554[36], 555[11, 123, 125], 556 [11, 240], 559[11, 64, 155], 562 [155], 564[155], 567[36, 186, 190], 568[71, 125], 569,64, 71, 125, 190], 570[64, 190], 571 [186], 573[64, 186], 574[190], *579, 580, 581, 583, 584, 585,*
[Siegl, G.]
587
Sikl, D., 45[23], *57*
Silagi, S., 412[231], *431*
Silverberg, H., 798[126], *842*
Silverman, C., 855[44], 872[178], *884, 887*
Silverstein, S., 800[137], 801[137], 842
Silverstein, S. C., 386[51], 398 [149], 400[156, 161], 402[51, 149, 156, 170, 179], 403[149, 179, 181], 404[51], 405[149, 156, 161], 406[149], 407[207], 408[51, 211], 409[212], 410, 411 [181, 212, 219], *426, 429, 430, 431*
Silvestre, D., 808[163], 835[163, 308, 311], *843, 847*
Siminovich, L., 851[15], 877[15], 879[15], *883*
Simizu, B., 147[160], 145[202], *164, 166*
Simmons, D. T., 114[11], 117[11, 34, 38], 126[11, 34], 127[92], 128[34, 92], 129, 131[34], 132 [34], 134[34], 135[38], 136, 137[38], 139[34], *160, 161, 162*
Simmons, T., 729[254], *765*
Simon, E. H., 365[227], *381*
Simon, J., 595[59], *654*
Simons, K., 114[8, 10], 118[8], 119[57], 120[58], 121[10], 122[10, 57, 69, 70], 123[10, 57, 72], 133[10, 72], 135[10, 72], 142[57], 143[10], *160, 161, 162*
Simpson, D. H., 385[12], 400[12], *424*
Simpson, D. I. H., 169, 174[29], 175[51], 183[29, 92], 215 [92], 216[51], 218[51], *271, 272, 274*
Simpson, R. W., 118[50, 51], *161*, 199[162], 224[162], 226[216],

[Simpson, R. W.]
227[219], 228[216], 264[279, 280], *276, 278, 279,* 293[65], 302[99], 306[99], 307[126], 313, 314[65, 163, 164, 228], 315[173], 318, 320[65], 336 [213, 219, 242], *343, 344, 345, 346, 347, 348,* 373[228], *381*
Sinarachatanant, P., 154[88], 155 [188], *165*
Sinclair, F., 880[260], *889*
Singer, I. I., 569[206], *585*
Singer, M. F., 609[208], *659*
Singh, K. R. P., 155[190], *165*
Singh, K. V., 351[229], *381*
Sinsheimer, R. L., 541[37], 561 [146], 565[146, 162], 574 [218], *580, 584, 586*
Sipe, J. D., 49[[73], *59*, 387 [54, 65], 390[84], 392[54], 394[65], 397[84], 400[84], 403[54], 405[84], 409[65], *426, 427*
Sirtori, C., 825[250], *846*
Sishido, A., 352[230], *381*
Sitnikov, B. S., 336[224], *347*
Sjogren, H. O., 648[559], *670*, 738[255], *765*
Skačianska, E., 353[231], *381*
Skehel, J. J., 45[33], *57*, 256 [256], *279*, 293[59, 63, 235], 297, 298[82, 235], 300[63], 301, 302[101], 306[101], 313 [160], 314[160], 320[160], 324[187], 325[96], 326, 333 [101], *343, 344, 346, 348,* 392 [112], 400[112], 401[112], 405[112], *428*[233], 705[234], 708[234], 709[234], *764*
Skinner, G. R. B., 813, *844*
Skipkowitz, N. L., 815[209], *844*
Skurkovich, S. V., 55[148], *61*
Skyler, J. S., 423[269], *432*
Slayter, H. S., 624[317], *662*
Slifkin, M., 634[456], *667*
Slor, H., 126[89], *162*
Slotnick, V. B., 595[49], *654*
Smadel, J. E., 45[21], *57*, 854 [31], *883*
Smale, C. J., 122[71], *162*, 195 [210], 197[145], 198[148], 199[148], 200[145, 167], 204 [167, 201], 221[210], 238 [167], *275, 276, 277*
Smelt, D., 361[190], *380*
Smiley, J. R., 893, *898*
Smith, A. E., 67[24], 80[24], 85 [24], 87[120, 121], 88, *106, 109,* 117[39], 139, *161*, 631 [410], 632[410], *665*
Smith, B. G., 55[143], *61*
Smith, C. A., 608[195], 610 [195], 611[195], *658*
Smith, H. O., 617[294], *662*, 749 [181], *762*
Smith, H. S., 638[494], 641[518], *668, 669*
Smith, J. B., 632[423], *666*
Smith, J. D., 601[121], *656*, 825 [253], *846*
Smith, J. E., 123[135], 141[135], *163*
Smith, J. F., 156[196], *165*
Smith, J. W., 352[109], *377*, 384 [9], *424*, 594[47], *654*
Smith, K. M., 15[104], 16[104, 119], *38, 39*
Smith, K. O., 387[72. 74], 391 [74], *426*, 541[31], 543[31, 75], 546, 547[55], 550[31, 75], 554[75, 117], 569[117], 578[233], *580, 581, 583, 586,* 691[256], *765*, 795[117], 797 [117], *841*, 866[125], 879 [125], *886*
Smith, L., 186[108], *274*
Smith, R., 20[1], *35*
Smith, R. E., 386[46], 387[46], 390[46], 391[46], 394[46], 395[46], 398[46], *426*
Smith, R. W., 476[64], *541*

Smoler, D., 100[156], *110*
Smythers, G., 469[52], 470 [52], *540*
Snowdon, W. A., 174[27], *272*
Snyder, H. W., 137[127], *163*
Snyder, R. M., 257[259], 258 [259, 261], *279*, 352[249], 360[250], *381, 382*
Söderlund, H., 114[6], 117[36], 118[49], 121[6], 131[36], 132[36], 137[36], *160, 161*
Söderlung, H., 720[373], *768*
Soehner, R. L., 773[45], *839*
Soeiro, R., 324[189], *346*, 410 [220], 420[220], *431*
Soergel, M., 387[76], 389[76], 402[173], 412[234], *426, 430, 431*
Soergel, M. E., 190[124], 248 [248], 269[248, 306], *275, 278, 280*
Sogawa, Y., 352[117], 364[117], *378*
Sohier, R., 283[16], *341*, 677 [257], *765*
Sokol, F., 194[134, 137, 138, 140, 141], 195[138, 141], 196[134, 137, 138, 141], 199[138, 141, 161], 201[137, 138, 141], 202 [137], 206, 218, 237[137], 258[137], *275, 276*, 353[231], 360[232], *381*, 598[92], 610 [215], 614[253, 254], 617 [279], 620[279], 622[279], 624[316], 636[474], *655, 659, 660, 661, 662, 667*
Solditt-Georgieff, M., 719[65], *759*
Solem, R., 706[363], 711[363], 718[363], 738[363], 753 [363], *768*, 893[29], 896[29], *898*
Solomon, J. J., 779[39], *839*
Soloviev, G. Ya., 313[157], *345*
Soloviev, V. D., 283[16], *341*, 438 [15], 457[15], 480[15], *539*
Somers, K., 649[581], *671*
Sommers, S., [374], *768*
Sonnabend, J. A., 50[86, 99], *59*, 126[82], *162*
Sopori, M. L., 51[115], 52[115, 119], *60*
Soreq, H., 132[113], *163*
Soria, M., 236[230], 251, 252 [255], 254[255], *278, 279*, 372[233], *381*
Soriano, E. Z., 186[109], *274*
Soriano, F., 595[58], *654*
Soule, H. D., 537[115], *542*
Southam, C. M., 778[19], 826 [19], *838*
Southern, E. M., 750, *765*, 787 [86], *840*
Southwick, C. H., 352[222], *381*
Spahn, G. F., 54[135], *61*
Spandidos, D. A., 417[245, 246], 421, 424[276, 277, 278], *432, 433*
Spanier Collins, B., 370[24], *375*
Spear, I., 565[167], *584*
Spear, P. G., 15[101], *38*, 772[25], 774[25, 67], 779[36], 780[67, 77], 787[77], 792[36], 793 [36, 97], 794, 795[97, 114, 116], 796[114], 797[116], 802[149], 810[116, 166, 167, 169], 811[149], 822[243], 826 [149], 828[116, 275], 832 [297], 833[116, 275, 297], 834[167], 836[116, 167], 837 [297], *838, 839, 840, 841, 842, 843, 845, 846, 847*
Spector, D. H., 68[36, 38, 43], 94 [36, 38, 43], 95[43], 96, 100 [155], 101[155], *106, 109, 110*, 132, *163*
Spence, L., 170[14], *271*
Spencer, J. H., 553[114], *583*
Spencer, M., 387[61], 392[61], *426*

Spendlove, R., 80[100], *108*
Spendlove, R. S., 387[63, 73], 391[96, 99, 106], 402[172, 178], 417[106], *426, 427, 429, 430*
Sperling, R., 4[105], *39*
Spickard, A., 384[7], 385[7], *424*
Spiegelman, S., 227[219], 270 [309], *278, 280,* 480[68], 535[68, 110], *541, 542,* 800, *842*
Spies, F., 729[302], *766*
Spigland, I., 385[17], *425*
Spillman, T., 632[429], 633[429], *666*
Spira, G., 592[18], *653*
Spomer, B., 632[429], 633[429], *666*
Sporn, M. B., 632[423], *666*
Spradbrow, P. B., 174[28], *272*
Sprecher-Goldberger, S., 541[35], 545[35], 550[35], 551[35], 552[35], 569[35], *580*
Spring, S. B., [230], *348*, 795 [111], 796[111], 811[170], 835[312], 836[312, 313, 314], *841, 843, 848*
Squire, R. A., 541[31], 543[31, 75], 554[75], *580, 581*
Squires, E. J., 879[232], *889*
Sreevalsan, T., 126[84], 130, 131 [99, 102, 108], 137[127], *162, 163*
Srinivasan, P. R., 644[540], 645 [540], *669*
Staal, S. P., 739[259], *765*
Stackpole, C. W., 778[27], 822, *838, 845*
Stampfer, M., 85[111], *108*, 201 [172], 203, 226[194], 239[237, 238], 243[237], 246[238], 248 [250], 257[238], 269[250], 270[250], *276, 277, 278,* 352 [8], 370[167], *374, 379*
Stancek, D., 49[79, 80], 50[80,
[Stancek, D.]
82, 83, 84], *59*, 89[124], 93 [124], *109*, 194[140], *275*
Standfast, H. A., 174[26], *272*
Stanley, N. F., 384[4, 5, 10], 385[20], 386[37], 387[68], 390[10], *424, 425, 426*
Stanley, P., 292[54], 298[80], 326[80, 198], 332[198], *342, 343, 347*
Stanners, C. P., 605[187], 610 [187], *658*
Stansy, J. T., 559[156], 562[156], *584*
Staples, S. E., 115[28], 143[28], 145[28], *160*
Stark, C. R., 595[55], *654*
Stark, G., 632[428], 633[428], *666*
Stark, G. R., 746[4, 220], *757, 763*
Stasny, J. T., 698[260], *765*
Staub, M., 629[380], *664*
Stebbing, N., 69[48], *106*
Steenbergh, P. H., 683[261], 728 [267], *765,* 895[43], *898*
Stehelin, D., 517[82], *541*, 614 [259], 625[346], 636[475], *660, 663, 667*
Stein, S., 795[118], 797[118], *841*
Steinberg, A. D., 54[135], *61*
Steiner-Pryor, A., 75[81, 82], *107*
Steinhart, W. L., 780[59, 60], 782[59, 60], *839*
Stephenson, J. R., 309[140], 312, 313, 316[151], 318, *345*, 474 [59], *540*
Stephenson, S., 881[263], *889*
Steplewski, Z., 648[570, 576], 650[570], *670, 671*
Stern, W., 865[100], *885*
Sterz, H., 780[79], 787[79], 789[79], *840*

Steve-Bocciarelli, D., 309[137], *345*
Stevely, W. S., 793[102], 830 [285], *841, 847*
Steven, A., 20[1], *35*
Stevens, A., 154[185], 155 [185], *165*
Stevens, D. A., 55[139], *61*
Stevens, J. G., 808[160], 820 [235], 822[235], 825[235], 826[235], *843, 845*
Stevens, T. M., 114[22], 117[22, 31], 118[22], 126[22], 131 [22, 104], 132[22], *160, 163*
Stevenson, D., 50[100], *59*
Stevenson, M., 239[236], *278*
Stewart, A. M., 595[56], *654*
Stewart, C. C., 143[139], *164*
Stewart, M. L., 85[113], *108*, 808[164], *843*
Stewart, S. E., 594[38, 39, 40], *653*
Stich, H. F., 825[258], *846*
Stinebaugh, S. E., 636[479], *668*
Stinebring, W. R., 43[16], 45[16, 18], 48[65], *57, 58*
Stirpe, F., 615[262], *661*
Stobbe, R., 541[35], 545[35, 94], 550[35], 551[35], 552 [35, 94], 555[94], 569[35], *580, 582*
Stoeckenius, W., 352[39], 359 [39], *375*, 387[62], *426*
Stoker, M., 601[116], 612[230], *656, 660*
Stoker, M. G. P., 2[20], *36*, 641 [517], 642[517, 526], *669*, 826, *846*
Stollar, B. D., 154[186], *165*
Stollar, V., 114[19, 22], 117[22], 118[22, 56], 119[19], 123 [19], 126[22], 127[172], 131 [22, 104], 132[22], 146[147], 147[155], 153[170, 173], 154 [172, 173, 174, 186], 155[155, 173, 174], *160, 161, 163, 164,*
[Stollar, V.] *165*
Stoltz, D. B., 177[61], *273*
Stoltzfus, C. M., 69[56], 71[56], *107*, 394[130], 395[130, 134], 401[168], 404, 410, *428, 429*
Stone, H. O., 69[59], *107*, 359 [234, 235, 236], 360[48, 234, 235, 236], 366[236], 367[47, 48, 236], 371[234, 237], *376, 381*
Stone, J. D., 308[132], *345*
Stone, R. H., 405[200], *430*
Stone, R. S., 591[8], *652*
Storz, J., 543[79], 551[79], 552 [79], 555[120, 126], 568[120], 570[120], *581, 583*
Stott, J., 68[45], *106*
Strand, M., 202[184], *277*, 472, 473[57], 535[108], *540, 542*
Strandberg, J. D., 596[65], *654*, 777, 778[12], 823[247], *838, 846*
Strander, H., 54[128], *60*
Strauss, E. G., 115[28], 140[145], 143[28, 141], 144[141], 145 [28], 146[145, 146], 147[146], 148[156], 149[156], 150[156], 151[156, 205], *160, 164, 166*
Strauss, J. H., 114[7, 11, 16], 115 [28, 29], 117[11, 34, 38], 118 [7, 29], 121[7, 16, 66], 124[7, 77], 126[11, 34], 127[92], 128 [34, 92], 129, 131[34], 132 [34], 134[7, 34], 135[38], 136, 137[7, 38], 138, 139[34], 140 [132, 134, 145], 143]28, 132, 141], 144[141], 145[28], 146 [145, 146], 147[146], 148 [156], 149[156], 150[156], *160, 161, 162, 163, 164,* 363 [238], *381*
Strauss, S. E., 575[223, 224], 576 [224], *586*
Strickland, J. E., 463[45], *540*

Strikland, J. E., [113], *542*
Strohl, W. A., 737[267], *765*
Strohmaier, K., 216[207], 217 [207], 269[207], *277*
Stubbs, G. J., 4[54], *37*
Studdert, M. J., 543[99], 552 [99], *582*, 773[46], *839*
Studier, F. W., 565[161], *584*
Suárez, D., 176[58], *273*
Suarez, F., 597[83], *655*
Subak-Sharpe, H., 592[28], *653*
Subak-Sharpe, J., 592[27], *653*
Subak-Sharpe, J. H., 793[98], 812 [183], 813[183, 188, 189], 815 [183], 828[183], *841, 843, 844,* 866[112], 881[112], *885*
Subramanian, K., 605[180, 182], 619[180], 620[180], *658,* 747 [38], *758*
Subramanian, K. N., 604[159, 163], 607[159, 163], *657*
Sudia, W., 154[179], *165*
Sugamura, K., 373[239], *381*
Sugawara, K., 715[72], 716[72], 725[263, 351], 727[124, 263], 738[351], [375], *759, 760, 765, 768,* 896[23], *898*
Sugden, B., 687[244], 748[237], *764*
Sugden, W., 604[157], 607[157], 615[263], 616[267], 617[157], 619[297], 620[297], 621[297], 623[297], 624[297], 644[538], 645[538], *657, 661, 662, 669*
Sugimoto, K., 628[372], *664*
Sugino, A., 628[371, 372], *664*
Sugino, Y., 719[365], *768*
Sugita, K., 353[240], *381*
Sugiura, A., 305[221], 336[214, 215, 220, 221], 337, *347*
Sugiura, M., 394[124], *428*
Suitor, E. C., Jr., 155[192], *165*
Sulimanovic, D., 174[39], *272*
Sulkin, S. E., 113[3], *159,* 180 [82], *273, 274*
Sulkowski, E., 47[57], 48[69], 49 [57, 69, 78], *58, 59*
Summers, D. F., 67[25], 75[76, 79], 76[79], 78[79, 91], 80[25, 76, 79], 83[103, 104], 85[113], 86 [117p, 89[124], 92[131], 93 [124], 95[142], 97[147], *106, 107, 108, 109,* 190[120], 194 [134], 196[134], 198[155], 202[182, 190], 233[229], 236 [228, 229], 237[190, 228], 239 [229], 257[120], 258[229, 263], *275, 276, 277, 278, 279,* 353[87], *377*, 808[164], *843*
Summers, J., 604[153], 606[153], 650[592], *657, 671*
Summers, M. R., 469[52, 53], 470 [52], *540*
Sundquist, B., 691[50], 697[50], [264], *758, 765*
Surgay, V. V., 868[150], 869[150], *886*
Surzycki, J. A., 732[265], *765*
Surzycki, S., 732[265], *765*
Sussenbach, J. S., 683[261], 685 [359], 690[359], 705[297], 707[297], 714[266], 726[268], 727[297], 728[266, 267, 268, 296], 301, 376, 377], *765, 766, 768,* 894[26], 895[43], *898*
Sutton, W. D., 787[87], *840*
Suzuki, E., 567[181], 575[181], *585*, 711[269], 712[123, 271], *760, 765*
Suzuki, K., 145[202], *166*, 605 [179], 631[179], *658*
Svedmyr, A., 835[309], *847*
Svejgaard, A., 351[188], 362[184], *380*
Swan, I. C., 578[237], *587*
Swango, L. J., 352[149], *379*
Swanson, R., 126[87], *162*
Swart, C., 720[174], *762*
Sweet, B. H., 594[46], 595[49], *654*
Swenson, C. C., 114[23], 118[23],

[Swenson, C. C.]
126[23], 131[23], *160*
Swetly, P., 614[252], 657, 650[604], *660, 670, 671*
Swift, H., 19[43], *36*, 773[7], 775 [7], 777[7], 778[7], 797[7], *838*
Sydiskis, R. J., 802[146, 146], 827, 828[145], 829[146], 831[146], *842*
Synder, R. M., 197[146], *276*
Sypowicz, D., 772[25], 774[25], *838*
Syrett, C., 618[362], 627[362], 629[377], *664*
Syverton, J. T., [14], *653*
Szilágyi, J. F., 200[169], 227[218], 238[169], 259[313], 260[218], 262[218], 264, *276, 278, 280,* 877[215], 881[215], *888*
Szymanski, M., 385[28], *425*

Tabachnik, N., 604[158], 607[158], *657*
Taber, L. H., 552[93], *582*
Taber, R., 424[274], *433*
Taber, R. L., 86, 87[115], *108*
Tagaya, I., 879[234], *889*
Taguchi, F., 631[394], 650[595], *665, 671*
Tai, H., 608[195], 610[195], 611 [195], *658*
Takagi, Y., 610[223], 631[223], *659*
Takahashi, K., 358[4], *374*
Takahashi, M., 701[272], *765*, 851 [6], *883*
Takano, T., 358[4], *374*, 494[75], *541*
Takayama, N., 147[160], *164*
Takemori, N., 699, 700[273, 274, 275], *765*
Takemoto, K. K., 15[106], 23[85, 86, 106], *38, 39,* 595[62], 596 [69], 611[227], 614[251], 628 [227], 631[391, 402], 632 [402], 634[443], 636[477, 484], 638[499], 639[402], 646[499], [587], *654, 655, 659, 660, 665, 666, 667, 668, 671*
Tal, J., 690[29], 718[29, 276], 751[29, 277], *757, 765*
Talbot, P., 70[64], 71, 72[64], *107*, 200[168], *276*
Talens, L., 798[128], 824[128],. *842*
Tamagawa, S., 363[94], *377*
Tamanoi, F., 628[373], *664*
Tamm, I., 75[74, 75], 79[98, 99], 80[74, 75], 92[74, 75], 102, *107, 108, 110,* 114[9], 115 [26], 118[9], 123[26], *160,* 386[49], 387[49, 58, 59], 390 [88, 94], 392[49, 58], 398 [92], 402[58], 403[182], 406 [182, 203], 411[182, 203, 226, 227, 228, 229], 412[58], *426, 427, 430, 431*
Tan, K. B., 148[158], 149, 151, *164,* 598[92], 614[253], 636 [474], *655, 660, 667,* 872 [180], *887*
Tan, Y. H., 46[43], *58*
Tanada, Y., 422[256], *432*
Tanaka, S., 865[99], *885*
Tanaka, Y., 174[24], *272*
Tang, J., 89[127], 90[127], *109*
Tannock, G. A., 88[122], 89 [122], 93[122], *109*
Tantravahi, R., 73[66], *107*
Tanzer, J., 593[34, 35], *653*
Tarragó, A., 69[53], *106*
Tarragó-Litvak, L., 69[53], *106*
Tattersall, P. J., 556[138], 557, 560, 567[185], 568[185], 570 [185, 189], 571[185], 574 [185, 220], *583, 585, 586*
Taube, S. E., 43[16], 45[16], *57*

Taylor, B. A., 505[87], 508[87], *541*
Taylor, G., 677[288], *766*
Taylor, J., 50[85], *59*
Taylor, J. A., 326[198], 332 [198], *347*
Taylor, J. M., 298[77], 324[188], 329[188], *343, 346*, 459[41], 461[42], *540*
Taylor, M. M., 270[309], *280*
Taylor, M. W., 75[86], *107*
Taylor, W. P., 182[90], 184 [90], 185[90], 215[90], *274*
Tegtmeier, G., 190[124], *275*
Tegtmeyer, D., 632[412], *665*
Tegtmeyer, P., 613[413, 414, 488], 630[389], 631[389, 403], 632 [304, 403, 413, 414, 431], 636 [413, 414, 473], 637[413, 414, 431, 488], 638[304, 389, 403, 414, 431, 503], 639[389, 503], *662, 665, 666, 667, 668*, 705 [279], 731[278, 279], 746 [278, 279], *765*, 823[246], *846*
Teich, N., 446[20], 520[20], *539*
Teich, N. M., 437[11], 521[11], *539*
Temin, H., 436, 437, 439, 477[8], 478[8], *538*
Temin, H. M., 436[2], 454[29], 483[2], *538, 539*
Teng, M.-H., 68[41], 94[41], *106*
Teninges, D., 185[97], *274*
Tennant, R. W., 567[183, 184, 188], 568[183, 188], 570[184], 571 [184, 188], 572[188], 574[188], 576[188], *585*
Tenser, R. B., 778[23], 779[23], 822[23], 824[23], *838*
Teramoto, Y. A., 456[32], *539*
Teresky, A. J., 610[217], *659*
Ter Meulen, V., 351[160], *379*
Terni, M., 771[5], 793[5], 837[316], *838, 848*
Tesh, R. B., 176[213], *273, 277*
Tessman, I., 15[47], *37*
Tevethia, M. J., 613[489], 637 [489], *668*
Tevethia, S. S., 613[489], 637 [489], 648[565], *668, 670*, 826[265], 827[265], 833[265], *846*, 88[240], *889*
Tewari, K. K., 780[66], *840*
Thach, R. E., 76[88], 77[88], 78 [88, 92], 79[92], 90[130], 94 [136], *108, 109*
Thach, S., 94, *109*
Thacore, H., 373[241], *381*
Theiler, S., 170[10], *271*
Thelander, L., [264], *765*
Thiel, J. T., 546[55], 547[55], *581*
Thomas, C. A., 678[92], 681[92], 682[92], *759*
Thomas, C. A., Jr., 549[70], *581*, 681[282], *765*
Thomas, D., 714[280], 716[280], 733[91], 738[91], *759, 765*
Thomas, D. C., 714[281], 716[219], *763, 765*
Thomas, G. F., 576[225], *586*
Thomas, J. O., 599[102], *655*
Thomas, M., 593[34, 35], *653*, 893[44], *898*
Thomas, M. T., 54[127], *60*
Thomas, V., 878[218], *888*
Thompson, R. L., 881[261, 262], *889*
Thomsen, M., 351[188], 362[184], *380*
Thomssen, R., 880[241], *889*
Thoren, M., 629[379], *664*
Thoren, M. M., 626[351], 627 [351], *663*
Thorne, H. V., 597[76], 603[140], 605[188, 189], 610[188, 189], 625[347], *655, 657, 658, 663*
Thornhill, T. S., 546[40], 551[40], *580*
Thouless, M. E., 812[186, 187],

[Thouless, M. E.]
813, *844*
Tibbetts, C., 688[283, 284], 707 [210], 716[210], 717[283], 718[210], 720[210], 728[367], 732[207], 736[367], 749[210, 284], 750[210, 283, 367], 751 [210], *763, 765, 768*
Tierkel, E. S., 175[46], *272*
Tierney, E. L., [230], *348*
Tiffany, J. M., 290[43], 292, *342*, 360[242], 361[242], *381*
Tignor, G. H., 181, 220, *274, 277*
Tikchonenko, T. I., 19[107], *39*
Tikhonenko, A. S., 22[108], *39*
Tillotson, J. R., 385[22], 390[89, 90], *425, 427*
Timbury, M. C., 793[98], *841*, 866 [112], 881[112], *885*
Tinsley, T. W., 540[7], 550[7], *579*
Tischer, E. G., 609[210], *659*
Titmuss, D. H. J., 852[18], 854 [18], *883*
Tobita, K., 298[83], 299[83], 300 [83], 305[221], 336[220, 221], 337[83, 221], *343, 347*, 362 [184], *380*
Tockstein, G., 721[285], *765*
Todaro, G., 436, 439[16], 501[5], *538, 539*, 642[528], *669*
Todaro, G. J., 437[7], 446[21], 463[47, 48], 476[61], 486[61], 487[61], 488[48, 70], 509[88], 511, 514, 516[48], 535[66, 109], 536[90], 537[66, 109, 117], *538, 539, 540, 541, 542*, 641 [518], [587], 651[605], *669, 671, 672*, 744[15], *757*
Todd, P., 795[118], 797[118], *841*
Tokuno, S., 636[485], *668*
Tolun, A., 690[286], 728[286], [377, 379], *765, 768*
Tomassian, N., 798[130], *842*
Tomassini, N., 614[254], *660*
Tomkins, J. K. N., 866[104, 108],
[Tomkins, J. K. N.]
885
Tonegawa, S., 616[274], 617[274, 281], 623[274], *661*
Toni, R., 457[34], 471[34], 480 [34], *539*
Toolan, H. W., 540[3, 6], 541 [3, 14, 15, 21], 543[3, 6, 21, 103, 171], 544[213], 546, 549[3, 63], 551[3], 552[3, 15], 553[3, 6, 14, 115], 554[3, 6], 555[54, 63, 130], 556[130], 557[130], 559 [63], 561[63], 566[115, 169, 170], 567[14, 21], 568[3], 569 [115, 205], 570[169, 170], 571 [6, 213], 576[226], 578[170], *579, 580, 581, 582, 583, 584, 585, 586*
Tooze, J., 445[19], 479, 497[19], 500[19], 511[19], 530[19], 538 [19], *539*, 599[96], 602[96], 614[96], *655*, 676[287], *766*
Torikai, K., 549[62], *581*
Torpier, G., 570[208], *586*
Torten, J., 608[203], *659*
Tournier, P., 2[20, 79], 18[113], 19[113], *36, 38, 39*, 389[78, 80], *427*, 636[478], *667*
Tovell, D. R., 862[66], *884*
Towle, H. C., 814[205], *844*
Tozawa, H., 361[244], 373[239], *381*
Traboni, C., 613[246], 615[246], *660*
Trafford, R., 186[110], *274*
Traub, A., 89[126], 96[126], *109*
Traynor, B. L., 397[145], *429*
Trent, D. W., 114[18, 21, 23], 117[45, 105], 118[23], 119[18, 21], 123 [18, 21], 124[18], 126[23], 131 [23, 105], 132[45], 139, *160, 161, 163*
Trentin, J. J., 545[73], 550[73], 569 [202], *581, 585*, 677[288], *766*
Tress, E., 452[27], 473[27], [62],

[Tress, E.]
539, 540
Trilling, D. M., 610[218], *659*
Trim, A. R., 6[56], *37*
Tripier, F., 869[155], *887*
Trkula, D., 636[485], *668*, 812 [179, 180], *843*, 872[176], *887*
Tromans, W. J., 15[109], *39*, 547[59], *581*
Tronick, S. R., 474[59], *540*
Trousdale, M. D., 691[256], *765*
Trukhmanova, L. B., 45[27], *57*
Tsai, K.-H., 202[178], *276*
Tsilinsky, Ya. Ya., 154[176], *165*
Tsubahara, H., 384[6], 385[6, 14], *424, 425*
Tsuchida, N., 459[39], 491[72], 508[72], 518[92], 519[72], *540, 541, 542*
Tsuei, D., 733[289], 735[289], *766*
Tsukamoto, K., 719[365], *768*
Tsuruhara, T., 877[208], *888*
Tsuruo, T., 729[338], *768*
Tuan, D. Y., 599[94], *655*
Tumova, B., 283[16], 288[30], 289[30], 307[30], 336, *341, 342*
Turner, G. S., 879[232], *889*
Turner, H. C., 437[9], 443[17], 502[17], 508[9], 509[9], *538, 539,* 631[390], 648[390], *665,* 677[116], 731[116], 744[114], *760*
Turner, T. B., 833[292], *847*
Tyrell, D., 42[3], *56*
Tyrrell, D. A. J., 47[47], *58*, 312 [149], *345*
Tytell, A. A., 43[4], 45[4, 39], *56, 57*

Uchida, S., 605[186, 190], 610 [186, 190, 223], 631[223],
[Uchida, S.]
658, 659
Ueda, M., 298[83], 299[83], 300 [83], 305[221], 336[215, 221, 222], 337[83, 221, 222], *343, 347,* 362[184], *380*
Ueda, Y., 879[234], *889*
Underdahl, N. R., 287[19, 20, 21], *341, 342*
Underwood, B., 351[245], *381*
Underwood, B. O., 195[210], 221 [210], *277*
Underwood, P. A., 779[38], *839*
Unger, J. T., 265[283], *279*
Unwin, P. N. T., 4[10, 111], *39*
Uomala, P., 135[120], 137[120], *163*
Upholt, W. B., 602[132], 603[139], *656, 657*
Urbano, C., 190[127], *275*
Urishibara, T., 863[77], *884*
Uryvayev, L., 200[169], 238[169], *276*
Usategui-Gomez, M., 549[63], 555 [63], 559[63], 561[63], *581*
Ustacelebi, S., 701[320], 705[318], 708[318], 709[290], [322], *766, 767*
Utermann, G., 122[69, 70], *162*

Vago, C., 540[2], 541[39], *579, 580,* 854[28], 877[210], *883, 888*
Vahlne, A. G., 792[95], 793[95], *841*
Vaidya, A. B., 535[111], *542*
Valentine, R. C., 293[58], 295 [58], 298[58], *343*, 546[51], *581*, 691[291], 692[291], 698 [291], *766*
VanAlstyne, D., 404[197], 407 [197], *430*
Vanaman, T. C., 395[141], 397 [141], 407[141], *429*, 476[64, 65], *541*

Van Bruggen, E. F. J., 681[293], *766*
van der Eb, A. J., 638[497], 642 [497], *668*, 676[78], 681[293], 686[78], 688[78], 690, 728 [292], 732[78], 753[78], *759, 766*
Van Der Maaten, M. J., 457[34], 471[34], 480[34], *539*
Vanderpool, A. E., 679[55], 680 [55], *758*
Van Der Vliet, P. C., 704[149], 705[149, 295, 297], 707 [149, 294, 295, 297], 715 [294], 726[268], 727[294, 295, 297], 728[267, 268, 296], 738[295], [378], *761, 765, 766, 768*
Van Der Voorde, A., 604[164], 605[185], *657, 658*
Van Der Westhuizen, B., 174[23], 184[23], *271*
Vandvik, B., 351[180, 246], *380, 381*
Van Herreweghe, J., 605[185], *658*
Van Heuverswyn, H., 605[185], *658*
Van Hoosier, G. L., 780[78], 787 [78], *840*
Van Mitchell, M., 386[53], 387 [53], 389[53], *426*
Van Praag Koch, H., 175[49], 181 [49], 220[49], *272*
Van Regenmortel, M. H. V., 15[40], 16[40], *36*
Vantis, J. T., 826, *846*
Van Vorstenbosch, C. J. A. H. V., 187[112], *274*
Van Wielink, P. S., 685[359], 690 [359], *768*, 894[26], *898*
Varich, N. L., 369[126], 371[124], *378*, 868[150], 869[150], *886*
Varma, A., 6[112], *39*
Varmus, H. E., 496[77], 517[82], 519[93], *541, 542*, 733[298],
[Varmus, H. E.] *766*
Varricchio, F., 707[369], *768*
Vasquez, C., 18[113], 19[113], *39*, 389[78, 80], 392[114], *427, 428*, 547, 555[58], 557 [58], *581*
Vassileva, V., 353[226], *381*
Vaughan, M. H., 68[35], 94[35], *106*
Vaughn, R. K., 818[231], *845*
Velicer, L., 716[299, 300], *766*
Vennström, B., 707[210], 716 [210], 718[210], 720[210, 373], 749[210], 750[210], 751[210], *763, 768*
Veres-Molnár, Z., 718[21], 736 [341], 751[341], *757, 768*, 891[5], *897*
Verma, I. M., 531, 532, *542*
Vernon, L., 437[13], 463[13], *539*
Vernon, S. K., 194[164], *276*, 361[177], 363[176], *379*, 559[156], 562, *584*
Verwoerd, D. W., 422[251, 255], *432*
Vilcek, J., 46[42, 44], 49[81], 50[91], *58, 59*
Villa-Komaroff, L., 67[19], 80 [19], 85[19], 100[155], 101 [155], *105, 110*, 139[129], 143[129], *163*, 404[189], 407 [189], *430*
Villarejos, V. M., 546[41], *580*
Villarreal, L. P., 202[185], 236 [185], 248[246], *277, 278*, 374[274], *382*
Vilner, L. M., 45[27], *57*
Vinograd, J., 592[25], 602[25, 126, 127, 130, 131, 132, 135], 603[126, 137, 138, 139], 608 [195], 610[195], 611[195], *653, 656, 657, 658*
Vlak, J. M., 728[267], 729[302], [301], *765, 766*

Vogel, A., 651[609, 616], *672*
Vogel, J. E., 880[238], *889*
Vogell, W., 779[31], *838*
Vogt, M., 610[121], 603[136], 625[322, 325], 629[376], 642[530], 643[531], 650 [591, 592], *656, 662, 663, 664, 669, 671,* 676[42], *758*
Vogt, P., 459[37], 460[37], 502, 504[37], *540*
Vogt, P. K., 356[62], *376*, 459 [38], 502, 503[38, 80], 504 [80], *540, 541,* 733[298], *766*
Vogt, V., 604[166], 605[170], *657, 658*
Vogt, V. M., 497[78], *541*
Volkaert, G., 605[185], *658*
Von Bonsdorff, C.-H., 118[53], 123[53], 143[53], *161*
Vonka, V., 591[4], *652*, 780[69, 73], 782[73], 819[69], *840*
Von Magnus, P., 306[119], 307 [119], 334, *344*
Von Mandelkow, E., 4[7], *35*
Von Muenchhausen, W., 48[69], 49[69], *58*
Vorkunova, G. K., 359[247], 365 [26, 27], *375, 381*
Vovk, T. S., 368[29], *375*

Wachsman, J. T., 401[165], *429*
Wachtel, E. J., 6[81, 82, 114], *38, 39*
Wachter, R. F., 120[160], *161*
Wadsworth, S. C., 772[62], 780 [61, 62, 65, 74, 76], 782[61, 62, 74, 76], 783, 784, 785 [61, 65, 76], 787, 790[65], 800[76], 801[76], 803[76], 806[76], 807[76], 808[76], 819[61, 76], *839, 840*
Wagley, P. F., 363[164], *379*
Wagner, E. K., 772[25], 774[25], 780, 800[140], 801[140, 143], 829[140, 277], 833[277], *838, 840, 842, 846*
Wagner, K., 351[248], *381*
Wagner, R. R., 43[15], 49[72], *57, 59,* 169, 189[115], 190[121], 191[2], 193[132], 194[134], 195[139], 196[134, 139], 197 [146], 198[2, 150, 151, 152], 199[139, 150, 151, 157, 159], 200, 201[139, 173, 175], 202 [189], 204[157], 214[115], 237[189], 238[173, 232, 233], 248[121], 257, 259], 258[189, 233, 259, 261], 260, 261, 267 [291], 268[150], 269[2, 115, 121, 173], *271, 275, 276, 277, 278, 279, 280,* 352[249, 251], 360[65, 115, 250], *376, 378, 381, 382*
Wainberg, M. A., 363[252], *382*
Waite, M. R. F., 118[52], 130, 135 [140], 143[52, 140], 146[143, 144], 152[94, 140], 153, *161, 162, 164*
Waldeck, W., 610[214], *659*
Walen, K. H., 877[202], *888*
Walker, D. H., Jr., 20[115], *39*
Walker, D. L., 595[60], 596[64, 70, 72], *654, 655*, 880[257, 258], *889*
Walker, G., 897[14], *897*
Walker, I. O., 562[154], *584*
Wall, R., 647[552], *670*, 719 [211], 733[304], 735[304], *763, 766*
Wall, R. L., 720[303], *766*
Wallace, J., 867[138], 877[138], *886*
Wallace, R. D., 714[305], *766*
Walling, M. J., 779[33], *839*
Wallis, C., 387[71, 74], 391[71, 74, 101], *426, 427*, 814[199], *844*

Walter, G., 199[165], *276*, 597 [77], 598, 616[274], 617 [274, 281], 623[274], 630 [77], 635[77], *655, 661,* [162], [306], 741, 742, 747 [307], 748[307], *762, 766,* 987[13], *897*
Walters, M. N.-I., 385[20, 33], 386 [37], *425*
Waner, J. L., 772[14], 805[14], *838*
Wang, L.-H., 503[80], *541*
Ward, D. C., 556[138], 557[138], 560[138], *583*
Ward, F. E., 180[83], *274*
Ward, R. L., 394[118], 400[160], 405[160], 406[118, 160, 204, 205], *428, 429, 431,* 808[160], *843*
Ward, T. G., 676[229], *764*
Warden, D., 597[76], 605[189], 610[189], *655, 658*
Warfield, M., 351[1], *374*
Warfield, M. S., 541[18], 543[18], *580*
Waring, M. J., 602[134], *656*
Warnaar, S. O., 616[269], 617 [269, 288], 638[497], 642 [497], *661, 668,* 676[78], 686 [78], 688[78], 732[78], 753 [78], *759*
Warner, H., 629[380], *664*
Warner, R. C., 780[66], *840*
Warren, G. S., 543[79], 551[79], 552[79, 555[126], *581, 583*
Warrington, R. C., 406[202], *430*
Wasserman, F. E., 385[17], *425*
Watanabe, K., 353[253], *382*, 394[124], *428*
Watanabe, M., 361[244], *381*
Watanabe, S., 605[186, 190], 610 [186, 190, 223], 631[223], *658, 659*
Watanabe, Y., 386[48], 389[81], 392[110, 111], 394[48, 126], 403[126, 185], 404[126, 185,
[Watanabe, Y.]
194], 409[216], 411[213], 412[185], 417[48], 418[194], 421[194], *426, 427, 428, 430, 431,* 647, *670*
Watanabi, Y., 865[99], *885*
Waterfield, M. D., 293[235], 298 [235], 326, *348*
Waters, D. J., 351[254], 358[32], 359[255], *375, 382*
Waterson, A. P., 7[3], 16[55, 116], *35, 37, 39,* 169, *271*, 289[34, 35], 294[67], *342, 343,* 351 [256], 358[95], *377, 382,* 690 [109], *760*
Watkins, J., 648[571], 649[571], 650[571], *670*
Watkins, J. F., 648[575, 578], 649, *671*, 835, *847*
Watson, C. J., 474[60], *540*
Watson, D., 793, *841*
Watson, D. H., 306[120], 332 [120], *344*, 779[35], 780[80], 787[80], 797[125], 802[153], 811[171], 812[181, 185], 813 [181, 197], 815[214], *839, 840, 842, 843, 844*
Watson, J. D., 7, 16[23], *36*, 592 [21], *653*, 676[309], 730[308], *766*
Watson, R., 602[126, 130], 603 [126], *656*
Way, H. J., 866[115], 880[115], 88[115], *885*
Weber, J., 712[10, 311], 713[310, 311], 725[310], 733[304], 735 [304], *757, 766,* 891[24, 45], *898*
Weber, K., 597[87], 598[87], 632 [425], 638[501], 639[425, 501], 648[425], *655, 666, 668,* 705 [190], 707[361], 717[361], 727[361], 731][90], *762, 768*
Weber, M., 634[439], 641[521], *666, 669*
Webster, R. G., 282[3, 4], 283

[Webster, R. G.]
[16], 288[4, 25, 26, 27, 28, 29, 31, 32], 289[4, 31, 32], 290, 298[81], 299[86, 90], 307[25, 26, 27, 28, 29, 31, 32], 308[81], 310[86], 323, 336[4, 32], 339, *341, 342, 343, 344, 346, 347*
Wecker, E., 117[43], *161*
Wei, C.-M., 67[30], *106*, 203[193], *277*, 719[312], *766*, 863[75, 76], *884*
Weil, R., 47[56, 59], 48[59], *58*, 597[82], 599[98, 99], 601[123], 602[99, 123], 603[137], 610[219], 612[231, 244], 614[258], 616[258], 617[258, 280], 622[280], 623[158, 311], 625[98, 99, 329, 331, 340], 630[82], 631[340], 635[82], 641[340], 652[258], *655, 656, 659, 660, 661, 662, 663*
Weinberg, R. A., 367[257], *382*, 616[269], 617[269, 286, 290], *661*
Weinberg, R. L., 635[466], *667*
Weiner, A., 726[239], *764*
Weiner, L., 595[57], *654*
Weingartner, B., 690[240], 728[240], [379], *764, 768*
Weinmann, R., 79[97], *108*, 718[313], [380], *767, 768*
Weintraub, S., 876[193], 879[237], *888, 889*
Weiser, B., 716[41], *758*
Weiser, R. S., 592[15], *653*
Weisman, S. M., 718[221], *763*
Weiss, A. F., 595[59], *654*
Weiss, B., 153[171], 154[171], *165*
Weiss, R. A., 267[298], 268[298], *280*, 446[22], 501, 504, 520,
[Weiss, R. A.]
539, 541
Weiss, S. R., 367[258], 370[24, 258], *375, 382*
Weissbach, A., 629, *664*, 729[314], *766*, 814, 815[211], *844*, 872[178, 179], 877[203], *887, 888*
Weissman, S., 604[159, 163], 605[180], 607[159, 163], 619[180], 620[180], *657, 658*, 718[188, 189], *762*
Weissman, S. M., 605[181, 182, 183, 184], 612[241], 624[184], 634[450], *658, 660, 667*, 747[38], *758*
Weissmann, C., 92[133], *109*, 126[89], *162*
Wellemans, G., 541[35], 545[35, 94, 239], 550[35], 551[35], 552[35, 94], 555[94], 569[35], *580, 582, 587*
Wells, H. J., 201[176], *276*
Wende, R. D., 184[94], *274*
Wendel, I., 602[128], 603[128], 630[128], *656*
Wengler, G., 117[37], 131[37], 135[122], 139, *161, 163*
Wenner, H. A., 385[32], 386[32], *425*
Wensink, P. C., 678[92], 681[92], 682[92], *759*
Werchau, H., 625[334], *663*
Werenne, J. J., 411[224], 412[224], *431*
Werner, J. H., 676, *760*
Wertz, G. W., 239[243], *278*
West, B., 390[86], *427*
West, R., 174[29], 183[29], *272*
Westaway, E. G., 114[13], 117[13], 119[13], 139, *160, 163*
Westphal, H., 600[115], 608[205], 617[115, 295], 624[115, 295, 315, 319], 625[334], 631[115], 634[454], 642, 644[540], 645[540], *656, 659, 662, 663, 667,*

[Westphal, H.]
669, 739[48], 752[47, 316], 756[315], *758, 766,* 893[31, 32, 33], *898*
Westphal, M., 697[17], 698[17], 699[17], 716[41], *757, 758,* 894[4], *897*
Westwood, J. C. N., 852[18], 854 [18, 34], 879[228], 880[242], 881[270], *883, 888, 889, 890*
Wetmur, J., 644[534], *669*
Wettstein, F. O., 320[244], 321 [244], *348*
Wever, G. H., 649[580], *671*
Wewerka, Y. L., 731[152], 749 [152], *761*
Wewerka-Lutz, J., 619[300], 620 [300], *662*
Wewerka-Lutz, Y., 724[54], 747 [54], 748[54], *758*
Wheeler, C. E., 831[291], *847*
Wheeler, T., 117[39], 139[39], *161*, 631[410], 632[410], *665*
Wheelock, E. F., 45[25], 54[126], *57, 60,* 365[259], *382*
Whipple, W. J., 174[44], *272*
Whitcomb, R. F., 387[63], *426*
White, C. K., 423[267], *432*
White, C. N., 374[273], *382*
White, D. O., 64[1], 74[1], 75 [1], 80[1], 93[1], *105*, 282 [7], 292[47], 298[47, 77, 80], 300[47], 312[146], 324[188], 326[80, 198], 329[188], 332 [198], *341, 342, 343, 345, 346, 347,* 351[69], 358[85], 361[260], *376, 377, 382,* 633 [438], 636[471], *666, 667,* 691[165, 166], 698[166], *762*
White, G. B. B., 423[262], *432*
White, H. B., Jr., 865[102, 103], *885*
White, M., 537[116], *542*, 600 [107], *656*
White, R. L., 893[44], *898*
White, R. M., 424[281], *433*
White, W. L., 555[122, 131],
556[131], 557[131], 559[122], 561[122], 562[122], 564[122], 567[193], 568[193], 570[193], 572[193], 574[193], 576[193], *583, 585*
Whitfield, S. G., 117[32], 146[32], 154[179], *160, 165,* 174[32], 182[90], 184[32, 90], 185[90], 215[90], *272, 274*
Whitford, W., 424[274], *433*
Whitney, E., 174[31], 184[31], *272*
Wicker, R., 631[408], 636[478], *665, 667*
Wickus, G. G., 123[75], 143[75], *162*
Widdows, D., 546[42], *580*
Wiebe, L., 404[191], *430*
Wiener, F. P., 361[177], *379*
Wiese, W..H., 634[451], *667*, 746 [154], *761*
Wiktor, T. J., 7[61], *37*, 178[66], 188[114], 194[138], 195[138], 196[138], 199[138], 201[138], 206[138], 218[138], 256[114], *273, 275, 276*
Wilbanks, G. D., 827[269], 833, *846*
Wilbert, S. M., 601[118], 650[603], *656, 671*
Wilcox, K., 617[294], *662*
Wilcox, W. C., 253[261, 262], *382,* 698[317], *766*
Wild, T. F., 248[253], *279*
Wilde, A., 618[362], 627[362], *664*
Wildy, P., 2[20], 15[117], 16[55, 117], *36, 37, 39,* 289[34], *342,* 358[95], *377*, 690[109], *760*, 778, 779[13], 780[80], 787[80], 797[125], 811[171], 812[187], 815[214], 826, *838, 840, 842, 843, 844, 846*

Wilkie, N. M., 705[318], 708[318], *766*, 780[72], 782[72], 819 [72], *840*
Wilkins, M. H. F., 387[61], 392[61, 108], *426, 427*
Willems, M., 75[80], 80[80], *107*
Williams, D., 489[71], *541*
Williams, J., 644[539], 649[539], *669*, 690[236], 701[101, 238], 707[98, 170, 238, 321], 709 [238], 713[98, 238], 732[236], 749[98], 753[236], 754[98, 238, 321], 755[98], *760, 762, 764, 767*
Williams, J. E., [322], *767*
Williams, J. F., 701[77, 320, 323, 334], 703[323], 705[232, 234, 318, 323], 707, 708[232, 234, 318, 323], 709[232, 234, 290, 323, 324], 712[323], 721[340], 726[323], 732[319], 749[319], 754[323], *759, 764, 766, 767, 768*, 866[112], 881[112], *855*
Williams, M. C., 174[29], 175[55], 183[29], *272, 273,* 385[12], 400[12], *424*
Williams, M. G., 15[2], *35*, 591 [9], *653*
Williams, R. C., 16[119], 20[118], *39*, 592[23], *653*
Williamson, J. D., 873[186, 187], *887*
Wills, C. G., 541, 543[10], *579*
Wills, E. J., 709[324], *767*
Wilsnak, R., 473[57], *540*
Wilson, G. L., 463[48], 488[48], 516[48], *540*
Wilson, R. G., 248[247], *278*
Wimmer, E., 67[13, 14, 17], 68 [13, 17, 34, 40, 41, 42], 94 [34, 40, 41, 42], 96[109], 100 [156], *105, 106, 109, 110*
Winer, B., 866[117], 881[117, 264, 265, 267, 272], *885, 889, 890*
Winget, C. A., 48[62], *58*
Winkler, W. G., 180[79], *273*
Winn, P., 390[93], 398[93], 401 [93], *427*
Winn, W. C., 179[71, 72], 180[71], 181[71, 72], *273*
Winnacker, E. L., 628[368], 629 [368], *664*, 690[240], 728[240], 729[325], [379], *764, 767, 768*
Winocour, E., 599[97], 605[191], 608[203, 204], 609[191, 206, 208, 209, 211], 612[238, 239], 616[269], 617[269, 276], 620 [276], 623[314], 624[320], 625[97, 238, 239], 635[467, 468], 643, 645[276], *655, 658, 659, 660, 661, 662, 667*
Winston, S. H., 373[263], *382*
Winterhoff, U., 697[17], 698[17], 699[17], 716[41], *757, 758,* 894[4], *897*
Wiseman, R. L., 6[28, 81, 82, 114], *36, 38, 39*
Witter, R. L., 779[28, 39], 827 [267], *838, 839, 846*
Witz, J., 4[76], *37*
Wold, W. S. M., 677[355], 688[326, 327], 715, 717, 719[328], 720 [326, 327], 723[102, 326], 725 [351], 727[124], 738[351], 749[326, 327], 750[327], [375], *760, 767, 768,* 896[23], *898*
Wolf, H., 772[52], 780[52], *839*
Wolf, K., 185[103], 186[103, 110], 187[103], *274*, 770[3], 824[3], *838*
Wolff, M. H., 802[318], 809[318], *848*
Wolford, R. G., 478[67], *541*
Wolfson, J., 564[159], *584*, 683 [329], *767*
Wöllert, W., 326[194], 328[194], 332[194], *346*
Wollman, S., 651[616], *672*
Wong, D., 351[1], *374*
Wong, P. K. Y., 259[269], 260

[Wong, P. K. Y.]
[269], 265[269, 282], *279*
Wood, H. A., 422[250], 423, *432*
Wood, M. L., 55[149], *61*
Woodall, J. P. 385[12], 400[12], *424*
Woode, G. N., 423[262, 263], *432*
Woodie, J. D., 180[80], *273*
Woodman, D. R., 552[83], 553 [83], *582*
Woodroofe, G. M., 851[7, 9], *883*
Woods, R. D., 6[44, 51, 88, 112], *36, 37, 38, 39*
Woodson, B., 851, 862[62], 867 [146], 868[146], 873[62], 881[271], *882, 884, 886, 890*
Work, T. H., 113[3], *159*
Worth, C. B., 174[21], 182[21], *271*
Warthall, A. E., 552[97], 553 [97], *582*
Wright, P., 351[128], *378*
Wright, P. J., 75[82, 83], *107*
Wright, S. E., 506[83], *541*
Wrigley, N. G., 15[120], 16[120], *39*, 298[82], *343*, 691[198], 692, 697[198], *763*
Wu, M., [381], *768*
Wungkobkiat, S., 154[187], 155 [187], *165*
Wunner, W. H., 194[136], 195 [136], 196[136], 239[234], 256[256], 267[290], *275, 278, 279, 280*
Wyatt, R. G., 385[29], *425*, 546 [40], 551[40], *580*
Wyke, J., 651[615], *672*
Wyke, J. A., 651[614], *672*

Yabe, Y., 677[288], *766*
Yakobson, E., 876[196], *888*
Yamaguchi, K., 726[249], 729 [252], *764*
Yamaguchi, N., 632[419, 427],
[Yamaguchi, N.]
636[427], 637[419], *666*
Yamamoto, H., 631[409], *665*, 700, *767*
Yamanouchi, K., 352[230], *381*
Yamashita, T., 726[331, 332, 333], 727[331, 332, 333], 728[331, 332], 729[332, 338], *767, 768*
Yamazaki, H., 677[63], 678[63], 732[63], *758*
Yamazaki, S., 49[72], *59*, 257 [259], 258[259], *279*, 360 [250], *382*
Yanagawa, R., 541[33], 545[33], 562[33], *580*
Yanagida, M., 22[121], *39*
Yang, R., 604[164], 605[185], *657, 658*
Yang, Y. J., 177[61], *273*
Yaniv, M., 605[174], *658*
Yasuda, I., 775[9], *838*
Yata, Y., 835[311], *847*
Yates, V. J., 545[73], 550[73], 569[202], *581, 585*
Yin, F. H., 74[72], *107*, 131 [102], 151, *163, 164*
Yogo, Y., 68[34, 40, 41, 42], 94 [34, 40, 41, 42], 96[145], *106, 109*
Yohn, D. S., 877[207], *888*
Yoshida, T., 361[112, 154], *378, 379*
Yoshida, Y., 650[595], *671*
Yoshiike, K., 605[179, 190], 610 [190, 220, 221, 223], 631 [179, 223], *658, 659*
Yoshimori, R. N., 617[293], *662*
Yot, P., 69[50, 51], *106*
Young, C. S., 701[323], 703[323], 705[323], 707[323], 708[323], 709[323], 712[323], 726[323], 754[323], *767*
Young, C. S. H., 676[352], 699 [352], 701[334, 352], *767, 768*

Young, G. A., 287[21], *342*
Young, H. A., 536[114], 537 [114], *542*
Young, J. C., 534[107], *542*
Young, L. J. T., 456[32], *539*
Young, M. R., 867[130], *886*
Young, R. J., 304[111], *344*
Younger, J. S., 386[44], *426*, 879 [123], *886*
Younghusband, H. B., 681[225, 335, 336], 683[225], 728[12], 730[225], *757, 764, 767,* 894 [37], *898*
Youngner, J. S., 43[16], 45[16, 18], 48[65], 55[140], *57, 58, 61,* 373[241], *381*
Yow, M. D., 552[93], *582*
Yu, Y.-H., 201[174], 238[174], *276*, 360[66], *376*
Yunker, C. E., 154, 155[191], *165*
Yusutake, W. T., 188[113], *274*

Zabielski, J., 625[340], 631[340], 641[340], *663*
Zaides, V. M., 359[265], 368 [264], *382*
Zain, B. S., 605[180, 181, 182, 183, 184], 619[180], 620 [180], 624[184], *658*
Zain, S., 604[159, 163], 607[159, 163], *657*, 747[38], *758*
Zajac, B., 798[130], *842*
Zakstelskaya, L. Ya., 283[16], 298[84], *341, 343*
Zandberg, J., [378], *768*
Zang, K. D., 595[59], *654*
Zarling, J. M., 880[240], *889*
Zaslavsky, V., 876[196], *888*
Závada, J., 267[294, 295, 296], 268[295, 296, 299], *280*
Závadova, Z., 268[299], *280*
Zbitnew, A., 351[59], *376*
Zdanov, V. M., 283[16], *341*
Zebovitz, E., 117[106], 126 [103], 131[103, 106], 147 [159], *163, 164*
Zee, Y. C., 798[128], 824[128], *842*
Zeigler, R. E., 546[241], *587*
Zeitlenok, N. A., 45[27], *57*
Zeleznick, L. D., 43[7], 54[7], *56*
Zemla, J., 620[302], *662*, 833 [295], *847*
Zhdanov, V. M., 365[27], 373 [266], *375, 382,* 438[15], 457[15], 480[15], *539*
Zhirnov, O. P., 359[265], 368 [265], *382*
Zilberstein, A., 52[118], *60*
Zimmer, S., 690[29], 718[29], 720[30], 723[30], 736[30], 748[30], 751[29, 30, 277], *757, 765*
Zimmer, S. G., [382], *768*
Ziola, T. R., 69[58], *107*
Zittelli, A., 825[252], *846*
Zubay, G., 15[62], 16[62], *37*
Zur Hausen, H., 789[92, 93], *841*
Zu Rhein, G. M., 595[60], 596 [64, 70], *654, 655*
Zwartovw, H. T., 852[18], 854 [18, 34], 855[37], 865[37], 879[228], 880[242], *883, 888, 889*
Zweerink, H. J., 386[46], 387 [46, 55], 389[55], 390[46, 55], 391[46], 392[113], 394 [46], 395[46, 140], 397[55, 146], 398[46, 113, 140], 402 [180], 404[180], 405[180], 406[113], 407[113, 140], 409 [218], 410[221], 411[113, 218], 420, 423[267, 268, 269], *426, 428, 429, 430, 431, 432*
Zwillenberg, H. H. L., 185[105], *274*
Zwillenberg, L. O., 174[39], 185 [105], *272, 274*

Subject Index

Adenoviruses, 673-768
adenovirus replicative cycle, 714-729
biogenesis of adenovirus mRNA, 718
biosynthesis of adenovirus proteins, 721
early viral proteins, 721
function of early proteins, 721
late adenovirus proteins, 724
cell transformation by adenoviruses, 730
adenovirus-induced proteins, 738
nature of mRNA in transformed cells, 733
T antigens, 738
transcription of viral DNA, 733
transplantation antigen, 738
viral DNA sequences in transformed cells, 732
amount, 732
in situ hybridization, 733
reassociation kinetics, 732
state of viral DNA integration, 733
DNA genome, 681
biological activity, 690
capsomeres, 691
conformation, 681
core, 698
DNA-DNA homology, 685
DNA regions encoding mRNA, 686
[Adenoviruses]
H strand, 686
infectivity, 690
interrelationships among human adenoviruses, 685
intramolecular structure, 681
L strand, 686
localization of transforming genes, 686
map, 736
mapping size, 688
molecular weight, 681
DNA replication, 726
current model, 729
enzymes involved in adenovirus DNA replication, 728
model for DNA replication in mammalian cells, 726
viral-coded proteins involved in replicative forms, 726
viral DNA intermediates, 727
early events, 714
fiber, 698
genetics, 699
Ad 2 mutants, 712
Ad 5 *ts* and *hr* mutants, 700
complementation groups, 702
early mutants, 705
late mutants, 708
recombination map, 706
Ad 12 *ts* mutants, 710

Volume 2 comprises pages 539-900.

[Adenoviruses]
Ad 31 *ts* mutants, 711
CELO *ts* mutants, 712
cyt mutants, 699
genetic source of, 699
hexon, 697
hybrids, adenovirus-SV40, 738-748
defective AD-SV40 hybrids, 744
defective Ad_2-SV40 hybrids, 745
$E46^+$, 744
enhancement of adenovirus replication by SV40, 738
isolation, properties, 744
nondefective Ad_2-SV40 hybrids, 746
messenger RNA, 717
early mRNA, 717
late mRNA, 717
methylation, 719
polyadenylation, 719
processing, 719
productive infection, 714
size classes, 735
switch from early to late viral gene expression, 721
transcription, 717
molecular biology of transformed cells, 731
morphology, 691
nonpermissive cells, adenovirus, 713
pathology, 676
penton base, 698
physical mapping of Ad *ts* mutants, 700
polypeptides components, 691
productive infection, 714
structure of the virion, 690
African horse sickness virus, 422
Alphaviruses, 112 (*see also* Togaviruses)
Arboviruses, 112 (*see also* Togaviruses)
Avipoxviruses, 852

B virus, 772
Barur virus, 172
BeAn virus, 172
BK virus, 595
Blue tongue virus, 422
Bolivar virus, 171
Bovine ephemeral fever virus, 172
Bovine mammalitis virus, 773

Calciviruses, 65
Canine herpesvirus, 773
Capripoxviruses, 852
Cardio viruses, 65
Catfish herpesvirus, 774
Chandipura virus, 171
Chikungunya virus, 112
Cocal virus, 117
Coital-exanthema virus, 773
Colorado tick fever virus, 422
Cormorant herpesvirus, 774
Coxsackieviruses, 65
Cytomegalovirus, 772
Cytoplasmic polyhedrosis virus, 422

Dendrobium virus, 173
Dengue virus, 113
Distemper virus, 351
Double-stranded RNA viruses, 383 (*see also* Reoviruses)
Duck-plague herpesvirus, 774
Duvenhagé virus, 172

Eastern equine encephalitis virus, 112
Echoviruses, 65
Egtved virus, 172
Encephalomyocarditis (EMC) virus, 65
Enders-Ruckle virus, 351
Enteroviruses, 65
Entomopoxviruses, 852
Epstein-Barr virus, 772
Equine abortion virus, 773
Equine rhinovirus, 65
European red mite virus, 173

Feline rhinotracheitis virus, 773
Fiji disease virus, 422
Flanders virus, 172
Flaviviruses, 112 (*see also* Togaviruses)
Foot-and-mouth disease virus, 64
Frog virus 4, 774

Glycoproteins
herpesviruses, 792
myxoviruses, 295
paramyxoviruses, 361
poxviruses, 859
rhabdoviruses, 193
RNA tumor viruses, 471
togaviruses, 121
Guinea pig cytomegalovirus, 774

Hart Park virus, 172
Herpesviruses, 770-848
alteration of cellular membrane, 831
change in immunologic specificity of plasma membrane, 834
social behavior of infected cells, 837
structural changes in plasma membrane, 833
viral proteins in plasma membrane, 833
alteration of host cell function during viral replication, 824
change in cytoplasm, 826
change in nucleus, 825
architectural components, 775
capsid, 778
core, 775
envelope, 779
tegument, 778
DNA synthesis, 817
characteristics, 818
requirements of viral DNA synthesis, 818
effect on host macromolecular synthesis, 827
host DNA, 827
host protein, 828
host RNA, 828
mechanism of inhibition, 829
function of virus-specific proteins, 811
deoxyribonucleases, 815
DNA polymerase, 814
kinases, thymidine and deoxycytidine, 811
[Herpesviruses]
proteases, 817
protein kinases, 816
ribonucleotide reductase, 816
transcriptase, 811
herpes simplex virus, 772
herpesvirus saimiri, 772
herpesvirus tamarinus, 772
herpesviruses in eukaryotes, 772
lipids, 795
morphogenesis of virion, 819
egress, 823
envelopment, 820
inefficiency of viral assembly, 823
morphogenesis of capsid and core, 819
polyamines, 795
replication of herpesvirus, 797
adsorption, 798
duration of replication cycle, 798
infectious unit, 797
penetration, 798
site of transcription, 801
viral RNA sequences in cytoplasm, 800
viral RNA sequences in nucleus, 801
transcription, 799
transcriptional and posttranscriptional processing, 807
translation, 802
coordinate and sequential synthesis of viral proteins, 803
glycosylation of viral proteins, 810
posttranslational modification, 808
regulation of viral proteins, 803
role of viral DNA, 808
role of viral RNA, 806
site of translation, 802
topological distribution of chemical components, 796
uncoating, 798
viral DNA, 779
arrangement of sequences, 782
buoyant density, 772

[Herpesviruses]
complexity, 780
composition, 779
conformation, 780
molecular weight, 780
relatedness of DNA among herpesviruses, 787
size, 780
viral proteins, 792
host-controlled modification, 793
major polypeptides, 793
minor polypeptides, 793

Iguana herpesvirus, 774
Inclusion-body rhinitis virus, 773
Infectious bovine rhinotracheitis virus, 773
Infectious hematopoietic necrosis virus, 173
Infectious laryngotracheitis virus, 774
Influenza viruses, 281 (*see also* Myxoviruses)
Interferon, 41-61
antiviral action, 50
transcription inhibition, 53
translation inhibition, 50
effect on host macromolecule synthesis, 54
inducers, 45
mechanism of induction, 46
synthetic polyribonucleotides, 45
viruses, 45
induction, 43
properties, 46
assaying purity, 49
biological, 47
chemical, 47
physical, 48
purification procedure, 49
unit of assay, 42

JC virus, 595
Japanese encephalitis virus, 112
Joinjaka virus, 172

K virus, 596
Kamese virus, 172
Kern Canyon virus, 172
Klamath virus, 172
Kotonkan virus, 171
Kunjin virus, 112
Kwatta virus, 172

Lagos bat virus, 171
Leporipoxviruses, 852
Leukovirus (*see* RNA tumor viruses)
Lucke virus, 774

Maize rough dwarf virus, 422
Marek's disease virus, 774
Marmoset herpesvirus, 772
Mouse Elberfeld (ME) virus, 65
Measles virus, 351
Mengovirus, 65
MM virus, 595
Mokola virus, 171
Mossuril virus, 172
Mt. Elgon virus, 172
Mouse cytomegalovirus, 774
Mumps virus, 351
Myxoviruses, 281-348
antigenic drift, 287
antigenic shift, 288
antigenic variation, 287
carbohydrates, 307
characteristics, 283
classification, 283
composition, 295
hemagglutinin, 295
M protein, 300
neuraminidase, 298
NP protein, 301
P protein, 301
virion proteins, 295
epidemics, 286
genetics, 336
mechanism of genetic exchange, 339
temperature-sensitive mutants, 337
genome, 302
biological implications, 307
complexity, 306
nucleotide sequence of RNA segments, 304

[Myxoviruses]
number of segments, 306
segmented nature of RNA, 302
host-virus interaction, 308
effect on host macromolecule synthesis, 309
host range, 308
infectious cycle, 310
attachment, penetration, uncoating, 310
pandemic strains, 285
pandemics, 286
pathogenesis, 286
processing of viral proteins, 329
budding, 329
modification of cell membrane, 322
processing of RNP, 333
RNA replication, 313
cell-associated virion-specific polymerase, 315
intracellular virus-specific RNA, 317
kinetics of RNA synthesis, 317
messenger RNA, 320
replicative (RI), transcriptive (TI) intermediates, duplexes (DS), 318
virion-associated transcriptase, 313
VRNA and CRNA synthesis, 317
structure, 289
lipid bilayer, 292
lipid-protein interaction, 293
membrane or matrix, 293
morphology, 289
ribonucleoprotein, 293
spikes, 290
virus-specific protein synthesis, 323
HA, 326
localization and development of viral antigens, 324
M protein, 328
NA, 328
NP, 329
NS, 329
P protein, 328
viral antigens, 324
virus-specific polypeptides, 325
Von Magnus virus, 334

[Myxoviruses]
characteristics, 334
mechanism of formation, 335

Navarro virus, 172
Neuraminidase, 298
myxoviruses, 298
paramyxoviruses, 361
Newcastle disease virus, 351 (*see also* Paramyxoviruses)
Nigerian horse virus, 171

Obodhiang virus, 171
Oncornavirus, 436 (*see also* RNA tumor viruses)
Orbivirus, 422 (*see also* Reoviruses)
Oregon sockeye disease virus, 173
Orthoreovirus, 385 (*see also* Reoviruses)

Papilloma viruses, 591
Papovaviruses, 590-672
architecture, 591
agglutination of red cells, 596
distribution in human population, 594
empty particles, 599
full particles, 599
histone-DNA complexes, 599
nucleoprotein complexes, 599
polypeptides, structural, 596
T antigen, 631
U antigen, 633
viral histones, 599
DNA
chain elongation of DNA, 628
circular, 602
component I and II, 602
deletion, 605
DNA synthesis, 625
foreign DNA, 608
mapping of DNA, 603, 606
nicked DNA, 603
rearrangements, 605
replicating DNA, 625
restriction cleavage of DNA, endonuclease, 606
substituted molecules of DNA, 608

[Papovaviruses]
substitution, 605
superhelical, 602
termination of DNA synthesis, 630
topology, 601
genetics, 636
helper function of SV40, 634
human papovaviruses, 595
BK virus, 595
JC virus, 595
regulation of transcription, 615
symmetrical transcription, 615
time course of synthesis of
viral DNA, 625
viral proteins, 630
viral RNA, 615
transcription, 615
transformation, 641
ts mutants, 636
virus-specific RNA, 615
early, 615
late, 615
lytic cycle, 615
nonpermissive cells, 645
permissive cells, 615
Parainfluenzavirus, 351 (*see also* Paramyxoviruses)
Paramyxoviruses, 350-382
disease aspects, 350
fusion, 363
the role of F protein, 363
genome RNA, 353
hemagglutination, 361
negative strand viruses, 350
distribution in nature, 351
members of the group, 351
nomenclature, 351
role in diseases, 350
neuraminidase, 361
nucleocapsid, 357
basic properties, helical symmetry, 357
functions of the polypeptides, 359
template for transcription, 367
positive and negative strand RNA molecules, 356
practical aspects of biochemical work with paramyxoviruses, 352
[Paramyxoviruses]
RNA replication, replicase, 371
virion characteristics, 350
virion envelope, 360
glycopolypeptide F, 363
glycopolypeptide HN, 361
polypeptide M, 364
virus replication, 365
encapsidation, 372
assembly, envelopment, 372
transcription, 366
lack of a temporal transcription, 367
nucleocapsid as transcription template, 367
virion transcriptase, primary transcription, 366
translation, 368
messenger RNA species, 368
regulation of polypeptide abundances, 370
Parapoxviruses, 852
Parvoviruses, 540-587
characteristics, 546
antigenic cross-reaction, 550
buoyant density, 547
capsid configuration, 547
hemagglutination, 549
host range, 551
latency, 553
molecular weight, 549
particle size, 546
pathogenicity, 551
sedimentation coefficient, 549
effect of helper viruses, 566
family, 540
members, 541
nomenclature, 541
interference, 577
effect on adenovirus replication, 577
effect on tumor induction, 578
nucleic acid, 560
configuration, 562
density, 562
enzymatic cleavage, 565

[Parvoviruses]
homology, 565
strands, 562
replication, 567
DNA synthesis, 574
infectious cycle, 570
permissive cells, 568
protein synthesis, 572
RNA synthesis, 576
ultrastructure studies, 569
structural proteins, 555
Picornaviruses, 63-110
assembly, 97
procapsid, 98
provirion, 98
characteristics, 67
classification, 65
defective interfering particles, 100
disease aspects, 66
genetic map, 86
experimental approach(es), 86
inhibition by guanidine, 102
inhibition by 2-α(hydroxybenzyl)-benzimidazole (HBB), 102
interference by defective particles, 100
replication, 73
absorption, 73
alteration of host cell function, 74
inhibition of cellular DNA, 79
inhibition of cellular RNA, 78
inhibition of host protein synthesis, 75
mechanism of inhibition, 75
penetration, 73
uncoating, 73
VP 4, role in adsorption, 74
replication of viral RNA, 91
polyadenylation of viral RNA, 94
replication complex, 92
replicative intermediate, 93
structure components, 67
viral proteins, 69
structural arrangement in virion, 69
viral RNA, 67
molecular weight, 67
[Picornaviruses]
termini (5′ and 3′ ends), 68
unsolved questions, 103
virus-specified proteins, 80
cleavage, proteolytic, 85
function of virus-specified proteins, 88
primary polypeptides, 80
processing, 80
synthesis, 80
Pigeon herpesvirus, 774
Pike fry rhabdovirus, 187
Piry virus, 176
PML virus, 595
Pneumonia (virus) of mice, 351
Polioviruses (*see* Picornaviruses)
Polyoma virus, 593
Poxvirus, 849-890
characteristics, 851
chemical composition, 854
deoxyribonucleic acid, 855
ribonucleic acid, 855
classification, 851
virion structure, 851
dissemination of poxvirus, 879
effect on host cells, 879
inhibition of macromolecular synthesis, 880
modification of cell membranes, 879
enzymes, 859
deoxyribonucleases, 864
guanylyl and methyl transferases, 863
nucleoside triphosphate phosphohydrolases, 863
poly(A)polymerase, 862
protein kinase, 865
RNA polymerase, 862
genetics, 866
growth cycle, 866
entry, 867
transcription, 867
uncoating, 867
host range mutants, 866
lipids, 865
proteins, 855

[Poxvirus]
localization of structural proteins, 859
modified structural proteins, 859
glycoproteins, 859
phosphoproteins, 859
protein synthesis, 870
identification of viral proteins, 870
posttranslational modifications, 874
amino acid modification, 876
cleavage, 874
regulation of protein synthesis, 876
structural proteins, 873
virus-induced enzymes, 872
early, 872
late, 873
replication, 876
assembly of DNA virus, 877
DNA, 876
specific inhibitors of poxvirus reproduction, 881
temperature-sensitive mutants, 866
PPV virus, 542
Pseudorabies herpesvirus, 773

Rabbit herpesvirus, 774
Rabbit vacuolating virus, 596
Rabies, 177
strains of rabies virus, 178
Rat cytomegalovirus, 774
Reoviruses, 383-433
characteristics, 384
biological features, 390
hemagglutination, 390
host range, natural occurrence and serology, 384
infectivity, 390
morphology, 387
pathobiology, 385
physical properties, chemical composition, 386
virion structure, 387
[Reoviruses]
genetics, 412
complementation groups, 414
deletion mutants, 417
physiology, 418
recombination frequency, 413
specific gene lesions, 421
temperature-sensitive mutants, 412
replication, 402
absorption, penetration, uncoating, 402
effect on the host cell, 411
infected cells, morphology, 402
inhibitors of host macromolecules, 412
morphogenesis, 410
one-step growth cycle, 403
replication of double-stranded RNA, 408
synthesis of oligonucleotides, 409
transcription in vitro, 405
transcription in vivo, 404
translation in vitro, 407
translation in vivo, 406
viral polypeptides, 395
localization within virions, 397
nonstructural polypeptides, 398
structural polypeptides, 395
virion enzymes, 400
characteristics, 400
other enzymes, 401
product analysis, 405
virion transcriptase, 400
virion RNA, 392
genome RNA, 392
molecular weight, 392
number of segments, 392
oligonucleotides, 394
Respiratory syncytial virus, 351
Retroviruses (*see* RNA tumor viruses)
Rhabdoviruses, 169-278
carbohydrates, 202
characteristics, 169
defective interfering particles, 269
autointerference, 269
genetic analyses, 259

[Rhabdoviruses]
isolation and characterization, 259
recombination between VSV Indiana *ts* mutants, 267
RNA synthesis of *ts* mutants, 265
temperature-sensitive mutants, 259
infectious process, a topography of infected cells, 256
isolates from vertebrates and invertebrates, 170
phenotypic mixing, pseudotypes, 267
rabies subgroup, 177
replication process, 221
model, 252
penetration, 223
RNA synthesis, 224
product analyses, 228
identification of the transcriptase proteins, model of primary transcription, 238
transcription capping, initiation, methylation, 228
transcription polyadenylation, termination, 236
serological relationships, 206
serological tests, 203
antigenic determinants, 203
complement-fixation tests, 213
neutralization tests, 207
structural components, 189
genome homology, 191
RNA genome, 190
shape, 189
structural proteins, 193
G proteins, 198
host constituents, 201
L proteins, 201
lipid membrane, 201
M proteins, 199
N proteins, 200
NS proteins, 201
transcriptase, 203
transcription in vitro, 226
transcription in vivo, 239
characterization of viral mRNA, 246

[Rhabdoviruses]
control of transcription, 243
primary transcription, 239
secondary transcription, 240
transmission and pathobiology, 175
bovine ephemeral fever virus, 184
Chandipura virus, 176
Duvenhagé and Bolivar viruses, 181
Flanders, Hart Park, and Navarro viruses, 183
Joinjakaka, Mossuril, Kamese, Kwatta, and BeAn 157575 viruses, 182
Klamath and Barur viruses, 183
Kotonkan and Obodhiang viruses, 182
Lagos bat and Mokola viruses, 180
Mount Elgon and Kern Canyon bat viruses, 183
Nigerian horse virus, 181
Oregon sockeye disease, Sacramento River, Chinook salmon disease, infectious hematopoietic necrosis viruses, 186
Pike fry rhabdovirus, 187
Piry, 176
rabies, 177
sigma, 185
spring viremia of carp virus, 186
vesicular stomatitis viruses, 176
VHS and Egtved viruses, 185
Rhinoceros beetle virus, 173
Rhinoviruses, 65
Ribodeoxyvirus, 435 (*see also* RNA tumor viruses)
Rice dwarf virus, 422
Rinderpest virus, 351
RNA tumor virus(es), 435-542
antigenic relationship, 462
interspecies and intraspecies relationship, 462
envelope glycoproteins, 471

[RNA tumor virus(es)]
molecular hybridization with DNA, 482
P30 determinants, 462, 467, 469
type-specific immunoassay, 466
intraspecies variability, 485
endogenous viruses, 499
interrelationship, mammalian viruses, 489
interrelationship, murine viruses, 485
interrelationship, primate viruses, 488
interspecies hybrid viruses, 489
bovine leukemia virus, 480
chick helper factor (chf), 446
classification, 438
type B, 440
type C, 438
COFAL test, 443
DNA intermediate, 492
endogenous viruses, 499
genetic control of viral expression in mouse, 521
chromosomal localization, 521
control of endogenous virus expression in chicken, 528
Fv-1 loci in mouse, 522
regulatory genes of MULV, 525
host-viral relationship, 492
defective viruses, 443
events in replication, 492
helper viruses, 445
leukemia viruses, 443
nondefective type C viruses, 492
proviral DNA synthesis, 496
recombination, 499
replication, 492
sarcoma sequences, 499
sarcoma viruses, 443
S+L- cells, 499
transcription of DNA, 492, 497

[RNA tumor virus(es)]
Mason Pfizer monkey virus, 480
mouse mammary tumor viruses (MMTV), 478
oncogene, 436, 506
oncornaviruses, 439
pathogenesis, 532
leukemia induction, 532
pseudotypes, 445
recombination, 499
cDNA sarc, 505
sarcoma-specific sequences, 506
replication, 492
reverse transcriptase, 477
structure, 449
genome complexity, 457
70 S RNA, 459
35 S RNA, 459
5 S RNA, 459
4 S RNA, 459
primer RNA, 461
glycoproteins, 452
internal core shell, 452
morphology, 449
MMTV, structural components, 456
nucleic acids, 457
molecular weight, 457
proteins, 452
reverse transcriptase, 436
RNA-dependent DNA polymerase, 436
spikes, knobs, 449
spleen necrosis virus, structural components, 456
studies in man, 535
virus-related sequences in human malignancy, 537
viral components in normal cells, 508
biological detection of viruses in normal cells, 508
DNA homology among primates, 515
induction by BudR and IudR, 519
RD114 viral sequences in

[RNA tumor virus(es)]
domestic cats, 511
transfection, 519
viral nucleic acids in normal cells, 510
viral proteins in normal cells, 508
viral mutants, 530
conditional mutants, 530
heat lability of reverse transcriptase, 531
nonconditional, 530
replication defective, 530
transcriptase in *ts* mutants, 532
temperature-sensitive, *ts*, 531
transformation defective, 530
xenotropic, 437

SA 6 virus, 772
SA 8 virus, 772
Sacramento River chinook disease virus, 173
St. Louis encephalitis virus, 112
Semliki Forest virus, 112
Sendai virus, 351 (*see also* Paramyxoviruses)
Sialidase, 298, 361 (*see also* Neuraminidase)
Sigma virus, 185
Simian virus-40 (SV40 (*see also* Papoviruses)
host range, 595
polyoma, 593
transformation, 641
integration site of viral DNA, 645
physical state of viral DNA in transformed cells, 644
rescue of virus from transformed cells, 648
transcription in transformed cells, 645
tumor-specific transplantation antigen, 648
viral genes for transformation, 642

[Simian virus]
viral proteins in transformed cells, 648
Sindbis virus, 112
Spider-monkey herpesvirus, 772
SV 5 virus, 351
Symmetry in virus architecture, 1-39
combined symmetry, 20
derivation, 24
dimer, trimer, multiple protein patterns, 26
triangulation numbers, 24
general architecture, 2
capsids, 2
capsomeres, 8
nucleocapsids, 3
viroids, 3
helical nucleocapsids, enveloped, 6
myxoviruses, 203
paramyxoviruses, 357
rhabdoviruses, 167
icosahedral symmetry, 7
capsomeres, 8
construction of icosahedron, 8
core structure, 19
dimer clustering, 17
hexamer, 16
icosahedron, 7
nonidentical structural units, 19
pentamer, 16
quasi-equivalent position, 16
structural units, 16
triangulation number, 9
trimer clustering, 19
tubular variants of viral capsomeres, 22
vertex (vertices), 7
viral architecture, 1

Togaviruses, 111-166
characteristics, 112
classification, 112
defective interfering particles, 153
genetics, 146
complementation, 149
functional defects of RNA^+

[Togaviruses]
mutants, 150
functional defects of RNA⁻ mutants, 152
isolation of mutants, 146
growth in arthropod cells, 154
carrier state, 155
growth cycle, 124
sensitive host cells, 124
receptors, 124
temperature optimum, 124
transcription of alphavirus RNA, 126
mechanism of transcription, 126
replicative forms (RF), 127
transcription of flavivirus RNA, 131
maturation of alphaviruses, 140
budding process, 141
interaction between glycoproteins and nucleocapsids, 142
maturation of flaviviruses, 146
translation, 131
identification of viral messenger, 131
processing of viral proteins, 133
translation of alphaviral messengers, 137

[Togaviruses]
translation of flaviviral messengers, 139
translation in vitro, 135
virion structure, 113
composition, 113
envelope proteins, 113
glycoproteins, 121
carbohydrate composition, 114
polysaccharide chains, 121
lipid bilayer, 119
nucleocapsid, 118
RNA, 117
triangulation number, 118
Tree shrew herpesvirus, 773
Turkey herpesvirus, 774

Varicella-zoster virus, 772
Venezuelan equine encephalitis virus, 112
Vesicular stomatitis virus, 170 (*see also* Rhabdoviruses)

Warts, 591
Western equine encephalitis virus, 112
Whirligig beetle virus, 173
Wildebeest herpesvirus, 773
Wound tumor virus, 422

Yucaipa virus, 351